T0269096

Genetics and the Behavior of Domestic Animals

Genetics and the Behavior of Domestic Animals

Second Edition

Edited by

Temple Grandin
Department of Animal Science
Colorado State University
Fort Collins, Colorado, USA

Mark J. Deesing
Grandin Livestock Handling Systems Inc.
Fort Collins, Colorado, USA

AMSTERDAM • BOSTON • HEIDELBERG • LONDON
NEW YORK • OXFORD • PARIS • SAN DIEGO
SAN FRANCISCO • SINGAPORE • SYDNEY • TOKYO
Academic Press is an imprint of Elsevier

Academic Press is an imprint of Elsevier

32 Jamestown Road, London NW1 7BY, UK
225 Wyman Street, Waltham, MA 02451, USA
525 B Street, Suite 1800, San Diego, CA 92101-4495, USA

Acquiring Editor: Kristi Gomez
Development Editor: Pat Gonzalez
Project Managers: Karen East and Kirsty Halterman
Designer: Alan Studholme

Second edition 2014

Copyright © 2014 Elsevier Inc. All rights reserved

No part of this publication may be reproduced, stored in a retrieval system or transmitted in any form or by any means electronic, mechanical, photocopying, recording or otherwise without the prior written permission of the publisher.

Permissions may be sought directly from Elsevier's Science & Technology Rights Department in Oxford, UK: phone (+44) (0) 1865 843830; fax (+44) (0) 1865 853333; email: permissions@elsevier.com. Alternatively, visit the Science and Technology Books website at www.elsevierdirect.com/rights for further information.

Notice
No responsibility is assumed by the publisher for any injury and/or damage to persons or property as a matter of products liability, negligence or otherwise, or from any use or operation of any methods, products, instructions or ideas contained in the material herein.

Because of rapid advances in the medical sciences, in particular, independent verification of diagnoses and drug dosages should be made.

British Library Cataloguing-in-Publication Data
A catalogue record for this book is available from the British Library

Library of Congress Cataloging-in-Publication Data
A catalog record for this book is available from the Library of Congress

ISBN: 978-0-12-810016-5

For information on all Academic Press publications
visit our website at elsevierdirect.com

Typeset by MPS Limited, Chennai, India
www.adi-mps.com

Printed and bound in United States of America

14 15 16 10 9 8 7 6 5 4 3 2 1

Contents

Preface

The purpose of this book is to bridge the gap between the field of behavior genetics and the research on behavior published in the animal science and veterinary literature. Fifteen years have passed since the first edition of this book was published. In that time considerable advances have been made in the study of the genetics and behavior. The second edition contains the addition of three new chapters and updates on the original chapters. Many of the early studies of behavior discussed by editors Temple Grandin and Mark J. Deesing (Chapter 1) are still relevant to our understanding of the genetic effects on behavior. The important lessons learned by these studies endure to this day and are included in the second edition. Well-done behavioral studies never become obsolete.

In Chapter 2, Per Jensen and Dominic Wright discuss domestication as a genetic process, whereby animals change phenotypically and genetically as a response to living under human supervision.

In Chapter 3, Alain Biossy and Hans Erhard provide an overview of the current behavioral and cognitive aspects of emotions in animals and explores the impacts of emotional experiences on the animal adaptation to its circumstances.

In Chapter 4, editors Temple Grandin and Mark J. Deesing discuss observations under field conditions and review research findings which affect behavior of livestock during handling, herding, and restraint.

In Chapter 5, Peter Chenoweth, Antonio J. Landeta-Hernandez, and Cornelia Florecki discuss maternal and reproductive behavior in livestock and emphasize the importance of incorporating maternal behavior in selection to ensure that mothering ability is not compromised by selection for production traits.

In Chapter 6, Kathryn Lord discusses genetic effects on the behavior of different dog breeds. Dogs and wolves share genotypes, which are nearly identical, but their behavioral phenotypes are very different.

In Chapter 7, editors Mark J. Deesing and Temple Grandin review the current literature on horse behavior and offer suggestions for further research on subjects such as early experience effects on behavior and temperament.

In Chapter 8, Jean-Michel Faure and Andrew Mills discuss their classic work on fearfulness and social reinstatement (separation distress) in Japanese quail. This chapter is a good example of a well-done behavioral study that is still relevant after many years.

In Chapter 9, William Muir and Heng Wei Cheng focused attention on genetic modification to improve chicken adaptability to its environment and consequently to enhance its welfare.

In Chapter 10, Anna Kukekova, Lyudmila Trut, Gregory Acland, discuss ongoing research focused on identification of molecular genetic mechanisms associated with selection for behavior and behavioral genetics of dogs, wolves, and foxes with the aim of providing insight into the complex structure of domesticated behavior in canids.

In Chapter 11, Lotta Rydhmer and Laurianne Canerio discuss the importance of behavioral genetics for pig welfare, and discuss the possibility to select for behavioral traits in order to improve welfare.

In Chapter 12, editors Temple Grandin and Mark J. Deesing discuss serious animal welfare problems caused by over-selection for production traits such as rapid growth, leanness, and high milk yield. Animals can be altered by genetic selection to such an extent that serious structural or neurological defects develop. It is not possible to have an adequate level of animal welfare if a selected trait becomes so extreme that it causes obvious mobility problems, or if it causes a condition that is known to be painful in humans.

Behavior genetics is a wide field and an integrated science. Understanding the relative influence of nature (genetics) and nurture (environment) in the development of behavior is the focus of behavioral geneticists. Basic genetic mechanisms producing changes in appearance and behavior of animals are described and key principles are explained in order to make it easier for the non-geneticist to read and understand the latest papers.

The contributing authors explain the relevance of their work and discuss the basic principles that apply to other animals. This book will be of interest to biologists, researchers, veterinarians, livestock producers, zoo curators, behavioral ecologists, evolutionary biologists, and people who own and train domestic animals. A wide range of literature is reviewed and presented in an easy-to-understand manner with a minimum of scientific jargon. When technical explanations are necessary, the subject is explained as clearly and simply as possible.

Temple Grandin
Mark J. Deesing

List of Contributors

Gregory M. Acland Baker Institute for Animal Health, Cornell University, Ithaca, NY, USA

Alain Boissy UMR INRA 1213 Herbivores, Centre de Theix, St-Genès Champanelle, France

Laurianne Canario French National Institute for Agricultural Research, Animal Genetics Division, Castanet-Tolosan, France

Heng Wei Cheng Purdue Agriculture, Livestock Behavior Research Unit, USDA-ARS, West Lafayette, IN, USA

Peter J. Chenoweth Charles Sturt University, Wagga Wagga, Australia

Lorna Coppinger School of Cognitive Science, Hampshire College, Amherst, MA, USA

Raymond Coppinger School of Cognitive Science, Hampshire College, Amherst, MA, USA

Mark J. Deesing Grandin Livestock Handling Systems, Inc., Fort Collins, Colorado, USA

Hans W. Erhard UMR INRA 791 Mosar, AgroParisTech, 16 rue Claude Bernard, Paris Cedex 05, France

Jean Michel Faure Station de Recherches Avicoles, INRA de Tours, Novzilly, France

Cornelia Flöercke Department of Animal Sciences, Colorado State University, Fort Collins, Colorado, USA

Temple Grandin Department of Animal Sciences, Colorado State University, Fort Collins, Colorado, USA

Per Jensen IFM Biology, Linköping University, Sweden

Anna V. Kukekova Department of Animal Sciences, University of Illinois at Urbana-Champaign, Urbana, IL, USA

Antonio J. Landaeta-Hernández Universidad del Zulia, Facultad de Ciencias Veterinarias, Unidad de Investigaciones Zootécnicas, Maracaibo, Venezuela

Kathryn Lord School of Cognitive Science, Hampshire College, Amherst, MA, USA

Andrew D. Mills Station de Recherches Avicoles, INRA, Centre de Tours, Nouzilly, France

William M. Muir Purdue Agriculture, West Lafayette, IN, USA

Lotta Rydhmer Department of Animal Breeding and Genetics, Swedish University of Agricultural Sciences, Uppsala, Sweden

Lyudmila N. Trut Institute of Cytology and Genetics of the Russian Academy of Sciences, Novosibirsk, Russia

Dominic Wright IFM Biology, Linköping University, Sweden

Chapter 1

Behavioral Genetics and Animal Science

Temple Grandin* **and Mark J. Deesing**[†]
*Department of Animal Science, Colorado State University, Fort Collins, Colorado, USA;
[†]Grandin Livestock Handling Systems, Inc., Fort Collins, Colorado, USA*

INTRODUCTION

A bright orange sun is setting on a prehistoric horizon. A lone hunter is on his way home from a bad day at hunting. As he crosses the last ridge before home, a quick movement in the rocks, off to his right, catches his attention. Investigating, he discovers some wolf pups hiding in a shallow den. He exclaims, "Wow . . . cool! The predator . . . in infant form."

After a quick scan of the area for adult wolves, he cautiously approaches. The pups are all clearly frightened and huddle close together as he kneels in front of the den . . . all except one. The darkest-colored pup shows no fear of the man's approach. "Come here you little predator! Let me take a look at you," he says. After a mutual bout of petting by the man and licking by the wolf, the man suddenly has an idea. "If I take you home with me tonight, maybe mom and the kids will forgive me for not catching dinner . . . again."

The opening paragraphs depict a hypothetical scenario of man first taming the wolf. Although we have tried to make light of this event, the fact is no one knows exactly how or why this first encounter took place. More than likely, the "first" encounter between people and wolves occurred more than once. Previous studies suggest that dogs were domesticated 14,000 years ago (Boessneck, 1985). However, Ovodov *et al.* (2011) reported finding dog fossils 33,000 years old in Siberia. Domestication of dogs may have begun before 35,000 years ago in what Galibert *et al.* (2011) described as a period of proto-domestication. Early hunter-gatherers may have captured wolf pups which became tame and habituated to living with human groups. Some wolves may have become aggressive as they matured and were killed or chased away. Others remained submissive and bred with less fearful wolves scavenging around human settlements. Analysis of mitochondrial DNA of 67 dog breeds and wolves from 27 localities indicates that dogs may have diverged from wolves over 100,000 years ago (Vita *et al.*, 1997). Other researchers question this finding and suggest that dogs were

Genetics and the Behavior of Domestic Animals. DOI: http://dx.doi.org/10.1016/B978-0-12-394586-0.00001-9
© 2014 Elsevier Inc. All rights reserved.

domesticated 5400 to 16,300 years ago from many maternal lines (Pang *et al.*, 2009; Savolainen *et al.*, 2002). Other evidence suggests that domestication occurred in several different regions (Boyko, 2011). Ancient breeds such as the Australian Dingo, Basenjis, and New Guinea singing dogs, all originated in areas where there were no wolves (Larsen *et al.*, 2012). Possibly the scenario at the beginning of this chapter happened many times.

Another scenario is that wolves domesticated themselves. The presumption is that calm wolves with low levels of fear were more likely to scavenge near human settlements. Both Coppinger and Smith (1983) and Zenner (1963) suggest that wild species which later became domesticants started out as camp followers. Some wolves were believed to have scavenged near human settlements or followed hunting parties. Modern dog breeds probably were selectively bred from feral village dogs (Boyko, 2011).

The human brain is biologically programmed to pay attention to animals. Electrical recordings from the human amygdala, a brain structure involved with emotion, showed that pictures of animals caused a larger response than pictures of landmarks, people or objects (Morman *et al.*, 2011). Both threatening and non-threatening animal pictures evoked the same response. Maybe this shows the importance of animals in our past.

Genetics Shapes Behavior

Genetic differences in animals affect behavior. A major goal of the 2nd edition of this book is to review both old and new research on individual differences within a breed and behavioral differences between breeds. Well-done behavioral studies never become obsolete. Scientists may discover new ways to interpret "why" a certain behavior occurs but well-done behavioral studies always retain their value. During our literature review, we found many new studies that verified older studies reviewed in the first edition. A good example is the classic work on fearfulness and social reinstatement (separation distress) in quail (Faure and Mills, 1998, Chapter 8). These emotional traits can be independently strengthened or weakened by selective breeding.

This book is aimed at students, animal breeders, researchers, and anyone who is interested in animal behavior. There has been a great increase in research on genetic mechanisms that affect both behavior and physical traits of animals. In this chapter, we review studies of behavioral differences with a focus on domestic animals such as dogs, cattle, horses, pigs, sheep, and poultry. Detailed reviews of genetic mechanisms and molecular biology are beyond the scope of this book.

GENETIC EFFECTS OF DOMESTICATION

Price (1984) defined domestication as:

"a process by which a population of animals becomes adapted to man and the captive environment by some combination of genetic changes occurring over

generations and environmentally induced developmental events recurring during each generation."

Major behavioral differences exist between domesticated animals and their wild relatives. For example, the jungle fowl is much more fearful of novel objects and strange people compared to the domestic white Leghorn chicken (Campler *et al.*, 2009). A strong genetic component underlies differences in fearfulness between jungle fowl and domestic chickens (Agnvall *et al.*, 2012).

Domestication may have been based on selection for tameness. In long-term selection experiments designed to study the consequences of selection for the "tame" domesticated type of behavior, Belyaev (1979) and Belyaev *et al.* (1981) studied foxes reared for their fur. The red fox (*Vulpes fulva*) has been raised on seminatural fur farms for over 100 years and was selected for fur traits and not behavioral traits. However, the foxes had three distinctly different behavioral responses to people. Thirty per cent were extremely aggressive, 60% were either fearful or fearfully aggressive and 10% displayed a quiet exploratory reaction without either fear or aggression. The objective of this experiment was to breed animals similar in behavior to the domestic dog. By selecting and breeding the tamest individuals, 20 years later the experiment succeeded in turning wild foxes into tame, border collie-like fox-dogs. The highly selected "tame" population of (fox-dog) foxes actively sought human contact and would whine and wag their tails when people approached (Belyaev, 1979). This behavior was in sharp contrast to wild foxes which showed extremely aggressive and fearful behavior toward man. Keeler *et al.* (1970) described this behavior:

Vulpes fulva *(the wild fox) is a bundle of jangled nerves. We had observed that when first brought into captivity as an adult, the red fox displays a number of symptoms that are in many ways similar to those observed in psychosis. They resemble a wide variety of phobias, especially fear of open spaces, movement, white objects, sounds, eyes or lenses, large objects, and man, and they exhibit panic, anxiety, fear, apprehension, and a deep trust of the environment. They are 1) catalepsy-like frozen positions, accompanied by blank stares, 2) fear of sitting down, 3) withdrawal, 4) runaway flight reactions, and 5) aggressiveness. Sometimes the strain of captivity makes them deeply disturbed and confused, or may produce a depression-like state. Extreme excitation and restlessness may also be observed in some individuals in response to many changes in the physical environment. Most adult red foxes soon after capture break off their canine teeth on the mesh of our expanded metal cage in their attempts to escape. A newly captured fox is known to have torn at the wooden door of his cage in a frenzy until he dropped dead from exhaustion.*

Belyaev (1979) and Belyaev *et al.* (1981) concluded that selection for tameness was effective in spite of the many undesirable characteristics associated with it. For example, the tame foxes shed during the wrong season and developed black and white patterned fur. Changes were also found in their hormone profiles and the monestrous (once a year) cycle of reproduction was disturbed. The tame foxes would breed at any time of the year. Furthermore, changes in behavior occurred simultaneously with changes in tail position and ear shape, and the appearance of a white muzzle, forehead blaze, and white shoulder hair. The white color pattern on the head is similar to many domestic animals (Belyaev, 1979) (Figures 1.1 and 1.2). The most dog-like foxes had white spots and patterns on their heads, drooping ears, and curled tails and looked more like dogs than the foxes that avoided people. The behavioral and morphological (appearance) changes were also correlated with corresponding changes in the levels of sex hormones. The

FIGURE 1.1 Wild-type fox before Belyaev started selective breeding for tameness. Reprinted from *Journal of Heredity* by permission of Oxford University Press.

FIGURE 1.2 Selecting for tameness for many generations altered the body shape and coat color pattern. Foxes selected for tameness resembled dogs. Reprinted from *Journal of Heredity* by permission of Oxford University Press.

tame foxes had higher levels of the neurotransmitter serotonin (Popova *et al.*, 1975). Serotonin is known to inhibit certain kinds of aggression (Belyaev, 1979). Serotonin levels are increased in the brains of people who take Prozac (fluoxetine).

The fox experiments started by Dmitry Belyaev and Lyudmila Trut in the late 1950s are still ongoing. Using molecular genetic techniques, Kukekova *et al.* (Chapter 10) are working to understand basic biological principles that guide social behavior. They suggest that domesticated behavior in dogs and foxes may have a similar genetic basis.

Basic Genetic Mechanisms

Since the first edition, there has been a huge increase of research on genetic mechanisms. By the time this book is published, some of the material on genetic mechanisms may be obsolete. However, behavior studies reviewed in this book will remain useful to scientists working to discover new genetic mechanisms. The classical genetic concepts of recessive and dominant traits discovered by Gregor Mendel explain only a small fraction of the genetic factors affecting inheritance. It is beyond the scope of this book to provide an in-depth review of genetic mechanisms, instead we describe key principles that will make it easier for the non-geneticist to read and understand the latest papers. Life is complicated. Increasingly complex non-Mendelian genetic mechanisms are being discovered, and they are best viewed as networks of information (Hayden, 2012). Below is an outline of some basic genetic mechanisms that produce changes in the appearance and behavior of animals.

SNPs (Single Nucleotide Polymorphisms)

An SNP is a single code change in a single base pair of DNA. Guryev *et al.* (2004) states that SNPs are a major factor in genetic variation. In some Mendelian diseases single SNPs or multiple SNPs are involved in disease inheritance (Kong, *et al.*, 2009; Shastry, 2002).

Repeats

Repeats are also called tandem repeats, single sequence repeats (SSRs), or genetic stutters. Repeats are sequences of DNA code repeated more than once, and the number of repeats can vary within a gene. The number of repeats can determine many traits ranging from the length of a dog's nose to variations in brain development (Fondon and Garner, 2004; Fondon *et al.*, 2008).

CNVs (Copy Number Variations)

CNVs are rearrangements of genetic code. It is likely CNVs contribute greatly to genetic variation by modifying genetic expression (Chaignat *et al.*, 2011;

Henrichson *et al.*, 2009). CNVs are often spontaneous (*de novo*) mutations not inherited form the parents. The many types of CNVs range from translocation of pieces of genetic code, to deletions of genetic code, or to extra copies of code. Twelve per cent of the human genome is in copy number variable regions of the genome (Redon *et al.*, 2006). CNVs are very numerous in the brain and the immune system.

Jumping Genes

Also called transposable elements (transposons), jumping genes are short segments of genetic material that transport themselves throughout the genome in a "cut and paste", or a "copy and paste" manner. Mikkelson *et al.* (2007) states that transposons are a "creative force" in the evolution of mammalian gene regulation." Jumping genes are more numerous in the brain than in liver or heart cells (Vogel, 2011).

Coding DNA

The very small percentage of the genome that specifically codes for proteins used in development of the animal. Until recently, only coding DNA was sequenced.

Non-Coding DNA, Also Called Regulatory DNA

In the 1980s, this was called junk DNA because it does not code for proteins. Researchers have discovered that non-coding DNA has a regulatory function and approximately 80% of the non-coding DNA is transcribed by RNA and has biochemical functions (ENCODE Project Consortium 2012). Non-coding DNA is the "computer operating system" that directs the coding DNA. Non-coding DNA may be the gene's "project managers" that orchestrate and direct the sequence of building proteins (Saey, 2011). Chakravarti and Kapoor (2012) state that to understand the genes that code for proteins, the regulatory non-coding DNA needs to be understood. Some portions of non-coding DNA are highly conserved, and similar sequences occur in many different animals. Other portions of non-coding genome may rapidly evolve (Maher, 2012). Coding regions of DNA that direct the development of basic patterning of the body (hoxgenes) are highly conserved across many species from arthropods to mammals (Linn *et al.*, 2008). Early pattern formation of the notochord and neural tube is also highly conserved (Richardson, 2012). In both plants and animals, the embryos of many species look similar during the mid-embryonic stage of development. The mid-stage of development is "dominated by ancient genes" (Quint *et al.*, 2012). A basic principle is that similar traits in a species originate from highly conserved genetic code. Traits that have recently changed originate from newer code. Research shows that changes in non-coding DNA are drivers of evolutionary change. The human neocortex has a higher percentage of young genes expressed during fetal development compared to mice

(Zhang *et al.*, 2011). Transposable bits of code can make changes in regulatory DNA (Mikkelson *et al.*, 2007). The fact that similar sections of non-coding DNA occur in many species indicates its important function. Changes in non-coding DNA during environmental changes cause stickleback fish to adapt by changing traits, such as body shape, skeletal armor or the ability to live in salt or fresh water (Hockstra, 2012). Sections of non-coding regulatory DNA evolve and change along with the coding DNA (Jones *et al.*, 2012). Studies in other animals also show a role for non-coding DNA mutations in the development of domestic animals (Anderson, 2012).

Exome

The exome is the DNA sequencing of all the protein-coding regions of the genome. The non-coding DNA is left out of the exome.

RNA Transcriptome

The RNA trancriptome is theDNA code that is read and transcribed by RNA. Sequencing the genome indicates that the animal's genome contains a particular piece of genetic code. The transcriptome indicates whether or not the code was transcribed by RNA and expressed as either a protein or involved in regulatory functions.

De Novo *Mutations*

De novo mutations are random mutations that are not inherited. Common *de novo* mutations are CNVs, SNPs, and other changes in DNA code.

Quantitative Trait Loci (QTLs)

QTLs are regions of DNA containing many nucleotide base pairs associated with continuous traits such as height or temperament. QTLs are not associated with simple discreet Mendelian traits such as hair and eye color. QTLs are associated with phenotypic traits influenced by many genes (polygenic).

Haplotypes

Haplotypes are a group of genes linked together and inherited as a group.

Epigenetics

An animal's DNA may contain a certain sequence of genetic code, but it may be locked out by epigenetic mechanisms. Environmental influences can either lock out sections of code or unlock them. For example, epigenetic mechanisms either upregulate (make more anxious) or downregulate the nervous system of rodents depending on how much the pregnant mother was stressed, or how often a rodent is attacked by another rodent (Nestler, 2011, 2012).

Lamarckism

Jean-Baptiste Lamarck (1744–1829) was a French naturalist remembered for a theory of inheritance of acquired characteristics, more commonly referred to as soft inheritance, Lamarckism, or the theory of use/disuse. Lamarck believed that animals could acquire a certain trait during its life-span and that the genetic code could be changed. For example, he believed that if a giraffe throughout its life stretched its neck to reach leaves high in the tree tops, its neck would become longer and these changes would be passed along to the offspring. Many of Lamarck's ideas were ridiculed. Today we know that a giraffe stretching its neck will not make it longer. However, his theory that genes can be changed during life and the changes can be passed on to offspring is actually now very credible. Part of Lamarck's argument is actually supported now by the emerging field of epigenetics. One part of this new branch of science is based on the proteins (histones) binding to the DNA, winding it into a small enough shape to fit into the cell (Probst *et al.*, 2009). Chemicals cause histones to bind either tighter or looser. Certain influences during life can influence the tightening or loosening of histones, these changes are then passed on (Sarma and Reinberg, 2005). If a piece of genetic code is loose, it will be easy to read, if it is tighter, it will not be read. This changes the expression of the genes and to a large extent explains the difference between twins at older ages, even when their genes are exactly the same (Fraga *et al.*, 2005; Poulsen *et al.*, 2007). Another epigenetic modification which serves to regulate gene expression without altering the underlying DNA sequence is DNA methylation. In simple terms, DNA methylation acts to "turn on" or "turn off" a gene. The methylation "lock" prevents a section of DNA code gene from being read and expressed. For example, maternal obesity before and during pregnancy in mice affects the establishment of body weight regulatory mechanisms in her baby. This is caused by methylation locks that lock out a section of genetic code. Overweight mothers give birth to offspring who become even heavier, resulting in amplification of obesity across generations (Champagne and Curley, 2009).

All of this has little to do with Lamarck, except to support his idea that individuals could evolve, not just populations as Darwin said. These traits are then passed on to their offspring. While no biologist believes that organisms can willfully change their physiology in response to their environment and pass those changes on to their offspring, some evidence suggests that the environment can make lasting changes to the expression of genes via epigenetic mechanisms—changes that may be passed on to future generations (Crews, 2010). Studies in rats show that epigenetics influences maternal behavior and the effect can be passed on from one generation to the next (Cameron *et al.*, 2008). The offspring of rat mothers who display high levels of nurturing behavior such as licking and grooming are less anxious and produce less stress hormones, compared to the offspring of less nurturing mothers. In turn, the female offspring of nurturing mothers become nurturing mothers themselves. The effects of maternal behavior are mediated in part through epigenetic mechanisms.

Brain Genetics More Complex Than Other Traits

For centuries, intensive selection in dogs has narrowed the gene pool for traits such as body shape and type of coat. Researchers have discovered that only a few genomic regions control many dog appearance traits (Boyko *et al.*, 2010). This is not true for behavior. Genetics that control brain development are much more complex. Selective breeding experiments in foxes and other experiments involving selection for appearance traits show that those traits are sometimes linked to behavior traits. Why does selecting for a calm temperament produce a black and white fox? When the first edition of this book was written, these unusually linked traits were unexplained. The long-running ENCODE project (2012) which is mapping the non-coding regions of DNA may help provide answers. Regions of non-coding DNA are not always located adjacent to the piece of code it regulates. There are long-range interactions. Sanyal *et al.* (2012) states that regulatory elements and coding DNA are in complex three-dimensional networks. Maybe when long strands of DNA are folded up, the temperament and coat color regions are folded up beside each other.

A BRIEF HISTORICAL REVIEW OF ANIMAL BEHAVIOR STUDY

This historical review is not intended to be completely comprehensive. Our objective is to discuss some of the early discoveries important for our current understanding of animal behavior, with particular emphasis on genetic influences on behavior in domestic animals.

Early in the 17th century, Descartes came to the conclusion "that the bodies of animals and men act wholly like machines and move in accordance with purely mechanical laws" (in Huxley, 1874). After Descartes, others undertook the task of explaining behavior as reactions to purely physical, chemical, or mechanical events. For the next three centuries, scientific thought on behavior oscillated between a mechanistic view that animals are "automatons" moving through life without consciousness or self-awareness and an opposing view that animals had thoughts and feelings similar to those of humans.

In "On the Origin of the Species" (1859), Darwin's ideas about evolution began to raise serious doubts about the mechanistic view of animal behavior. He noticed animals share many physical characteristics and was one of the first to discuss variation within a species, both in behavior and in physical appearance. Darwin believed that artificial selection and natural selection were intimately associated. Darwin (1868) cleverly outlined the theory of evolution without any knowledge of genetics. In "The Descent of Man" (1871)

Darwin concluded that temperament traits in domestic animals are inherited. He also believed, as did many other scientists of his time, that animals had subjective sensations and could think. Darwin wrote: "The differences in mind between man and the higher animals, great as it is, is certainly one of degree and not of kind."

Other scientists realized the implications of Darwin's theory on animal behavior and conducted experiments investigating instinct. Herrick (1908) observed the behavior of wild birds in order to determine, first, how their instincts are modified by their ability to learn, and second, the degree of intelligence they attain. On the issue of thinking in animals, Schroeder (1914) concluded: "The solution, if it ever comes, can scarcely fail to illuminate, if not the animal mind, at least that of man." It is evident that by the end of the 19th century, scientists studying animal behavior in natural environments learned that the mechanical approach could not explain all behavior.

BEHAVIORISM

During the middle of the 20th century, scientific thought again reverted to the mechanical approach and behaviorism reigned throughout America. The behaviorists ignored both genetic effects on behavior and the ability of animals to engage in flexible problem solving. The founder of behaviorism, J.B. Watson (1930), stated, "differences in the environment can explain all differences in behavior." He did not believe that genetics had any effect on behavior. In "The Behavior of Organisms," the psychologist B.F. Skinner (1958) wrote that all behavior could be explained by the principles of stimulus–response and operant conditioning.

The first author visited with Dr Skinner at Harvard University in 1968. Skinner responded to a question about the need for brain research by saying, "We don't need to know about the brain because we have operant conditioning" (Grandin and Johnson, 2005). Operant conditioning uses food rewards and punishments to train animals and shape their behavior. In a simple Skinner box experiment, a rat can be trained to push a lever to obtain food when a green light turns on, or to push a lever very quickly to avoid a shock when a red light appears. The signal light is the "conditioned stimulus." Rats and other animals can be trained to perform a complex sequence of behaviors by chaining together a series of simple operant responses. Skinner believed that even the most complex behaviors can be explained as a series of conditioned responses.

However, in a Skinner box a rat's behavior is very limited. It's a world with very little variation, and the rat has little opportunity to use its natural behaviors. It simply learns to push a lever to obtain food or prevent a shock. Skinnerian principles explain why a rat behaves a certain way in the sterile confines of a 30×30 cm Plexiglas box, but they don't reveal much about the behavior of a rat in the local dump. Outside of the laboratory, a rat's behavior is more complex.

Instincts *Versus* Learning

Skinner's influence on scientific thinking slowed a bit in 1961 following the publication of "The Misbehavior of Organisms" by Brelands and Brelands. Their paper described how Skinnerian behavioral principles collided with instincts. The Brelands were trained Skinnerian behaviorists who attempted to apply strict principles of operant conditioning to animals trained at fairs and carnivals. Ten years before this classic paper, Brelands (1951) wrote, "we are wholly affirmative and optimistic that principles derived from the laboratory can be applied to the extensive control of animal behavior under non laboratory condition." However, by 1961, after training more than 6000 animals as diverse as reindeers, cockatoos, raccoons, porpoises, and whales for exhibition in zoos, natural history museums, department store displays, fairgrounds, trade convention exhibits, and television, the Brelands wrote a second article featured in the American Psychologist (1961), which stated, "our background in behaviorism had not prepared us for the shock of some of our failures."

One of the failures occurred when the Brelands tried to teach chickens to stand quietly on a platform for 10–12 seconds before they received a food reward. The chickens would stand quietly on a platform in the beginning of training. However, once they learned to associate the platform with a food reward, half (50%) started scratching the platform, and another 25% developed other behaviors, such as pecking the platform. The Brelands salvaged this disaster by developing a wholly unplanned exhibit involving a chicken that turned on a jukebox and danced. They first trained chickens to pull a rubber loop which turned on some music. When the music started, the chickens would jump on the platform and start scratching and pecking until the food reward was delivered. This exhibit made use of the chicken's instinctive food-getting behavior. The first author remembers as a young adult seeing a similar exhibit at the Arizona State Fair of a piano-playing chicken in a little red barn. The hen would peck the keys of a toy piano when a quarter was put in the slot and would stop when the food came down the chute. This exhibit also worked because it was similar to a Skinner box in the laboratory. It also utilized the natural pecking behavior of the chicken.

The Brelands experienced another classic failure when they tried to teach raccoons to put coins in a piggy bank. Because raccoons are adept at manipulating objects with their hands, this task was initially easy. As training progressed, however, the raccoons began to rub the coins before depositing them in the bank. This behavior was similar to the washing behavior raccoons do as instinctive food-getting behavior. The raccoons at first had difficulty letting go of the coin and would hold and rub it. However, when the Brelands introduced a second coin, the raccoons became almost impossible to train. Rubbing the coins together "in a most miserly fashion," the raccoons got worse and worse as time went on. The Brelands concluded that the innate behaviors were suppressed during the early stages of training and sometimes for long into the training, but as training progressed, instinctive food-getting behaviors gradually replaced the conditioned behavior. The animals were unable to override their instincts and thus a conflict between conditioned and instinctive behaviors occurred.

ETHOLOGY

While Skinner and his fellow Americans were refining the principles of operant conditioning on thousands of rats and mice, ethology was being developed in Europe. Ethology is the study of animal behavior in natural environments. The primary concern of ethologists is instinctive or innate behavior (Eibl-Eibesfeldt and Kramer, 1958). Essentially, ethologists believed that the secrets to behavior are found in the animal's genes, and the way genes were modified during evolution to deal with particular environments. The ethological trend originated with Whitman (1898), who regarded behavioral reactions to be so constant and characteristic for each species that, like morphological structures, they may be of taxonomic significance. A similar opinion was held by Heinroth (1918, 1938). He trained newly hatched fledglings in isolation from adults of their own species and discovered that instinctive movements, such as preening, shaking, and scratching, were performed by young birds without observing other birds.

Understanding the mechanisms and programming of innate behavioral patterns and the motivation underlying behavior is the primary focus of ethologists. Konrad Lorenz (1939, 1965, 1981) and Niko Tinbergen (1948, 1951) catalogued the behavior of many animals in natural environments. Together they developed the ethogram. An ethogram is a complete listing of the behaviors an animal performs in its natural environment. The ethogram includes both innate and learned behaviors.

An interesting contribution to ethology came from studies on egg-rolling behavior in the graylag goose (Lorenz, 1965, 1981). When a brooding goose notices an egg outside her nest, Lorenz observed that an instinctive program triggers the goose to retrieve it. The goose fixates on the egg, rises to extend her neck and bill out over it, then gently rolls it back to the nest. This behavior is performed in a highly mechanical way. If the egg is removed as the goose begins to extend her neck, she still completes the pattern of rolling the nonexistent egg back to the nest. Lorenz (1939) and Tinbergen (1948) termed this a "fixed action pattern." Remarkably, Tinbergen also discovered that brooding geese can be stimulated to perform egg rolling on such items as beer cans and baseballs. The fixed action pattern of rolling the egg back to the nest can be triggered by anything outside the nest that even marginally resembles an egg. Tinbergen realized that geese possess a genetic-releasing mechanism for this fixed action pattern. Lorenz and Tinbergen called the object that triggers the release of a fixed action pattern "sign stimuli." When a mother bird sees the gaping mouth of her young, it triggers the maternal feeding behavior and the mother feeds her young. The gaping mouth is another example of sign stimuli acting as a switch that turns on the genetically determined program (Herrick, 1908; Tinbergen, 1951).

Ethologists also explained the innate escape response of newly hatched goslings. When goslings are tested with a cardboard silhouette in the shape of a hawk moving overhead, it triggers a characteristic escape response. The goslings will crouch or run. However, when the silhouette is reversed to look like a goose, there is no effect (Tinbergen, 1951). Several members of the research community doubted the existence of such a hard-wired instinct because other scientists failed to repeat these experiments (Hirsch *et al.*, 1955). Canty and Gould (1995), repeated the classic experiments and explained why the other experiments failed. First, only goslings under 7 days old respond to the silhouette. Second, a large silhouette, which casts a shadow, must be used. Third, goslings respond to the perceived predator differently depending upon the circumstances. For example, birds tested alone try to run away from the hawk silhouette and birds reared and tested in groups tend to crouch (Canty and Gould, 1995). Nevertheless, fear is likely to be the basis of the response. Ducklings were shown to have higher heart rate variability when they saw the hawk silhouette (Mueller and Parker, 1980). Research by Balaban (1997) indicates that species-specific vocalizations and head movements in chickens and quail are controlled by distinct cell groups in the brain. To prove this, Balaban transplanted neural tube cells from developing quail embryos into chicken embryos. Chickens hatched from the transplanted eggs exhibited species-specific quail songs and bobbing head movements.

Do similar fixed action patterns occur in mammals? Fentress (1973) conducted an experiment on mice which clearly showed that animals have instinctive species-specific behavior patterns which do not require learning. Day-old baby mice were anesthetized and had a portion of their front legs amputated. Enough of the leg remained that the mice could easily walk. The operations were performed before the baby mice had fully coordinated movements so there was no opportunity for learning. When the mice became adults, they still performed the species-specific face-washing behavior; normal mice close their eye just prior to the foreleg passing over the face, and in the amputees, the eye still closed before the nonexistent paw hit it. The amputees performed the face-washing routine as if they still had their paws. Fentress (1973) concluded that the experiment proved the existence of instincts in mammals.

Two years after the Breland's article, Jerry Hirsh (1963) at the University of Illinois, wrote a paper emphasizing the importance of studying individual differences. In it he wrote, "Individual differences are no accident. They are generated by properties of organisms as fundamental to behavior science as thermodynamic properties are to physical science."

Ethology and Behaviorism Provide Tools to Study Emotions and Behaviors

Both the behaviorists and the ethologists avoided the question of whether or not animals had emotions. They both developed a strictly functional approach to the motivations of behavior (deWaal, 2011). Until relatively recently, most behaviorists and ethologists did not get involved with neuroscience. A review of the neuroscience literature makes it clear that emotional systems in the brain drive behavior. The research tools provided by the disciplines of both ethology and behaviorism are essential to further our understanding of animal behavior. Ethology provides the methodology for studying animals in complex environments. Bateson (2012) discusses the need to study freely moving animals. Animal behavior is more complex in natural settings or on a farm. Lawrence (2008) reviewed the behavior literature and determined that domestic animal research was changing from studying the basic biology of domestic animal behavior to studying animal behavior related to specific animal welfare issues. Lawrence (2008) warns that too narrow a focus on specific welfare concerns may be detrimental to answering broader welfare issues such as the subjective state of animals.

Neuroscience and Behavior

Modern neuroscience supports Darwin's view on emotions in animals. All mammal brains are constructed with the same basic design. They all have a brainstem, limbic system, cerebellum, and cerebral cortex. The cerebral cortex is the part of the brain used for thinking and flexible problem solving. The major difference between the brains of people and animals is in the size and complexity of the cortex. The emotional systems serving as drivers for behavior are located in the subcortex, and are similar in all mammals (Panksepp, 2011). Primates have a larger and more complex cortex than a dog or a pig. Pigs have a more complex cortex than a rat or a mouse. Furthermore, all animals possess innate species-specific motor patterns which interact with experience and learning in determining behavior. Certain behaviors in both wild and domestic animals are governed largely by innate (hard-wired) programs. Behaviors for copulation, killing prey, nursing young, and nest building tend to be more instinctual and hard wired. Experience and learning play a larger role in behaviors that require more flexibility such as finding food, social interactions, and hunting.

Another basic principle to remember is that animals with large, complex brains are less governed by innate behavior patterns. For example, bird behavior is governed more by instinct than that of a dog, whereas an insect would have more hard-wired behavior patterns than a bird. This principle was clear to Yerkes (1905) who wrote:

Certain animals are markedly plastic or voluntary in their behavior, others are as markedly fixed or instinctive. In the primates, plasticity has reached its highest known stage of development; in the insects fixity has triumphed, instinctive action is predominant. The ant has apparently sacrificed adaptability to the development of ability to react quickly, accurately, and uniformly in a certain way. Roughly, animals might be separated into two classes: those which are in high degree capable of immediate adaptation to their conditions, and those that are apparently automatic since they depend upon instinct tendencies to action instead of upon rapid adaptation.

Emotional Systems Motivate Behavior

Great strides have been made in understanding how genetic factors affect behavior when scientists started by looking at brain systems that control emotions. This is the starting point for making it possible to sort out many conflicting results in behavioral studies.

The neuroscientist Jaak Panksepp outlined the major emotional systems located in the subcortical areas of the brain. The four main emotions are FEAR, RAGE, PANIC (separation distress), and SEEKING (novelty seeking) (Panksepp, 2005, 1998; Morris *et al.*, 2011). He also listed three additional emotional systems of LUST, CARE (mother–young nurturing behavior) and PLAY. Each primary system is associated with a genetically based subcortical brain network. Panksepp (2011) defined the basic emotional circuits of mammalian brains:

FEAR: An emotion induced by a perceived threat that causes animals to move quickly away from the location of the perceived threat, and sometimes hide. Fear should be distinguished from the emotion anxiety, which typically occurs without any certain or immediate external threat. Some examples of fear are reactions to exposure to sudden novelty, startle responses, and hiding from predators. Fear is sometimes referred to as behavioral reactivity, behavioral agitation, or a highly reactive temperament.

PANIC: Separation distress is an emotional condition in which an individual experiences excessive anxiety regarding separation from either home or from other animals that the individual has a strong emotional attachment to. One example of separation distress is a puppy or lamb vocalizing when it is separated from its mother. The PANIC system may also be activated when a single cow is separated from her herd. Sometimes referred to as social isolation stress or high social reinstatement behavior.

RAGE: A feeling of intense anger. Rage is associated with the fight-or-flight response and is activated in response to an external cue such as frustration or attempts to curtail an animal's activity. RAGE is the emotion that enables an animal to escape when it is in the jaws of a predator.

SEEKING: The seeking system (novelty seeking) in the brain motivates animals to become extremely energized to explore the world, but is not restricted to the narrow behavioristic concept of approach, or the pleasure/reinforcement system. Seeking is a broad action system in the brain that helps coordinate feelings of anticipation, eagerness, purpose and persistence, wanting, and desire.

This system promotes learning by urging animals to explore and to find resources needed for survival. A dog that excitedly sniffs and explores every room when turned loose in a strange house is an example of high SEEKING in an animal.

LUST: The lust system in the brain controls sexual desire or appetite. Sexual urges are mediated by specific brain circuits and chemistries that overlap but are distinct between males and females and are aroused by male and female hormones.

CARE: The maternal nurturing system that assures that parents take care of their offspring. Hormonal changes at the end of pregnancy activate maternal urges that promote social bonding with the offspring.

PLAY: A key function of the play system is to help young animals acquire social knowledge and refine subtle social interactions needed to thrive. One motivation for PLAY is the dopamine energized SEEKING system.

The existence of these emotional structures is well documented and review articles on this research can be found in Morris *et al.* (2011), Burgdorf and Panksepp (2006), and Panksepp (2011). Direct electrical or chemical stimulation of specific subcortical structures elicit emotional responses. The brain circuits controlling fear and seeking have been extensively mapped (LeDoux, 2000; Reynolds and Berridge, 2008).

Confusion of Emotional Systems May Confound Studies

In the behavior literature, many inconsistencies exist in papers on novelty seeking and fear. Novelty seeking may be confused with other emotional systems when terms such as activity level or emotional reactivity are used. Confusion may also exist between FEAR and PANIC when the term reactivity is used. The FEAR and PANIC (separation distress) systems have totally different functions. Fear keeps an animal away from danger and the PANIC system prevents the offspring from getting separated from its mother and helps to keep social groups together. Some scientists assume that an open-field test only evaluates fearfulness in animals (see Chapter 4). An animal alone in an open field may be reacting to separation from its mother or the social group. Confusing fear and separation distress is more likely when herding and flocking animals are tested in an open field, compared to animals that live a more solitary life. Panksepp's framework of the seven emotional systems may help sort out conflicting results in the scientific literature.

Research by Reynolds and Berridge (2008) has shown that the emotional traits of seeking novelty and fear are both controlled in a structure in the brain called the nucleus accumbens. When one end of the nucleus accumbens is stimulated, the animal becomes fearful. Stimulating the other end turns on seeking (Faure *et al.*, 2008). There is a mixture of fear and seeking receptors in the middle portion of the nucleus accumbens. The discovery of this function for the nucleus accumbens may explain the "curiously afraid" behavior we observed in cattle. Cattle curiously approach a novel clipboard laid on the ground, but when the wind flaps the paper, they fearfully jump back. When the paper stops moving, the cattle approach it again. The nucleus accumbens may be in SEEK mode when cattle voluntarily approach and switches into fear mode when the paper suddenly moves.

Genetics and Emotional Systems

Research clearly shows that genetic factors have a very strong effect on both fearfulness and novelty seeking (Campler *et al.*, 2008; Clinton *et al.*, 2007; Stead *et al.*, 2006; Yukihide *et al.*, 2005). Maternal factors had little effect on novelty seeking in rats (Stead *et al.*, 2005). Cross-fostering also had little effect, which shows that novelty seeking is highly heritable. Animals that are high seekers have more dopaminergic activity in the nucleus accumbens (Dellu *et al.*, 1996). In addition to the heritable component to fearfulness, environmental influences also have an effect on fearfulness. Stressful treatment of either a pregnant mother or her offspring can upregulate the fear system. Lemos *et al.* (2012) propose that severe stress can disable the appetitive system in the nucleus accumbens. To state it more simply, stress can break the animal's SEEK function and it will no longer explore. Other emotional systems are also subject to both genetic and environmental influences. The PANIC (separation distress) system is highly heritable. In sheep, separation distress measured by isolating a single animal and measuring how many times it bleats (vocalizes) shows that vocalization is highly heritable (Boissy *et al.*, 2005). Strength of the sex drive is also heritable. When the first Chinese pig breeds were imported into the U.S., caretakers observed that boars were more highly motivated to mate compared to U.S. and European commercial pigs. The Chinese pig was bred for large litter size and low levels of meat production.

INTERACTIONS BETWEEN GENETICS AND EXPERIENCE

Some behavior patterns are similar between different species, and some are found only in a particular species. For example, the neural programs that enable animals to walk are similar in most mammals (Melton, 1991; Grillner, 2011). On the other hand, courtship rituals in birds are very species-specific (Nottebohm, 1977). Some innate behavior patterns are very rigid and experience has little effect on them. Other instinctive behaviors can be modified by learning and experience. The flehmann, or lip curl response of a bull when he smells a cow in estrus, and the kneel-down (lordois) posture of a rat in estrus are examples of rigid behaviors. Suckling by newborn mammals is another example of a hard-wired behavioral system. Suckling behavior does not vary. Newborn mammals suckle almost everything put in their mouth.

An example of an innate behavior affected by learning is burrowing behavior in rats. Boice (1977) found that wild Norway rats and albino laboratory rats both dig elaborate burrows. Learning has some effect on the efficiency of burrowing, but the configuration of the burrows was the same for both the wild and domestic rats. Albino laboratory rats dug excellent burrows the first time they were exposed to an outdoor pen. Nest building in sows is another example of the interaction between instinct and learning. When a sow is having her first litter, she has an uncontrollable urge to build a nest. Nest building is hard-wired and hormonally driven. Widowski and Curtis (1989) showed that injections of prostaglandin F_2a induce nest building in sows. However, sows learn from experience how to build a better nest with each successful litter.

Other behaviors are almost entirely learned. Seagulls are known to drop shellfish on rocks to break them open, while others drop them on the road and let cars break them open (Grandin, 1995). Many animals ranging from apes to birds use tools to obtain food. New Calendonian crows use complicated tools to obtain food and can solve problems other birds cannot (Weir and Kacelnik 2006). Griffin (1994) and Dawkins (1993) provide many examples of complex learned behaviors and flexible problem solving in animals.

Innate behaviors used for finding food, such as grazing, scavenging, or hunting, are more dependent on learning than behaviors used to consume food. Sexual behavior, nesting, eating, and prey-killing behaviors tend to be governed more by instinct (Gould, 1977). The greater dependence on learning to find food makes animals in the wild more flexible and able to adapt to a variety of environments. Behaviors used to kill or consume food can be the same in any environment. Mayr (1974) called these different behavioral systems "open" or "closed" to the effects of experience. A lion hunting her prey is an example of an open system. The hunting female lion recognizes her prey from a distance and carefully stalks her approach. Herrick (1910) wrote, "the details of the hunt vary every time she hunts. Therefore, no combination of simple reflex arcs laid down in the nervous system will be adequate to meet the infinite variations of the requirements for obtaining food."

Interactions Between Instinctual Hardwired Behavior and Experience

Some of the interactions between genetics and experience have very complex effects on behavior. In birds, the chaffinch learns to sing its species-specific song even when reared in a sound-proof box where it is unable to hear other birds (Nottebohm, 1970, 1979). However, when chaffinches are allowed to hear other birds sing, they develop a more complex song. The basic pattern of canary song emerges even in the absence of conspecific (flock-mate) auditory models (Metfessel, 1935; Poulsen, 1959). Young canaries imitate the song of adult canaries they can hear, and when reared in groups they develop song patterns that they all share (Nottebohm, 1977). Many birds, such as the white crowned sparrow, chaffinch, and parrot, can develop local song dialects (Nottebohm *et al.*, 1976; Adler, 1996). Sparrows are able to learn songs by listening to recordings of songs with either pure tones or harmonic overtones. Birds trained with harmonic overtones learned to sing songs with harmonic overtones, but 1 year later, 85% of their songs reverted back to innate pure tone patterns (Nowicki and Marler, 1988). Experiments by Mundinger (1995) attempted to determine the relative contribution of genetics and learning in bird song. Inbred lines of roller and border canaries were used in this study along with a hybrid cross of the two. The rollers were cross-fostered to border hens and vice versa to control for effects of maternal behavior. The roller and border males preferred to sing innate song patterns instead of copying their tutors. The hybrids preferred to learn some of both songs. Furthermore, canaries are capable of learning parts of an alien song but have a definite preference for their own songs. Comparing these animals to those in Brelands and Brelands (1961) exhibits, birds can be trained sing a different song, but genetically determined patterns have a strong tendency to override learning. In reviewing all this literature, it became clear that innate patterns in mammals can be overridden. Unfortunately, the animals tend to revert back to innate behavior patterns.

THE PARADOX OF NOVELTY

Novelty is anything new or strange in an animal's environment. Novelty is a paradox because it is both fear-provoking and attractive. Paradoxically, it is most fear-provoking and attractive to animals with a nervous, excitable temperament. Skinner (1922) observed that pronghorn antelope, a very flighty animal, will approach a person lying on the ground waving a red flag. Kruuk (1972) further observed attraction and reaction to novelty in Thompson's gazelles in Africa. In small groups, Thompson's gazelles are most watchful for predators (Elgar, 1989). Animals that survive in the wild by flight are more attentive to novelty than more placid animals. Gazelles can also distinguish between a dangerous hunting predator and one that is not hunting. In Thompson's gazelles, the most dangerous predators attract the highest degrees of attraction. They often move close to a cheetah when the cheetah is not hunting. Furthermore, when predators walk through a herd of Thompson's gazelles, the size of the flight zone varies depending on the species of predator.

Reaction to Novelty

Highly reactive animals are more likely to have a major fear reaction when confronted with sudden novelty. In domestic animals, examples of sudden novelty include being placed in a new cage, transport in a strange vehicle, an unexpected loud noise, or being placed in an open field. Using various experimental environments, Hennessy and Levine (1978) found that rats show varying degrees of stress and stress hormone levels proportional to the degree of novelty of the environment they are placed in. A glass jar is totally novel in appearance compared to a lab cage the animal is accustomed to. Being placed in a glass jar was more stressful for rats than a familiar lab cage with no bedding.

Studies of reactions to novelty in farm animals have been conducted by Moberg and Wood (1982), Stephens and Toner (1975), and Dantzer and Mormede (1983). Calves show the highest degrees of stress when placed in an open-field test arena very dissimilar from their home pen (Dantzer and Mormede, 1983). Calves raised indoors were more stressed by an outdoor arena and calves raised outdoors were more stressed by an indoor arena. The second author is familiar with similar responses in horses. When horses are taken to the mountains for the first time, a well-trained riding horse accustomed to different show rings may panic when it sees a butterfly or hears a twig snapping on a mountain trail.

Genetic Factors and the Need for Novelty

In mammals and birds, normal development of the brain and sense organs requires novelty and varied sensory input. Nobel prize winning research of Hubel and Wiesel (1970) showed that the visual system is permanently damaged if kittens do not receive varied visual input during development. Dogs are more excitable when raised in barren and non-stimulating environments (Melzack and Burns, 1965; Walsh and Cummins, 1975). Schultz (1965) stated; "when stimulus variation is restricted central regulation of threshold sensitivities will function to lower sensory thresholds." Krushinski (1960) studied the influence of isolated conditions of rearing on the development of passive defense reactions (fearful aggression) in dogs and found that the expression of a well-marked fear reaction depends on the genotype of the animal. In this experiment, Airedales and German shepherds were reared under conditions of freedom (in homes) and in isolation (in kennels). Krushinski (1960) found that the passive defense reaction developed more acutely and reached a greater degree in the German shepherds kept in isolation compared to the Airdales. In general, animals reared in isolation become more sensitive to sensory stimulation because the nervous system attempts to readjust for the previous lack of stimulation.

In an experiment with chickens, Murphy (1977) found that chicks from a flighty genetic line were more likely to become highly agitated when a novel ball was placed in their pen, but were more attracted to a novel food than birds from a calm line. Cooper and Zubeck (1958) and Henderson (1968) found that rats bred to be dull greatly improved in maze learning when housed in a cage with many different objects. However, enriched environments had little effect on the rats bred for high intelligence. Greenough and Juraska (1979) found that rearing rats in an environment with many novel objects improves learning and results in increased growth of dendrites (nerve endings in the brain).

Pigs raised in barren concrete pens also actively seek stimulation (Grandin, 1989a,b; Wood-Gush and Beilharz, 1983; Wood-Gush and Vestergaard, 1991). Piglets allowed to choose between a familiar object and a novel object prefer the novel object (Wood-Gush and Vestergaard, 1991). Pigs raised on concrete are strongly attracted to objects to chew on and manipulate. The first author has observed that nervous, excitable hybrid pigs often chew and bite vigorously on boots or coveralls. This behavior is less common in placid genetic lines of pigs. Although hybrid pigs are highly attracted to novelty, tossing a novel object into their pen will initially cause a strong flight response. Compared to calm genetic lines, nervous-hybrid pigs pile up and squeal more when startled. Pork producers report that nervous, fast-growing, lean hybrid pigs also tail-bite other pigs more often than calmer genetic lines of pigs (see Chapter 11). Tail biting occurs more often when pigs are housed on a concrete slatted floor which provides no opportunity for rooting.

Practical experience by both authors suggests that highly reactive horses are more likely to engage in vices such as cribbing or stall weaving when housed in stalls or runs where they receive little exercise. Denied variety and novelty in their environments, highly reactive animals adapt poorly compared to animals from calmer genetic lines (Huck and Price, 1975).

In summary, in both wild and domestic animals, novelty is both highly feared and necessary. Novelty is more desirable when animals can approach it voluntarily. Unfortunately, novelty is also fear-provoking when animals are suddenly confronted with it.

Temperament is Not Just About Fear

Early research showed that animals as diverse as rats, chickens, cattle, pigs, and humans, genetic factors influence differences in temperament (fearfulness) (Royce *et al.*, 1970; Bilzard, 1971; Broadhurst, 1975; Fordyce *et al.*, 1988; Fujita *et al.*, 1994; Grandin, 1993b; Hemsworth *et al.*, 1990; Kagan *et al.*, 1988; Murphy, 1977; Murphey *et al.*, 1980a; Reese *et al.*, 1983; Tulloh, 1961). Trillmich and Hudson (2011) suggest that a broader area of animal personality needs to be studied. Today the concept of temperament includes all the neurotransmitter systems that affect the strength of different emotional systems. Barr (2012) contains an outline of how behaviors such as fear and exploring are affected by different neurotransmitter systems. Some individuals are wary and fearful and others are calm and placid. Boissy (1995) stated, "fearfulness is a basic psychological characteristic of the individual than predisposes it to perceive and react in a similar manner to a wide range of potentially frightening events." In all animals, genetic factors influence reactions to situations which cause fear (Boissy and Bouissou, 1995; Davis, 1992; Kagan *et al.*, 1988; Murphey *et al.*, 1980b). Therefore, temperament is partially determined by an individual animal's fear response. Rogan and LeDoux (1996) suggest that fear is the product of a neural system that evolved to detect danger and causes an animal to make a response to protect itself. Plomin and Daniels (1987) found a substantial genetic influence on shyness (fearfulness) in human children. Shy behavior in novel situations is considered a stable psychological characteristic of certain individuals. Shyness is also suggested to be among the most heritable dimensions of human temperament throughout the life-span.

In an experiment designed to control for maternal effects on temperament and emotionality, Broadhurst (1960) conducted cross-fostering experiments on Maudsley Reactive (MR) and Non-Reactive (MNR) rats. These lines of rats are genetically selected for high or low levels of emotional reactivity. The results showed that maternal effects were not great enough to completely mask the temperament differences between the two lines (Broadhurst, 1960; Eysenck and Broadhurst, 1964). Maternal effects can affect temperament, but they are not great enough to completely change the temperament of a cross-fostered animal which has a temperament very different from that of the foster mother. In an extensive review of the literature, Broadhurst (1975) examined the role of heredity in the formation of behavior and found that differences in temperament between rats persist when the animals are all raised in the same environment.

In their study of genetic effects on dog behavior in a large sample, Fuller and Thompson (1978) found that "simply providing the same defined controlled environment for each genetic group is not enough. Conditions must not only be uniform for all groups, but also favorable to the development of the behavior of interest." In wartime Russia, Krushinski (1960) investigated the ability of dogs to be trained for the antitank service or as trail dogs trained to track human scent. The dogs were tied to a spike driven into the ground and the person who regularly looked after them let them lick from a bowl of food for a few moments then summoned the dog to follow the man as he moved 10–15 meters away. Activity of each dog was measured with a pedometer for the next 2 minutes. The most active dogs were found to be the best antitank dogs. They were also fearless. In the antitank service, dogs were trained to run up to a tank and either run alongside of it or penetrate under the caterpillars of the tank. In order to do this, the dogs had to overcome their natural fear of a tank moving toward them at high speed. The less active dogs (as measured by the pedometer) were found to make the best trailer dogs. They slowly followed a trail and kept their noses carefully to the scent while negotiating the corners and turns on the trail. The more active dogs trailed at too high a speed and often jumped the corners and turns in the trail, which sometimes resulted in switching to another trail.

Mahut (1958) demonstrated an example of differences in fear responses between beagles and terriers. When frightened, beagles freeze and terriers run around frantically. In domestic livestock, measuring fear reactions during restraint or in an open field test reveals differences in temperament both between breeds and between individuals within a breed (Dantzer and Mormede, 1983; Grandin, 1993a; Murphey *et al.*, 1980b, 1981; Tulloh, 1961). Fearful, flighty animals become more agitated and struggle more violently when restrained for vaccinations and other procedures (Fordyce *et al.*, 1988; Grandin, 1993a). Fear is likely to be the main cause of agitation during restraint in cattle, horses, pigs, and chickens. Recent studies on the genetic effects on behavior during transport handling, and restraint of these animals are discussed in Chapter 4.

Species Differences in Emotional Reactions to Similar Tests

In an open-field test, a single animal is placed in an arena (Hall, 1934). Rodents often stay close to the arena walls whereas cattle may run around wildly and attempt to escape. Possibly the reaction of the rodent is motivated by fear and the reaction of a single bovine may be PANIC (separation distress). The rodent has an instinctual fear of open spaces and stays close to the walls. The motivation is to avoid open spaces where it likely to be seen by a predator. The bovine is motivated to rejoin its herdmates. The open-field test may be measuring different emotional systems in these two species. When fear is the motivator, a frightened animal may react in two different ways. It may run around frantically and try to escape or in another situation, it may freeze and stay still. Chickens often freeze when handled by humans. Jones (1984) called this "tonic immobility." The chickens become so frightened they cannot move. In cattle, brahman *Bos indicus* cattle are more likely to go into tonic immobility than *Bos Taurus* breeds such as Hereford. In wildlife, forceful capture can cause enough fear to sometimes inflict fatal heart damage. Wildlife biologists call this capture myopathy. In summary, much is known about the complex phenomenon of fear, but many questions still remain.

BIOLOGICAL BASIS OF FEAR

Genetic factors influence the intensity of fear reactions. Genetic factors can also greatly reduce or increase fear reaction in domestic animals (Flint *et al.*, 1995; Parsons, 1988; Price, 1984). Research in humans has clearly revealed some of the genetic mechanisms governing the inheritance of anxiety (Lesch *et al.*, 1996). LeDoux (1992) and Rogan and LeDoux (1996) state that all vertebrates can be fear-conditioned. Davis (1992) recently reviewed studies on the biological basis of fear. Overwhelming evidence points to the amygdala as the fear center in the brain. A small bilateral structure located in the limbic system, the amygdala is where the triggers for "flight or fight" are located. Electrical stimulation of the amygdala is known to increase stress hormones in rats and cats (Matheson *et al.*, 1971; Setckleiv *et al.*, 1961). Destroying the amygdala can make a wild rat tame and reduce its emotionality (Kemble *et al.*, 1984). Destroying the amygdala also makes it impossible to provoke a fear response in animals (Davis, 1992). Blanchard and Blanchard (1972) showed that rats lose all of their fear of cats when the amygdala is lesioned. Furthermore, when a rat learns that a signal light means an impending electric shock, a normal response is to freeze. Destroying the amygdala eliminates this response (Blanchard and Blanchard, 1972; LeDoux *et al.*, 1988, 1990). Finally, electrical stimulation of the amygdala makes humans fearful (Gloor *et al.*, 1981). Animal studies also show that stimulation of the amygdala triggers a pattern of responses from the autonomic nervous system similar to that found in humans when they feel fear (Davis, 1992). Different types of fear are processed in separate neural circuits. Fear of predators, learned fear, and fear of conspecifics are some examples (Gross and Canteras, 2012). In birds, fear of a novel object is different from fear during tonic immobility. Fear can either elicit a response to freeze or flight to escape. Selection of quail for either high or low tonic immobility had little effect on reaction to a novel object (Saint-Dizier *et al.*, 2008). Since the publication of the first edition of this book, scientists have learned that fear is not a single emotional trait. Further research shows that adding a third variable of human handling indicates that fear has more than one independent dimension (Richard *et al.*, 2010).

Heart rate, blood pressure, and respiration also change in animals when the flight or fight response is activated (Manuck and Schaefer, 1978). All these autonomic functions have neural circuits to the amygdala. Fear can be measured in animals by recording changes in autonomic activity. In humans, Manuck and Schaefer (1978) found tremendous differences in cardiovascular reactivity in response to stress, reflecting a stable genetic characteristic of individuals.

Fearfulness and Instinct

Fearfulness and instinct can conflict. This principle was observed first hand by the second author during his experience raising Queensland Blue Heeler dogs. Annie's first litter was a completely novel experience because she had never observed another dog giving birth or nursing pups. She was clearly frightened when the first pup was born. It was obvious she did not know what the pup was. However, as soon as she smelled it, her maternal instinct took over and a constant uncontrollable licking began. Two years later, Annie's daughter Kay had her first litter. Kay was more fearful than her mother and her highly nervous temperament overrode her innate licking program. When each pup was born, Kay ran wildly around the room and would not go near them. The second author had to intervene and place the pups under Kay's nose. Otherwise, they may have died, Kay's nervous temperament and fearfulness were a stronger motivation than her motherly instinct.

NERVOUS SYSTEM REACTIVITY CHANGED BY THE ENVIRONMENT

Raising young animals in barren environments devoid of variety and sensory stimulation will have an effect on development of the nervous system. It can cause animals to be more reactive and excitable as adults. This is a long-lasting, environment-induced change in how the nervous system reacts to various stimuli. Effects of deprivation during early development are also relatively permanent. Melzack and Burns (1965) found that puppies raised in barren kennels developed into hyperexcitable adults. In one experiment, deprived dogs reacted with "diffuse excitement" and ran around a room more than control dogs raised in homes by people. Presenting novel objects to the deprived dogs also result in "diffuse excitement." Furthermore, the EEGs of the kennel-raised dogs remained abnormal even after they were removed from the kennel (Melzack and Burns, 1965). Research by Simons and Land (1987) show the somatosensory cortex in the brains of baby rats do not develop normally if sensory input is eliminated by trimming their whiskers. A lack of sensory input made the brain hypersensitive to stimulation. The effects persisted even after the whiskers had grown back.

Emotional reactivity develops in the nervous system during early gestation. Denenberg and Whimbey (1968) showed that handling a pregnant rat can cause her offspring to be more emotional and explore less in an open field compared to control animals. This experiment is significant because it shows that handling the pregnant mother had the opposite effect on the behavior of the infant pups. Handling and stressing pregnant mothers changed the gestational environment of the fetus resulting in nervous offspring. However, handling newborn rats by briefly picking them up and setting them in a container reduced emotional reactivity when the rats became adults (Denenberg and Whimbey, 1968). The handled rats developed a calmer temperament.

The adrenal glands are known to have an effect on behavior (Fuller and Thompson, 1978). The inner portions of the adrenals secrete the hormones, adrenaline and noradrenaline, while the outer cortex secretes the sex hormones androgens and estrogens (reproductive hormones); and various corticosteroids (stress hormones). Yeakel and Rhoades (1941) found that Hall's (1938) emotional rats had larger adrenals and thyroids compared to the non-emotional rats. Richter (1952, 1954) found a decrease in the size of the adrenal glands in Norway rats accompanied by domestication. Several line and strain differences have been found since these early reports. Furthermore, Levine (1968) and Levine *et al.* (1967) showed that brief holding of baby rats reduces the response of the adrenal gland to stress. Denenberg *et al.* (1967) concluded that early handling may lead to major changes in the neuroendocrine system.

Taming Does Not Change Nervous System Reactivity

Adult wild rats can be tamed and become accustomed to handling by people (Galef, 1970). This is strictly learned behavior. Taming full-grown wild animals to become accustomed to holding by people will not diminish their response to a sudden novel stimulus. This principle was demonstrated by Grandin *et al.* (1995) in training wild antelope at the Denver Zoo for low-stress blood testing. Nyala are African antelope with a hair-trigger flight response used to escape from predators. During handling in zoos for veterinary treatments, nyala are often highly stressed and sometimes panic and injure themselves. Over a period of three months, Grandin *et al.* (1995) trained nyala to enter a box and stand quietly for blood tests while being fed treats. Each new step in the training had to be done slowly and carefully. Ten days were required to habituate the nyala to the sound of the doors on the box being closed.

All the training and petting by zoo keepers did not change the nyala's response to a sudden, novel stimulus. When the nyala saw repairmen on the barn roof, they suddenly reacted with a powerful fear response and crashed into a fence. They had become accustomed to seeing people standing at the perimeter of the exhibit, but the sight of people on the roof was novel and very frightening. Sudden movements, such as raising a camera up for a picture, also caused the nyala to flee.

Domestic *Versus* Wild

Wild herding species show much stronger fear responses to sudden novelty compared to domestic ruminants such as cattle and sheep. Domestic ruminants have attenuated flight responses due to years of selective breeding (Price, 1984). Wild ruminants learn to adapt in captivity and associate people with food, but are more likely to become agitated and injure themselves when frightened by novel stimulus (Grandin, 1993b, 1997). Fear is more likely when they are prevented from fleeing by a fence or other barrier. Principles for training and handling all herding animals are basically similar. Training procedures used on flighty antelope or placid domestic sheep are the same. The only difference is the amount of time required. Grandin (1989c) demonstrated this by training placid Suffolk sheep to voluntarily enter a tilting restraining device in one afternoon. The nyala at the Denver zoo took three months to train. Each new procedure had to be introduced in smaller increments to prevent an explosive flight reaction.

NEOTENY

Neoteny is the retention of the juvenile features in an adult animal. Genetic factors influence the degree of neoteny in individuals. Neoteny is manifested both behaviorally and physically. In the forward to "The Wild Canids" (Fox, 1975), Conrad Lorenz adds a few of his observations on neoteny and the problems of domestication:

The problems of domestication have been an obsession with me for many years. On the one hand, I am convinced that man owes the life-long persistence of his constitutive curiosity and explorative playfulness to a partial neoteny which is indubitably a consequence of domestication. In a curiously analogous manner does the domestic dog owe its permanent attachment to its master to a behavioral neoteny that prevents it from ever wanting to be a pack leader? On the other hand, domestication is apt to cause an equally alarming disintegration of valuable behavioral traits and an equally alarming exaggeration of less desirable ones.

Infantile characteristics in domestic animals are discussed by Price (1984), Lambooij and van Putten (1993), Coppinger and Coppinger (1993), Coppinger and Scheider (1993), and Coppinger *et al.* (1987). The shortened muzzle in dogs and pigs is an example. Domestic animals have been selected for a juvenile head shape, shortened muzzles, and other features (Coppinger and Smith, 1983). Furthermore, retaining juvenile traits makes animals more tractable and easy to handle. The physical changes are also related to changes in behavior.

Genetic studies point to the wolf as the ancestor of domestic dogs (Isaac, 1970). During domestication, dogs retained many infant wolf behaviors. For example, wolf pups bark and yap frequently, but adult wolves rarely bark. Domestic dogs bark frequently (Fox, 1975; Scott and Fuller, 1965). Wolves have hard-wired instinctive behavior patterns that determine dominance or submission in social relationships. In domestic dogs, the ancestral social behavior patterns of the wolf are fragmented and incomplete. Frank and Frank (1982) observed that the rigid social behavior of the wolf has disintegrated into "an assortment of independent behavioral fragments." Malamutes raised with wolf pups fail to read the social behavior signals of the wolf pups. Further comparisons found that the physical development of motor skills is slower in the malamute. Goodwin *et al.* (1997) studied 10 different dog breeds, ranging from German shepherds and Siberian huskies to bulldogs, cocker spaniels, and terriers. They found that breeds which retained the greatest repertoire of wolf-like social behaviors were breeds that physically resembled wolves, such as German shepherds and huskies. Barnett *et al.* (1979) and Price (1984) both conclude that experience may also cause animals to retain juvenile traits. Gould (1977) found that the effects of neoteny are determined by changes in a few genes that determine the timing of different developmental stages. On average, wolves are smarter than dogs on a test involving spatial orientation and pulling rope. All the adult wolves passed the most complex version of this test, but only five out of 40 German shepherds could do it (Hiestand, 2011).

OVERSELECTION FOR SPECIFIC TRAITS

There are countless examples in the medical literature of serious problems caused by continuous selection for a single trait (Dykman *et al.*, 1969; Steinberg *et al.*, 1994). People experienced in animal husbandry know that overselection for single traits can ruin animals. Good dog breeders have known this for centuries. Some traits that appear unrelated are in fact linked. Wright (1922, 1978) demonstrated this clearly by continuous selection for hair color and hair patterns in inbred strains of guinea pigs. Selection for hair color and patterns resulted in decreased reproduction in all the strains. Furthermore, differences in temperament, body conformation, and the size and shape of internal organs were found. Belyaev (1979) showed that continuous selection for tameness in foxes reduced maternal behavior and caused neurological problems. Graded changes occured in many traits over several years of continuous selection for tame behavior. Physiological and behavioral problems increased with each successive generation. In fact, some of the tamest foxes developed abnormal maternal behavior and cannibalized their pups. Belyaev *et al.* (1981) called this "destabilizing selection," in contrast to "stabilizing selection," found in nature (Dobzhansky, 1970; Gould, 1977).

There are countless examples in the veterinary medical literature of abnormal bone structure and other physiological defects caused by over selecting for appearance traits in dog breeds (Ott, 1996). The abnormalities range from bulldogs with breathing problems to German shepherds with hip problems. Scott and Fuller (1965) reported the negative effects of continuous selection for a certain head shape in cocker spaniels:

In our experiments, we began with what were considered good breeding stocks, with a fair number of champions in their ancestry. When we bred these animals to their close relatives for even one or two generations, we uncovered serious defects in every breed . . . Cocker spaniels are selected for a broad forehead with prominent eyes and a pronounced "stop" or angle, between the nose and forehead. When we examined the brains of some of these animals during autopsy, we found that they showed a mild degree of hydroencephaly, that is, in selecting for skull shape, the breeders accidentally selected for a brain defect in some individuals. Besides all this, in most of our strains, only about 50% of the females were capable of rearing normal, healthy litters, even under nearly ideal conditions of care.

Overselection in Livestock

In pigs and cattle, indiscriminate selection for production traits such as rapid gain and leanness result in more excitable temperaments in pigs (Grandin, 1994). The first author has observed thousands of modern lean hybrid pigs. Lean hybrids have a greater startle response and are more excitable and difficult to drive through races compared to older genetic lines with more back fat. Separating single animals from the group is also more difficult. Research shows that cattle with an excitable temperament have lower weight gains and more meat quality problems (Café *et al.*, 2011; Silveira *et al.*, 2012; Voisinet *et al.*, 1997a, b). This research illustrates that selection away from a very excitable temperament would be beneficial. However, ranchers report that mothering ability is reduced by overselecting for extremely calm temperaments. There are always trade offs.

Links Between Different Traits

Casual observations by the first author indicate that the most excitable pigs and cattle have long slender bodies and fine bones. Some lean hybrid pigs have weak legs and normally brown-eyed pigs now have blue eyes. Blue eyes are often associated with neurological problems (Bergsma and Brown, 1971; Schaible, 1963). Furthermore, pigs and cattle with large bulging muscles often have calmer temperaments compared to lean animals with less muscle definition. However, animals with double muscling (hypertrophy) have more excitable temperaments (Holmes *et al.*, 1972). Double muscling is abnormal and may have opposite effects on temperament compared to normal muscling.

Deafness in pointer dogs selected for nervousness is another example of apparently unrelated traits being linked (Klein *et al.*, 1987; 1988). There appears to be a relationship between thermoregulation and aggressiveness. Sluyter *et al.* (1995) found that wild mice selected for aggressiveness used larger amounts of cotton to build their nests than mice selected for low aggression. This effect occurred in both laboratory and wild strains of mice.

The complexity of genetic interactions continues to frustrate researchers using high-tech "knockout" gene procedures. Genes are knocked out in a gene-targeting procedure, and prevented from performing their normal functions. Knockout experiments show that blocking different genes can have unexpected effects on behavior. In one experiment, super-aggressive mice were created when genes involved with learning were inactivated (Chen *et al.*, 1994). The mutant mice had little or no fear and fought until they broke their backs. In another experiment, knockout mutants demonstrated normal behavior until they had pups, which they failed to care for (Brown *et al.*, 1996). In still another experiment, Konig *et al.* (1996) disabled the gene that produces encephalin (a brain opioid substance) and found unexpected results. Enkephalin is a substance normally involved in pain perception. Mice deficient in this substance were very nervous and anxious. They ran frantically around their cages in response to noise. Traits are linked in unexpected ways and completely isolating single gene effects may be impossible. Researchers suggest using caution and being careful not to jump to conclusions about claims by those who say they found an "aggression gene" or a "maternal gene" or an "anxiety gene." To use an engineering analogy, one cannot conclude the "picture center" in a television set was found after cutting one circuit inside the set used to create the pictures. Gerlai (1996) and Crawley (1996) also warn that knocking out the same gene in two different species may have different effects on behavior. This is due to the complex interactions between many different genes.

Thirty-five years ago behavior geneticists found that inheritance of behavior is complex. Fuller and Thompson (1978) concluded, "It has been found repeatedly that no one genetic mechanism accounts exclusively for a particular kind of behavior."

Transgenic Mice

Since the first edition of this book was published, hundreds of papers have been published on mouse models of conditions such as autism and various behavioral extremes (Roullet and Crawley, 2011; Silverman *et al.*, 2010). Models of high-novelty-seeking mice, timid mice, fearful mice, and obsessive-compulsive mice are used to study these conditions in humans (Lipkind *et al.*, 2004). There are different behavioral tests for the hundreds of types of special mice. Crawley (2007, 2008) devoted an entire book to behavioral testing of transgenic mice. Eisener-Dorman *et al.* (2009) stated that knock out procedures do not always work correctly to produce the desired effects and is concerned that transgenic mice are not always what they were originally bred to be.

Random Factors

Behavioral geneticists find it impossible to completely control variation in some traits. Gartner (1990) found that breeding genetically similar inbred lines of rats failed to stop weight fluctuations. Even under highly standardized laboratory conditions, body weights continued to fluctuate between animals. Pig breeders also observe that commercially bred hybrid lines of pigs do not gain weight at the same rate. Even in genetically identical animals, random unknown factors affect variability. *In utero* factors may be one cause, other causes are unknown. Dr Daryl Tatum and students at Colorado State University found both body conformation and meat quality variation in 50% English (*Bos taurus*) and 50% Brahman (*Bos indicus*) cattle. Some animals showed more Brahman characteristics, with larger humps and longer ears. However, the body conformation was not half English and half Brahman, and meat characteristics varied as well. Cattle that looked more Brahman had tougher meat. The animals had about 10% variation from the body shape and meat characteristics of Brahman half-bloods.

Gartner (1990) concluded that up to 90% of the cause of random variability cannot be explained by differences in the physical environment. In both mice and cattle, random factors affected body weights. Gartner (1990) believes that random factors influence development either before or shortly after fertilization. The interactions between environmental and genetic factors are complex. Genetic makeup and environmental conditions influence behavior. In subsequent chapters in this book, the interactions of genetics and environment will be discussed in greater detail.

CONCLUSIONS

There is a complex interaction between genetic and environmental factors which determines how an animal will behave. The animal's temperament is influenced by both genetics and learning. Another principle to remember is that changes in one trait, such as temperament, can have unexpected effects on other apparently unrelated traits. Overselection for a single trait may result in undesirable changes in behavioral and physical traits.

REFERENCES

Adler, A., 1996. How songbirds get their tunes. Sci. News. 149, 280–281.

Agnvall, B., Jongren, M., Strandberg, E., Jensen, P., 2012. Heritability and genetic correlations of fear related behavior in Red Jungle fowl—possible implications for early domestication. PLoS ONE 7 (4), 35162.

Anderson, L., 2012. How selective sweeps in domestic animals provide new insight into biological mechanisms. J. Intern. Med. 271, 1–14.

Balaban, E., 1997. Changes in multiple brain regions underlie species differences in complex, congenital behavior. Proc. Natl. Acad. Sci. USA 94, 2001–2006.

Barnett, S.A., Dickson, R.G., Hocking, W.E., 1979. Genotype and environment in the social interactions of wild and domestic "Norway" rats. Aggress. Behav. 5, 105–119.

Barr, C.S., 2012. Temperament in animals. In: Zentner, M., Shiner, R.L. (Eds.), Handbook of Temperament. Guildford Press, pp. 251–272.

Bateson, P., 2012. Behavioral biology: past and a future. Ethology 118, 216–221.

Belyaev, D.K., 1979. Destabilizing selection as a factor in domestication. J. Hered. 70, 301–308.

Belyaev, D.K., Ruvinsky, A.O., Trut, L.N., 1981. Inherited activation–inactivation of the star gene in foxes. J. Hered. 72, 267–274.

Bergsma, D.R., Brown, K.S., 1971. White fur, blue eyes, and deafness in the domestic cat. J. Hered. 62, 171–185.

Blanchard, D.C., Blanchard, R.J., 1972. Innate and conditioned reactions to threat in rats with amygdaloid lesions. J. Comp. Physiol. Psychol. 81, 281–290.

Blizard, D.A., 1971. Autonomic reactivity in the rat: effects of genetic selection for emotionality. J. Comp. Physiol. Psychol. 76, 282–289.

Boessneck, J., 1985. (Domestication and its sequelae) Die Domestikation und irhe Folgen. Tieraerzil. Prax. 13 (4), 479–497, In German.

Boice, R., 1977. Burrows of wild and albino rats: effects of domestication, outdoor raising, age, experience and maternal state. J. Comp. Physiol. Psych. 91, 649–661.

Boissy, A., 1995. Fear and fearfulness in animals. Q. Rev. Biol. 70 (2), 165–191.

Boissy, A., Bouissou, M.F., 1995. Assessment of individual differences in behavioral reactons of heifers exposed to various fear-eliciting situations. App. Aim. Behav. Sci. 46, 17–31.

Boissy, A., Bouix, J., Orgeur, P., Paindron, P., Bibe, B., LeNeindre, P., 2005. Genetic analysis of emotional reactivitiy of sheep, effects of genotypes of lambs and their dams. Genet. Sel. Evol. 37, 381–401.

Boyko, A.R., Quignon, P., Li, L., Schoenebeck, J.J., Degenhardt, J.D., Lohmueller, K.E., et al., 2010. A simple genetic architecture underlies morphological variation in dogs. PLoS Biol. doi: 10.1371/journal.pbio.1000451.

Boyko, A.R., 2011. The domestic dog: man's best friend in the genomic era. Genome. Biol. 12, 216–226.

Brelands, K., Brelands, M., 1951. A field of applied animal psychology. Am. Psychol. 6, 202–204.

Brelands, K., Brelands, M., 1961. The misbehavior of organisms. Am. Psychol. 16, 681–684.

Broadhurst, P.L., 1960. Analysis of maternal effects in the inheritance of behavior. Anim. Behav. 9, 129–141.

Broadhurst, P.L., 1975. The Maudsley reactive and non-reactive strains of rats: a survey. Behav. Genet. 5, 299–319.

Brown, J.R., Ye, H., Bronson, R.T., Dikkes, P., Greenberg, M.E., 1996. A defect in nurturing in mice lacking he immediate early gene fos B. Cell 86, 297–309.

Burgdorf, J., Panksepp, J., 2006. The neurology of positive emotions. Neurosci. Biobehav. Rev. 30, 173–187.

Café, L.M., Robinson, D.L., Ferguson, D.M., McIntyre, B., Beesink, G.H., Greenwood, P.L., 2011. Cattle temperament: persistence of assessments and associations with productivity, carcass and meat quality traits. J. Anim. Sci. 89, 1452–1465.

Cameron, N.M., et al., 2008. Epigenetic programming of phenotypic variations in reproductive strategies in the rat through maternal care. J. Neuroendocrinol. 20, 795–801.

Campler, M., Jongren, N., Jensen, P., 2008. Fearfulness in red jungle fowl and domesticated white leghorn chickens. Behav. Process 61, 39–43.

Canty, N., Gould, J., 1995. The Hawk/Goose experiment: sources of variability. Anim. Behav. 50, 1091–1095.

Chaignat, E., Yahya-Graison, E., Henrichsen, C.N., Chrast, J., Schultz, F., Pradervand, S., et al., 2011. Copy number variation modifies expression time courses. Genome. Res. 21, 106–113.

Chakravarti, A., Kapoor, A., 2012. Mendelian puzzles. Science 335, 930–931.

Champagne, F.A., Curley, J.P., 2009. Epigenetic mechanisms mediating the long-term effects of maternal care on development. Neurosci. Biobehav. Rev. 33 (4), 593–600.

Chen, C., Rainnie, D.G., Greene, R.W., Tonegawa, S., 1994. Abnormal fear response and aggressive behavior n mutant mice deficient for a-calcium-calmodin kinease II. Science 266, 291–294.

Clinton, S.M., Vazquez, D.M., Kabbaj, M., Kabbaj, M.H., Watson, S.J., Akil, H., 2007. Individual differences in novelty seeking and emotional reactivity correlate with variation in maternal behavior. Horm. Behav. 51, 655–664.

Cooper, R.M., Zubek, J.P., 1958. Effects of enriched and restricted early environments on learning ability of bright and dull rats, Can. J. Psychol. 12, 159–164.

Coppinger, R., Schneider, R., 1993. Evolution of working dog behavior. In: Serpell, J. (Ed.), The Domestic Dog: Its Evolution, Behavior and Interactions with People. Cambridge University Press, Cambridge, UK.

Coppinger, L., Coppinger, R., 1993. Dogs for herding livestock. In: Grandin, T. (Ed.),Livestock Handling and Transport, pp. 179–196. CABIInternational, Wallingford, UK.

Coppinger, R., Glendinning, J., Torop, E., Mattthay, C., Sutherland, M., Smith, C., 1987. Degree of behavioral neoteny differentiates canid polymorphy. Ethology 75, 85–108.

Coppinger, R.P., Smith, C.K., 1983. The domestication of evolution. Environ. Conserv. 10, 283–292.

Crawley, J.N., 1996. Unusual behavioral phenotypes of inbred mouse strains. Trends. Neurosci. 19 (5), 181–182.

Crawley, J.N., 2007. What's Wrong with My Mouse? Wiley and Sons, New Jersey.

Crawley, J.N., 2008. Behavioral phenotyping strategies for mutant mice. Neuron 57, 809–818.

Crews, D., 2010. Epigenetics, brain, behavior, and the environment. Hormones 9, 40–50.

Dantzer, R., Mormede, P., 1983. Stress in farm animals: a need for re-evaluation. J. Anim. Sci. 57, 6–18.

Darwin, C.R., 1859. On the Origin of Species. Oxford University Press (Published in 1958 by Mentor, New York).

Darwin, C.R., 1868. The Variation of Plants and Animals Under Domestication, vols. 1 and 2. John Murray, London.

Darwin, C., 1871. The Descent of Man and Selection in Relation to Sex. Modern Library, New York.

Davis, M., 1992. The role of the amygdala in fear and anxiety. Annu. Rev. Neursci. 15, 353–375.

Dawkins, M.S., 1993. Through our Eyes Only: The Search for Animal Consciousness. Freeman, New York.

Dellu, F., Piazza, P.V., Mayo, W., LeMoal, M., Simm, H., 1996. Novelty seeking in rats, behavioral characteristics and possible relationships with sensation seeking in man. Neuropsycholobiology 34, 136–145.

Denenberg, V.H., Brumaghim, J.T., Haltmeyer, G.C., Zarrow, M.X., 1967. Increased adreno-cortical activity in the neonatal rat following handling. Endocrinology (Baltimore) 81, 1047–1052.

Denenberg, V.H., Whimbey, A.E., 1968. Experimental programming of life histories: towards an experimental science of individual differences. Dev. Psychobiol. 1 (1), 55–59.

DeWaal, F.B., 2011. What is animal emotion. Ann. N. Y. Acad. Sci. 122, 191–206.

Dobzhansky, T., 1970. Genetics of the Evolutionary Process. Columbia University Press, New York.

Dykman, R.A., Morphee, O.D., Peters, J.E., 1969. Like begets like: behavior tests, classical autonomic and motor conditioning in two strains of pointer dogs. Ann. N. Y. Acad. Sci 159, 976–1007.

Eisener-Dorman, A.F., Lawrence, D.A., Bolivar, V.J., 2009. Cautionary insights on knockout mouse studies: the gene or not gene. Brain Behav. Immun. 23, 318–324.

Elgar, M.A., 1989. Predator vigilance and group size in mammals and birds, A critical review of the empirical evidence. Biol. Rev. Camb. Philos. Soc. 64, 13–33.

Eibi-Eibesfeldt, I., Kramer, S., 1958. Ethology, the comparative study of animal behavior. Q. Rev. Biol. 33, 181–211.

ENCODE Project Consortium, 2012. An integrated encyclopedia of DNA elements in the human genome. Nature 489, 57–74.

Eysenck, H.J., Broadhurst, P.L., 1964. Experiments with animals: introduction. In: Eysenck, H.J. (Ed.), Experiments in Motivation. Macmillan, New York, pp. 285–291.

Faure, A.S.M., Reynolds, J.M., Richard, Berridge, C., 2008. Mesolimbic dopamine in desire and dread enabling motivation to be generated by localized glutamate disruption of the nucleus accumbens. J. Neurosci. 28, 7184–7192.

Faure, J.M., Mills, A.D., 1998. Improving the adaptability of animalsby selection. In: Grandin, T. (Ed.), Genetics and the Behavior of DomesticAnimals. Academic Press.

Fentress, J.C., 1973. Development of grooming in mice with amputated forelimbs. Science 179, 204–205.

Flint, J., Corley, R., DeFries, J.C., Fulker, D.W., Gray, J.A., Miller, S., et al., 1995. A simple genetic basis for a complex physiological trait in laboratory mice. Science 269, 1432–1435.

Fondon, J.W. III, Garner, H.R., 2004. Molecular origins of rapid and continuous morphological evolution. Proc. Natl. Acad. Sci. 101, 18058–18063.

Fondon, J.W. III, Hammack, E.A.D., Hannak, A.J., King, D.G., 2008. Simple sequence repeats: genetic modulators of brain function and behavior. Trends. Neurosci. 31, 328–334.

Fordyce, G., Dodt, R.M., Wythes, J.R., 1988. Cattle temperaments in extensive herds in northern Queensland. Aust. J. Exp. Agric. 28, 683–687.

Fox, M.W. (Ed.), 1975. The Wild Canids: Their Systematics, Behavioral Ecology and Evolution. Van Nostrand-Reinhold, New York.

Fraga, M.F., et al., 2005. Epigenetic differences arise during the lifetime of monozygotic twins. PNAS 102, 10604–10609.

Frank, H., Frank, M.G., 1982. On the effects of domestication on canine social development and behavior. Appl. Anim. Ethol. 8, 507–525.

Fujita, O., Annen, Y., Kitaoka, A., 1994. Tsukuba high and low emotional strains of rats (*Rattus norvegicus*): an overview. Behav. Genet. 24, 389–415.

Fuller, J.L., Thompson, W.R., 1978. Foundations of Behavior Genetics. Mosby, St. Louis, MO.

Galef Jr., B.G., 1970. Aggression and timidity: responses to novelty in feral Norway rats. J. Comp. Physiol. Psychol. 70, 370–381.

Galibert, F., Quignon, P., Hitte, C., Andrew, C., 2011. Towards understanding dog evolutionary and domestication history. C. R. Biol. 334, 190–196.

Gartner, K., 1990. A third component causing random variability besides environment and genotype. A reason for the limited success of a thirty year long effort to standardize laboratory animals. Lab. Anim. 24, 71–77.

Gerlai, R., 1996. Gene-targeting studies of mammalian behavior: is it the mutation or the background genotype? Trends. Neurosci. 19 (5), 177–181.

Gloor, P., Oliver, A., Quesney, L.F., 1981. The role of the amygdala in the expression of psychic phenomenon in temporal lobe seizures. In: Ben-Ari, Y. (Ed.), The Amygdaloid Complex. Elsevier/North-Holland, New York, pp. 489–507.

Goodwin, D., Bradshaw, J.W.S., Wickens, S.M., 1997. Paedomorphosis affects visual signals of domestic dogs. Anim. Behav. 53, 297–304.

Gould, S.J., 1977. Ontogeny and Phylogeny. Harvard University Press (Belknap Press), Cambridge, MA and London.

Grandin, T. (1989a) Effects of Rearing Environment and Environmental Enrichment on Behavior and Neural Development in Young Pigs, Ph.D. Dissertation, University of Illinois, Urbana.

Grandin, T., 1989b. Environmental causes of abnormal behavior. Large. Anim. Vet. 4 (3), 13–16, June 13–16.

Grandin, T., 1989c. Voluntary acceptance of restraint by sheep. Appl. Anim. Behav. Sci. 23, 257–261.

Grandin, T., 1993a. Behavioral agitation during handling of cattle is persistent over time. Appl. Anim. Behav. Sci. 36, 1–9.

Grandin, T., 1993b. Behavioral principles of cattle handling under extensive conditions. In: Grandin, T. (Ed.), Livestock Handling and Transport. CAB International, Wallingford, UK, pp. 43–57.

Grandin, T., 1994. Solving livestock handling problems. Vet. Med. 89, 989–998.

Grandin, T., 1995. Thinking in Pictures. Doubleday, New York.

Grandin, T., 1997. Assessment of stress during handling and transport. J. Anim. Sci. 75, 249–257.

Grandin, T., Johnson, C., 2005. Animals in Tanslation. Scribner, New York NY.

Grandin, T., Rooney, M.B., Phillips, M., Canibre, R.C., Irlbeck, N.A., Graffam, W., 1995. Conditioning of nyala (*Tragelaphus angasi*) to blood sampling in a crate with positive reinforcement. Zoo. Biol. 14, 261–273.

Greenough, W.T., Juraska, J.M., 1979. Experience induced changes in fine brain structure: their behavioral implications. In: Hahn, M.E., Jensen, C., Dudek, B.C. (Eds.), Development and Evolution of Brain Size: Behavioral Implications. Academic Press, New York, pp. 295–320.

Griffin, D., 1994. Animal Minds. University of Chicago Press, Chicago, IL.

Grillner, S., 2011. Human locomotion circuits. Science 334, 912–913.

Gross, C.T., Canteras, N.S., 2012. The many paths of fear. Nat. Rev. Neurosci. 13, 651.

Guryev, V., Berezikov, E., Malik, R., Plasterk, R.H.A., Cuppen, E., 2004. Single nucleotide polymorphisms associated with rat expressed sequences. Genome. Res. 14, 1438–1443.

Hall, C.S., 1934. Emotional behavior in the rat, I. Defecation and urination as measures of individual differences in emotionality. J. Comp. Psychol. 18, 385–403.

Hall, C.S., 1938. The inheritance of emotionality. Sigma. Xi. Q. 26, 17–27.

Hayden, E.C., 2012. Life is complicated. Nature 464, 664–667.

Heinroth, O., 1918. Reflektorische Bewegungen bei Voegeln. J. Ornithol. 66 (1 and 2).

Heinroth, O., 1938. Aus dem Leben der Vogel. Berlin.

Hemsworth, P.H., Barnett, J.L., Treacy, D., Madgwick, P., 1990. The heritability of the trait fear of humans and their association between this trait and subsequent reproductive performance of gilts. Appl. Anim. Behav. Sci. 25, 85–95.

Henderson, H.D., 1968. The confounding effects of genetic variables in early experience research: can we ignore them? Dev. Psychobiol. 1, 146–152.

Henrichsen, C.N., Chaignet, E., Reymond, A., 2009. Copy number variants diseases and gene expression. Hum. Mol. Genet. 18, R1–R8.

Hennessy, M.G., Levine, S., 1978. Sensitive pituitary–adrenal responsiveness to varying intensities of psychological stimulation. Physiol. Behav. 21, 295–297.

Herrick, C.J., 1910. The evolution of intelligence and its organs. Science 31, 7–18.

Herrick, F.H., 1908. The relation of instinct to intelligence in birds. Science 27, 847–850.

Hiestand, L., 2011. A comparison of problem solving and spatial orientation in the wolf (*Canis lupis*) and dog (*Canis familiaris*). Behav. Genet. 41, 840–857. 10.1007/510519-011-9455-4.

Hirsh, J., 1963. Behavior genetics and individuality understood. Science 142, 1436–1442.

Hirsch, J., Lindley, R.H., Tolman, E.C., 1955. An experimental test of an alleged sign stimulus. J. Comp. Physiol. Psychol. 48, 278–280.

Hockstra, H.E., 2012. Stickleback catch of the day. Nature 484, 46–47.

Holmes, J.H.G., Robinson, D.W., Ashmore, C.R., 1972. Blood lactic acid and behavior of cattle with hereditary muscular hypertrophy. J. Anim. Sci. 55, 1011–1014.

Hubel, D.H., Wiesel, T.N., 1970. The period of susceptibility to the physiological effects of unilateral eye closure in kittens. J. Physiol. London. 206, 419.

Huck, U.W., Price, E.O., 1975. Differential effects of environmental enrichment on the open field behavior of wild and domestic Norway rats. J. Comp. Physiol. Psychol. 89, 892–898.

Huxley, T.H. (1874) On the hypothesis that animals are automata and its history, In: "Collected Essays, vol. 1, Methods and Results: Essays," p. 218 (Published in 1901 by MacMillan, London).

Isaac, E., 1970. Geography and Domestication. Prentice–Hall, Englewood Cliffs, NJ.

Jones, R.B., 1984. Experimental novelty and tonic immobility in chickens (*Gallus domesticus*). Behav. Processes. 9, 155–260.

Jones, F.C., Grabberr, M.G., Chan, Y.F., Russell, P., et al., 2012. The genomic basis of adoptive evolution in three spine stickbacks. Nature 484, 55–62.

Kagan, J., Reznick, J.S., Snidman, N., 1988. Biological bases of childhood shyness. Science 240, 167–171.

Keeler, C., Mellinger, T., Fromm, E., Eade, I., 1970. Melanin, Adrenalin and the legacy of fear. J. Hered. 61, 81–88.

Kemble, E.D., Blanchard, D.C., Blanchard, R.J., Takushi, R., 1984. Taming in wild rats following medial amygdaloid lesions. Physiol. Behav. 32, 131–134.

Klein, E., Steinberg, S.A., Weiss, S.R.B., Matthews, D.M., Uhde, T.W., 1988. The relationship between genetic deafness and fear-related behaviors in nervous pointer dogs. Physiol. Behav. 43, 307–312.

Kong, A., Steinthorsdottir, V., Masson, G., Thorleifsson, G., 2009. Parental origin of sequence variants associated with complex diseases. Nature 462, 868–874.

Konig, M., Zimmer, A.M., Steiner, H., Holmes, P.V., Crawley, J.M., Brownstein, M.J., et al., 1996. Pain responses, anxiety and aggression in mice deficient in preproenkephalin. Nature (London) 383, 535–538.

Krushinski, L.V., 1960. Animal behavior—its normal and abnormal development. In: Wortis, J. (Ed.), International Behavioural Sciences Service. Consultants Bureau, New York (Original Russian version published by Moscow University Press).

Kruuk, H., 1972. The Spotted Hyena. University of Chicago Press, Chicago, IL.

Lambooij, E., van Putten, G., 1993. Transport of pigs. In: Grandin, T. (Ed.), Livestock Handling and Transport. CAB International, Wallingford, UK.

Larsen, G., Karlsson, E.K., Perri, A., Webster, M.T., Ho, S.Y.W., Peters, J., et al., 2012. Rethinking dog domestication by integrating genetics, archeology, and biogeography. Proc. Natl. Academ. Sci. 109, 8878–8883.

Lawrence, A.B., 2008. Applied animal behavior Science: past, present, and future prospects. Appl. Anim. Behav. Sci. 115, 1–24.

LeDoux, J.E., 1992. Brain mechanisms of emotion and emotional learning. Curr. Opin. Neurobiol. 2 (2), 191–197.

LeDoux, J.E., 2000. Emotion circuits in the brain. Annu. Rev. Neurosci. 23, 155–184.

LeDoux, J.E., Ciccheti, P., Nagoraris, A., Romanski, L.M., 1990. The lateral amygdaloid nucleus: sensory interface of the amygdala in fear conditioning. J. Neurosci. 10, 1062–1069.

LeDoux, J.E., Iwata, J., Ciccheti, P., Reis, D.J., 1988. Different projections of the central amygdaloid nucleus mediate autonomic and behavioral correlates of conditioned fear. J. Neurosci. 8, 2517–2529.

Lemos, J.C., Wanata, M.J., Smith, J.S., Reyes, B.A.S., Hollon, N.G., Van-Bockstaele, E.J., et al., 2012. Severe stress switches CRF action in the nucleus accumbens from appetitire to aversive. Nature 490, 402–406.

Lesch, K.P., Bengel, D., Heils, A., Sabol, D.Z., Greenburg, B.D., Perri, S., et al., 1996. Association of anxiety-related traits with a polymorphism in the serotonin transporter gene regulatory region. Science 274, 1527–1531.

Levine, S., 1968. Influence of infantile stimulation on the response to stress during preweaning development. Dev. Psychobiol. 1 (1), 67–70.

Levine, S., Haltmeyer, G.C., Karas, G.G., Denenburg, V.H., 1967. Physiological and behavioral effects of infantile stimulation. Physiol. Behav. 2, 55–59.

Linn, Z., Ma, H., Nei, M., 2008. Ultra conserved coding regions outside the homeobox in mammalian box genes. Evol. Biol.

Lipkind, D., Sakov, A., Kafkafi, N., Elmer, G.I., Benjamin, Y., Golani, 2004. New replicable anxiety-related measures of wall vs. center behavior of mice in the open field. J. Appl. Physiol. 97, 347–359.

Lorenz, K.S., 1939. Vergleichende Verhaltensforschung Zool. Anz.(12), 69–109.

Lorenz, K.S., 1965. Evolution and Modification of Behavior. University of Chicago Press, Chicago.

Lorenz, K.Z., 1981. The Foundations of Ethology. Springer-Verlag, New York.

Maher, B., 2012. The human encyclopedia. Nature 489, 46–48, 8:260. doi:10.1186/1471-2148-8-260.

Mahut, H., 1958. Breed differences in the dog's emotional behavior. Can. H. Psycho. 12 (1), 35–44.

Manuck, S.B., Schaefer, D.C., 1978. Stability of individual differences in cardiovascular reactivity. Physiol. Behav. 21, 675–678.

Matheson, B.K., Branch, B.J., Taylor, A.N., 1971. Effects of amygdaloid stimulation on primary-adrenal activity in conscious cats. Brain. Res. 32, 151–167.

Mayr, E., 1974. Behavioral programs and revolutionary strategies. Am. Sci. 62, 650–659.

Melton, D.A., 1991. Pattern formation during animal development. Science 252, 234–241.

Melzack, R., Burns, S.K., 1965. Neurophysiological effects of early sensory restriction. Exp. Neurol. 13, 163–175.

Metfessel, M., 1935. Roller canary song produced without learning from external source. Science 81, 470.

Mikkelson, T.S., Wakefield, M.J., Aken, B., et al., 2007. Genome of the marsupial Monodephis domestica reveals innovation in non-coding sequence. Nature 447, 167–177.

Moberg, G.P., Wood, V.A., 1982. Effects of differential rearing on the behavioral adrenocortical response of lambs to a novel environment. Appl. Anim. Ethol. 8, 269–279.

Momozawa, Y., Yuari, Y., Kusunose, R., Kikusui, T., Mori, Y., 2005. Association of equine temperament and polymorphism in dopamine D4 gene. Mamm. Genome. 16, 538–544.

Morman, F., Dubois, J., Korblith, S., Milosovljevic, M., Cerf, M., et al., 2011. A category specific response to animals in the human right amygdala. Nat. Neurosci. 14, 1247–1249.

Morris, C.L., Grandin, T., Irlbeck, N.A., 2011. Companion animal symposium: environmental enrichment for companion exotic and laboratory animals. J. Anim. Sci. 89, 4227–4238.

Mueller, H.C., Parker, P., 1980. Cardiac responses of domestic chickens to hawk and goose models. Behav. Processes. 7, 255–258.

Mundinger, P.C., 1995. Behavior analysis of canary song interstrain differences in sensory learning and epigenetic rules. Anim. Behav. 50, 1491–1511.

Murphy, L.B., 1977. Responses of domestic fowl to novel food and objects. Appl. Anim. Ethol. 3, 335–349.

Murphey, R.M., Moura Durate, F.A., Coelho, Novaes, W., Torres Penedo, M.C., 1980a. Age group differences in bovine investigatory behavior. Dev. Psychobiol. 14 (2), 117–125.

Murphey, R.M., Moura Duarte, F.A., Torres Penedo, M.C., 1980b. Approachability of bovine cattle in pastures: breed comparisons and a breed X treatment analysis. Behav. Genet. 10, 171–181.

Murphey, R.M., Moura Duarte, F.A., Torres Penedo, M.C., 1981. Responses of cattle to humans in open spaces: breed comparisons and approach-avoidance relationships. Behav. Genet. 11 (1), 37–48.

Nestler, E., 2011. Hidden switches in the mind. Sci. Am. December, 77–83.

Nestler, E.J., 2012. Stress leaves its molecular mark. Nature 490, 171–172.

Nottebohm, F., 1970. Ontogeny of bird song. Science 1678, 950–956.

Nottebohm, F., 1977. Asymmetries n neural control of vocalization in the canary. In: Harnard, S. (Ed.), Lateralization of the Nervous System. Academic Press, New York and London.

Nottebohm, F., 1979. Origins and mechanisms in the establishment of cerebral dominance. In: Gazzaniza, M. (Ed.), Handbook of Behavioral Neurobiology. Plenum, New York.

Nottebohm, F., Stokes, T.M., Leonard, C.M., 1976. Central control of song in the canary (*Serinus canaria*). J. Comp. Neurol. 165, 457–486.

Nowicki, S., Marler, P., 1988. How do birds sing? Music. Percept. 5, 391–421.

Ott, R.S., 1996. Animal selection and breeding techniques that create diseased populations and compromise welfare. J. Am. Vet. Med. 208, 1969–1974.

Ovodov, N.D., Crockford, S.J., Kuzmin, Y.V., Higham, T.F.G., Hodgins, G.W.L. and der Plicht, J. (2011) PLOS ONE 6 (7), e22821. doi:w0.1371/Journal.pone.0022821.

Pang, J.F., Kleutsch, C., Zau, X., Zhang, A.B., Luo, L.Y., et al., 2009. mtDNA data indicated a single origin of dogs south of the Yangtzee River less than 16,300 years ago from numerous wolves. Mol. Biol. Evol. 26, 2849–2864.

Panksepp, J., 1998. Affective Neuroscience. Oxford University Press, New York NY.

Panksepp, J., 2005. Affective consciousness: core emotional feelings in animals and humans. Conscious. Cogn. 14, 30–80.

Panksepp, J., 2011. The basic emotional circuits of mammalan brains: do animals have inner lives? Neurosci. Biobehav. Rev. 35, 1791–1804.

Parsons, P.A., 1988. Behavior, stress and variability. Behav. Genet. 18 (3), 293–308.

Plomin, R., Daniels, D., 1987. Why are children in the same family so different from one another? Behav. Brain. Sci. 10, 1–60.

Popova, N.K., Voitenko, N.N., Trut, L.N., 1975. Changes in serotonin and 5-hydroindoleacetic acid content in the brain of silver foxes under selection for behavior. Proc. Acad. Sci. USSR. 233, 1498–1500 (In Russian).

Poulsen, P., Esteller, M., Vaag, A., Fraga, M.F., 2007. The epigenetic basis of twin discordance in age-related diseases. Pediatr. Res. 61, 38R–42R.

Poulsen, H., 1959. Song learning in the domestic canary. Z. Tierpsychol. 16, 173–178.

Price, E.O., 1984. Behavioral aspects of animal domestication. Q. Rev. Biol. 59, 1–32.

Probst, A.V., Dunleavy, E., Almouzni, G., 2009. Epigenetic inheritance during the cell cycle. Nat. Rev. Mol. Cell. Biol. 10, 192–206.

Quint, M., Drost, H.G., bagel, A., Ullrich, K.K., Bonn, M., Gross, I., 2012. A transcript ? hourglass in plant embryogenesis. Nature 490, 98–108.

Redon, R., et al., 2006. Global variation in copy number in the human genome. Nature 444, 444–454.

Reese, W.G., Newton, J.E.O., Angel, C., 1983. A canine model of psychopathology. In: Krakowski, A.J., Kimball, C.P. (Eds.), Psychosomatic Medicine. Plenum, New York, pp. 25–31.

Reynolds, S.M., Berridge, K.C., 2008. Emotional environmental retune the balance between appetitive versus fearful functions in nucleus accumbens. Nat. Neurosci. 11, 423–425.

Richard, S., Lend, N., Saint-Dizier, H., Letemer, C., Faure, J.M., 2010. Human handling and presentation of novel objects evoke independent dimensions of fear in quail. Behav. Processes. 85, 18–23.

Richardson, M.K., 2012. A phylotypic stage for all animals? Dev. Cell. 22, 903–904.

Richter, C.P., 1952. Domestication of the Norway rat and its implications for the study of genetics of man. Am. J. Hum. Genet. 4, 273–285.

Richter, C.P., 1954. The effects of domestication and selection on the behavior of Norway rat. J. Nat. Cancer Inst. (U.S.) 15, 727–738.

Rogan, M.T., LeDoux, J.E., 1996. Emotion: systems cells, and synaptic plasticity. Cell (Cambridge Mass) 83, 369–475.

Roullet, F.I., Crawley, J.N., 2011. Mouse models of autism: Testing hypotheses about molecular mechanisms. Curr. Top. Behav. Neurosci. 7, 187–212.

Royce, J.R., Carran, A., Howarth, E., 1970. Factor analysis of emotionality in ten strains of inbred mice. Multidiscip. Res. 5, 19–48.

Saey, T.M., 2011. Missing link. Sci. News.22–25, December, 17.

Saint-Dizier, H., Letemier, C., Levy, F., Richard, S., 2008. Selection for tonic immobility duration does not affect response to novelty. Appl. Anim. Behav. Sci. 112, 297–306.

Sanyal, A., Lajole, B.R., Jain, G., Dekker, J., 2012. The long-range interaction landscape of gene promoters. Nature 489, 109–113.

Sarma, K., Reinberg, D., 2005. Histone variants meet their match. Nat. Rev. Mol. Cell Biol. 6, 139–149.

Savolainen, P., Zhang, Y.-P., Lup, J., Lundeberg, J., Leitner, T., 2002. Genetic evidence for an East Asian origin of domestic dogs. Science 298, 610–613.

Schaible, R., 1963. Clonal distribution of melanocytes n piebald spotting and variegated mice. J. Explor. Zool. 172, 181–200.

Schroeder, C., 1914. Thinking animals. Nature (London) 94, 426–427.

Schultz, D., 1965. Sensory Restriction. Academic Press, New York.

Scott, J.P., Fuller, J.L., 1965. Genetics and the Social Behavior of the Dog. University of Chicago Press, Chicago.

Setckliev, J., Skaug, O.E., Kaada, B.R., 1961. Increase of plasma 17-hygroxy-corticosteroids by cerebral cortisol and amagdaloid stimulation in the cat. J. Endocrinol. 22, 119–129.

Shastry, B.S., 2002. SNP alleles in human disease and evolution. J. Hum. Genet. 47 (11), 0561–0566.

Silveira, I.D.B., Fischer, J., Farinatti, H.E., Restle, J., Filho, D.C.A., deMenezes, L.F.G., 2012. Relationship between temperament with performance and meat quality of feedlot steers with predominantly Charolais or Nellose breed. Rev. Bras. de. Zootechnia. 41.

Silverman, J.L., Yang, M., Lord, C., Crawley, J.N., 2010. Behavioral phenotyping assays for mouse models of autism. Nat. Rev. Neurosci. 11, 490–502.

Simons, D., Land, P., 1987. Dearly experience of tactile stimulation influences organization of somatic sensory cortex. Nature (London) 326, 694–697.

Skinner, F.F., 1958. Behavior of Organisms. Appleton–Century–Crofts, New York.

Skinner, M.P., 1922. The pronghorn. J. Mammal. 3, 82–106.

Sluyter, F., Bult, A., Lynch, C.B., VanOortmerssen, G.A., Kookhaus, J.S., 1995. Comparison between house mouse lines selected for attack latencies or nest building. Evidence for a genetic basis of alternative behavioral strategies. Behav. Genet. 25 (3), 247–252, May 1995.

Stead, J.D.H., Clinton, S., Neal, C., Schneider, J., 2006. Selective breeding for divergence in novelty seeking traits: heritability and enrichment in spontaneous, anxiety-related behaviors. Behav. Genet. 36, 697–712.

Steinberg, S.A., Klein, E., Killens, R.L., Udhe, T.W., 1994. Inherited deafness among nervous pointer dogs. J. Hered. 85, 56–59.

Stephens, D.E., Toner, J.M., 1975. Husbandry influences on some physiological parameters of emotional responses in calves. Appl. Anim Ethol. 1, 233–243.

Tinbergen, N., 1948. Social releasers and the experimental method required for their study. Wilson. Bull. 60, 6–52.

Tinbergen, N., 1951. The Study of Instinct. Clarendon Press, Oxford University Press, New York.

Trillmich, F., Hudson, R., 2011. The emergence of personality in animals: The need for a developmental approach. Dev. Psychobiol. 53, 505–509.

Tulloh, N.M., 1961. Behavior of cattle in yards, II. A study of temperament. Anim. Behav. 9, 25–30.

Vita, C., Savolainen, P., Maldoado, J.E., Amorim, I.R., Rice, J.E., Honeycutt, R.L., et al., 1997. Multiple and ancient origins of the domestic dog. Science 276, 1687–1689.

Vogel, G., 2011. Do jumping genes spawn diversity. Science 332, 300–301.

Voisinet, B.D., Grandin, T., Tatum, J.D., O'Conner, S.F., Struthers, J., 1997a. Feedlot cattle with calm temperaments have higher average daily gains than cattle with excitable temperaments. J. Anim. Sci. 75, 892–896.

Voisinet, B.D., Grandin, T., O'Connor, S.F., Tatum, J.D., Deesing, M.J., 1997b. Bos-indicus cross feedlot cattle with excitable temperaments have tougher meat and a higher incidence of borderline dark cutters. Meat. Sci. (In press).

Walsh, R.N., Cummins, R.A., 1975. Mechanisms mediating the production of environmental induced brain changes. Psychol. Bull. 82, 986–1000.

Watson, J.B., 1930. Behaviorism. W.W. Norton, New York.

Weir, A.A.S., Kacelnik, A., 2006. New Caledonian crows (Corvus moneduloides) creatively re-design tools by bending or unbending metal strips according to needs. Anim. Cogn. 9 (4), 317–334.

Whitman, C.O., 1898. "Animal Behavior" Biol. Lec. Marine Biological Laboratory, Woods Hole, MA.

Widowski, T.M., Curtis, S.E., 1989. Behavioral response of per parturient sows and juvenile pigs to prostaglandin F_2a. J. Anim. Sci. 67, 3266–3276.

Wood-Gush, D.G.M., Beilharz, R.G., 1983. The enrichment of a bare environment for animals in confined conditions. Appl. Anim. Ethol. 10, 209.

Wood-Gush, D.F.M., Vestergaard, K., 1991. The seeking of novelty and its relation to play. Anim. Behav. 42, 599–606.

Wright, S., 1922. The effects of inbreeding and crossbreeding on guinea pigs. U.S. Dep. Agric. Bull., 1090.

Wright, S., 1978. The relation of livestock breeding to theories of evolution. J. Anim. Sci. 46 (5), 1192–1200.

Yeakel, E.H., Rhoades, R.P., 1941. A comparison of the body and endocrine gland (adrenal, thyroid and pituitary) weights of emotional and nonemotional rats. Endocrinol (Baltimore) 28, 337–340.

Yerkes, R.M., 1905. Animal psychology and criteria of the psychic. J. Comp. Neurol. Psychol. 15:137.

Zenner, F.E., 1963. A History of Domesticated Animals. Harper & Row, New York.

Zhang, Y.E., Landback, P., Vibranovski, M.D., Long, M., 2011. Accelerated recruitment of new brain development genes into the human genome. PLoS Biol. 9 (10), e1001179. doi: 10.1371/journal.pbio.1001179.

Chapter 2

Behavioral Genetics and Animal Domestication

Per Jensen and Dominic Wright
IFM Biology, Linköping University, Sweden

DEFINITION OF DOMESTICATION

When Darwin first formulated his ideas on evolution, domesticated plants and animals played a central role. The first chapter of *On the Origin of Species* was devoted to a discussion about the rapid selection responses occurring during domestication, affecting widely different aspects of animal phenotypes. Darwin noted that size, color and shape could all be modified in few generations when the selection pressure was sufficiently intense, and he also rightly noted that the behavior of domesticated animals differs from wild relatives in a number of aspects. Hence, Darwin saw domestication as a rapid evolutionary process, something that has received massive support from research during the last century, and behavior was conceived as a phenotype amongst others.

The basic principles underlying any evolutionary process are few. In essence, evolutionary development of a trait relies on three conditions: the trait varies between individuals within the population (the variation principle), the variation has a genetic component (the heritability principle) and individuals with different variants of the trait contribute unequally to the gene pool of the next generation (the principle of natural selection). The phenotypes of animals undergoing domestication develop according to the same principles. What makes domestication special is the fact that humans exert a large part of the selection pressure in deciding which individuals will be most likely to have their genes propagated to the next generation.

Attempts to define domestication are often based on the different selection pressure as compared to the natural situation. Hence, Price (2002) suggested it should be defined as the process whereby captive animals adapt to man and the environment he provides. Price further argued that this is achieved through genetic changes occurring over generations and through environmental stimulation and experiences during an animal's lifetime. This has become the most influential definition and is widely cited by researchers in the field.

Genetics and the Behavior of Domestic Animals. DOI: http://dx.doi.org/10.1016/B978-0-12-394586-0.00002-0
© 2014 Elsevier Inc. All rights reserved.

However, in this chapter we will not consider the experiential effects, since these are shared with any animal in captivity (for example, habituation to humans and other aspects of the captive environment), and since they are not transmitted to the next generation, they should not be included in a formal definition of domestication. However, epigenetic effects may partly bridge the gap between genetics and experience; this will be considered more closely later.

Hence, we suggest that a more fruitful definition should focus on the effects, which transfer across generations. Domestication is the process whereby populations of animals change genetically and phenotypically in response to the selection pressure associated with a life under human supervision. Hence, domestication is an evolutionary process, and it acts through three different pathways, in addition to natural selection, which of course continues also in captivity. Firstly, there is a relaxation of specific natural selection pressures, such as predation and starvation; secondly, there is an intensified human selection for preferred traits, for example, growth, appearance, and reproduction; thirdly, there is a development of traits which are genetically or functionally correlated to the ones selected, for example, increased relative gut length in fast-growing broilers (Jackson and Diamond, 1995, 1996).

These three processes have also been discussed at length by (Price, 2002). Our definition leads to the view that domestication should be viewed as a genetic process. Alleles, which ultimately increase the reproductive output—the fitness—of an individual, are favored and will therefore be over-represented in coming generations. But due to the third process, certain alleles may also "hitch-hike" by being genetically linked to the selected locus.

In this chapter, we will therefore focus on behaviorally relevant genes and mutations which may have been selected during domestication, and discuss their functions and the mechanisms whereby they affect behavior. But we will also discuss genetic architecture, i.e. the way in which genes are physically organized in the genome, as this will be an important piece of information for understanding correlated selection and how the different aspects of the domestication phenotype are controlled genetically.

THE DOMESTICATED PHENOTYPE

A striking aspect of domestication is the emergence in various, unrelated species, of very similar phenotypic changes. The complex of convergent traits is often referred to as "the domestic phenotype" (Price, 2002).

The first part of this complex relates to the appearance of the domesticated animals in relation to their ancestors. For example, domesticates are generally smaller and have modified and often reduced pigmentation. The size reduction can be seen both as an overall smaller body mass. This can however be reversed by selective breeding, and many domestic breeds are in fact considerably larger than their wild progenitors. For example, some dog breeds exceed the weight and wither height of wolves (Mosher *et al.*, 2007), and many horse breeds are considerably larger than their presumed ancestor (Clutton-Brock, 1999). Nevertheless, in the absence of specific selection for the opposite, the general pattern is that domestication causes animals to decrease in size.

There is also a disproportionate size modification of certain body parts. For example, legs tend to be shorter in domesticated animals, referred to as chondrodystrophy, and the skull tends to show a more or less drastic shortening from nose to neck, so called brachycephaly (Clutton-Brock, 1999). Further examples of typical aspects of the domesticated phenotype are floppy ears, curled tails, and modifications of fur structure. Also in birds, similar analogous changes occur, including alterations of size, skull shape, leg length, pigmentation, and plumage morphology.

Behavior and physiology have undergone a range of changes. Most prominently, domesticated animals generally exhibit a greatly reduced fear of humans (Price, 2002), which seems to generalize to other aspects of fear as well (Campler *et al.*, 2009). Furthermore, domesticates tend to have an overall reduced activity level and perform less explorative behavior, and they also show less rigid social structures and often less agonistic behavior (Schütz and Jensen, 2001). For example, free-range dogs are known often to form dynamically changing packs of 30–50 individuals, while wolves live in strictly closed kinship groups of usually fewer than ten. Domesticated animals usually develop faster and reach sexual maturity earlier than their wild relatives. For example, dogs usually become sexually mature before one year of age, whereas wolves take twice that time (Boitani and Ciucci, 1995). Red Junglefowl females start laying eggs at about 25 weeks of age, whereas modern layers are reproductive already seven or eight weeks earlier (Schütz *et al.*, 2002).

Thus, although the variation is large, mainly due to a high variety of selection pressures for different, human-controlled purposes, a relatively uniform set of phenotypic changes reoccur in many species as a consequence of domestication. So this begs the question: why do these changes keep occurring? Are they part of an adaptive process or solely caused by human preferences? In any case, which are the genetic mechanisms involved?

EVOLUTIONARY MECHANISMS IN DOMESTICATION

Although natural selection does not end when a population is domesticated—diseases, parasites, competition, and predation still exert a strong effect on which individuals will reproduce—a strong selection is carried out by humans. Still, the traits, which can be selected will depend on the trait variation and the heritability of this, just like in the natural population. Hence, variation, heritability, and selection remain the same driving forces during domestication as they are in the wild.

It is therefore reasonable to claim that domestication is an evolutionary process whereby populations overall adapt to the prevailing selection pressures. In that sense, the domesticated phenotype may largely be caused by an adaptive response to the human protection. For example, body size of the wild ancestors has been optimally adjusted to the natural selection pressures in their wild habitats, such as food availability and predation risk. When predation pressure decreases and food availability goes up, it may be an adaptive response to reduce the body size. This will in turn enable faster development and increase the lifetime reproductive success, something that may also have been appreciated and exacerbated by early farmers. It has been argued that many of the different parts of the domestic phenotype are typical of adaptive responses to r-selection, as opposed to k-selection (Price, 2002).

In the same way, several behavioral responses appear to be adaptive. For example, reduced rigidity in tolerance of variations of social structures can save energy and reduces the risk of injuries in situations where conditions are more likely to be crowded with many individuals in a limited area. It can also be adaptive to save energy by not engaging in excessive exploratory behavior when the net benefits from this are outweighed by the fact that food is provided by humans.

In a systematic comparison between domesticated White Leghorn laying hens and their ancestors, the Red Junglefowl, Schütz and Jensen (2001) studied groups of birds reared and kept under identical conditions. The domesticates were in general less active, displayed less social behavior, and were less investigative. The latter was shown, for example, by a decreased level of so-called contrafreeloading; given a choice between hidden and freely available food, the domesticates ate less than the Red Junglefowl from the food source requiring some searching, digging, and scratching. The behavior of the White Leghorns was interpreted as an adaptive response to life under human protection, where food is abundant, predation is less of a risk, and crowding is common.

It has been argued that the reduced natural selection pressure under domestication should lead to an increase in trait variation (Price, 2002). The logic of the argument is that, in absence of directed selection, the fitness of individuals carrying sub-optimal traits should be relatively higher, and the population distribution should be flattened. There is little actual evidence to support this when considering specific domestic populations. However, when directed selection is applied on a specific trait, between-breed variation may be dramatic. Dog breeds, for example, vary in adult size from about 700 g to 70 kg, and chickens come in dwarf versions of less than 500 g adult size, as well as giant breeds, reaching 7–9 kg. However, within breeds the variation is less pronounced, and mostly does not appear to be larger than in a population of the wild ancestors. Hence, the directed, human-controlled selection under domestication may be as strong as the natural selection in the wild, causing trait distributions to change in domesticated populations without significantly changing the variation.

Over the last decade or so, researchers have started to investigate the specific adaptive properties of domesticated behavior. In particular, the cognitive and communicative abilities of dogs have received a large amount of attention. In general, dogs have developed traits which are absent in wolves, and which make them more suited to interacting and communicating with humans (Miklosi, 2008). For example, they are better at responding to human cues such as pointing and gazing, and they can learn and understand large numbers of spoken words and put these into categories. This probably represents exceptional adaptations in a species, which has been selected for many thousands of years for their ability to cooperate and interact properly with humans, but it demonstrates that the specific selection pressures under domestication may lead to novel adaptations within a limited evolutionary time. The extent of this adaptation is further shown by the fact that dogs, when released of the human selection (such as during feralization and in stray dogs), wolf-like social behavior does not re-emerge. Free-ranging dogs have a very different social and reproductive system than their wolf ancestors; they mate promiscuously and all individuals reproduce (unlike in wolves, where normally only one pair in a pack raises offspring) and they do not form highly cooperative, stable packs (Boitani and Ciucci, 1995).

A widely discussed problem is how many genes are actually involved in causing the domesticated phenotype. It has been suggested that the reoccurrence of similar phenotypic changes in many different domesticated species could be indicative of a limited number of "master genes", perhaps only one, controlling a range of traits (Stricklin, 2001). As selection acts on this gene, for example by reducing stress and flightiness, several or all of the domesticated phenotypes would develop as correlated responses.

An important study sparkling suggestions such as these, is the one carried out by Belyaev and co-workers in Russia during the 1960s (Trut, 1999). They selected farm foxes (silver foxes, descendants of the red fox) for tameness in a simple test, measuring the propensity to flee or attack when approached by the hand of an observer. The response in the selected trait was rapid, and within a few generations, the selected foxes were tame and behaved in a dog-like manner towards humans.

Interestingly, several other domestication-related traits increased in frequency among the tame foxes. This included loss of pigmentation, faster ontogenetic development, curly tails, floppy ears, and compressed jaws. Belyaev himself suggested that this may have been the result of a released selection pressure on many genes, controlled by one or few "master genes" which were directly selected upon (Trut, 1999).

Similar results were found in an experiment where the Belyaev approach was used to select rats of high and low fearfulness towards handling. The tamer rats had a generally dampened stress response, as shown in both physiological and behavioral variables (Albert *et al.*, 2008). However, even though several genetic loci seem to affect more than one of the phenotypic traits, Albert *et al.* (2009, 2011) did not find any evidence of a major, domestication gene. Rather, genetic architecture and the combined effects of several physically linked genes may explain the correlated selection responses.

In a very similar study, focusing on Red Junglefowl (ancestors of all domestic chickens), Agnvall and colleagues measured correlated behavior responses to selection for increased tameness against humans over three generations (Agnvall *et al.*, 2012). They found a moderate but significant heritability for reduced fear of humans ($h^2 = 0.17$), and also significant genetic correlations between fear of humans and, for example, foraging and exploration, and hatch weight. This indicates that selection for tameness is likely to cause correlated behavioral effects also in chickens, and again suggests that genetic architecture of few and pleiotropic genes may control a large part of the domesticated phenotype.

What is interesting in the study of the genetic architecture of domestication is that when these differentially selected, or indeed wild and domestic, populations are then inter-crossed, these trait correlations begin to erode (Albert *et al.*, 2009; Wright *et al.*, 2010, 2012). The implications of this are discussed in the section on genetic architecture, below.

GENETIC MECHANISMS IN DOMESTICATION

All the standard genetic mechanisms that act on natural populations will also act on domestic populations, though some to a much greater and others to a much lesser degree. Genetic mechanisms affecting gene frequency can be largely broken down in to systematic processes and dispersive processes (Falconer and Mackay, 1996).

Systematic processes affect gene frequency in both a predictable manner and direction, and include migration, mutation, and selection. Dispersive processes have greater effects in smaller populations, and are often unpredictable in their effects on populations. This is principally drift, but also inbreeding and assortative mating can fall under this category. In terms of the latter two (inbreeding and assortative mating) these affect not only gene frequencies, but more importantly genotype frequencies, as discussed below. Of these processes, selection, inbreeding, and drift are perhaps the factors with the greatest effect on domestic populations (Price and King, 1968). Migration (i.e. new gene flow into the population) is generally extremely tightly controlled in domestic animals, whilst mutation may provide some novel discrete Mendelian (single gene) phenotypes that are then selected upon. Quantitative traits will once again be relatively unaffected in the relatively short time frame (evolutionarily) of domestication (especially when one considers that some of the largest changes in phenotype have occurred in the most recent selection history of domestication). Assortative mating (whereby phenotypically similar animals mate/are mated with one another) will have an effect in domestic animals, though this will generally be due to selection for a specific phenotype, and will also often be closely related to inbreeding.

Selection

Selection can be further subdivided into artificial (both intentional and inadvertent), natural, and relaxed in the case of domestic animals.

Artificial Selection

This can probably be thought of as both the strongest and most defining mechanism of domestication (Price and King, 1968). The effects of this can be extremely rapid, and even within even just a few generations of selection, large differences can occur. For example, both silver foxes (Belyaev *et al.*, 1981, 1984) and rats (Albert *et al.*, 2008, 2009) selected for tameness show a variety of behavioral and physiological responses after just a few generations, as discussed above. Chickens selected for differential growth show a similar marked response to selection (Carlborg *et al.*, 2006). These are examples of intentional artificial selection, though unintentional artificial selection can also occur. Examples of this would be the coonstripe shrimp, *Pandalus danae*, inadvertently selected over 10 generations for a decrease in escape tail-flip response (this response lead to injury during handling (Marliave *et al.*, 1993) and for trout inadvertently selected for stress reduction to increase fecundity (Fevolden *et al.*, 1991). One important difference to consider between artificial and natural selection has been put forward by Price (Price, 1999), in that artificial selection is "goal-oriented", whereby individuals with a desirable phenotype are then bred from (i.e. selection occurs prior to reproduction), whereas with natural selection an individual is only "assessed" through its representation in future generations (i.e. the number of offspring it produces).

Relaxed Selection

As mentioned above, certain behavioral adaptations, principally foraging ability and predator avoidance, are no longer relevant to the survival of an animal in captivity. In the case of antipredation behaviors in particular, there are associated costs with exhibiting these traits and relaxed pressure may therefore drive the traits in the opposite direction (especially considering that wild populations with lower predation pressure exhibit these behaviors less). (Kronenberger and Medioni, 1985) showed that domestic mice preferred novel saccharin-water more than their wild counterparts. Such neophobia is often seen in wild-derived populations, but less so in domestic populations (Schütz *et al.*, 2001; Wright *et al.*, 2006c).

This relaxation of selection could be expected to lead to an increase in phenotypic and genetic variability, however this is often confounded with the limited genetic variability found in certain domestic populations. For example, laboratory strains often have limited genetic polymorphism. Animals tend to come from a limited founder population (e.g., mice (Bonhomme and Guenet, 1984) and therefore have greater genetic homogeneity. Using a new series of crosses between animals recently caught from the wild (Koide *et al.*, 2000) derived a series of mouse strains that exhibited much greater polymorphism than standard laboratory strains, with several strains also diverging in assorted behavioral assays.

Natural Selection

Although domestic animals are obviously subject to strong artificial selection, this does not mean that natural selection does not occur. Indeed, any selection that occurs in the captive environment that is not artificial is by definition natural (Hale, 1969). Even those animals that are artificially selected for breeding may nevertheless fail to reproduce. This can be due to a myriad of factors, including social stress, parasitism, and in general a failure to cope with the captive environment. Those animals that cope best with this new environment will increase their reproductive potential. For example, wild-caught *Drosophila* populations that were maintained in captivity were found to exhibit rapid adaptation to the captive environment over nine generations (Frankham *et al.*, 1999).

Not only that, but the artificial selection process itself, acting on certain traits, may also lead to a reduction in reproductive potential. For example, dog breeds have long been known to show negative fitness effects due to selected characteristics (e.g., the ridge in ridgebacks is associated with sinus formation (Salmon Hillbertz *et al.*, 2007), hip dysplasia in alsatians (Hedhammar *et al.*, 1979)), growth is linked with decreases in reproduction in broilers (Siegel and Dunnington, 1985), and the red hair color in rabbits leads to metabolic problems, disease susceptibility, and increased mortality (Muntzing, 1959). Generally, the extent of the occurrence of natural selection can be thought of as determined by the extent to which the environment allows the expression of species-typical traits (Spurway, 1955). Animals with very few pre-adaptations to this environment will therefore naturally be under greater natural selection as they are subjected to a strong, novel environment.

Inbreeding

Inbreeding, the process whereby related individuals assortatively breed together, leads to an increase in alleles identical by descent (Falconer and Mackay, 1996). This can be typically measured by assessing the excess numbers of homozygotes in the population, due to substructure. In domestic populations, inbreeding is relatively common, especially in more modern selection practises, as fewer numbers of animals tend to get a far larger percentage of the matings. This is especially true when selection indices (BLUP and the like) are used to calculate the most desirable individuals to breed from, and therefore tend to increase the rate of inbreeding (Belonsky and Kennedy, 1988; Bijma and Woolliams, 2000; reviewed in Kristensen and Sørensen, 2005). This can be a greater problem in certain domestic breeds than others.

For instance, the level of inbreeding in dogs has been shown to be extremely high (Calboli *et al.*, 2008), whereas chicken breeds remain remarkably diverse, even in domestic breeds (Rubin *et al.*, 2010). The general effects of inbreeding are commonly agreed to lead to inbreeding depression. In many appearances the opposite of heterosis (or outbreeding vigor), this is manifest in a decrease in general fitness (Lynch and Walsh, 1998). Although many such examples exist for a variety of animal models with rather variable results, as concerning domestication, sheep (Wiener *et al.*, 1992, 1994) and dairy cows (Miglior *et al.*, 1995; Smith *et al.*, 1998) have been shown to exhibit effects in primary fitness. Inbreeding can also have negative effects on behavior. Courtship behavior in guppies (Mariette *et al.*, 2006) and *Drosophila* (Sharp, 1984) have both been shown to be negatively affected by inbreeding. Parental behavior has also been found to be affected in mice (Margulis, 1998). The detrimental effects of inbreeding can also be seen in the numerous mechanisms to avoid inbreeding (for example non-selfing mechanisms in plants (Charlesworth and Charlesworth, 1979) and behavioral mechanisms in animals (Pusey and Wolf, 1996)).

The mechanism through which the negative effects of inbreeding act have traditionally been thought of as being due to the effects of dominance between loci (Lynch and Walsh, 1998). If the effects were purely additive, they would be independent of the genetic background and there should be no effect of increased homozygosity. There are three general hypotheses for describing inbreeding depression: the (erroneously named) dominance hypothesis, the over-dominance hypothesis, and the breakdown of epistasis hypothesis. The dominance hypothesis is in fact nothing to do with dominance, but is related to the expression of deleterious recessives that are more likely to arise due to identity by descent (Bruce, 1910; Davenport, 1908). The overdominance hypothesis relies on the fact that heterosis causes an increase in fitness, therefore a reduction in heterozygosity (as occurs during inbreeding) will lead to a reduction in fitness (Bruce, 1910; Davenport, 1908). This hypothesis means that variation is maintained in the absence of mutation, whereas with the dominance hypothesis a flux of deleterious mutations arise and then become fixed in the inbred individuals. The breakdown of epistasis is a more recent theory and relates to the web of epistatic interactions that potentially exist between alleles (Templeton and Read, 1994). As these are disrupted in inbred individuals (especially if heterozygosity is required for these epistatic interactions) this could explain both heterosis (hybrid vigor) and inbreeding depression.

Drift

Drift is the name given to random fluctuations in gene frequency due to stochastic processes (Falconer and Mackay, 1996), and is generally most important in small populations (with domestic populations frequently having low effective population sizes). It is also most pronounced when substructure, for example inbreeding, has occurred to further decrease the effective population size. In this instance it can readily lead to a loss of allelic diversity, and is therefore similar in effect to inbreeding. Founder effects (when limited numbers of individuals are used to generate the initial population) and bottlenecks (whereby only a limited number of individuals are used to generate the next generation at certain points) will also decrease this initial diversity (in the case of founder effects) or reduce it during their occurrence (in the case of bottlenecks) (Hedrick *et al.*, 2001; Mickett *et al.*, 2003; Sunden and Davis, 1991).

Allelic diversity is important to long-term adaptation (Allendorf, 1986) and is lost more readily than heterozygosity. However, the same measures to avoid the loss of heterozygosity should also help with drift, for instance, with the use of large founder populations with efforts made to maintain diversity (Ferguson *et al.*, 1991) or back-crossing back to wild individuals (Vilà *et al.*, 2005). Effects of this loss of diversity are potentially somewhat equivocal, though generally thought to be negative. A study with bottle-necked *Drosophila* populations found a decreased tolerance to NaCl compared to control populations (Frankham *et al.*, 1999), whilst effects on within-line variance (Avery and Hill, 1977) and variance in response (Wray and Hill, 1989) have been shown. However, Carson (1990) found no effects on variance after bottleneck events in *Drosophila* and house-fly populations.

GENETIC ARCHITECTURE OF DOMESTICATION

The term "genetic architecture" refers to the location, number and effect size of the number of loci (Quantitative Trait Loci, or QTL for short) that underpin the variation in domestication phenotypes (Falconer and Mackay, 1996). A QTL is the name given to a genetic locus that causes variation in a phenotypic trait. For example, one allele at this particular locus may cause an increase in the expression of a trait, whereas another may cause a decrease. The effect of the locus may be caused by a single base change within the coding region of a gene (referred to as a SNP, single nucleotide polymorphism), a single base change in a non-coding (regulatory) region, or indeed any number of different changes in the genetic code (duplications, deletions, etc.). The strong directional selection associated with domestication can potentially lead to a distinctive genetic architecture (Andersson and Georges, 2004), whilst the form of this genetic architecture appears to be remarkably consistent in both plant and animal species (Wright *et al.*, 2010). This architecture has been principally analyzed through QTL studies, whereby wild populations are crossed with domestic breeds, with the resulting hybrids (termed F1s) then either inter-crossed together again or back-crossed to one of the parental populations. These individuals are then analyzed genetically, allowing the major effect loci to be identified positionally on the genome, and the relative strength of effect (i.e. how much they influence the phenotype) estimated.

What all these diverse domesticated species appear to have in common is that the domestication traits themselves all seem to be grouped together in clusters or modules in the genome, rather than being randomly distributed (i.e. a QTL for increased growth may lie near a QTL for decreased anxiety behavior, rather than being very dispersed). Starting with domesticated plant species, beans (Koinange *et al.*, 1996; Pérez-Vega *et al.*, 2010), rice (Cai and Morishima, 2002), maize (Doebley and Stec, 1991, 1993), and sunflowers (Burke *et al.*, 2002) all show this pattern. In animals such modules are also once again seen when using similar wild × domestic crosses. For example, in the chicken various behavioral, morphological and life-history traits have all been shown to be grouped together in clusters (Karlsson *et al.*, 2011a; Schütz *et al.*, 2002; Wright *et al.*, 2006b, 2008; 2010, 2012). Similarly, a study using rat populations selected for high and low aggression also found similar modules composed of QTL affecting physiological and behavioral traits (Albert *et al.*, 2009). Silver foxes differentially selected for tame and aggressive behavior have also been shown to exhibit correlations with other dog-like characteristics (including skull shape and color) (Hare *et al.*, 2005; Trut, 1999). In a QTL analysis of these animals, which was restricted to behavior, overlaps between different behavioral measures were found (Kukekova *et al.*, 2011). A cross between wild and laboratory zebrafish populations found correlations between growth and boldness behavior in the inter-cross generation (Wright *et al.*, 2006a, c).

When such modules do occur, a pertinent question is of course whether each module represents one general "domestication-locus" (i.e. all the domestication effects arise from a single pleiotropic locus), or whether they are in fact comprised of many different, but closely physically linked, loci. In the case of plants, there have been some examples where the actual causative genes and mutations have been identified, allowing this question to be addressed more accurately. Where individual genes have been identified in plants, these do indeed appear to have pleiotropic effects (pleiotropy here refers to when a gene or QTL affects multiple different phenotypes). The *Q* gene in wheat (Faris *et al.*, 2003; Simons *et al.*, 2006) and the gene *tb1* in maize (Clark *et al.*, 2004; Doebley and Stec, 1993) are classic examples. However, with each of these examples, additional mutations and causative genes have also been identified in separate, but physically linked, genes (Doebley and Stec, 1991, 1993; Ji *et al.*, 2006). In the animal examples it is therefore tempting to suggest that these modules, similar to the pleiotropic plant genes, represent genuine pleiotropy. However, this may not be the case—one of the main issues with QTL analysis is the large confidence intervals when an F2 or backcross (BC) format are used, due to the relatively low number of recombinations that may have occurred between closely linked genes (Lander and Botstein, 1989; Lynch and Walsh, 1998). In this case, such large confidence intervals will be unable to distinguish close linkage from pleiotropy without very large sample sizes.

As an illustration, in the chicken study by Wright *et al.* (Wright *et al.*, 2010, 2012), which analyzed a wide variety of behavioral (anxiety and fear–avoidance), morphological (growth and bone density) and life-history (onset of sexual maturity and fecundity) traits, a very large sample size and multivariate statistical analysis were used to attempt to dissect pleiotropy from linkage. Whereas the central loci in the modules were shown to be indistinguishable from pleiotropy, those on the periphery were linked, rather than pleiotropic. A follow-up study using an advanced inter-cross (to increase the number of recombinations present in the cross) examined one module, which affected comb mass and bone allocation, and this was genuinely pleiotropic (Johnson *et al.*, 2012). Pooling these different results from domesticated species, it appears that these modules may contain central "cores" of pleiotropic loci, surrounded by more loosely linked loci.

This idea of loose-linkage was first alluded to by Grant (1981). He noticed that even with a strong domestication phenotype, multiple correlations between different aspects of the domestication phenotype would rapidly break down when reverse selected or crossbred. Although he noticed this with several plant-based examples, the same is true with the correlations between different domestication characteristics breaking down rapidly during inter-crossing (Albert *et al.*, 2008, 2009; Wright *et al.*, 2010). The uses of this in domestication would be to enable new phenotypes to be rapidly bred into a certain background (far harder if traits are generally all pleiotropic), and the pre-existing domestic phenotype regained. Though this may be an advantage, as it allows desirable characteristics to be rapidly bred into the population and then the previous domesticated phenotype to be rapidly regained, it is unclear whether this ability would be specific to domesticated species (potentially explaining the rather limited numbers of domesticated species) or is a general feature of strong directional selection on phenotypic evolution, which is then exploited in the domestication process.

In addition to modularity being observed in the organization of loci affecting different aspects of the domestic phenotype, modularity can also be seen in genome organization and gene expression in general. Using the whole genome expression variation, it appears that the differences between wild and domestic populations are due to hundreds or thousands of gene expression changes (Lai *et al.*, 2008; Natt *et al.*, 2012; Rubin *et al.*, 2007). Once again, these differentially expressed genes tend to be grouped together into modules that are once again close to one another. In effect, what this means is that genes that lie physically close to one another tend to show the same expression pattern.

Examples come from microarray data (whole-genome expression profiles) using inbred lines, or Recombinant Inbred Lines (RILs). What is striking is that in many of the microarray studies comparing wild and domestic populations (that show large-scale differential expression), the QTL architecture appears to be relatively simplistic. Litvin found that the total number of transcriptional differences could be reduced down to a relatively small number of modules (Litvin *et al.*, 2009), whilst Chesler *et al.* found that a small number of QTL modulated large transcriptional gene expression sets that also correlated with behavior (Chesler *et al.*, 2005). These gene modules may therefore represent pleiotropy or close linkage, with a single gene potentially leading to numerous changes in gene expression. In effect the large-scale changes would be downstream effects of the initial causative change. Wild inbred lines also show this pattern, with differentially expressed genes grouped together in clusters in the genome (Ayroles *et al.*, 2009). In these examples it is once again hard to know if these represent multiple linked loci, genuine pleiotropy, or a combination of the two. Pleiotropic hotspots have also been discovered in other eQTL studies (see a review in Breitling *et al.* (2008)), though there is some debate as to whether these are genuine or not, and how common they are (Breitling *et al.*, 2008).

WHAT TYPES OF MUTATIONS CAUSE THE DOMESTIC PHENOTYPE?

Having seen how gene expression differences can lead to variation in a phenotype and their architecture in the genome, it remains to be seen what kind of mutations actually cause these gene expression differences and hence the domestication phenotypes. The genome itself is usually made up of between 13–30 k genes, but the actual size of the genome can vary tremendously between species. Some genomes are rather compact (for example the chicken is around 1.09 Gb), whilst others are larger (humans have ~3 Gb and mice ~3.4 Gb). The genes themselves are made up of exons (which are actively transcribed and often translated to form proteins) and introns (the regions between exons which are spliced out of the gene during transcription).

Promoters are regions at the start of a gene which help determine how the gene is regulated, whilst other enhancers and transcription factor binding sites can be present around the gene (commonly referred to as a cis effect—when the mutation is close to the gene it is affecting) or even much further away (trans effect), and also regulate the degree of expression. Therefore mutations can either affect the coding regions themselves (resulting in altered proteins, or even the absence of a particular protein), or the non-coding regions that regulate expression. Coding DNA is the proportion of the genomic DNA that is translated into proteins. For each gene (which is composed of both exons and introns) the introns are spliced out and the exons are spliced together, and are then transcribed into messenger RNA (mRNA). The mRNAs are then translated into actual proteins. All the other parts of the genomic DNA that are not translated are therefore non-coding DNA (so introns, intergenic DNA and genic DNA that codes for non-coding RNA). Examples of coding-change mutations are generally more abundant, though this is ascertainment bias, due to the increased ease of identification of this type of mutation. This is due to the greater ease of showing that a protein is either truncated or altered after such a mutation, whereas for non-coding changes a multitude of candidates may be present in a region and the exact location is far harder to prove without transgenic evidence or similar advanced techniques. See Carroll *et al.* (2005) for an overview of genome organization and the different mutation effects.

Examples of coding changes affecting domestic animals include examples from coat coloration in chickens (insertion/deletion polymorphisms (Kerje *et al.*, 2004)) and horses (missense mutations (Brunberg *et al.*, 2006)), muscling in dogs (in this case a premature stop codon (Mosher *et al.*, 2007)) and cattle (Grobet *et al.*, 1998). There are growing numbers of examples of cis mutations, whereby non-coding changes affect a trait. In domestic animals examples exist for pig fatness (Van Laere *et al.*, 2003), sheep muscling (the creation of a microRNA binding site (Clop *et al.*, 2006)), color (Eriksson *et al.*, 2008) and comb shape (Wright *et al.*, 2009) in chickens, with more occurring all the time. Indirect evidence may be taken by the large numbers of gene expression variation seen between wild and domestic animals (reviewed above), with this more likely to occur through regulatory changes (though this may be harder to prove, with coding changes not detected in such screens).

Although to date the debate has been regarding which mutation types are more common, it is certain that both these two types play very important parts in the expression of a phenotype. What is perhaps more interesting is whether the non-coding (cis or trans) mutations are qualitatively distinct in terms of their relevance to evolution and development (Wray, 2007). In principle, regulatory mutations may be more pliable for causing quantitative changes due to the nature of pleiotropy and co-dominance. For example, certain phenotypes may be easier to modify using cis mutations than coding mutations. Given the relatively small numbers of genes present in the genome, a toolkit approach occurs, especially with developmental genes (Carroll *et al.*, 2005), with genes used in a multitude of different circumstances.

Cis mutations can cause specific changes in a given tissue at a given time point, whilst expression in other tissues will be unaffected. For example, many of the genes identified in the above studies (*IGF2* for pig fatness, myostatin in sheep, *BCDO2* in chickens, etc.) are all expressed in a wide variety of tissues at a wide variety of times. Therefore, any changes in a coding region will lead to that effect being manifested in every tissue and time point that protein is expressed. In that way, such protein coding changes may be far more pleiotropic, whereas regulatory mutations will allow more specific effects to be manifested without relying on potentially large trade-offs.

Furthermore, most regulatory mutations are co-dominant, whereas protein-coding changes are recessive (i.e. two copies are required to manifest an effect). This enables selection to act more efficiently on heterozygotes in the case of regulatory mutations (as there will be a difference between the heterozygotes and the two different homozygotes).

Copy Number Variation

Copy number variation (CNV) refers to the situation whereby a region contains one or more duplications (or indeed deletions) in one individual relative to other individuals. Technically, these regions are greater than 1 kb (1 kb = 1000 bases) and can be up to several megabases, or even greater, in length (Henrichsen *et al.*, 2009), though more recently much smaller duplications are now being detected and classed as CNVs.

Initially, the analysis of such CNVs rose to prominence around 2006/2007 (Feuk *et al.*, 2006), though now they are becoming more and more a part of standard analysis for the detection of the genetic basis of phenotypic diversity. In standard analysis they are typically detected using a Competitive Genomic Hybridization Array (CGH) (Pinkel *et al.*, 1998), whereby the intensity of regions are compared between two different individuals, pools, or populations, for example, and the intensity gives an indication of the degree of duplication. More recently Whole Genome Resequencing can also be used for this. In this case, individuals have their entire transcriptome (all transcribed genes) sequenced using resequencing technology. The transcriptome is first amplified and then randomly broken into smaller segments. These segments then have the first (and in some cases last) parts "read", with these reads then aligned against the reference genome. This allows both a very accurate distinction of which precise allele is being expressed, and the relative number of reads for each given region can then be compared between multiple individuals or pools of individuals.

Such CNVs have been found to typically occur in regions of segmented duplication (SD; regions of DNA with almost identical sequences) and also in genes affecting immunity, sensory perception and response to stimuli (Clop *et al.*, 2012; de Smith *et al.*, 2008). There can even be large differences between individuals—for example, even within mouse inbred strains (thought to be largely isogenic—meaning essentially clones of one another) differences in CNVs occur. Domestic animals appear no different to the other species looked at, with SD regions and immune genes harboring the greatest numbers of CNVs, with such CNVs found in cattle, pigs, chickens, and dogs amongst others (Chen *et al.*, 2009; Clop *et al.*, 2012; Fadista *et al.*, 2008; Liu *et al.*, 2010; Wang *et al.*, 2010).

The effects of CNVs can be extremely varied, as can their mechanisms. Confining ourselves to domestic animals (Clop *et al.*, 2012), effects have been found on color in horses (premature graying (Rosengren Pielberg *et al.*, 2008)), goats (Fontanesi *et al.*, 2009), and sheep (Norris and Whan, 2008) (both with a CNV in the *ASIP* gene), morphology in chickens (causing the classic pea comb variant (Wright *et al.*, 2009)), and also late feathering in the same species (Elferink *et al.*, 2008). In dogs CNVs are related to the ridge mutation in ridgeback dogs (Salmon Hillbertz *et al.*, 2007) and chondrodysplasia in a variety of small dog breeds (Parker *et al.*, 2009). These examples are all discrete mutations. However one study did find an association between production traits and CNVs in cattle (Seroussi *et al.*, 2010), though in this case the exact mechanism has not been conclusively proven.

Moving away from domestic animals, other examples of correlations between complex diseases and CNVs have been identified (Stankiewicz and Lupski, 2010; Zhang *et al.*, 2009). Associations with neurocognitive defects (Brunetti-Pierri *et al.*, 2008; Golzio *et al.*, 2012), autism (Sanders *et al.*, 2011), and schizophrenia (Sebat *et al.*, 2007) are of particular note in a behavioral setting. The actual mechanisms by which CNVs affect the expression of a trait can be extremely varied and range from the simple to the complex. At the most basic level, deletions (a form of CNV) can remove one or more exons from one or more genes (or indeed even entire genes), leading to large changes in the protein that is expressed. More subtly, dosage effects can be found, with the greater the number of copies of a gene, the greater the expression (especially if the gene promoter is also duplicated). Further mechanisms can be more complex still, with transcriptional sites that up- or down-regulate a gene altered through dissociation between these sites and their promoters, alteration of chromatin structure and the like (see Henrichsen *et al.* (2009) for a brief overview), though little is yet known about such regulation.

MAPPING GENES FOR BEHAVIOR—TOP DOWN APPROACHES

The mapping of genes for behavior is similar to mapping other quantitative traits, only with the exception that the trait is often harder to define and measure reliably. The general methods either use a top down or a bottom up approach. Essentially this refers to whether one starts at the phenotype and attempts to work down to the gene level, or whether one starts at a gene level and attempts to work up to the phenotype (Boake *et al.*, 2002).

Pedigree Studies and Heritability Analysis

Initially this uses statistical analysis of pedigrees to demonstrate an actual heritable component to the trait of interest, through breeding designs, artificial selection or the like. The requirements at this stage are only the behavioral phenotypes for all individuals and knowledge of their relatedness (the pedigree). Heritability studies can further dissect traits down to distinguish between broad-sense and narrow-sense heritability, with the genetic component further dissected down to additive and dominance components (additive in this case refers to those loci which give a cumulative effect, whereas dominance represents the interactions between alleles at the same locus) (Falconer and Mackay, 1996; Lynch and Walsh, 1998).

Quantitative Trait Loci (QTL) and Association Mapping

QTL analysis is nowadays a very well-established technique, which became increasingly popular in the 1990s (Lander and Botstein, 1989). In recent times it has enjoyed something of a renaissance, first with the use of microsatellite markers, and now with SNP markers and resequencing technology. Microsatellite markers are regions of the DNA that are non-coding and contain small motifs (two, three or four bases long) that are repeated anywhere from three to a hundred or more times. The nature of these repeats means they are much more likely to expand than other regions (they are typically in non-conserved regions, hence this expansion is rarely a problem), thereby generating a large degree of variability, and making them excellent as molecular markers. In contrast, SNPs are any single-base changes that are present in the genome. They are obviously by far the most prevalent marker in the genome, and the ability to identify these markers has truly opened up the world of genomics, not least with the ability to reliably saturate any given genome with a wealth of markers. These markers are finally enabling the step to gene discovery to be performed (previously very much a weakness with this approach).

At its most basic, this is the crossing of two distinct populations that differ for one or more traits of interest. As such, this technique is particularly amenable for domestic animals, where extremely large phenotypic differences frequently occur between wild and domestic populations (Andersson and Georges, 2004). The inter-cross individuals are then genotyped for multiple markers spread throughout the genome, with the genotype information correlated with the phenotypic data. This enables the identification of the number of loci affecting the traits that differ between the two populations, their effect size and their genomic location. The loci can be defined as their additive and dominance components, whilst pleiotropy and epistasis (multiple genes interacting with one another) can also be identified. Two-way epistasis is the only realistic form of detectable epistasis, due to the massive increase in the number of tests performed (essentially an exponential increase).

As mentioned, the major issue is one of resolution of the detected loci (with very large confidence intervals generally the norm) and the sample sizes required for the detection of smaller effect loci (Beavis, 1998).

Although this is an extremely flexible technique, the resolution of the detected loci is limited by the number of recombinations present in the test cross (Darvasi, 1998; Lynch and Walsh, 1998). Typically this means that the resolution of each QTL is rather low, and relatively few QTL have been refined to the causative gene (often referred to as Quantitative Trait Genes or Quantitative Trait Nucleotides if the exact mutation is identified) though this is beginning to change.

Association mapping is a similar technique, but frequently uses a single outbred population and a high-density SNP analysis, to identify QTL. The difference with this technique is that as outbred or single large populations are used, many more recombinations are present and therefore the resolution is far greater. In contrast, however, many more markers are required per individual and population substructure can be an issue. Additionally, in such large populations, the genetic architecture tends to be very complicated (as far more polymorphisms and alleles are being considered than in a standard QTL cross which utilizes inbred populations/individuals and most commonly only two different populations). Results with association mapping in large populations in humans tend to show that very few QTL with large effects are present, and very little of the variance present can be explained by the loci which are discovered (Carlson *et al.*, 2004). However, in the case of domestic animals, they possess many features that make them more amenable to this analysis (Goddard and Hayes, 2009). Firstly, the strong directional effects of domestication selection can mean that fewer markers are required as haplotype blocks are larger (for instance in the case of the dog (Karlsson *et al.*, 2007)) and the genetic architecture is less complex than more outbred populations, in terms of the numbers of alleles present which affect the trait at any given QTL. This does have the drawback that only large regions are identified, however, when combined with multiple breeds very narrow regions can be identified. Jones *et al.* used this to tentatively map QTL for distinct breed characteristics, including pointing, herding, boldness and trainability (Jones *et al.*, 2008).

Selective Sweep Mapping

Linkage disequilibrium and the identification of selective sweeps is potentially an extremely powerful tool for the dissection of genetic architecture, and shows great potential in domestic populations (Andersson and Georges, 2004). The basis of selective sweeps is that when a new mutation arises in a population and is then subjected to strong directional selection, not only it but the haplotype it arose in will go to fixation. The surrounding SNP markers will therefore "hitch-hike" alongside the mutation (Smith and Haigh, 1974).

Once fixation is reached, depending on the strength of selection and the time that has elapsed since fixation, this signature of selection will be slowly eroded. In the case of domestic populations, the strength of selection is often extremely high, increasing the likelihood of such sweeps occurring, though this is also dependent on the genetic architecture of the trait (Pritchard and Di Rienzo, 2010). Such sweeps have been seen in the chicken, where around 50 40-kb regions were putatively identified as being under selection in domestic layer and broiler populations (Rubin *et al.*, 2010), the dog (with blocks of around 106 kb in length (Goldstein *et al.*, 2006)), and even the stickleback (sweep sizes between 20 kb and 90 kb in size (Makinen *et al.*, 2008)). This approach is particularly powerful when multiple domestic populations are analyzed, with shared regions of identity-by-descent identified, and has been used to locate both discrete mutations (for example the pea comb (Wright *et al.*, 2009) and yellow skin (Eriksson *et al.*, 2008) mutations in chickens) and mutations affecting quantitative traits (QTNs), for example pig fatness (Van Laere *et al.*, 2003), cattle twinning rate (Meuwissen *et al.*, 2002) and milk production (Riquet *et al.*, 1999), and comb size in chickens (Johnson et al., 2012).

MAPPING GENES FOR BEHAVIOR—BOTTOM-UP APPROACHES

With this approach, research starts at the gene level and is built-up to the behavioral phenotype. At its purest form, gene knock-outs, knock-downs (whereby the expression of a gene is reduced using a variety of different techniques), and transgenics (the insertion of a novel gene into a genome) (Flint and Mott, 2001) can be used to directly study the effects of individual genes, though one must still find which genes should be perturbed prior to analysis. Often such research therefore starts with a mutagenesis screen, whereby large numbers of mutant individuals are generated and then bulk screened for alterations to the behavior of interest (Nadeau and Frankel, 2000).

Such an approach has produced many of the early successes with behavioral gene identification, though these were restricted to circadian rhythm in *Drosophila* (Sawyer *et al.*, 1997; Tully, 1996), social aggregation in *C. elegans* (Coates and de Bono, 2002; de Bono *et al.*, 2002) and larval foraging in *Drosophila* (de Belle and Sokolowksi, 1989; Sokolowski, 1998), so the relevance to domesticated animals may be lower, especially considering the size and generation times of many of the domestic animals under investigation.

Moving up, global gene expression analysis, traditionally through microarray analysis though more recently through RNA-sequencing (whole-genome sequencing of the transcriptome) gives more precise measures of transcription (Wang *et al.*, 2009). The issue with this approach is the amount of data generated and interpreting it. For instance, what tissue does one look in and at what time point? A particular gene may be limited in its expression to a very narrow window and even then relatively modest differences in expression may cause the phenotypic change. Additionally, if you find differentially expressed genes between populations for example, many hundreds or even thousands can be identified, but determining the critically important genes from those which are due to downstream changes of such key regulators can be extremely complicated (Verdugo *et al.*, 2010).

Merging both a top-down and a bottom-up approach, expression QTL (eQTL) analysis uses a combination of expression microarrays run on each individual from a standard QTL study (Gibson and Weir, 2005). In essence, the microarray gene expression values are used as phenotypes to be correlated with the genotypic marker information for each individual. The difference here is that the physical position of the gene is known, so what is mapped is the causative element leading to variation in the gene expression. Obviously, the same issue with tissue and time point choice is still present, but this does allow the correlation with a behavioral phenotype, which can be much more persuasive evidence of causative gene identification (Mehrabian *et al.*, 2005).

On a note regarding model organisms (*Drosohila, C.elegans*, mice, etc.), though these model systems may obviously be less directly relevant to domestic animals (especially in the case of *Drosophila* and *C.elegans*), the number of actual known mutations that affect behavioral traits is extremely low, and the vast majority (indeed virtually all) of the mutations that have been identified have been done so in these model organisms. They therefore represent a powerful resource of what one should expect and what systems may be involved in genes affecting behavioral variation in domestic animals. The greater availability and more advanced genomic tools present in these model organisms also make it useful to more finely dissect any putative candidate genes that affect domestic animals.

EFFECTS OF SPECIFIC MUTATIONS

Social Aggregation in *C. Elegans*

The identification of the actual mutations that lead to variation in behavior have been notoriously hard to identify, whilst none have been truly identified for domestication behavior in a domestic animal. Having said that, the lessons learnt from the few examples identified in model organisms can give important insights to the type and effect such mutations will no doubt have.

Beginning with social aggregation, one of the classical examples for this actually comes from the nematode *C. elegans*. In the wild, there are two naturally occurring variants of worms feeding on *E. coli*—social and solitary foragers (de Bono and Bargmann, 1998; Hodgkin and Doniach, 1997). Greater than 50% of social foragers are found in groups, while less than 2% of solitary foragers are found in such states, irrespective of age and sex. The social foragers move more rapidly (de Bono and Bargmann, 1998), and accumulate at the richest food regions. However, social foragers do not aggregate in the absence of food (de Bono and Bargmann, 1998), implying that ecological factors are also important to the manifestation of this behavior.

Both morphs are associated with natural variation at the *npr-1* (neuropeptide receptor resemblance) locus, although several other loci may affect this behavior, albeit in a lesser manner. This locus was identified using a mutagenesis screen, and encodes a neuropeptide receptor, similar to neuropeptide Y (NPY) receptors (G-proteins, regulating food consumption, anxiety, and pain in mammals). These similarities suggest that *npr-1* has a neuropeptide ligand, however no clear NPY homolog has been found. Null mutations in *npr-1* transform solitary feeders into strong social feeders, with a specific single amino acid change in *npr-1* also found in 17 natural isolates (de Bono and Bargmann, 1998).

The gene *npr-1* regulates this aggregation through stimulating aversion to noxious stimuli (de Bono *et al.*, 2002), utilizing specific neuronal pathways (the ASH and ADL neurons). If these neurons are ablated, the social aggregation response is severely decreased. Aggregation is regulated by food in *C. elegans* (de Bono and Bargmann, 1998), therefore food in this case is the aversive stimulus, not surprising given that soil bacteria can kill *C. elegans* (Marroquin *et al.*, 2000). In mammals NPY has also been shown to have sedative effects and decreases the sensitivity to nociceptive stimuli (Naveilham, 2001), showing a similar trend to the *C. elegans* model system with *npr-1* activity suppressing the behavior induced by noxious stimuli.

Foraging in *Drosophila* Larvae and *Apis mellifera*

Another classic mutation affecting behavior, causing extreme effects and largely attributable to a single gene locus is found in the *for* gene (de Belle and Sokolowksi, 1989; Sokolowksi, 1980). Individuals with what is termed the "rover" allele move greater distances whilst feeding than those homozygous for the "sitter" allele, with neither showing any variation in behavior in the absence of food (Pereira and Sokolowksi, 1993). Both of these behavioral morphs exist at appreciable frequencies in the wild (70% rover; 30% sitter (Sokolowksi, 1980, 1982). The *for* gene itself is in fact the gene *dg2*, one of two cGMP-dependent protein kinase genes (PKG) in *Drosophila*, with the cGMP signal transduction pathway implicated in foraging behavior in *D. melanogaster* (Osborne *et al.*, 1997). PKGs function in naturally occurring variation in larval foraging behavior, with gene levels correlating with larval foraging behavior (Sokolowski, 1998). Sokolowski *et al.* (1997) used an empirical study comprising of three separate base populations, each selected for either high- or low-population density. Those individuals in a low-density selection environment had shorter foarging path lengths, whilst those from high-density environments had longer path lengths.

The evolutionary benefits of these different phenotypes are therefore apparent in varying-density environments. At high densities larvae are required to navigate around other individuals to reach food patches, whilst under low-density conditions this locomotion behavior is an unnecessary waste of energy, with food being continuously distributed and in greater quantity. *For* is found to be expressed in a diverse set of tissue types, principally those involved in taste, olfaction, gut, and central brain function (closely paralleling rat PKG1 expression (Kroner *et al.*, 1996)), and thought to be involved in the response to aversive stimuli. The association between social behavior and neurons that respond to aversive conditions in this instance therefore provide a mechanistic basis for these observations from behavioral ecology. Such adverse environmental conditions often provide the stimulus for group formation (for instance antipredation behavior in fish (Krause and Ruxton, 2002)).

Pigmentation and Behavior

As seen in ancient wall pictures and as is obvious from recent selection experiments, loss of pigmentation seems to be one of the first and most conspicuous phenotypic changes occurring in most domesticated species (Clutton-Brock, 1999; Price, 2002). This leads to a rapid increase in frequency of spotted and white individuals in the population. It is of course quite possible that the fast emergence of hypo-pigmented individuals is a result of directed human selection of rare, spontaneous mutations. For example, white and spotted individuals may have been easier to find and recognize, and to distinguish from the wild ancestors, which would have been abundant around the early farm villages, and hence may have been perceived as more attractive. However, it is also possible that loss of pigmentation is related to tameness and reduced stress sensitivity, as we discussed earlier in the perspective of the Belyaev experiment on foxes. In this case, non-pigmented animals may have developed spontaneously and could have been favored during selection either because of their visual attractiveness or because of their tameness, or both.

Several studies of different species indicate a clear correlation between pigmentation on the one hand and stress sensitivity or fearfulness on the other. For example, non-pigmented mink are generally easier to handle than their colored conspecifics (Trapezov *et al.*, 2008) and in wild vertebrates, darker and more pigmented species and populations are generally more aggressive and more sexually active (Ducrest *et al.*, 2008). Salmon selected for low cortisol response in a stress test have fewer pigmented skin spots than more stress-prone individuals (Kittilsen *et al.*, 2009).

Skin and hair pigment come in two different variants: black eumelanin, and red phaeomelanin. The phenotypic expression of both is mediated by the melanocortin receptor, which, in turn, exists in five different variants, present in different tissues. The most important one for the expression of pigmentation is melanocortin receptor 1 (MC1R), a G-protein-coupled receptor in melanocytes. When bound to the agonist α-MSH (melanin stimulating hormone) the receptor causes a cascade of reactions leading to the deposition of black eumelanin in the melanosomes, whereas the receptor antagonist, the Agouti signaling protein (ASIP) leads to a deposition of phaeomelanin instead.

The key to how pigmentation may act pleiotropically to behavior and stress sensitivity lies in the fact that the melanocortin receptors are spread in many different tissues and that their ligands cause a variety of effects. A mutation in the receptor gene may thus inhibit or modify pigment expression, while simultaneously altering the stress response system. This is further supported by observations that cortisol levels in dog hairs vary systematically with pigmentation: phaeomelanistic (red) hairs have higher concentrations than eumelanistic (black), while banded hairs (agouti) are intermediate (Bennett and Hayssen, 2012). Furthermore, in rodents, agouti-colored animals are more aggressive and less tractable than the non-agouti conspecifics (Hayssen, 1997) (Cottle and Price, 1987).

Fang and co-workers studied mutations in *MC1R* of pigs from China and Europe and observed nine different protein-changing modifications in the gene in different populations (Fang *et al.*, 2009). This is a striking difference to the case in the ancestor, the wild boar, where all observed mutations were silent (i.e. synonymous substitutions). The authors conclude that pigmentation in the wild ancestor is under strong purifying selection, whereas different phenotypically active mutations have been selected by different human populations. They argue that this shows that color variation during domestication is a result of direct selection for attractive color phenotypes, rather than a side-effect of selection for tameness. However, it is also possible that different mutations in pigmentation-related genes, such as *MC1R*, may cause similar pleiotropic effects on behavior. Behavioral differences between different pig mutants have not been studied, so it remains a possibility that selection has originally been on tameness rather than on color. As mentioned above, the two explanations are not mutually exclusive.

The phenotypic expression of pigmentation is of course also dependent on further downstream genetic processes, which may be related to behavior in different ways. In a large-scale QTL-study of a cross between domesticated White Leghorn chickens and ancestral Red Junglefowl, it was observed that the risk of being exposed to feather pecking, a detrimental behavior disorder, was related to a mutation in the *PMEL17*-gene (Keeling *et al.*, 2004). The same mutation was identified as the causative mutation in the Dominant White-locus, and causes a lack of all melanin expression in feathers and skin (Kerje *et al.*, 2004). The mutation protects against being exposed to feather pecking, and the pigmented birds are more vulnerable to becoming victims when they are relatively common.

Follow-up studies showed that the specific mutation, a nine-base-pair insertion in the transmembrane region, was related to differences in social and exploratory behavior, also when the birds were raised without the possibility to interact physically with conspecifics (Karlsson *et al.*, 2011a; Karlsson *et al.*, 2011b). In general, pigmented birds were more aggressive and explorative. This shows a clear pleiotropic effect on behavior of a specific pigment-related mutation, but it remains unclear which mechanisms may be involved. Some hypotheses may, however, be discussed.

PMEL17 is a melanocyte-specific protein, which is not directly involved in the synthesis of melanin, but rather affects the maturation of melanosomes, the melanocytic vesicles where melanin is deposited. The protein forms amyloid fibers on the inside of the melanosome membranes, and these serve as the substrate where melanin particles fasten (Yasumoto *et al.*, 2004). The mutation inhibits the formation of a transmembrane region, and the further development of the necessary amyloid fibers, and thus leaves the melanosomes without any substrate for melanin deposition.

Melanin is formed from tyrosine and DOPA as precursors. DOPA is also a precursor for the synthesis of the potent neurotransmitter dopamine and the chatecholamines norepinephrine and epinephrine. As a hypothesis, Nätt and coworkers suggested that when melanin cannot be deposited, its synthesis may be down-regulated, and the concentration of DOPA may consequently increase (Nätt *et al.*, 2007). This could in turn make more DOPA available for the dopamine synthesis pathway. Although this mechanism is highly speculative, the fact that there is a close biochemical connection between melanin synthesis and behaviorally important pathways makes a connection between pigmentation and stress responsiveness quite logic.

Neurotransmitters and Social Behavior

Among the more prominent behavior changes occurring as a response to domestication is the modified social behavior. Whereas the wild ancestors of our domesticates usually live in groups with very little flexibility concerning size, dynamics, and home ranges, most domestic species tolerate very variable group sizes and a high degree of dynamic changes in group composition. For example, wolves live in packs which rarely exceed 15–20 members, and where all individuals mostly are closely related, whereas free-ranging dogs form variable-sized groups of up to dozens of dogs where membership of a group may change from one week to the next (Boitani and Ciucci, 1995); Wild boars form tight family groups with little exchange of individuals, while domestic pigs tolerate being regrouped at regular intervals during rearing (Gonyou, 2001); Red Junglefowl live in small harem groups with strict territorial boundaries, while chickens under commercial conditions can cope with thousands of other birds in a limited space (Mench and Keeling, 2001). Mutations affecting sociality and social tolerance must have been important selection targets during domestication.

During recent times, there has been a growing interest in the role of neuropeptides, primarily vasopressin and oxytocin, in shaping the social behavior of an individual (Donaldson and Young, 2008). While the actual hormones are evolutionary well conserved, variations in the expression of their receptors have been experimentally connected to variations in social behavior both between and within species. For example, the expression of the vasopressin receptor gene *AVPR1a* is highly correlated with the degree of gregariousness across different finch species and with the degree of mate attachment in voles (Donaldson and Young, 2008). The expression differences appear to be largely connected to variations in the length of a highly polymorphic microsatellite (a stretch of repetitive DNA-sequences) located at the 5′ flanking region of the gene. It seems likely that the length of this microsatellite may affect the transcription level of the gene. Correlations between the length of the microsatellite allele and sociality have been demonstrated in birds, voles, and humans.

The other of the two closely related nine-amino-acid neuropeptides, oxytocin, has received a large interest for its role in social bonding and nursing and nurturing in humans. It exerts basically the same effects across different mammalian species, where increased levels of circulation are associated with a stronger bonding and nurturing tendency (Donaldson and Young, 2008). It appears that oxytocin mainly exerts its effects on females while vasopressin chiefly is the male effective hormone.

Selection for favorable variants of the receptor genes of these neuropeptides would hypothetically be a plausible pathway for causing the domesticated social behavior, which seems to be characterized by increased social tolerance and an increased ability to bond with humans. However, there has been little direct research into this possibility. Meanwhile, some suggestive observations lend a bit of support to the hypothesis.

In a large-scale QTL-analysis of a cross between Red Junglefowl and domesticated White Leghorn layers, a major QTL affecting growth and several other domestication-related traits, was discovered on chromosome 1 (Kerje *et al.*, 2003; Schütz *et al.*, 2002). The QTL was named Growth 1, in recognition of its immense effect on growth rates—the genotype at this locus explains more than 20% of the difference in adult body weight between the breeds. In addition, Growth 1 also affects several other domestication related traits, such as egg production and fearfulness, as measured in open-field tests.

A closer analysis of the gene content in the region showed that *AVPR1a* is located centrally in the QTL. Also another vasopressin receptor *AVPR2*, as well as the oxytocin gene, are well within the confidence interval of Growth 1. A plausible hypothesis would therefore be that social behavior could be modified by these loci as a correlated response to selection for increased growth.

Wirén and colleagues bred birds from an advanced inter-cross line between Red Junglefowl and White Leghorns to be homozygous for alternative alleles at a microsatellite marker in Growth 1, closely linked to AVPR1a (Wirén and Jensen, 2011). These birds, and birds from the pure lines, were tested in different social setups. Red Junglefowl, as expected, were more aggressive towards intruders and less inclined to interact with strangers. In the inter-cross birds, the difference in genotype on the selected marker locus close to *AVPR1a* caused the birds to show breed typical differences in males, but not females, indicating that the gene may be a major player in causing the domestication induced differences in social behavior. As a further support to this, the gene is differentially expressed in the part of the brain containing the ventral pallium when comparing Red Junglefowl and White Leghorn males (Wirén, 2011).

Of course, these studies are purely based on correlations, and the effects could just as well be caused by any other linked gene in the same region. However, it is intriguing that several socially potent neuropeptide genes are located in the same small genomic region. It remains to be studied which mutations may underlie the different expression rates and it is also as yet uncertain to what degree this locus is involved in shaping social behavior in chickens. Furthermore, neither of the neuropeptides or their receptors has been studied with respect to domestication in mammals, so the story remains a thought-provoking idea at this time.

Different Means to the Same Ends

The traditional domesticated animals have a history of thousands of years of living with humans, making it very difficult to pinpoint early genetic responses to the altered selection. However, studies of recently added members to the domesticated fauna may be interesting in this respect. Many fish species have been domesticated only in the last hundred or so years, while at the same time being under similar selection pressures as other domesticates. For example, they are bred to grow faster than their ancestors in an environment where food is abundant and predation is virtually absent.

Tymchuk and colleagues studied gene expression profiles in domesticated rainbow trout and their wild conspecifics in different tissues, and further compared this to other domesticated salmonids (Tymchuk *et al.*, 2009). They found that transcription profiles could differ substantially between different species, but taken together they target the same physiological end-point. The common denominator appears to be to affect the Growth Hormone gene, and the closely associated *IGF1*. This suggests that although the phenotypic target is the same, the genetic networks are sufficiently complex and flexible to cause a similar phenotypic effect by means of different pathways. This would be an important argument against the previously mentioned hypothesis that domestication is a single-gene event.

The fish studies also demonstrate the existence of important pleiotropic effects, which conversely adds some support to the idea that broad phenotypic alterations can be caused by selection of a single trait. For example, Johnsson *et al.* (1996) studied brown trout (*Salmo trutta*) which were either hatchery raised or injected with growth hormone. Both treatments caused a similar increase in growth rate. Irrespective of whether growth had been increased by selective breeding or by hormonal manipulations, the correlated behavioral responses were very similar. Both types of fish had a reduced anti-predator behavior, explained by the authors as a response to a reallocation of resources from vigilance and alertness to growth. Similarly, Wright *et al.* (2006c) found correlations between growth and fear–avoidance behavior in an inter-cross population between wild and laboratory zebrafish.

EPIGENETICS

So far, we have discussed how different mutations and variants in DNA-sequence may be involved in causing the phenotypes associated with domestication. According to current genetic and evolutionary theories, spontaneous and randomly occurring mutations are the only substrate for natural and artificial selection. As we have seen, many of the identified mutations are mainly regulatory, and do not affect protein structures. So the strongest driving forces behind evolutionary change appear to be related to the timing of expression of protein coding genes, or in other words, how the genome is orchestrated during ontogeny.

This has sparked an increased interest in understanding the mechanisms involved in gene regulation. Some of these mechanisms are purely genetic in their nature, as explained above when we discussed eQTL-studies, in which case the expression profile of a particular gene is contingent on the DNA-sequence on another locus. However, a substantial part of the expression variation is due to chemical modifications, which do not affect or relate directly to DNA-sequence. These are the factors we refer to as epigenetic (Richards, 2006), and they represent a recently discovered and as-yet poorly understood bridge between the environment and the genome.

At the molecular level, mainly two chemical changes have attracted research so far. The first consists of a modification of cytosine bases, where a methyl group is added to the fifth carbon, changing C to 5 mC. Most methylation of cytosine occurs when this nitrogen base is positioned next to a guanine base, a so called CpG-position, and throughout the genome, repetitive sequences of this combination occur mainly in regulatory regions, in so called CpG-islands. Increased methylation is usually associated with decreased expression of the gene.

The second important group of epigenetic modification affects histones, the proteins responsible for packing DNA in the nucleus. The histone tails may receive additions of, for example, methyl- or acetyl-groups, which will change the density with which DNA is packed. Consequently, the more densely packed the DNA, the less transcription of the genes in the region. It is important to remember that although both DNA-methylation and histone modifications are important regulators of genome activity, they do not affect the DNA-sequence itself. A novel term to capture the status of the genome with respect to its epigenetic modifications is “epigenome”.

A large proportion of the epigenetic variation is constitutional—it varies vastly between cell types and organs, but within a particular cell type, variation is much smaller between individuals (Richards, 2006). This constitutional variation depends on genetic differences, either in the promoter regions or in translocated loci. However, some of the epigenetic variation is dynamic and modifiable, so loci involved in this may be highly responsive to environmentally mediated input—hence, the epigenome can be thought of as a bridge between nature and nurture.

Epigenetic Changes and Long-Time Behavior Modifications

In a pioneering series of experiments, Meaney and co-workers studied the long-term effects of early experience on later behavior, and particularly investigated the epigenetic mechanisms involved (Weaver *et al.*, 2004). In a cross-fostering experimental design, rats were raised either by attentive and caring mothers (which licked their pups a lot and performed much arched-back nursing) or by less careful females. Offspring of the more caring mothers were more resistant to restraint stress when they were adults, and displayed a more attentive and careful maternal behavior when they themselves gave birth. This was found to be due to an induced difference in methylation status of a promoter region of the glucocorticoidreceptor gene (*GR*), particularly the binding site associated with the transcription factor *NGFI-A*, also known as *EGR1*. The methylation differences between differently nurtured rats also caused histone modifications in the same region, leading to a modified expression of *GR*.

This experiment was one of the first clearly demonstrating that experiences gained during ontogeny can alter the epigenetic state of the genome, and subsequently give rise to behavioral differences. A number of other experiments have demonstrated similar effects after this. For example, maternal care has been shown to affect methylation states of estrogen receptor alpha in rodents (Champagne and Curley, 2008), and epigenetic factors affect cognitive and behavioral traits in a variety of species, including humans (Gräff and Mansuy, 2008).

Stress plays a crucial role in the inter-relationship between environmental factors and epigenetic modifications. Although different definitions of stress are ubiquitous, there is a general consensus that steroid hormones, such as cortisol (or corticosterone), testosterone, and estrogens, are usually elevated under conditions perceived by an individual as stressful. Steroids act as DNA transcription modifiers when the steroid–receptor complex binds with binding proteins to specific DNA-regions, and this provides an open gate from stress experiences to modifications of gene expression (Murgatroyd *et al.*, 2009).

Epigenetic Changes in Domestication

Long-term effects of epigenetic modifications can also be transferred across generations, by changes in the germ-line epigenomes. Skinner and coworkers exposed rats to the environmental pollutant vinclozolin, which caused a dramatic effect on the methylation states and expression levels of many genes, and also affected their stress responses and anxiety-related behavior (Skinner *et al.*, 2008). Both the epigenetic effects and the behavioral modifications were clearly measureable in the third, untreated offspring generation. This shows that epigenetic effects may alter phenotypes in ways similar to those of DNA-mutations (although perhaps often in a reversible manner), and offers a hypothetical novel route for rapid evolutionary changes in response to, for example, environmental stress. A broad review of studies supporting this has been provided by (Jablonka, 2009).

Since domestication is an example of precisely such a rapid evolutionary process, causing wide phenotypic modifications in a short time, it is quite likely that epigenetic mechanisms may play an important role here. Lindqvist and coworkers studied this in chickens, by exposing groups of ancestral Red Junglefowl and domesticated White Leghorns to chronic stress during rearing (using an unpredictable light scheme as the stressor), and measured the outcome on a spatial learning task. In both breeds, stressed birds showed a reduced learning ability compared to control groups (Lindqvist *et al.*, 2007). However, only in domesticated chickens was this decreased learning mirrored in the offspring. Using microarray analysis, the authors found that stress-induced brain gene expression differences were correlated across generations in domesticates, but again, not in Red Junglefowl. Hypothetically, it was suggested that domestication could have selected animals with an increased capacity to respond epigenetically to environmental stress, and to transfer these responses to the next generation. In effect, the suggestion is that domestication may have selected animals with a higher evolvability (Carter *et al.*, 2005), which would have speeded up the pace of evolution considerably.

This suggestion is as yet hypothetical and speculative, but it has received a certain experimental support lately. Nätt and colleagues used microarrays to analyze differential gene expression in the hypothalamus regions of a small number of families of Red Junglefowl and White Leghorn, and in addition studied the methylation differences in the promoter regions of more than 3000 of the same genes (Natt *et al.*, 2012). The families were selected based on divergent results in a fear test, so within each breed there were highly fearful and less fearful individuals. As in the previous studies by Lindqvist *et al.* (2007), fearfulness was significantly correlated between generations in White Leghorns, but not in Red Junglefowl, again indicating that the ability for transgenerational effects may have been favored during domestication.

Comparing the breeds, close to 300 genes were differentially expressed in the parents, and more than 1600 in the offspring. Gene Ontology (GO) analysis showed that genes belonging to pathways associated with stress responses, memory consolidations, neural differentiation and reproduction were enriched among the differentially expressed ones. Looking at promoter regions, over 200 of the 3600 selected genes were differentially methylated in parents, and more than 800 in the offspring. Among genes carrying differential methylation, 79% were hypermethylated in White Leghorns, indicating that this breed has accumulated methylations during domestication. This suggests that selection for favorable epigenetic variation may have been an important aspect of domestication.

A particularly interesting aspect of the experiment is the fact that differential gene expression profiles, as well as methylation differences, were reliably inherited by the offspring, so there was high transgenerational stability in the epigenetic effects. This further supports the importance of epigenetic selection as a driver of phenotypic changes during domestication. The last piece of evidence along this line from the same study is the observation that differentially methylated loci were significantly more common in previously identified selective sweep regions associated with domestication of layer breeds. This strongly suggests that a significant proportion of the observed differential methylations may have been subject to directed selection.

CLOSING REMARKS

We are in the middle of one of the most dynamic developmental phases in biology ever experienced. Full-genome sequencing is becoming faster and cheaper every year, and epigenetic screening is developing at an unprecedented rate. Hence, we can foresee a rapid increase in knowledge regarding how genetic variation is related to phenotypic differences. This will of course also be the case for animal behavior.

The challenge faced by scientists will be to provide reliable phenotypic measurements. Among all possible phenotypes, behavior stands out as particularly difficult. It is the first and fastest way in which an animal will react to any stimulus, and will therefore vary with almost any environmental factor (unlike, for example, color or size). The obvious solution to this is to standardize behavioral tests as much as possible, and this introduces the risk of distancing the recordings more and more from natural conditions. This is a highly sensitive balance, and it will take a high degree of scientific skill from the researchers of the future.

The ultimate goal of the research will be to find the causative genetic polymorphisms responsible for the domesticated behavior. This will be of enormous importance to anyone interested in the biology and welfare of domesticated animals, but equally so for those mainly concerned with the mechanisms of evolution.

REFERENCES

Agnvall, B., Jöngren, M., Strandberg, E., Jensen, P., 2012. Heritability and genetic correlations of fear-related behaviour in Red Junglefowl—Possible implications for early domestication. PLoS ONE 7, e35162.

Albert, F.W., Carlborg, O., Plyusnina, I.Z., Besnier, F., Hedwig, D., Lautenschlager, S., et al., 2009. Genetic architecture of tameness in a rat model of animal domestication. Genetics 109.102186.

Albert, F.W., Hodges, E., Jensen, J.D., Besnier, F., Xuan, Z., Rooks, M., et al., 2011. Targeted resequencing of a genomic region influencing tameness and aggression reveals multiple signals of positive selection. Heredity 107, 205–214.

Albert, F.W., Shchepina, O., Winter, C., Römpler, H., Teupser, D., Palme, R., et al., 2008. Phenotypic differences in behavior, physiology and neurochemistry between rats selected for tameness and for defensive aggression towards humans. Horm. Behav. 53, 413–421.

Allendorf, F.W., 1986. Genetic drift and the loss of alleles versus heterozygosity. Zoo. Biol. 5, 181–190.

Andersson, L., Georges, M., 2004. Domestic-animal genomics: deciphering the genetics of complex traits. Nat. Rev. Genet. 5, 202–212.

Avery, P.J., Hill, W.G., 1977. Variability in genetic parameters among small populations. Genet. Res. 29, 193–213.

Ayroles, J.F., Carbone, M.A., Stone, E.A., Jordan, K.W., Lyman, R.F., Magwire, M.M., et al., 2009. Systems genetics of complex traits in Drosophila melanogaster. Nat. Genet. 41, 299–307.

Beavis, W.D., 1998. QTL Analyses: Power, Precision and Accuracy, 49th Annual Corn and Sorghum Industry Research Conference. ASTA, Washington DC, pp. 250–266.

Belonsky, G.M., Kennedy, B.W., 1988. Selection on individual phenotype and best linear unbiased predictor of breeding value in a closed swine herd. J. Anim. Sci. 66, 1124–1131.

Belyaev, D.K., Plyusnina, I.Z., Trut, L.N., 1984. Domestication in the silver fox (*Vulpes fulvus* desm.) changes in physiological boundaries of the sensitive period of primary socialization. Appl. Anim. Behav. Sci. 13, 359–370.

Belyaev, D.K., Ruvinsky, A.O., Trut, L.N., 1981. Inherited activation–inactivation of the star gene in foxes. Its bearing on the problem of domestication. J. Hered. 72, 267–274.

Bennett, A., Hayssen, V., 2012. Measuring cortisol in hair and saliva from dogs: coat color and pigment difference. Domest. Anim. Endocrinol. 39, 171–180.

Bijma, P., Woolliams, J.A., 2000. Prediction of rates of inbreeding in populations selected on best linear unbiased prediction of breeding value. Genetics 156, 361–373.

Boake, C.R.B., Arnold, S.J., Breden, F., Meffert, L.M., Ritchie, M.G., Taylor, B.J., et al., 2002. Genetic tools for studying adaptation and the evolution of behavior. Am. Nat. 160, 143–159.

Boitani, L., Ciucci, P., 1995. Comparative social ecology of feral dogs and wolves. Ethol. Ecol. Evol. 7, 49–72.

Bonhomme, F., Guenet, J.-L., 1984. The laboratory mouse and its wild relatives. In: Lyon, M.F., Rastan, S., Brown, S.D.M. (Eds.), Genetic Variants and Strains of the Laboratory Mouse. Oxford University Press, Oxford, pp. 1577–1596.

Breitling, R., Li, Y., Tesson, B.M., Fu, J., Wu, C., Wiltshire, T., et al., 2008. Genetical genomics: spotlight on QTL hotspots. PLoS Genet 4, e1000232.

Bruce, A.B., 1910. University and educational news. Science 32, 627.

Brunberg, E., Andersson, L., Cothran, G., Sandberg, K., Mikko, S., Lindgren, G., 2006. A missense mutation in PMEL17 is associated with the Silver coat color in the horse. BMC. Genet. 7, 46.

Brunetti-Pierri, N., Berg, J.S., Scaglia, F., Belmont, J., Bacino, C.A., Sahoo, T., et al., 2008. Recurrent reciprocal 1q21.1 deletions and duplications associated with microcephaly or macrocephaly and developmental and behavioral abnormalities. Nat. Genet. 40, 1466–1471.

Burke, J.M., Tang, S., Knapp, S.J., Rieseberg, L.H., 2002. Genetic analysis of sunflower domestication. Genetics 161, 1257–1267.

Cai, H.W., Morishima, H., 2002. QTL clusters reflect character associations in wild and cultivated rice. Theor. Appl. Genet. 104, 1217–1228.

Calboli, F.C.F., Sampson, J., Fretwell, N., Balding, D.J., 2008. Population structure and inbreeding from pedigree analysis of purebred dogs. Genetics 179, 593–601.

Campler, M., Jöngren, M., Jensen, P., 2009. Fearfulness in red junglefowl and domesticated White Leghorn chickens. Behav. Processes. 81, 39–43.

Carlborg, O., Jacobsson, L., Ahgren, P., Siegel, P., Andersson, L., 2006. Epistasis and the release of genetic variation in response to selection. Nat. Genet. 38, 418–420.

Carlson, C.S., Eberle, M.A., Kruglyak, L., Nickerson, D.A., 2004. Mapping complex disease loci in whole-genome association studies. Nature 429, 446–452.

Carroll, S.B., Grenier, J.K., Weatherbee, S.D., 2005. From DNA to Diversity. Blackwell, Oxford.

Carson, H.L., 1990. Increased genetic variance after a population bottleneck. Trends Ecol. Evol. 5, 228–230.

Carter, A.J., Hermisson, J., Hansen, T.F., 2005. The role of epistatic gene interactions in the response to selection and the evolution of evolvability. Theor Popul Biol 68, 179–196.

Champagne, F.A., Curley, J.P., 2008. Maternal regulation of estrogen receptor alpha methylation. Curr. Opin. Pharmacol. 8, 735–739.

Charlesworth, D., Charlesworth, B., 1979. The evolutionary genetics of sexual systems in flowering plants. Proc. R. Soc. Lond. B. Biol. Sci. 205, 513–530.

Chen, W.-K., Swartz, J.D., Rush, L.J., Alvarez, C.E., 2009. Mapping DNA structural variation in dogs. Genome. Res. 19, 500–509.

Chesler, E.J., Lu, L., Shou, S., Qu, Y., Gu, J., Wang, J., et al., 2005. Complex trait analysis of gene expression uncovers polygenic and pleiotropic networks that modulate nervous system function. Nat. Genet. 37, 233–242.

Clark, R.M., Linton, E., Messing, J., Doebley, J.F., 2004. Pattern of diversity in the genomic region near the maize domestication gene tb1. Proc. Natl. Acad. Sci. USA 101, 700–707.

Clop, A., Marcq, F., Takeda, H., Pirottin, D., Tordoir, X., Bibe, B., et al., 2006. A mutation creating a potential illegitimate microRNA target site in the myostatin gene affects muscularity in sheep. Nat. Genet. 38, 813–818.

Clop, A., Vidal, O., Amills, M., 2012. Copy number variation in the genomes of domestic animals. Anim. Genet. 43, 503–517.

Clutton-Brock, J., 1999. A Natural History of Domesticated Mammals. Cambridge University Press, Cambridge.

Coates, J.C., de Bono, M., 2002. Antagonistic pathways in neurons exposed to body fluid regulate social feeding in *C.elegans*. Nature 419, 925–929.

Cottle, C.A., Price, E.O., 1987. Effects of the nonagouti pelage-color allele on the behavior of captive wild Norway rats (Rattus norwegicus). J. Comp. Psychol. 101, 390–394.

Darvasi, A., 1998. Experimental strategies for the genetic dissection of complex traits in animal models. Nat. Genet. 18, 19–24.

Davenport, C.B., 1908. Recessive characters. Science 28, 454.

de Belle, J.S., Sokolowksi, M.B., 1989. Genetic localization of *foraging* (*for*): a major gene for larval behavior in *Drosophila melanogaster*. Genetics 123, 157–163.

de Bono, M., Bargmann, C.I., 1998. Natural variation in a neuropeptide y receptor homolog modifies social behavior and food response in *C.elegans*. Cell 94, 679–689.

de Bono, M., Tobin, D.M., Davis, M.W., Avery, L., Bargmann, C.I., 2002. Social feeding in *C.elegans* is induced by neurons that detect aversive stimuli. Nature 419, 899–903.

de Smith, A.J., Walters, R.G., Froguel, P., Blakemore, A.I., 2008. Human genes involved in copy number variation: mechanisms of origin, functional effects and implications for disease. Cyogenet. Gen. Res. 123, 17–26.

Doebley, J., Stec, A., 1991. Genetic analysis of the morphological differences between maize and teosinte. Genetics 129, 285–295.

Doebley, J., Stec, A., 1993. Inheritance of the morphological differences between maize and teosinte: comparison of results for two F(2) populations. Genetics 134, 559–570.

Donaldson, Z.R., Young, L.J., 2008. Oxytocin, vasopressin, and the neurogenetics of sociality. Science 322, 900–904.

Ducrest, A.-L., Keller, L., Roulin, A., 2008. Pleiotropy in the melanocortin system, coloration and behavioural syndromes. Trends. Ecol. Evol. 23, 502–510.

Elferink, M., Vallee, A., Jungerius, A., Crooijmans, R., Groenen, M., 2008. Partial duplication of the PRLR and SPEF2 genes at the late feathering locus in chicken. BMC Genomics 9, 391.

Eriksson, J., Larson, G., Gunnarsson, U., Bed'hom, B., Tixier-Boichard, M., Strömstedt, L., et al., 2008. Identification of the yellow skin gene reveals a hybrid origin of the domestic chicken. PLoS Genetics 4, e1000010.

Fadista, J., Nygaard, M., Holm, L.-E., Thomsen, B., Bendixen, C., 2008. A snapshot of CNVs in the pig genome. PLoS ONE 3, e3916.

Falconer, D.S., Mackay, T.F.C., 1996. Introduction to Quantitative Genetics. Prentice Hall.

Fang, M., Larson, G., Soares Ribeiro, H., Li, N., Andersson, L., 2009. Contrasting mode of evolution at a coat color locus in wild and domestic pigs. PLoS Genetics 5, e1000341.

Faris, J.D., Fellers, J.P., Brooks, S.A., Gill, B.S., 2003. A bacterial artificial chromosome contig spanning the major domestication locus Q in wheat and identification of a candidate gene. Genetics 164, 311–321.

Ferguson, M.M., Ihssen, P.E., Hynes, J.D., 1991. Are Cultured Stocks of Brown Trout (Salmo Trutta) and Rainbow Trout (Oncorhynchus Mykiss) Genetically Similar to their Source Populations ? National Research Council of Canada, Ottawa, ON, Canada.

Feuk, L., Carson, A.R., Scherer, S.W., 2006. Structural variation in the human genome. Nat. Rev. Genet. 7, 85–97.

Fevolden, S.E., Refstie, T., Roed, K.H., 1991. Selection for high and low cortisol stress response in atlantic salmon (*Salmo salar*) and rainbow-trout (*Oncorhynchus mykiss*). Aquaculture 95, 53–65.

Flint, J., Mott, R., 2001. Finding the molecular basis of quantitative traits: successes and pitfalls. Nat. Rev. Genet.

Fontanesi, L., Beretti, F., Riggio, V., Gómez González, E., Dall'Olio, S., Davoli, R., et al., 2009. Copy number variation and missense mutations of the agouti signaling protein (ASIP) gene in goat breeds with different coat colors. Cyogenet. Gen. Res. 126, 333–347.

Frankham, R., Lees, K., Montgomery, M.E., England, P.R., Lowe, E.H., Briscoe, D.A., 1999. Do population size bottlenecks reduce evolutionary potential? Anim. Conserv. 2, 255–260.

Gibson, G., Weir, B., 2005. The quantitative genetics of transcription. Trends. Genet. 21, 616–623.

Goddard, M.E., Hayes, B.J., 2009. Mapping genes for complex traits in domestic animals and their use in breeding programmes. Nat. Rev. Genet. 10, 381–391.

Goldstein, O., Zangerl, B., Pearce-Kelling, S., Sidjanin, D.J., Kijas, J.W., Felix, J., et al., 2006. Linkage disequilibrium mapping in domestic dog breeds narrows the progressive rod-cone degeneration interval and identifies ancestral disease-transmitting chromosome. Genomics 88, 541–550.

Golzio, C., Willer, J., Talkowski, M.E., Oh, E.C., Taniguchi, Y., Jacquemont, S., et al., 2012. KCTD13 is a major driver of mirrored neuroanatomical phenotypes of the 16p11.2 copy number variant. Nature 485, 363–367.

Gonyou, H.W., 2001. The social behaviour of pigs. In: Keeling, L.J., Gonyou, H.W. (Eds.), Social Behaviour in Farm Animals. CABI Publishers, Wallingford.

Gräff, J., Mansuy, I.M., 2008. Epigenetic codes in cognition and behaviour. Behav. Brain. Res. 192, 70–87.

Grant, G., 1981. Plant Speciation. Columbia University Press, New York.

Grobet, L., Poncelet, D., Royo, L., Brouwers, B., Pirottin, D., Michaux, C., et al., 1998. Molecular definition of an allelic series of mutations disrupting the myostatin function and causing double-muscling in cattle. Mamm. Genome 9, 210–213.

Hale, E.B., 1969. Domestication and the evolution of behaviour. In: Hafez, E.S.E. (Ed.), The Behaviour of Domestic Animals. Baillieire, Tindall &Cassell, London, pp. 22–42.

Hare, B., Plyusnina, I., Ignacio, N., Schepina, O., Stepika, A., Wrangham, R., et al., 2005. Social cognitive evolution in captive foxes is a correlated by-product of experimental domestication. Curr. Biol. 15, 226–230.

Hayssen, V., 1997. Effects of the nonagouti coat-color allele on behavior of deer mice (Peromysceus maniculatus). J. Comp. Psychol. 111, 419–423.

Hedhammar, A.A., Olsson, S.-E., Andersson, S.A., Persson, L., Pettersson, L., Olausson, A., et al., 1979. Canine hip dysplasia: study of heritability in 401 litters of German Shepherd dogs. J. Am. Vet. Med. Assoc. 174, 1012–1016.

Hedrick, P.W., Gutierrez-Espeleta, G.A., Lee, R.N., 2001. Founder effect in an island population of bighorn sheep. Mol. Ecol. 10, 851–857.

Henrichsen, C.N., Chaignat, E., Reymond, A., 2009. Copy number variants, diseases and gene expression. Hum. Mol. Genet. 18, R1–R8.

Hodgkin, J., Doniach, T., 1997. Natural variation and copulatory plug formation in *C.elegans*. Genetics 146, 149–164.

Jablonka, E., 2009. Transgenerational epigenetic inheritance: prevalence, mechanisms, and implications for the study of heredity and evolution. Q. Rev. Biol. 84, 134–176.

Jackson, S., Diamond, J., 1995. Ontogenetic development of gut function, growth, and metabolism in a wild bird, the Red Jungle Fowl. Am. J. Physiol.—Reg. Int. Comp. Physiol 269, R1163–R1173.

Jackson, S., Diamond, J., 1996. Metabolic and digestive responses to artificial selection in chickens. Evolution 50, 1638–1650.

Ji, H.-S., Chu, S.-H., Jiang, W., Cho, Y.-I., Hahn, J.-H., Eun, M.-Y., et al., 2006. Characterization and mapping of a shattering mutant in rice that corresponds to a block of domestication genes. Genetics 173, 995–1005.

Johnson, M.I., Gustafsson, C.-J., Rubin, A.-S., Sahlqvist, K.B., Jonsson, S., Kerje, O., Ekwall, O., Kämpe, L., Andersson, Jensen P., Wright D., 2012. A sexual ornament in chickens is affected by pleiotropic alleles at HAO1 and BMP2, selected during domestication. PLoS Gen. 8, e1002914.

Johnsson, J.I., Petersson, E., Jonsson, E., Bjornsson, B.T., Jarvi, T., 1996. Domestication and growth hormone alter antipredator behavior and growth patterns in juvenile brown trout, Salmo trutta. Can. J. Fish. Aquat. Sci. 53, 1546–1554.

Jones, P., Chase, K., Martin, A., Davern, P., Ostrander, E.A., Lark, K.G., 2008. Single-nucleotide-polymorphism-based association mapping of dog stereotypes. Genetics 179, 1033–1044.

Karlsson, A.-C., Mormede, P., Kerje, S., Jensen, P., 2011a. Genotype on the pigmentation regulating PMEL17 gene affects behavior in chickens raised without physical contact with conspecifics. Behav. Genet. 41, 312–322.

Karlsson, A.C., Kerje, S., Mormède, P., Jensen, P., 2011b. Genotype on the pigmentation regulating PMEL17 gene affects behavior in chickens raised without physical contact with conspecifics. Behav. Genet. 41, 312–322.

Karlsson, E.K., Baranowska, I., Wade, C.M., Salmon Hillbertz, N.H.C., Zody, M.C., Anderson, N., et al., 2007. Efficient mapping of mendelian traits in dogs through genome-wide association. Nat. Genet. 39, 1321–1328.

Keeling, L., Andersson, L., Schutz, K.E., Kerje, S., Fredriksson, R., Carlborg, O., et al., 2004. Chicken genomics: feather-pecking and victim pigmentation. Nature 431, 645–646.

Kerje, S., Carlborg, O., Jacobsson, L., Schutz, K., Hartmann, C., Jensen, P., et al., 2003. The twofold difference in adult size between the red junglefowl and White Leghorn chickens is largely explained by a limited number of QTLs. Anim. Genet. 34, 264–274.

Kerje, S., Sharma, P., Gunnarsson, U., Kim, H., Bagchi, S., Fredriksson, R., et al., 2004. The Dominant white, dun and smoky color variants in chicken are associated with insertion/deletion polymorphisms in the PMEL17 gene. Genetics 168, 1507–1518.

Kittilsen, S., Schjolden, J., Beitnes-Johansen, I., Shaw, J.C., Pottinger, T.G., Sörensen, C., et al., 2009. Melanin-based skin spots reflect stress responsiveness in salmonid fish. Horm. Behav. 56, 292–298.

Koide, T., Moriwaki, K., Ikeda, K., Niki, H., Shiroishi, T., 2000. Multi-phenotype behavioral characterization of inbred strains derived from wild stocks of *Mus musculus*. Mamm. Genome. 11, 664–670.

Koinange, E.M.K., Singh, S.P., Gepts, P., 1996. Genetic control of the domestication syndrome in common bean. Crop. Sci. 36, 1282–1291.

Krause, J., Ruxton, G., 2002. Living in Groups. Oxford University Press, Oxford.

Kristensen, T.N., Sørensen, A.C., 2005. Inbreeding—lessons from animal breeding, evolutionary biology and conservation genetics. Anim. Sci. 80, 121–133.

Kronenberger, J.-P., Medioni, J., 1985. Food neophobia in wild and laboratory mice (*Mus musculus domesticus*). Behav. Processes 11, 53–59.

Kroner, C., Boekoff, I., Lohmann, S.M., Genieser, H.-G., Breer, H., 1996. Regulation of olfactory signalling via cGMP-dependent protein kinase. Eur. J. Biochem. 236, 632–637.

Kukekova, A., Trut, L., Chase, K., Kharlamova, A., Johnson, J., Temnykh, S., et al., 2011. Mapping loci for fox domestication: deconstruction/reconstruction of a behavioral phenotype. Behav. Genet. 41, 593–606.

Lai, Z., Kane, N.C., Zou, Y., Rieseberg, L.H., 2008. Natural variation in gene expression between wild and weedy populations of helianthus annuus. Genetics 179, 1881–1890.

Lander, E.S., Botstein, D., 1989. Mapping mendelian factors underlying quantitative traits using RFLP linkage maps. Genetics 121, 185–199.

Lindqvist, C., Jancsak, A.M., Nätt, D., Baranowska, I., Lindqvist, N., Wichman, A., et al., 2007. Transmission of stress-induced learning impairment and associated brain gene expression from parents to offspring in chickens. PLoS ONE. 10.1371/journal.pone.0000364.

Litvin, O., Causton, H.C., Chen, B.-J., Pe'er, D., 2009. Modularity and interactions in the genetics of gene expression. Proc. Natl. Acad. Sci. 106, 6441–6446.

Liu, G.E., Hou, Y., Zhu, B., Cardone, M.F., Jiang, L., Cellamare, A., et al., 2010. Analysis of copy number variations among diverse cattle breeds. Genome. Res. 20, 693–703.

Lynch, M., Walsh, B., 1998. Genetics and Analysis of Quantitative Traits. Sinauer Associates, Sunderland, MA.

Makinen, H.S., Shikano, T., Cano, J.M., Merila, J., 2008. Hitchhiking mapping reveals a candidate genomic region for natural selection in three-spined stickleback chromosome VIII. Genetics 178, 453–465.

Margulis, S.W., 1998. Relationships among parental inbreeding, parental behaviour and offspring viability in oldfield mice. Anim. Behav. 55, 427–438.

Mariette, M., Kelley, J.L., Brooks, R., Evans, J.P., 2006. The effects of inbreeding on male courtship behaviour and coloration in guppies. Ethology 112, 807–814.

Marliave, J.B., Gergits, W.F., Aota, S., 1993. F10 pandalid shrimp: sex determination; DNA and dopamine as indicators of domestication and outcrossing for wild pigmentation pattern. Zoo. Biol., 12.

Marroquin, L.D., Elyassnia, D., Griffiths, J.S., Feitelson, J.S., Aroian, R.V., 2000. Bt Toxin susceptibility and isolation of resistance mutants in the nematode *C.elegans*. Genetics 155, 1693–1699.

Mehrabian, M., Allayee, H., Stockton, J., Lum, P.Y., Drake, T.A., Castellani, L.W., et al., 2005. Integrating genotypic and expression data in a segregating mouse population to identify 5-lipoxygenase as a susceptibility gene for obesity and bone traits. Nat. Genet. 37, 1224–1233.

Mench, J., Keeling, L., 2001. The social behaviour of domestic birds. In: Keeling, L.J., Gonyou, H.W. (Eds.), Social Behaviour in Farm Animals. CABI Publishers, Wallingford.

Meuwissen, T.H., Karlsen, A., Lien, S., Olsaker, I., Goddard, M.E., 2002. Fine mapping of a quantitative trait locus for twinning rate using combined linkage and linkage disequilibrium mapping. Genetics 161, 373–379.

Mickett, K., Morton, C., Feng, J., Li, P., Simmons, M., Cao, D., et al., 2003. Assessing genetic diversity of domestic populations of channel catfish (Ictalurus punctatus) in Alabama using AFLP markers. Aquaculture 228, 91–105.

Miglior, F., Burnside, E.B., Dekkers, J.C.M., 1995. Nonadditive genetic effects and inbreeding depression for somatic cell counts of holstein cattle. J. Dairy Sci. 78, 1168–1173.

Miklosi, A., 2008. Dog Behaviour, Evolution and Cognition. Oxford University Press, Oxford.

Mosher, D.S., Quignon, P., Bustamante, C.D., Sutter, N.B., Mellersh, C.S., Parker, H.G., et al., 2007. A mutation in the myostatin gene increases muscle mass and enhances racing performance in heterozygote dogs. PLoS Genet 3, e79.

Muntzing, A., 1959. Darwin's view on variation under domestication in the light of present-day knowledge. Proc. Am. Philos. Soc. 103, 190–220.

Murgatroyd, C., Patchev, A., Wu, Y., Micale, V., 2009. Dynamic DNA methylation programs persistent adverse effects of early-life stress. Nat. Neurosci. 12, 1559–1566.

Nadeau, J.H., Frankel, W.N., 2000. The roads from phenotypic variation to gene discovery: mutagenesis versus QTLs. Nat. Genet. 25, 381–384.

Nätt, D., Andersson, L., Kerje, S., Jensen, P., 2007. Plumage color and feather pecking—behavioral differences associated with PMEL17 genotypes in chicken (Gallus gallus). Behav. Genet. 37, 399–407.

Natt, D., Rubin, C.-J., Wright, D., Johnsson, M., Belteky, J., Andersson, L., et al., 2012. Heritable genome-wide variation of gene expression and promoter methylation between wild and domesticated chickens. BMC Genomics 13, 59.

Naveilham, P., 2001. Reduced antinociception and plasma extravasation in mice lacking a neuropeptide Y receptor. Nature 409, 513–517.

Norris, B.J., Whan, V.A., 2008. A gene duplication affecting expression of the ovine ASIP gene is responsible for white and black sheep. Genome. Res. 18, 1282–1293.

Osborne, K.A., Robichon, A., Burgess, E., Butland, S., Shaw, R.A., Coulthard, A., et al., 1997. Natural behavior polymorphism due to a cGMP-dependent protein kinase of *Drosophila*. Science 277, 834–836.

Parker, H.G., VonHoldt, B.M., Quignon, P., Margulies, E.H., Shao, S., Mosher, D.S., et al., 2009. An expressed Fgf4 retrogene is associated with breed-defining chondrodysplasia in domestic dogs. Science 325, 995–998.

Pereira, H.S., Sokolowksi, M.B., 1993. Mutations in the larval foraging gene affect adult locomotory behavior after feeding in *Drosophila melanogaster*. Proc. Natl. Acad. Sci. USA 90, 5044.

Pérez-Vega, E., Pañeda, A., Rodríguez-Suárez, C., Campa, A., Giraldez, R., Ferreira, J., 2010. Mapping of QTLs for morpho-agronomic and seed quality traits in a RIL population of common bean (Phaseolus vulgaris L.). Theor. Appl. Genet.2010, online first.

Pinkel, D., Segraves, R., Sudar, D., Clark, S., Poole, I., Kowbel, D., et al., 1998. High resolution analysis of DNA copy number variation using comparative genomic hybridization to microarrays. Nat. Genet. 20, 207–211.

Price, E.O., 1999. Behavioral development in animals undergoing domestication. Appl. Anim. Behav. Sci. 65, 245–271.

Price, E.O., 2002. Animal Domestication and Behaviour. CABI Publishing, Wallingford, UK.

Price, E.O., King, J.A., 1968. Domestication and adaptation. In: Hafez, E.S.E. (Ed.), Adaptations of Domestic Animals. Lea & Febiger, Philadelphia.

Pritchard, J.K., Di Rienzo, A., 2010. Adaptation—not by sweeps alone. Nat. Rev. Genet. 11, 665–667.

Pusey, A., Wolf, M., 1996. Inbreeding avoidance in animals. Trends. Ecol. Evol. 11, 201–206.

Richards, E., 2006. Inherited epigenetic variation—revisiting soft inheritance. Nat. Rev. Genet..

Riquet, J., Coppieters, W., Cambisano, N., Arranz, J.-J., Berzi, P., Davis, S.K., et al., 1999. Fine-mapping of quantitative trait loci by identity by descent in outbred populations: application to milk production in dairy cattle. Proc. Natl. Acad. Sci. USA 96, 9252–9257.

Rosengren Pielberg, G., Golovko, A., Sundstrom, E., Curik, I., Lennartsson, J., Seltenhammer, M.H., et al., 2008. A cis-acting regulatory mutation causes premature hair graying and susceptibility to melanoma in the horse. Nat. Genet. 40, 1004–1009.

Rubin, C.-J., Lindberg, J., Fitzsimmons, C., Savolainen, P., Jensen, P., Lundeberg, J., et al., 2007. Differential gene expression in femoral bone from red junglefowl and domestic chicken, differing for bone phenotypic traits. BMC Genomics 8, 208.

Rubin, C.-J., Zody, M.C., Eriksson, J., Meadows, J.R.S., Sherwood, E., Webster, M.T., et al., 2010. Whole-genome resequencing reveals loci under selection during chicken domestication. Nature 464, 587–591.

Salmon Hillbertz, N.H.C., Isaksson, M., Karlsson, E.K., Hellmen, E., Pielberg, G.R., Savolainen, P., et al., 2007. Duplication of FGF3, FGF4, FGF19 and ORAOV1 causes hair ridge and predisposition to dermoid sinus in Ridgeback dogs. Nat. Genet. 39, 1318–1320.

Sanders, S.J., Ercan-Sencicek, A.G., Hus, V., Luo, R., Murtha, M.T., Moreno-De-Luca, D., Chu, S. H., Moreau, M.P., 2011. Multiple recurrent de novo CNVs, including duplications of the 7q11.23 Williams syndrome region, are strongly associated with autism. Neuron 70, 863–885.

Sawyer, L.A., Hennessy, J.M., Pexioto, A.A., Kyriacou, P., 1997. Natural variation in a drosophila clock gene and temperature compensation. Science 278, 2117–2120.

Schütz, K., Jensen, P., 2001. Effects of resource allocation on behavioural strategies: a comparison of red junglefowl (*Gallus gallus*) and two domesticated breeds of poultry. Ethology 107, 753–765.

Schütz, K., Kerje, S., Carlborg, Ö., Jacobsson, L., Andersson, L., Jensen, P., 2002. QTL analysis of a red junglefowl x White Leghorn intercross reveals trade-off in resource allocation between behaviour and production traits. Behav. Genet. 32, 423–433.

Schütz, K.E., Forkman, B., Jensen, P., 2001. Domestication effects on foraging strategy, social behaviour and different fear responses: a comparison between the red junglefowl (*Gallus gallus*) and a modern layer strain. Appl. Anim. Behav. Sci. 74, 1–14.

Sebat, J., Lakshmi, B., Malhotra, D., Troge, J., Lese-Martin, C., Walsh, T., et al., 2007. Strong association of de novo copy number mutations with autism. Science 316, 445–449.

Seroussi, E., Glick, G., Shirak, A., Yakobson, E., Weller, J., Ezra, E., et al., 2010. Analysis of copy loss and gain variations in Holstein cattle autosomes using BeadChip SNPs. BMC Genomics 11, 673.

Sharp, P.M., 1984. The effect of inbreeding on competitive male-mating ability in *Drosophila melanogaster*. Genetics 106, 601–612.

Siegel, P.B., Dunnington, E.A., 1985. Reproductive complications associated with selection for broiler growth. In: Hill, W.G., Manson, J.M., Hewitt, D. (Eds.), Poultry Genetics and Breeding. British Poultry Science Ltd, Harlow, pp. 59–72.

Simons, K.J., Fellers, J.P., Trick, H.N., Zhang, Z., Tai, Y.-S., Gill, B.S., et al., 2006. Molecular characterization of the major wheat domestication gene Q. Genetics 172, 547–555.

Skinner, M., Anway, M., Savenkova, M., Gore, A., 2008. Transgenerational epigenetic programming of the brain transcriptome and anxiety behavior. PLoS ONE 3, e3745.

Smith, J.M., Haigh, J., 1974. The hitch-hiking effect of a favourable gene. Genet. Res. 23, 23–35.

Smith, L.A., Cassell, B.G., Pearson, R.E., 1998. The effects of inbreeding on the lifetime performance of dairy cattle. J. Dairy Sci. 81, 2729–2737.

Sokolowksi, M.B., 1980. Foraging strategies of *Drosophila melanogaster*: a chromosomal analysis. Behav. Genet. 10, 291–302.

Sokolowksi, M.B., 1982. Rover and sitter larval foraging patterns in a natural population of *D.melanogaster*. Drosophila Inform. Serv 58, 138–139.

Sokolowski, M.B., Pereira, H.S., Hughes, K., 1997. Evolution of foraging behaviour in *Drosophila* by density-dependent selection. Proc. Nat. Acad. Sci. USA 94, 7373–7377.

Sokolowski, M.B., 1998. Genes for normal behavioural variation: recent clues from flies and worms. Neuron 21, 463–466.

Spurway, H., 1955. The causes of domestication: an attempt to integrate some ideas of Konrad Lorenz with evolution theory. J. Genet. 53, 325–362.

Stankiewicz, P., Lupski, J.R., 2010. Structural variation in the human genome and its role in disease. Annu. Rev. Med. 61, 437–455.

Stricklin, W.R., 2001. The evolution and domestication of social behaviour. In: Keeling, L.J., Gonyou, H.W. (Eds.), Social Behaviour in Farm Animals. CABI, Wallingford, pp. 83–110.

Sunden, S.L.F., Davis, S.K., 1991. Evaluation of genetic variation in a domestic population of Penaeus vannamei (Boone): a comparison with three natural populations. Aquaculture 97, 131–142.

Templeton, A.R., Read, B., 1994. Inbreeding: one word, several meanings, much confusion. In: Loeschcke, V., Tomiuk, J., Jain, S.K. (Eds.), Conservation Genetics. Birkhauser. Basel.

Trapezov, O., Trapezova, L., Sergeev, E., 2008. Effect of coat color mutations on behavioral polymorphisms in farm populations of American minks (Mustela vison Schreber, 1744) and sables (Martes zibellina Linnaeus, 1758). Russ. J. Genet. 44, 444–450.

Trut, L.N., 1999. Early canid domestication: the farm fox experiment. Am. Sci. 87, 160–169.

Tully, T., 1996. Discovery of genes involved with learning and memory: an experimental synthesis of hirschian and benzerian perspectives. Proc. Natl. Acad. Sci. USA 93, 13460–13467.

Tymchuk, W., Sakhrani, D., Devlin, R., 2009. Domestication causes large-scale effects on gene expression in rainbow trout: analysis of muscle, liver and brain transcriptomes. Gen. Comp. Endocrinol. 164, 175–183.

Van Laere, A.S., Nguyen, M., Braunschweig, M., Nezer, C., Collette, C., Moreau, L., et al., 2003. A regulatory mutation in IGF2 causes a major QTL effect on muscle growth in pigs. Nature 425, 832–836.

Verdugo, R., Farber, C., Warden, C., Medrano, J., 2010. Serious limitations of the QTL/Microarray approach for QTL gene discovery. BMC Biol. 8, 96.

Vilà, C., Seddon, J., Ellegren, H., 2005. Genes of domestic mammals augmented by backcrossing with wild ancestors. Trends. Genet. 21, 214–218.

Wang, X., Nahashon, S., Feaster, T., Bohannon-Stewart, A., Adefope, N., 2010. An initial map of chromosomal segmental copy number variations in the chicken. BMC Genomics 11, 351.

Wang, Z., Gerstein, M., Snyder, M., 2009. RNA-Seq: a revolutionary tool for transcriptomics. Nat. Rev. Genet. 10, 57–63.

Weaver, I.C.G., Cervoni, N., Champagne, F.A., D'Alessio, A.C., Sharma, S., Seckl, J.R., et al., 2004. Epigenetic programming by maternal behavior. Nat. Neurosci. 7, 847–854.

Wiener, G., Lee, G.J., Wooliams, J.A., 1992. Effects of rapid inbreeding and crossing of inbred lines on the body weight growth of sheep. Anim. Prod. 55, 89–99.

Wiener, G., Lee, G.J., Wooliams, J.A., 1994. Consequences of inbreeding for financial returns from sheep. Anim. Prod. 59, 245–249.

Wirén, A., 2011. Correlated selection responses in animal domestication: the behavioural effects of a growth QTL in chickens. Linköping studies in Science and Technology Dissertation No 1413.

Wirén, A., Jensen, P., 2011. A growth QTL on chicken chromosome 1 affects emotionality and sociality. Behav. Genet. 41, 303–311.

Wray, G., 2007. The evolutionary significance of cis-regulatory mutations. Nat. Rev. Genet. 8, 206–216.

Wray, N.R., Hill, W.G., 1989. Asymptotic rates of response from index selection. Anim. Prod. 49, 217–227.

Wright, D., Butlin, R.K., Carlborg, Ö., 2006a. Epistatic regulation of behavioural and morphological traits in the zebrafish (*Danio rerio*). Behav. Genet. 36, 914–922.

Wright, D., Kerje, S., Lundström, K., Babol, J., Schutz, K., Jensen, P., et al., 2006b. Quantitative trait loci analysis of egg and meat production traits in a red junglefowl x White Leghorn cross. Anim. Genet. 37, 529–534.

Wright, D., Nakamichi, R., Krause, J., Butlin, R.K., 2006c. QTL analysis of behavioural and morphological differentiation between wild and laboratory zebrafish (*Danio rerio*). Behav. Genet. 36, 271–284.

Wright, D., Kerje, S., Brändström, H., Schütz, K., Kindmark, A., Andersson, L., et al., 2008. The genetic architecture of a female sexual ornament. Evolution 62, 86–98.

Wright, D., Boije, H., Meadows, J.R.S., Bed'hom, B., Gourichon, D., Vieaud, A., et al., 2009. Transient ectopic expression of SOX5 during embryonic development causes the Pea-comb phenotype in chickens. PLoS Genetics 5 (6), e1000512.

Wright, D., Rubin, C.J., Martinez Barrio, A., Schütz, K., Kerje, S., Brändström, H., et al., 2010. The genetic architecture of domestication in the chicken: effects of pleiotropy and linkage. Mol. Ecol. 19, 5140–5156.

Wright, D., Rubin, C., Schutz, K., Kerje, S., Kindmark, A., Brandström, H., et al., 2012. Onset of sexual maturity in female chickens is genetically linked to loci associated with fecundity and a sexual ornament. Reprod. Domest. Anim. 47, 31–36.

Yasumoto, K., Watabe, H., Valencia, J.C., Kushimoto, T., Kobayashi, T., Appella, E., et al., 2004. Epitope mapping of the melanosomal matrix protein gp100 (PMEL17): rapid processing in the endoplasmic reticulum and glycosylation in the early Golgi compartment. J. Biol. Chem. 279, 28330–28338.

Zhang, F., Gu, W., Hurles, M.E., Lupski, J.R., 2009. Copy number variation in human health, disease, and evolution. Annu. Rev. Genomics. Hum. Genet. 10, 451–481.

Chapter 3

How Studying Interactions Between Animal Emotions, Cognition, and Personality Can Contribute to Improve Farm Animal Welfare

Alain Boissy* and Hans W. Erhard†

**UMR INRA 1213 Herbivores, Centre de Theix, St-Genès Champanelle, France; †UMR INRA 791 Mosar, AgroParisTech, 16 rue Claude Bernard, Paris Cedex 05, France*

INTRODUCTION

The welfare of animals used for food production is a major concern for society which stems from the recognition that animals are not only reactive to their environments but also sentient. Since the Brambell Committee (1965), it is clearly stated that animal welfare embraces both physical and mental well-being. Consideration of the mental well-being of animals implies that animals have emotional capacities, such that they attempt to minimize negative emotions (e.g., fear and frustration) and to seek positive emotions (e.g., pleasure and joy) (Boissy *et al.*, 2007b; Dawkins, 1990; Duncan, 1996). Nevertheless, while it is now well accepted that animals express emotional reactions, there is still a persistent belief that we can do no more than observe how the animals behave and that we could never know how they feel. In addition, while the concept of welfare is based on a balance between negative and positive experiences (Dawkins, 1990; Duncan, 1996; Mellor, 2012; Spruijt *et al.*, 2001), current research in animal welfare is mainly focused on identifying and preventing negative animal emotions and bad welfare. However, preventing negative welfare in animals is not the same as providing them with opportunities to experience positive emotions and positive welfare (Boissy *et al.*, 2007b).

In addition to emotions, which are by definition fleeting, welfare implies more persistent affective states that influence the way in which the individual reacts to his environment (Lazarus, 1991). Repeated exposure to negative

Genetics and the Behavior of Domestic Animals. DOI: http://dx.doi.org/10.1016/B978-0-12-394586-0.00003-2
© 2014 Elsevier Inc. All rights reserved.

events is known to induce long-term stress-related responses such as depression-like behaviors. For instance, chronic social stress induces behavioral changes, such as decrease in locomotor activity, self- and allo-grooming, feeding behaviors (both appetitive and consummatory components), and physiological changes, such as altered circadian rhythm, body temperature or body weight in various animal species (Boissy *et al.*, 2001; Carere *et al.*, 2001; Coutellier *et al.*, 2007; Veenema *et al.*, 2003). Nevertheless, as noted by Dawkins (2001) and Dantzer (2002), research on stress has generally been limited to indicators of stress and has not linked these indicators to the existence of long-lasting affective states. Therefore, it is essential to move beyond the simple description of the emotions/stress-related reactions in farm animals towards a scientific approach of their own emotional experiences (Désiré *et al.*, 2002; Mendl and Paul, 2004). Recently, attention has been paid to animal cognition and appraisal as a promising approach to scientifically assess sentience in animals (Boissy *et al.*, 2007a; Paul *et al.*, 2005).

The main objective of this chapter is to highlight the emotions and the longer-lasting affective states of an animal, and the individual characteristics by which its emotions are influenced. This chapter follows four complementary threads. Firstly, we review emotions defined as resulting from their perception of their own environment, and we discuss the difficulties involved in the assessment of fear and anxiety, the most commonly studied emotions in farm animals, and the need for better validation of the experimental approaches. Secondly, we review the interactions between emotions and cognition and present the relevance of taking into account the cognitive abilities of animals to better approach their emotions and long-lasting affective states. Thirdly, we discuss the relevance of the personality concept, as resulting from both genetics and developmental experience, for better embracing the animal's individuality in emotional behaviors and stress. Finally, we explore positive emotions in animals and their interests not only to alleviate fear but also to induce long-lasting positive states close to the welfare concept and to attempt to mitigate detrimental stress-induced effects on the welfare and health status.

DO ANIMALS FEEL EMOTIONS?

What is the Nature of Animal Sentience? What is an Emotion? What is Stress?

There is now a consensus in the scientific community that most animals that we use for our own purposes, whether for food, for work or for animal experimentation, are acknowledged as sentient beings (Dawkins, 2001). Sentient animals are

> *those who experience emotions associated with pleasure and suffering and who are motivated to promote their evolutionary fitness not as part of a*

well-planned, long-term strategy to ensure the well-being of future generations but through the simpler, but no less intense, need to feel good about themselves. Webster (2005).

This notion was included in the European Union Treaty of Amsterdam (European Union, 1997) highlighting the fact that because animals are sentient, their welfare matters (Duncan, 1993; Veissier and Boissy, 2007). This requires recognition that animals have emotional capacities, and that they attempt to minimize exposure to situations eliciting negative emotions like fear, frustration, distress and anxiety, and seek situations eliciting positive ones like pleasure and joy (Dawkins, 1990; Duncan, 1996; Spruijt *et al.*, 2001). Improving our understanding of the range and depth of emotions that animals can experience is essential in order to safeguard and to improve their welfare. One of the key factors explaining the evolutionary success of emotions is that they would favor adaptive cognition and action. Emotions refer to processes which are likely to have evolved from basic mechanisms that gave the animals the ability to avoid harm/punishment or to seek valuable resources/rewards (Panksepp, 1994; Paul *et al.*, 2005). Animals should thus be capable of assigning affective values to their environment.

Although there is no single general definition, an emotion can be defined as an intense but short-lived affective response to an event associated with specific body changes (Dantzer, 1988). An emotion is classically described by a subjective component, which is, strictly speaking, the emotional experience, and two expressive components, one motor (e.g., facial expressions, movements) and the other physiological (e.g., cardiac and cortisol reactions) (Dantzer, 1988). Emotions differ from sensations which are simply physical consequences of exposure to particular stimuli (e.g., heat, pressure), and from feelings which only designate internal states with no specific reference to external reactions. The emotional experience of animals is inferred from the behavioral and/or physiological components. There has been a growing interest in the study of emotions in animals over the last few decades, resulting in the emergence of a discipline referred to as Affective Neuroscience (Panksepp, 1998). Scientists have made huge progress in understanding how animals perceive their environment and the feelings prompted by this perception. First, in the 1970s, it became clear that the stress response, which was initially considered as a physiological concept (Selye, 1936), is triggered by psychological factors: it is the animal's representation of an event rather than the event itself that determines its stress reactions. This was inferred from studies on fasting monkeys who were either separated from or in the presence of normally fed counterparts (Mason, 1971). When separated from their normally fed counterparts, fastened animals were not aware of the threat imposed on their body and were not stressed, while they showed signs of stress in the presence of normally fed animals. Mason concluded that the non-specificity of stress responses, reported earlier by Selye, was due to the common emotion—felt by the animal—that triggers stress responses.

It is now well known that stress responses vary in form depending on how the individual perceives the situation and its coping possibilities, such as whether the animal engages in a passive response (e.g., immobility) or an active one (e.g., fight or flight) (Dantzer and Mormède, 1983; Veissier and Boissy, 2007). Likewise, it became clear that the ability to predict the occurrence of a stressful event—painful or not—and to control the termination of that event also affects stress responses (Weiss, 1972). This prompted scientists to suggest that animal welfare was closely linked to cognitive processes such as an awareness of some internal state (being hungry, being diseased...), expectancies about the environment (which help animals to detect whether something is absent or not), and the ability to predict or control the animal's environment (Duncan and Petherick, 1989; Wiepkema, 1987). From an adaptation perspective, the ability to perceive its own emotions enables the animal to detect and assess a discrepancy between its requirements and environmental conditions and the efficiency of its subsequent action to regain homeostasis. In addition, the subjective emotional experiences represent motivational urges or drives that are predominantly negative and include breathlessness, thirst, hunger, and pain (Mellor, 2012). Therefore, in order to illustrate the variety of approaches in emotion studies, we focus the following part on fear and anxiety because they are the most studied and the potentially most damaging of the emotions (Jones and Boissy, 2010).

Fear and Anxiety

Fear is generally defined as a response to the perception of actual danger, whereas anxiety is regarded as the reaction to a potential threat (Boissy, 1998). Fear-related reactions are characterized by physiological and behavioral components that prepare the animal to deal with the danger. Such defensive reactions promote fitness: an animal's life expectancy is obviously increased if it can avoid sources of danger such as predators (Panksepp and Burgdorf, 2003). Although captive animals have few natural predators, the behavioral and physiological emotional responses persist (Dwyer, 2004). Moreover, domestic animals reared on ranges may still experience severe predation by wild animals or dogs (Asheim and Mysterud, 2005). Farm animals also show predator-avoidance reactions to human contact, even though reduced fear of human beings has been a major component of domestication (e.g., in herbivores: Price, 1984). Routine management procedures like shearing, castration, tail docking, beak trimming, dehorning, vaccination, harvesting, herding, and transportation also cause fear and distress in cattle, sheep, and poultry (Gentle *et al.*, 1990; Hargreaves and Hutson, 1990; Manteca *et al.*, 2009; Wohlt *et al.*, 1994). In addition, intense fear can cause chronic stress which may compromise the expression of fundamental behaviors (social, sexual, parental) and reduce productivity and product quality in cattle, sheep, pigs, and poultry (Bouissou *et al.*, 2001; Faure and Jones, 2004; Fisher and Matthews, 2001; Hemsworth and Coleman, 1998).

Diversity of Fear-Eliciting Events

According to Gray (1987), the fear-eliciting properties of an event reflect its general characteristics, for example novelty, movement, intensity, duration, suddenness, or proximity. Fear may also be elicited by specific stimuli, such as height and darkness or bright light, in relation to the evolutionary history of the species (ancestral fears). Additionally, a stimulus can elicit fear through association with previous experience of another fear-eliciting event (conditioned fear).

In modern farming systems, animals may be exposed to a large variety of potentially stressful events from hatching or weaning through to puberty and adulthood. These include social mixing, transport, transfer from a rearing or nursing environment to a growing one, dietary change, new social partners, exposure to different stockpersons, harvesting, etc. Gregariousness is common to most domestic animals (Keeling and Gonyou, 2001) and the variety of social stimuli that accompany social cohesion and structure may elicit or modulate fear reactions. Social signals may represent particular cases of the types of fear-eliciting stimulation mentioned above. Some social signals are characterized by their unfamiliarity, for example the novelty of the neonate that affects maternal behavior in primiparous females (Poindron *et al.*, 1984), alarm calls can spontaneously elicit fear (Boissy *et al.*, 1998) and certain social odors can reduce fear and distress in the recipients (Madec, 2008). Some fear-eliciting social signals, for example threatening behaviors, may also be acquired (Bouissou *et al.*, 2001). Under farming conditions, animals are frequently exposed to changes in their social environment. For instance, sanitary processes often require isolation, and sorting animals into new groups generally entails breaking relationships with familiar conspecifics as well as encountering unfamiliar ones. Encountering unfamiliar animals or simple isolation are causes of marked stress reactions in the majority of domestic species that are known to be gregarious (Boissy *et al.*, 2001). Additionally, social isolation is one of the most stressful components of many fear tests when members of a social species are tested individually. Most domestic species are strongly motivated to rejoin conspecifics when isolated, and may consequently suffer more from separation anxiety than from the presumed fear-eliciting event itself.

Diversity in Fear-Related Responses

Fear-related behaviors vary greatly depending on the nature of the threat. They can sometimes seem contradictory, as both active and passive strategies may be observed in a challenging situation: these strategies include active defense (attack, threat), active avoidance (flight, hiding, escape) or passive avoidance (immobility) (Erhard and Mendl, 1999). Subtle expressions can also be regarded as fear indicators; for example, expressive movements like head postures and facial expressions, alarm calls, and the release of alarm odors or pheromones. These fear-related responses play important social roles by serving as alerting signals for conspecifics. Few studies that have been recently carried out in ruminants provide some insight that ear postures may be useful in assessing emotions in farm animals. In cattle, a high occurrence of pendulous ear postures was used as an indicator of the animals' positive rating of their favorite grooming sites (Schmied *et al.*, 2008). We found that sheep point their ears backward when they face an unfamiliar or an unpleasant event and they react to a sudden event by an asymmetric posture of their ears (Boissy *et al.*, 2011). These first results offer a promising step in the characterization of specific emotion-related facial expressions in farm animals since measuring ear postures constitutes a reliable non-invasive method for monitoring the emotional reactions in animals.

Fear-eliciting events can also affect the activity in which the animal is engaged: low levels of fear can enhance activity (e.g., attention, investigation) whereas intense fear can disturb or terminate an ongoing activity. Indeed, fear inhibits all other behavior systems including feeding, exploration, sexual and social interactions (Jones and Boissy, 2010). Finally, conflict between a negative emotional state and a positive motivation may generate a compulsive behavior; for example, a disturbed or hungry pig bites a chain or bar. Activation of the sympathetic nervous system, including the adrenal medulla, and the hypothalamic–pituitary–adrenal system are major neuroendocrine responses associated with negative emotions (Mormède *et al.*, 2007; von Borrell *et al.*, 2007) but they also accompany positive rewards such as food delivery, sexual interaction. Frightening stimulation also elicits a range of complex changes in central nervous mechanisms, such as neural pathways and neurotransmitters (Gray, 1987; Phillips *et al.*, 2003; Rosen, 2004).

Various Ways of Assessing Fear and Anxiety

Varied experimental situations have been designed to study fear in farm animals and most of them were originally developed for laboratory species (for a review, see Forkman *et al.*, 2007). Since Hall's classic work (1936), the open-field or novel arena test has been extensively used in rodents. Generally, a single animal is placed in a large novel arena and the amount of defecation and activity is interpreted as reflecting the emotional response to novelty. Later work showed that this test also incorporates many other threatening stimuli, such as absence of shelter and landmarks, human contact, social isolation, and bright light. Many other paradigms devised to assess fear in rodents, for example exposure to a predator or a novel object, confinement, handling, inescapable noxious stimulation, and passive or active avoidance conditioning (Ramos *et al.*, 1997) have also been used in farm animals from the 1970s onwards (Forkman *et al.*, 2007) though the designs were generally modified to suit the circumstances. Fear of novelty is evaluated using the open-field test or exposure to a novel object while few specific tests have been developed to assess an animal's fear of humans, such as the forced approach test or the voluntary approach test. The forced approach test is more likely to elicit an active response whereas the likelihood of observing either no response or a passive one is greater in the voluntary approach test (Waiblinger *et al.*, 2006). Restraint tests are also commonly used, for example by restraining the animal in a chute or, more commonly, by inducing a tonic immobility reaction (Forkman *et al.*, 2007). Finally, fear can be induced and evaluated in terms of the animals' reactions when exposed either to a natural predator, or to a sudden event (i.e. startle test), or to a signal that had been previously associated with a nociceptive event such as electric shock (i.e. conditioned fear test).

A Need for a Refined and Structured Methodology in Fear Studies

Many tests of fear were originally designed for laboratory rodents and, unfortunately, several of these have been used for farm animals with insufficient regard for their biological relevance. For example, while laboratory rodents are nocturnal and wall-seeking (wild rats build and live in burrows) most farm animals are diurnal and, apart from poultry, their ancestors generally occupied open areas/ranges. Cattle may actually perceive the open-field test as an enclosed area. Furthermore, most farm animals are highly gregarious and react heavily to being separated from group-mates, many have exclusive mother–young relationships and the young are generally precocious. In view of such species differences in ecological experience and motivation we must avoid testing animals in inappropriate environments that may elicit motivational states unrelated to the one under study, thereby leading to inaccurate estimations of fear and inconsistencies between studies. Reconsideration of the ecological contexts of domestic species may help us develop more reliable and valid tests and measures of fear.

Despite these cautionary notes, studies of laboratory animals may still guide the interpretation of a range of emotional reactions observed when farm animals are exposed to aversive situations. From an ecological perspective, suddenness and novelty, which underpin many fear tests (Boissy, 1998), are key features of predatory attack and we must remember that domestic ungulates and poultry kept free-range may still experience predation by wild animals and dogs (Shelton and Wade, 1979). The ability to cope with an aversive event can also strongly influence the animal's emotional experience. For instance, cows with the strongest tendency to approach a novel object voluntarily are also the most reactive to humans, but the opposite is true if they are forced to move toward that novel object (Murphey *et al.*, 1981).

The expressions of fear, like those of all the emotions, are brief reactions that are often difficult to measure, particularly in farm animals kept in commercial conditions. According to Forkman *et al.* (2007), research in this area would benefit substantially if more effort was devoted to overcoming three particular weaknesses. First, many tests used to measure fear need more rigorous examination of validity and repeatability. Validation through evaluation of inter-test consistency should thus be given more attention in future studies (Forkman *et al.*, 2007). Second, there is a pressing need for standardized protocols to ensure harmonization of methods across research teams and countries. This would reduce unnecessary duplication of effort while greatly facilitating meaningful interpretation of the results and their general acceptance. Third, the protocols have to be robust and practical: it must be possible to carry them out in an acceptable time frame, at the farm level, and without causing too much disturbance to the farmers' work. Therefore, it is not enough simply to transfer a laboratory test to the field without making appropriate modifications to fit the animals' biology and needs, the environmental conditions and the farmers' requirements.

A recent comprehensive review of the methodologies used to assess fear (Forkman *et al.*, 2007) in farm animals was a very useful first step in bringing together some of the above-mentioned issues; it describes the various methods used for cattle, pigs, horses, sheep, goats, chickens, and quail, and gives indications of their repeatability and validity. Similarly, fear tests involving exposure to humans were discussed in a review of human animal interactions (Waiblinger *et al.*, 2006). The Welfare Quality® project (2009), a multi-disciplinary EU-funded study, also made substantial progress in developing, refining, validating, and standardizing numerous animal-based measures of welfare, including fear-related behaviors. However, there is still a need for continued development and refinement of fear-related measures.

CONTRIBUTION OF COGNITIVE PSYCHOLOGY TO ACCESS ANIMAL EMOTIONS

Emotion is any mental experience with high intensity and high hedonic content (i.e. the negative or positive valence). In humans, the emotional experience is generally inferred from verbal self-reports. Since conceptual-psychological scales cannot be used in animals that have no verbal language, their emotional experience can thus only be inferred from their behavioral and physiological reactions. Emotions have both arousal and valence dimensions. However, behavioral and physiological reactions generally provide, as it has been reported in the previous section, a quantitative assessment of the emotional activation (i.e. the arousal of the emotional response), but they do not allow defining the exact nature of the emotion (i.e. the negative or positive valence) felt by the animal. For example, plasmatic concentrations of glucocorticoid can be increased both in response to acute negative emotional stimuli (Mason, 1971) or to the expectation of a positive situation such as the sexual partner (Colborn *et al.*, 1991). Likewise, common behavioral reactions such as startle, offensive or defensive postures, freezing, or approach, only provide information about the intensity of the underlying emotion (for a review, see Jones and Boissy, 2010). Therefore, the study of emotions in animals is mostly addressed by means of defining the response of a subject to an emotionally arousing situation. In this way, the emotional reaction is measured by behavioral and physiological parameters without requiring any preconceived theory about what emotions really are. Conceptual frameworks are thus needed to allow a convenient and reliable assessment of emotion in animals.

Among possible conceptual frameworks, appraisal theories originally developed in cognitive psychology to investigate human emotions can be helpful to better access the emotional experience in animals (Boissy *et al.*, 2007a; Désiré *et al.*, 2002). Furthermore, there is increasing evidence that causal links between emotions and cognition occur in both directions, with not only cognitive processes determining felt emotions but also with emotional manipulations more or less persistently influencing cognitive processes (Mendl *et al.*, 2009: Paul *et al.*, 2005). Below we provide a brief review of the interactions between emotions and cognition in farm animals. Firstly we examine the cognitive processes involved in the development of emotions. Then we explore how emotions may in turn alter these cognitive processes and how short-lived processes such as emotions can lead to prolonged affective states.

Influence of Cognitive Processes on Emotions

Since the pioneering work of Magda Arnold, appraisal theories have been developed in psychology to access emotions in humans: emotions result from how an individual evaluates a triggering situation *per se* followed by his responses to that situation (for a review, see Kappas, 2006; Lazarus *et al.*, 1991; Scherer, 1999). According to the pragmatic framework developed by Scherer (2001), this evaluation is based upon a limited number of elementary characteristics, from the most simple to the more complex, including (i) the characteristics of the event (its suddenness, familiarity, predictability, and pleasantness), (ii) the consistency of the event with the individual's expectations, (iii) the coping potential, mainly the ability of the individual to react and to control the event, and (iv) the normative significance of his response, including his internal standards (i.e. his self-esteem) and his social standards (i.e. the expectation of the social group). The outcomes of these elementary characteristics determine the emotional experience. For instance, fear is elicited by exposure to an unpleasant event that is sudden, unfamiliar, unpredictable and inconsistent with the expectations of the individual, whereas rage is experienced in similar situations, except that the individual's evaluation is that he can control the challenging situation (Sander *et al.*, 2005).

The framework based on appraisal theories was recently transposed to animals (Désiré *et al.*, 2002). We developed various experimental situations in sheep that were designed to activate one by one the evaluative characteristics, in order to ascertain which ones are relevant to animals (Boissy *et al.*, 2007a; Veissier *et al.*, 2009). Cardiac and behavioral reactions were recorded to probe the links between presumed appraisal and measurable emotional outcomes. As already reported in many species, it was found that sheep reacted to sudden and novel events (Désiré *et al.*, 2004, 2006). The evaluation of these elementary characteristics appears rather automatic, and may not require the animal to be aware of its evaluation of the situation. More interestingly, it was found that the emotional responses of sheep (e.g., behavioral and cardiac responses) are also affected by the predictability of a situation, its controllability, or its consistency with the animal's expectations (Greiveldinger *et al.*, 2007, 2009, 2011). More recently, we found that the emotional responses to a threatening event are influenced by the animal's social context (Greiveldinger *et al.*, 2012). Exposed to a sudden event (a panel falling down behind a trough when the animal is eating), sheep are more likely to display overt external responses (stepping back from the trough) when they are accompanied by a subordinate group-mate, but internal responses (tachycardia) when they are accompanied by a dominant group-mate. Therefore, it is clear that sheep are responsive to the main appraisal characteristics defined in human studies. Although this approach has not been completed for many farm animals, it is now widely accepted that not only mammals but also poultry can feel emotions (Mormède *et al.*, 2007; Valance *et al.*, 2008).

Since animals can use appraisal processes similar to humans, we can consider that they not only express emotional responses that in that case could be considered reflexes, but they really do feel emotions (Veissier *et al.*, 2009). This is in line with several studies that highlight the existence of mutual neural circuits underlying experience and expression of emotions in both human and animal (Boissy *et al.*, 2007a; Spruijt, 2001). Referring to the studies conducted in sheep and based on the elementary characteristics sheep use to evaluate their environment, it is suggested that they can experience a wide range of emotions: (i) fear and anger, as they are sensitive to suddenness, unpredictability, controllability, and social norms; (ii) rage, as they respond to suddenness, unfamiliarity, unpredictability, discrepancy from expectations, controllability, and social norms; (iii) despair, as they react to suddenness, unfamiliarity, unpredictability, discrepancy from expectations, and controllability; and (iv) boredom, as they are sensitive to suddenness, unfamiliarity, unpredictability, discrepancy from expectations, and controllability (Veissier *et al.*, 2009). Using a similar pragmatic framework in various farm animal species, comparisons across phyla could provide critical insights into the kind of emotions they can feel and thereby help to refine regulations designed to safeguard animal welfare.

Alteration of the Judgment and Decision Making by Emotions

While cognitive processes are at the origin of emotions, emotions can in turn influence cognitive processes. Here again, a significant body of work in human psychology has shown how emotions can temporarily bias the processing of information coming from the situation that is attention, memory, and judgment. For example, anxiety induces a shift in attention towards potential threats (Bradley *et al.*, 1997), emotionally charged events are more readily remembered than neutral ones (Reisberg and Heurer, 1995), and people exposed to strongly negative events tend to judge all subsequent ambiguous events negatively (Wright and Bower, 1992). Such cognitive biases induced by an emotion may have adaptive value by helping individuals to pay attention and to memorize threatening circumstances. They are not restricted to humans (Paul *et al.*, 2005). In rodents, the startle response induced by exposure to a sudden event is faster and larger under negative emotional states (Lang *et al.*, 1998). Heifers subjected to a potent stressor were unable to abandon a previously learned behavior that was no longer rewarded; this prevented them acquiring a new, more appropriate behavior (Lensink *et al.*, 2006). Sheep experiencing repeated aversive events demonstrated learning deficits (Destrez *et al.*, 2013). In contrast, rats given a catecholamine injection mimicking the physiological component of a moderate emotion are more attentive and display improved memory (Sandi *et al.*, 1997).

The evidence that an emotion has an immediate and temporary repercussion for cognitive functions suggests that it would be worthwhile to study how an accumulation of emotions modifies an animal's cognitive functions in a long-lasting manner. Rats or mice previously subjected to repeated frightening events respond less readily to an ambiguous stimulus signaling the delivery of a positive event long after the last exposure to a frightening event, thus revealing a persistent reduction of the capacities for judgment and decision-making (Harding *et al.*, 2004; Pardon *et al.*, 2000). Similarly, rats housed under social stress show less behavioral agitation after the presentation of a conditioned stimulus predicting the delivery of sucrose reward, suggesting reduced anticipation due to impaired judgment (von Frijtag *et al.*, 2000). Likewise, sheep submitted to repeated unpredictable aversive events for several weeks are less prone to respond to a stimulus signaling the delivery of a positive or negative event, especially when the signal is ambiguous, i.e. between a positive and a negative signal (Doyle *et al.*, 2011; Harding *et al.*, 2004). These findings suggest that the accumulation of negative experiences affects the way the animal appraises the environment to which it is exposed, making it less aware of positive cues and more susceptible to negative ones. Such persistent biases (optimistic versus pessimistic) guide the animal's decisions when appraising new situations or events, especially if there is a degree of ambiguity in their potentially rewarding or punishing consequences (Mendl *et al.*, 2009). If an animal lives in an environment where it has experienced fear several times, it would make adaptive sense to appraise ambiguous events as more likely to be negative and hence to take safety-first avoidance action. Such persistent biases that could be used as a non-invasive indicator of comfort or distress merit further investigations. Establishing the nature and frequency of cognitive processes used to evaluate the environment may help us to understand why chronic stress sometimes results in apathy or blunted emotion while in other cases it leads to heightened emotional reactivity (Boissy *et al.*, 2001). Apathy would likely develop when the animal has no way of altering negative events, whereas hyper-reactivity would occur when it thinks it can control such events.

In conclusion, the study of complex interactions between emotions and cognitive functions offers a new impulse to the study of affective states in animals. Few integrative and multidisciplinary (behavior, psychology, physiology, and neurobiology) approaches exist now to objectively assess affective states of animals in terms of welfare. In addition to studying the role of cognition in the generation of emotions, studies on the effects of emotions on cognitive biases should be promoted to provide deeper insight into the relationships between emotions and more persistent affective states close to welfare. Taken together, all these studies demonstrate how cognitive approaches can be used in animals to probe emotions as short-term affective experiences and welfare as persistent affective experiences.

ANIMAL INDIVIDUALITY IN EMOTIONS: THE CONCEPT OF PERSONALITY

Not all individuals will react in the same way to challenging situations. They will be more or less afraid, more or less curious, more or less difficult to handle, and will react more or less strongly to frightening situations. For instance, there were marked individual differences in the reactions of beef cattle to frightening situations (Kilgour *et al.*, 2006). Nobody who has ever worked with animals will question the fact that there is a considerable variation in how individuals behave in a given situation, i.e. in the "state" they are in, and that this variation is, to some extent, stable across time and situations. When we find such stability, we can call it a "trait," such as fearfulness, docility, and aggressiveness. We can then look for ways of assessing them and for underlying causes for the variation, such as genetic or developmental effects.

In the last 20 years or so, animal scientists have tried to catch up with what stockpeople have known since domestication began. And yet, they have not managed to agree on a label for this variation (Gosling, 2001). In this chapter, we will use the term "personality," since it has been shown repeatedly that consistent individual differences in behavior of human and non-human animals are sufficiently similar to be described by the same label (Gosling and John, 1999; Gosling, 2001; Mehta and Gosling, 2006). There are many definitions of "personality"; the one proposed by Pervin and John (1997; cited by Gosling, 2001), "consistent patterns of feeling, thinking, and behaving", incorporates three different elements, namely emotion, cognition, and behavior. Likewise, whereas dimensions of personality are generally discussed in terms of coping style or fearfulness, it has been proposed that there is currently no single concrete method for assessing a propensity to feel and express emotions (Gosling, 2001).

Individual Variability in Emotional Behaviors

To qualify for the label "personality," individual differences have to be consistent over time and across situations (Erhard and Schouten, 2001), which are typically social isolation, novelty, suddenness, predator response (in many cases humans may evoke the same reactions as predators). Comprehensive reviews of the literature on fear tests used on farm animals (Boissy, 1998; Dodd *et al.*, 2012; Forkman *et al.*, 2007; Jones and Boissy, 2010) already exist and we will just present some newer findings in the field in this section. Individual differences in fearfulness are not just interesting from a theoretical or animal welfare point of view. High levels of fear can make animals difficult and dangerous to handle.

Consistency of Fear-Related Responses Across Different Frightening Situations

In animal experimentation, this element of personality is referred to as the 'validity' of a behavior test (Forkman *et al.*, 2007). It is based on the notion that a personality trait will be expressed in a similar way in different situations within the same context. For example, sheep that were prepared to move away from their group to explore an unknown environment in a behavior test were found to have a lower group-cohesion nine weeks later when grazing undisturbed. They were less easily frightened, recovered more easily from a fright (Sibbald *et al.*, 2009), and split into sub-groups more easily (Michelena *et al.*, 2009) than sheep that explored less. While not all behavior variables may be correlated across tests, there is often sufficient evidence to suggest a common trait, in that individuals that react more strongly in one situation will also react more strongly in others. In horses, for instance, the frequency of licking/nibbling a novel object, the time to put one foot on a novel area and to eat from a bucket placed just behind it, and the flight distance and the time to eat under an opened umbrella were found to be interrelated and consistent over time (Lansade *et al.*, 2008a).

Not all reports show a strong consistency, however. This can have several reasons: maybe the test is not appropriate, or the situations belong to different contexts. Another possibility is that the same underlying fear may result in different behaviors, depending on situation. When horses were confronted with a range of frightening stimuli, such as a traffic cone, white noise and the smell of eucalyptus oil, the resulting heart-rate was correlated between situations while the behaviors were not. The authors explained this by a non-differentiated activation of the sympathetic nervous system (heart-rate), and a stimulus-specific link of the behavior (Christensen *et al.*, 2005).

Many situations give rise to several aspects of personality. Calves placed into an open-field (social isolation in a novel environment) and confronted with novel objects showed avoidance and cortisol responses that were interrelated. These were interpreted as reflecting underlying fearfulness. Locomotion and vocalization in the open-field, however, were independent, and interpreted as reflecting other personality traits, such as activity and sociality (Van Reenen *et al.*, 2005). Fear of people may also increase a female's motivation to defend its offspring, with dangerous consequences for the stockperson (Turner and Lawrence, 2007).

Consistency of Fear-Related Responses Over Time

The concept of personality includes an element of consistency over time. In horses that were put in social isolation, the frequency of neighing was found to be well correlated with other behaviors displayed in the same test, but also over time, from 8 months to 2.5 years of age (Lansade *et al.*, 2008b). A similar result was found for the horses' reaction to humans. Various behavioral reactions—such as the frequency of licking/nibbling a person, the time taken to touch the horse and to fit a halter—were significantly correlated across time (Lansade and Bouissou, 2008).

Beef cattle going through a handling system (squeeze chute) they were familiar with showed consistent differences in flight speed when tested twice on the same day, but also one month apart (Müller and Von Keyserlingk, 2006). In another study, the flight speed showed consistency over time, the behavior while in the crush did not (Cafe *et al.*, 2011). In another study (Benhajali *et al.*, 2010), however, the behavior during constraint in a chute was repeatable across years. The cattle used were of very different genetic backgrounds and from different production systems (Canada, Australia, and France). It is, therefore, not always possible to generalize tests from one system/breed to another. While we are recording the behavior of the animals, we must not forget that this behavior is a result of underlying emotions, which are dependent on how the animals interpret the situation (see the previous section). The reaction of a dairy cow and of a beef cow from an extensive range system towards a human will not be the same. While in one case the human is familiar and may be associated with positive experiences, in the other he might be regarded as a predator.

Personality as a Complex Intermediate Variable

As suggested throughout this chapter, behavioral reactions and physiological patterns observed in animals under emotional challenges are only indicators of fear/anxiety and cannot be considered as direct measures of a fear/anxiety state. Fear must be evaluated by studying not only the intensity of a unique response but also the overall strategy. The different mechanisms underlying fear-related responses probably depend on other motivational systems which may modulate these responses. The strength of reaction in response to a stressor is often used to assess the strength of the underlying emotion. The classic example is the fight/flight response that assumes that a higher level of activity indicates a higher level of fear. However, an absence of activity, immobility, also indicates fear. This might explain the negative correlation between activity while confined and in social isolation and the fear of humans in sheep (Beausoleil *et al.*, 2008). The interaction of several personality traits, such as fearfulness, activity and coping strategies (active/passive) can explain these complex results (Erhard *et al.*, 1999; Koolhaas *et al.*, 1999; van Reenen *et al.*, 2005).

Vocalizations are often considered to be an indicator of fear: the more an animal vocalizes, the more stressed it probably is. Vocalizations, however, are also a means of communication, and could be an indicator of sociability. Because of the complexity of the phenomena that can affect emotional behavior, it is not possible to just consider the results isolated from their context. A measurement used as a fear indicator in one situation cannot be directly extrapolated to others, and it is impossible to simply assess the magnitude of concepts like fear and anxiety on the basis of a single "objective and perfect measurement."

Pre-existing Characteristics of Temperament

Genetic Background and Genetic Models of Emotional Reactivity

Although farm animals have been domesticated over many centuries, fear-related behaviors still vary considerably both within and between populations (Boissy *et al.*, 2005b; Hemsworth, 2003). An animal's genetic background plays a major role in determining its fear levels and its ability to adapt to environmental changes and challenges. Large differences in emotional reactivity are also apparent within and between farm animal breeds (Boissy and Bouissou, 1995; Boissy *et al.*, 2005a; Faure *et al.*, 2003). The approaches go from comparison of breeds (Hansen *et al.*, 2001) to estimation of genetic correlations and heritability within the same breed (Benhajali *et al.*, 2010), selection lines (mink: Malmkvist and Hansen, 2002; sheep: Beausoleil *et al.*, 2008, 2012) and the study of quantitative trait loci (QTL) (Gutiérrez-Gil *et al.*, 2008; Hazard *et al.*, 2012; Jensen *et al.*, 2008). Estimates of heritability of fear in domestic animals seem sufficiently high to allow further selection against fearfulness (Boissy *et al.*, 2005a). Observed heritabilities also revealed a genetic component in calf reactivity (behavior, heart rate, cortisol) to a fear test; fearful calves became fearful heifers, and high fearfulness and a poor milking temperament were associated with low milk yield (Jones and Manteca, 2009; Van Reenen *et al.*, 2005; 2009). For a further discussion of the genetic aspect of fearfulness, see the section on *Genetics and Selective Breeding for Alleviating Negative Emotions*.

Developmental Aspects

While there is a strong genetic component to fearfulness, the environment plays a very important role as well. The role of the prenatal environment for the development of farm animals has recently been reviewed (Roussel *et al.*, 2007; Rutherford *et al.*, 2012). The role of the hypothalamus–pituitary–adrenal axis is of particular interest (Kapoor *et al.*, 2006; Matthews, 2002). The main aspects are maternal stress and nutritional state during pregnancy, and the effects depend on the severity of the stressor as well as on the developmental stage of the fetus and the time of the stressor. Peri-conceptional undernutrition in sheep (Hernandez *et al.*, 2010), as well as undernutrition in early (Erhard *et al.*, 2004) and late gestation (Laporte-Broux *et al.*, 2012), can affect the development of fearfulness in the offspring to affect them throughout their lives. Mammals are not the only ones affected by prenatal stress. Maternal stress in birds (i.e. pre-egg laying) results in more fearful offspring (Henriksen *et al.*, 2011; Janczak *et al.*, 2007). Even in farmed fish (salmon), there is evidence for the deleterious effects of maternal stress on the development of the offspring (Eriksen *et al.*, 2011; Espmark *et al.*, 2008).

The development of fearfulness does not end at birth. More fearful quail, for instance, can still increase their chicks' fearfulness post-hatching (Richard-Yris *et al.*, 2005). There are periods when young animals are specifically sensitive to long-term changes in fearfulness, for instance around weaning (Lansade *et al.*, 2004).

Personality and Health

There is evidence of a link between personality and health in animals (Koolhaas, 2008; Mehta and Gosling, 2008). Individuals scoring high on emotional reactivity and fearfulness are more easily stressed and, as a consequence, may have a weaker immune system. For instance, fear-related reactions to humans in cattle entering a feedlot are negatively associated with immune function (Fel *et al.*, 2003). The reality can, however, be more complicated. Pigs that react more actively to restraint-stress at a young age, show a stronger immune response several weeks later, compared to counterparts that are less active to the restraint test. The latter, however, seem to be more strongly affected by their housing environment at the time of the immune challenge (Bolhuis *et al.*, 2003).

In conclusion, there is overwhelming evidence that stable individual characteristics exist in animals, in the strength of emotions they show and in how they react to environmental challenges. These characteristics are influenced by genetics, but also by developmental factors, pre- and post-natally. To get animals that are easy and safe to work with and that have good welfare, the genetic background, but also the environment of the parent generation, must not be neglected.

A CONCEPT OF POSITIVE WELFARE BASED ON POSITIVE EXPERIENCES

Should we adapt the environment to suit the animal or change the animal to suit the environment? Both approaches can be helpful. On the one hand, we know that appropriate manipulation of environmental factors can profoundly change an animal's behavioral and physiological capabilities. Indeed, good stockmanship and environmental enrichment can improve the animals' quality of life (Boissy *et al.*, 2007b; Duncan, 1996; Fraser and Duncan, 1998; Hemsworth and Coleman, 1998; Jones, 2004). On the other hand, given the substantial genetic diversity within and between populations, selective breeding probably represents the quickest, most reliable method of reducing fear and other welfare problems. The examples of heritabilities for fearfulness reported in the previous section suggest sensitivity to genetic manipulation in farm animals (Boissy *et al.*, 2005b; Faure *et al.*, 2003; Rodenburg and Turner, 2012). In reality, an integrated approach involving targeted breeding programs, environmental manipulation and improved management is likely to be the most effective strategy (Boissy *et al.*, 2005b; Faure and Jones, 2004; Jones, 2004). For instance, while depression-like behaviors are expected after chronic stress, a significant proportion of animals (up to a third or half of samples) are resistant and do not develop these symptoms. Therefore, the emotional state of an animal seems to result from both its environment (especially previous stressful experiences) and its own individual characteristics.

We briefly report the genetic approaches for alleviating negative emotions, and then we present the common strategies of enrichment to induce positive emotions in livestock animals before introducing promising behavioral strategies based on evaluative abilities of the animals to improve their welfare and possibly their health.

Genetics and Selective Breeding for Alleviating Negative Emotions

Farm animal welfare management generally focused on minimizing disturbances that generate unpleasant emotions and longer-lasting affective states.

Breeding and genetics has played an important role in such an approach. Negative emotions and defensive behaviors might be reduced or eliminated by using genetic stocks that either do not show these behaviors or exhibit them so little or so infrequently as to be of trivial importance (Craig and Swanson, 1994). Genetic selection could be a powerful tool for decreasing the incidence of behaviors associated with welfare problems, such as fear. Indeed, selective breeding programs for reduced fear or dampened stress responsiveness have already been established in turkeys (Brown, 1974) and Japanese quail (Jones and Satterlee, 1996; Mills and Faure, 1991). For example, divergent selection of Japanese quail either for short tonic immobility (TI) fear reactions (Mills and Faure, 1991) or a decreased plasma corticosterone response to restraint (Jones and Satterlee, 1996) led to reductions in their fear and physiological stress responses to a wide range of potentially traumatic situations (Jones and Satterlee, 1996; Mills and Faure, 1991). Likewise, marker-assisted selection may be a useful tool for manipulating emotional states (Boissy *et al.*, 2005b). For example, quantitative trait loci (QTL) for fear (Schmutz *et al.*, 2001), for reactivity to humans (Davis and Denise, 1998), and for social reactivity (Hazard *et al.*, 2012) were identified in cattle and sheep using linked markers. Such linked markers could, in turn, be used for marker-assisted selection within sire families once the relationship between the marker and the gene has been determined. Genomic tools could thus facilitate selection for complex behavioral traits, which are frequently impossible to measure on a large number of animals. Identifying and manipulating fundamental behavioral traits that underlie adaptation to the physical and social farming environment might be an effective strategy for improving farm animal welfare in a broad sense as well as safeguarding or enhancing production, product quality, and profitability.

Some concerns must be addressed before genetic selection is widely accepted as a welfare-friendly tool. Firstly, as selection for one trait may also modify other characteristics, it is essential to ensure that there are no associated undesirable effects on welfare; new breeding tools can indeed enhance the rate of undesirable change in correlated traits of welfare significance (Rodenburg and Turner, 2012). Secondly, it is necessary to take into account the ethical objections that are raised regarding genetic selection. However; it is uncertain whether these objections make the distinction between selective breeding (practiced since animals were first domesticated) and genetic manipulation (e.g., gene insertion or deletion). Of course, interactions between the animals' genotype and their environment are critical determinants of animal welfare that may help reconcile production with ethical concerns (Boissy *et al.*, 2005b). Briefly, we regard genetic selection for reduced fearfulness as ethically sound because it could increase the animal's ability to interact successfully with its physical and social environment by dampening the inhibitory effects of fear and alleviating distress.

This not only reduces the animals' stress but also facilitates their management (better human–animal relationship), and consequently increases their welfare.

Eliciting Positive Emotions by Enriching the Environment and Management

Traditionally, the focus has been on the negative side of animal welfare that is acute fear and chronic stress (see above). This strategy produced animal welfare benefits, but at best it could only lift a poor welfare status to a neutral one. However, welfare is based on a balance between negative and positive experiences (Dawkins, 1990; Duncan, 1996; Mellor, 2012; Spruijt *et al.*, 2001), and it is now widely accepted that positive welfare is not simply the absence or reduction of negative experiences but needs also the expression of positive ones such as pleasure and comfort (for a review, see Boissy *et al.*, 2007b). Therefore, in addition to the assessment of negative experiences it is now necessary to consider both positive expectations (what an animal "likes") and resources that an animal is motivated to obtain (what an animal "wants"). However, despite the efforts of pioneering authors such as Colin Allen, Jaak Panksepp, or Kent Berridge, relatively little had been done until recently to understand positive experiences in comparison with their negative counterparts. As an emerging field in animal welfare, there is now a growing interest in positive emotions and enrichment in captive animals. We present the current efforts to induce positive animal welfare based on the environmental enrichment and the improvement of human–animal relationships.

Environmental Enrichment

Although it is often recommended that the animals' environment should be predictable and controllable (Bassett and Buchanan-Smith, 2007) an invariable and overly predictable one should be avoided as it may result in boredom (Van Rooijen, 1984). Environmental enrichment generally involves incorporating new, putatively interesting stimuli in the home environment and thereby increasing its complexity and stimulus value. This promotes desirable behaviors like foraging, exploration, social play, and grooming in pigs (Van de Weerd *et al.*, 2003; Bracke *et al.*, 2006; Spinka *et al.*, 2001), cattle (Wilson *et al.*, 2002), and poultry (Jones, 2004), reduces excitability and fear in pigs (Grandin *et al.*, 1987) and sheep (Vandenheede and Bouissou, 1998), as well as reduces neophobia in poultry (Jones, 2004). There is mounting evidence that animals housed in enriched conditions have more positive affective states compared to those housed in more barren environments. This belief is supported by the cognitive bias assessments, that report more optimistic cognitive biases in laboratory rats currently housed in more enriched or more stable cages (Brydges *et al.*, 2011; Harding *et al.*, 2004). Similar results have been reported in farm animals. For instance, pigs that have spent time in an enriched environment have more optimistic judgment biases, indicative of a more positive affective state, in comparison with counterparts reared in a barren environment (Douglas *et al.*, 2012).

However, there are several critical requirements, particularly in the farming sector. Practicality is paramount, e.g., enrichment devices must be safe (no sharp edges or protrusions), hygienic, affordable, durable, accessible to all members of the group, and retain interest as long as possible to reduce the need for frequent change. Ideally, we should determine what sorts of stimuli the animals find attractive and interesting (Bracke *et al.*, 2006; Jones, 2004; Wilson *et al.*, 2002). Important questions include: Are there unlearned preferences? Are movement and change important? Is physical interaction with the device essential? At what rates of introduction and stages of the animals' development would enrichment be most beneficial? Identifying the factors governing an animal's perception of a safe and interesting world would greatly facilitate the development of effective environmental enrichment.

Positive Human Contact and Training Programs

Poor human–animal relationships cause the animals to be fearful of humans (Manteca *et al.*, 2009). In some production systems (e.g., extensive farming), the animals are less likely to experience human contact which could in turn make them even more fearful of people through a lack of habituation (Boissy *et al.*, 2005b). Remedial efforts for more tractable animals focus on two main approaches, which can also be integrated in a complementary fashion: (i) regular and positive human contact and (ii) stockmanship training programs.

Picking an animal up regularly, stroking it or giving it food reliably reduces fear of people in farm mammals (Coulon *et al.*, 2012; Hemsworth *et al.*, 1993). Nevertheless, it is clearly not a practical option in modern agriculture with the tendency to increase the herd size and by contrast to reduce the labor force. Encouragingly though, simply allowing regular visual contact with people reduced fear of humans in lambs (Tallet *et al.*, 2008). Training programs are the main strategy for achieving and maintaining a good human–animal relationship (Hemsworth, 2003). A team of European and Australian researchers in the EU-funded Welfare Quality® project (www.welfarequality.net) developed a multimedia-based training package (courses, software, manuals, newsletters, etc.) designed to help stockpersons improve the quality of their interactions with cattle, pigs, and chickens, and hence enhance the human–animal relationship (Boivin and Ruis, 2011).

Cognitive Enrichment: A New Approach Based on the Animals' Appraisal Abilities

The conceptual framework that has been developed from recent advances in cognitive psychology and that is reported above (see the section, *Contribution of Cognitive Psychology to Access Animal Emotions*) is particularly useful to scientifically explore positive feelings in animals and to develop strategies in animal husbandry for enhancing positive emotional experiences. We outline three specific cognitive processes for eliciting positive experiences: signaling a reward in advance, giving a higher reward than expected, and enabling the animals to cope with or to control a wanted event (Boissy *et al.*, 2007b).

Investigations of what animals find positively reinforcing and of the animal's behavioral expression during anticipation or expectation of the rewarding result, often called positive anticipation, provide a basis for assessing positive emotional states in animals (Spruijt *et al.*, 2001). When anticipating food rewards, rats (van der Harst *et al.*, 2003), mink (Vinke *et al.*, 2004), poultry (Moe *et al.*, 2009), and pigs (Dudink *et al.*, 2005) show increased locomotor activity and frequent behavioral transitions, i.e. anticipatory hyperactivity, which reflect positive emotional experience. In pigs, signaled positive stimuli such as extra space, food, or straw lead to increased activity due to orientation toward the place where the reward is offered and frequent behavioral transitions, most often combined with play markers, i.e. hopping, scampering, pivoting, pawing, flopping, and head tossing (Dudink et al., 2005).

In a typical positive contrast experiment an animal is first trained to perform a task by giving it a reinforcer of a certain size (for a review, see Flaherty, 1996). The animal is then given a larger reinforcer than they had been given before. If the animal changes its behavior so that its response is faster/more vigorous than that of a control group given the larger reinforcer from the start, a successive positive contrast has occurred. Positive contrast can be found for both appetitive and consummatory responses. It is the difference between the expected value and the actual value that influences a subject's behavior rather than the exact level of reinforcement. In his original study, Crespi (1942) reported that rats for which the quantity of reward has just been increased run faster towards the food reward than do rats having always received the large quantity of reward. Similarly, sheep showed transient hyperactivity if a reward was of greater hedonic value than expected (Greiveldinger et al., 2011). Nevertheless, due to the relatively small number of studies, further research is needed to confirm the value of a positive contrast model in the study of potential positive emotions and their behavioral expressions.

Enabling the animals to cope with or to control a wanted event (controllability of reward) may be another source of positive emotions. Social and physical challenges can be perceived as positive experiences provided there is an adequate coping behavior. Meehan and Mench (2007) suggested that successful dealing with adequate cognitive challenges can induce positive emotion. Hence, frequent challenges that can always be successfully mastered and eventually enable the animal to reach a desired and rewarding goal may be suitable means to regularly elicit positive emotions. Using such a cognitive approach, a program of research on the effects of rewarded cognitive processes has been developed (Ernst *et al.*, 2005): each animal out of a group of pigs had to learn an individual acoustic signal as a call to work for food by pressing a button. Using this approach, Zebunke *et al.* (2011) report transient cardiac reaction revealing positive emotional experiences in pigs. Interestingly, pigs reared for several weeks under such a complex but predictable environment presenting positive challenges with which individuals are able to cope, present modifications in the reward-sensitive brain opioid receptors compared to conventionally housed pigs, indicating frequently occurring positive experiences (Kalbe and Puppe, 2010). In conclusion, this brief review on cognitive environmental enrichment provides valuable insights into the eliciting of positive emotions in animals. Based on the principle of animal-based measurements, adequate cognitive challenges are evaluated as emotionally positive by the tested animals.

Promoting Positive Experiences to Mitigate Negative Experiences and to Improve Animal Welfare and Health

Cognitive enrichment, such as anticipatory behavior, might be useful in developing a behavioral therapy to counteract the deleterious effects of stress. For instance, rats submitted to prolonged stress did not develop anhedonia—a major symptom of depression—if they received repeated food rewards announcements, suggesting that positive experiences can counteract the deleterious consequences of negative experiences (van der Harst *et al.*, 2005). Thus, signals of reward might form the basis for a new enrichment strategy. Likewise, reduction of fearfulness reduces pessimistic-like judgment in lambs compared with stressed counterparts (Destrez *et al.*, 2012). Furthermore, induction of anticipation via the announcement of certain alarming events might also be a way to reduce the negative emotion induced by such events. For instance, sheep for which the appearance of a sudden event was preceded by a light signal expressed fewer fear reactions than their counterparts that could not anticipate (Greiveldinger *et al.*, 2007). Therefore, the implementation of cognitive challenges into animal housing, based on positive anticipation, positive contrast, and positive control, is a promising approach to induce long-lasting positive effects on animal welfare.

Although welfare and health are distinct concepts, they influence each other. Considering health as a state of complete physical, mental, and social well-being and not merely the absence of disease or infirmity,[1] it is essential to reconsider the management of animal health through an integrated view, particularly by taking into consideration animal emotions and welfare/stress. There are permanent interactions between the immune system and the brain, notably through the inflammatory cytokines and the autonomic nervous and neuroendocrine systems. Although research in this area is recent, there is evidence that chronic stress results in glucocorticoid receptor resistance that induces exaggerated local release of inflammatory cytokines, resulting in increased disease risk because inflammation plays an important role in the onset and progression of a wide range of diseases (for a review, see Cohen *et al.*, 2012). Likewise, there is evidence in humans showing that, in contrast, experiencing positive emotions may improve health. Pioneering researchers suggest that there are health benefits in humans of humor and laughter as positive-emotions promoters, especially in coping with a diagnosis of cancer (Mahony *et al.*, 2002). Although the scientific evidence for this claim is inconclusive (Martin, 2001), humor is being increasingly used as a mean of reducing stress and making patients feel better (Penson *et al.*, 2005). Obviously, there are no such studies in animals. Nevertheless, an increase in stress responses is often associated with an increased vulnerability to infectious, metabolic, and production diseases. It is logical to suggest, therefore, that giving increased opportunities for positive experiences might also have a positive effect on animal health. For instance, pigs given the opportunity to show successful adaptation by rewarded cognitive processes (i.e. cognitive enrichment) express a higher immune reactivity and recover more quickly from a standard biopsy (Ernst *et al.*, 2006). Although psychological factors influence the immune system at any point in the life-span, it has been shown in laboratory animals that the immaturity of a young infant's immune responses makes it more vulnerable, especially during the fetal and neonatal stages (Coe and Lubach, 2003). Various studies in humans show clearly that perinatal stress has a programming effect resulting in vulnerability to the development of chronic diseases in adulthood (Lupien *et al.*, 2009). We can therefore hypothesize that proper welfare management of the young animal (either directly after birth or throughout the welfare state of the dam before birth) would improve its ability to cope with various forms of health challenges encountered not only during early postnatal life but also during its whole production life.

Although relatively recent, the concept of cognitive enrichment based on the evaluative abilities of the animals, allows defining behavioral strategies that have the potential to sustainably improve animal welfare and maybe health. Such behavioral strategies such as animal-friendly housing should be

1. World Health Organization (https://apps.who.int/aboutwho/en/definition.html).

easily implemented in farm husbandry and management since they are safe, relatively easy to implement, and economically affordable.

CONCLUSION

If we agree that farm animals are sentient and that they can more particularly feel emotions, then we should follow our representation of the sentient animals and behave towards them in the way that has been developed throughout the present chapter. The starting point of the welfare of an animal is its emotions, which may or may not evolve into prolonged affective states (Veissier *et al.*, 2012). We particularly emphasized in this chapter the cognitive aspects that can be useful to access emotions in animals. New conceptual frameworks based on cognitive sciences offer relevant tools to investigate positive affective states animals are capable of experiencing and to achieve real practical improvements in the welfare of the animals that we use for our own ends. Emotions are not merely determined by the characteristics of the challenges, but also by the personality of the individual animal. Avoiding the rearing of overly vulnerable individuals through appropriate genetic selection and careful consideration of the maternal and early environment will help to improve the welfare of both animals and stockpeople. We saw that the implementation of cognitive challenges into animal housing (based on positive anticipation, positive contrast, and positive control) is a promising and practical approach to induce long-lasting positive effects on animal welfare and speculatively on animal health. Further research is necessary to fully understand how appraisal processes and ongoing cognitive coping influence the immune system before developing relevant behavioral strategies to enhance animal welfare and its health. Now, scientific information reported throughout this chapter should be enough convincing to promote livestock management practices, based on the animal's emotions, that ensure a better match between the animals' sentience, their own expectations, and their environment, in order to achieve a positive animal welfare.

REFERENCES

Asheim, L.J., Mysterud, I., 2005. External effects of mitigating measures to reduce large carnivore predation on sheep. J. Farm Manag. 12, 206–213.

Bassett, L., Buchanan-Smith, H.M., 2007. Effects of predictability on the welfare of captive animals. Appl. Anim Behav. Sci. 102, 223–245.

Beausoleil, N.J., Blache, D., Stafford, K.J., Mellor, D.J., Noble, A.D.L., 2008. Exploring the basis of divergent selection for "temperament" in domestic sheep. Appl. Anim Behav. Sci. 109, 261–274.

Beausoleil, N.J., Blache, D., Stafford, K.J., Mellor, D.J., Noble, A.D.L., 2012. Selection for temperament in sheep: domain-general and context-specific traits. Appl. Anim Behav. Sci. 139, 74–85.

Benhajali, H., Boivin, X., Sapa, J., Pellegrini, P., Boulesteix, P., Lajudie, P., et al., 2010. Assessment of different on-farm measures of beef cattle temperament for use in genetic evaluation. J. Anim. Sci. 88, 3529–3537.

Boissy, A., 1998. Fear and fearfulness in determining behavior. In: Grandin, T. (Ed.), Genetics and the Behaviour of Domestic Animals. Academic Press, San Diego, pp. 67–111.

Boissy, A., Bouissou, M.F., 1995. Assessment of individual differences in behavioural reactions of heifers exposed to various fear-eliciting situations. Appl. Anim. Behav. Sci. 46, 17–31.

Boissy, A., Terlouw, C., Le Neindre, P., 1998. Presence of cues from stressed conspecifics increases reactivity to aversive events in cattle: evidence for the existence of alarm substances in urine. Physiol. Behav. 63, 489–495.

Boissy, A., Veissier, I., Roussel, S., 2001. Emotional reactivity affected by chronic stress: an experimental approach in calves submitted to environmental instability. Anim. Welf. 10, S175–S185.

Boissy, A., Bouix, J., Orgeur, P., Poindron, P., Bibé, B., Le Neindre, P., 2005a. Genetic analysis of emotional reactivity in sheep: effects of the genotypes of the lambs and of their dams. Genet. Select. Evol. 37, 381–401.

Boissy, A., Fisher, A.D., Bouix, J., Hinch, G.N., Le Neindre, P., 2005b. Genetics of fear in ruminant livestock. Livest. Prod. Sci. 93, 23–32.

Boissy, A., Arnould, C., Chaillou, E., Désiré, L., Duvaux-Ponter, C., Greiveldinger, L., et al., 2007a. Emotions and cognition: a new approach to animal welfare. Anim. Welf. 16, 37–43.

Boissy, A., Manteuffel, G., Jensen, M.B., Moe, R.O., Spruijt, B., Keeling, L., et al., 2007b. Assessment of positive emotions in animals to improve their welfare. Physiol. Behav. 92, 375–397.

Boissy, A., Aubert, A., Desiré, L., Greiveldinger, L., Veissier, I., 2011. Cognitive science to relate ear postures to emotions in sheep. Anim. Welf. 20, 47–56.

Boivin, X., Ruis, M., 2011. "Quality handling" a training program to reduce fear and stress in farm animals. In: Goby, L. (Ed.), The Fourth Boehringer Ingelheim Expert Forum on Farm Animal Well-Being. Boehringer Ingelheim Animal Health GmbH, Seville, pp. 21–26.

Bolhuis, J.E., Parmentier, H.K., Schouten, W.G.P., Schrama, J.W., Wiegant, V.M., 2003. Effects of housing and individual coping characteristics on immune responses of pigs. Physiol. Behav. 79, 289–296.

von Borell, E., Langbein, J., Desprès, G., Hansen, S., Leterrier, C., Marchant-Forde, J., et al., 2007. Heart rate variability as a measure of autonomic regulation of cardiac activity for assessing stress and welfare in farm animals a review. Physiol. Behav. 92, 293–316.

Bouissou, M.F., Boissy, A., Le Neindre, P., Veissier, I., 2001. The social behaviour of cattle. In: Gonyou, H., Keeling, L. (Eds.), Social Behaviour in Farm Animals. CAB International, Cambridge, pp. 113–146.

Bracke, M.B.M., Zonderland, J.J., Lenskens, P., Schouten, W.G.P., Vermeer, H., Spoolder, H.A.M., et al., 2006. Formalised review of environmental enrichment for pigs in relation to political decision making. Appl. Anim. Behav. Sci. 98, 165–182.

Bradley, B.P., Mogg, K., Lee, S.C., 1997. Attentional biases for negative information in induced and naturally occurring dysphoria. *Behav. Res. Ther.* 35, 911–927.

Brown Jr., W.G., 1974. Some aspects of beef cattle behaviour as related to productivity. *Dissert. Abstr. Int.* B. 34, 1805.

Brydges, N., Leach, M.C., Nicol, K., Wright, R., Bateson, M., 2011. Environmental enrichment induces optimistic cognitive bias in rats. Anim. Behav. 81, 169–175.

Cafe, L.M., Robinson, D.L., Ferguson, D.M., McIntyre, B.L., Geesink, G.H., Greenwood, P.L., 2011. Cattle temperament: persistence of assessments and associations with productivity, efficiency, carcass and meat quality traits. J. Anim. Sci. 89, 1452–1465.

Carere, C., Welink, D., Drent, P.J., Koolhaas, J.M., Groothuis, T.G.G., 2001. Effect of social defeat in a territorial bird (Parus major) selected for different coping styles. Physiol. Behav. 73, 427–433.

Christensen, J.W., Keeling, L.J., Nielsen, B.L., 2005. Responses of horses to novel visual, olfactory and auditory stimuli. Appl. Anim Behav. Sci. 93, 53–65.

Coe, C.L., Lubach, G.R., 2003. Critical periods of special health relevance for psychoneuroimmunology. Brain. Behav. Immun. 17, 3–12.

Cohen, S., Janicki-Devertsa, D., Doyle, W.J., Miller, G.E., Frank, E., Rabin, B.S., et al., 2012. Chronic stress, glucocorticoid receptor resistance, inflammation, and disease risk. PNAS 109, 5995–5999.

Colborn, D.R., Thompson Jr., D.L., Roth, T.L., Capehart, J.S., White, K.L., 1991. Responses of cortisol and prolactin to sexual excitement and stress in stallions and geldings. J. Anim. Sci. 69, 2556.

Coulon, M., Nowak, R., Andanson, S., Ravel, C., Marnet, P.G., Boissy, A., et al., 2012. Human–lamb bonding: oxytocin, cortisol and behavioural responses of lambs to human contacts and social separation. Psychoneuroendocrinology <http://dx.doi.org/10.1016/j.psyneuen.2012.07.008>.

Coutellier, L., Arnould, C., Boissy, A., Orgeur, P., Prunier, A., Veissier, I., et al., 2007. Pig's responses to repeated social regrouping and relocation during the growing-finishing period. Appl. Anim. Behav. Sci. 105, 102–114.

Craig, J.V., Swanson, J.C., 1994. Review: welfare perspectives on hens kept for egg production. Poultry Sci. 73, 921–938.

Crespi, L.P., 1942. Quantitative variation in incentive and performance in the white rat. American J. Psychol. 40, 467–517.

Dantzer, R., 1988. Les Émotions. Presses Universitaires de France, Paris, p.121.

Dantzer, R., 2002. Can farm animal welfare be understood without taking into account the issues of emotion and cognition? J. Anim. Sci. 80, E1–E9.

Dantzer, R., Mormède, P., 1983. Stress in farm animals: a need for reevaluation. J. Anim. Sci. 57, 6–18.

Davis, G.P., Denise, S.K., 1998. The impact of genetic markers on selection. J. Anim. Sci. 76, 2331–2339.

Dawkins, M.S., 1990. From an animal's point of view: motivation, fitness, and animal welfare. Behav. Brain Sci. 13, 1–61.

Dawkins, M.S., 2001. How can we recognise and assess good welfare? In: Broom, D.M. (Ed.), Coping with Challenge: Welfare in Animals Including Humans. Dahlem University Press, Berlin, pp. 63–76.

Désiré, L., Boissy, A., Veissier, I., 2002. Emotions in farm animals: a new approach to animal welfare in applied ethology. Behav. Process. 60, 165–180.

Désiré, L., Veissier, I., Després, G., Boissy, A., 2004. On the way to assess emotions in animals: do lambs evaluate an event through its suddenness, novelty or unpredictability? J. Comp. Psychol. 118, 363–374.

Désiré, L., Veissier, I., Després, G., Delval, E., Toporenko, G., Boissy, A., 2006. Appraisal process in sheep: interactive effect of suddenness and unfamiliarity on cardiac and behavioural responses. J. Comp. Psychol. 120, 280–287.

Destrez, A., Deiss, V., Belzung, C., Lee, C., Boissy, A., 2012. Does reduction of fearfulness tend to reduce pessimistic-like judgment in lambs? Appl. Anim. Behav. Sci. 139, 233–241.

Destrez, A., Deiss, V., Leterrier, C., Boivin, X., Boissy, A., 2013. Long-term exposure to unpredictable and uncontrollable aversive events alters fearfulness in sheep. Animal. 7, 476–484.

Dodd, C.L., Pitchford, W.S., Hocking Edwards, J.E., Hazel, S.J., 2012. Measures of behavioural reactivity and their relationships with production traits in sheep: a review. Appl. Anim Behav. Sci. 140, 1–15.

Douglas, C., Bateson, M., Walsh, C., Bédué, A., Edwards, S.A., 2012. Environmental enrichment induces optimistic cognitive biases in pigs. Appl. Anim. Behav. Sci. 139, 65–73.

Doyle, R.E., Lee, C., Deiss, V., Fisher, A.D., Hinch, G.N., Boissy, A., 2011. Measuring judgement bias and emotional reactivity in sheep following long-term exposure to unpredictable and aversive events. Physiol. Behav. 102, 503–510.

Dudink, S., De Jonge, F.H., Spruijt, B.M., 2005. Announcing the arrival of enrichment increases play behavior of piglets directly after weaning. In: KTBL, D. (Ed.), Current Research in Applied Ethology. Kuratorium für Technik und Bauwesen, Darmstadt, pp. 212–221.

Duncan, I.J.H., 1993. Welfare is to do with what animals feel. J. Agricult. Environment. Ethics. 6, 8–14.

Duncan, I.J.H., 1996. Animal welfare defined in terms of feelings. Acta Agric. Scand. Sect. A., 29–35.

Duncan, I.J.H., Petherick, J.C., 1989. Cognition: the implications for animal welfare. Appl. Anim. Behav. Sci. 24, 81–86.

Dwyer, C.M., 2004. How has the risk of predation shaped the behavioural responses of sheep to fear and distress? Anim. Welf. 13, 269–281.

Erhard, H.W., Mendl, M., 1999. Tonic immobility and emergence time in pigs: more evidence for behavioural strategies. Appl. Anim. Behav. Sci. 61, 227–237.

Erhard, H.W., Schouten, W.G.P., 2001. Individual differences and personality. In: Keeling, L.J., Gonyou, H.W. (Eds.), Social Behaviour in Farm Anmals. CABI Publishing, Wallingford, pp. 333–352.

Erhard, H.W., Mendl, M., Christiansen, S.B., 1999. Individual differences in tonic immobility may reflect behavioural strategies. Appl. Anim Behav. Sci. 64, 31–46.

Erhard, H.W., Boissy, A., Rae, M.T., Rhind, S.M., 2004. Effects of prenatal undernutrition on emotional reactivity and cognitive flexibility in adult sheep. Behav. Brain Res. 151, 25–35.

Eriksen, M.S., Faerevik, G., Kittilsen, S., McCormick, M.I., Damsgård, B., Braithwaite, V.A., et al., 2011. Stressed mothers—troubled offspring: a study of behavioural maternal effects in farmed Salmo salar. J. Fish Biol. 79, 575–586.

Ernst, K., Puppe, B., Schön, P.C., Manteuffel, G., 2005. A complex automatic feeding system for pigs aimed to induce successful behavioral coping by cognitive adaptation. Appl. Anim. Behav. Sci. 91, 205–218.

Ernst, K., Tuchscherer, M., Kanitz, E., Puppe, B., Manteuffel, G., 2006. Effects of attention and rewarded activity on immune parameters and wound healing in pigs. Physiol. Behav. 89, 448–456.

Espmark, Å.M., Eriksen, M.S., Salte, R., Braastad, B.O., Bakken, M., 2008. A note on pre-spawning maternal cortisol exposure in farmed Atlantic salmon and its impact on the behaviour of offspring in response to a novel environment. Appl. Anim Behav. Sci. 110, 404–409.

European Union 1997. The Amsterdam treaty modifying the treaty on European Union, the treaties establishing the European communities, and certain related facts. Official journal. C 340.

Faure, J.M., Bessei, W., Jones, R.B., 2003. Direct selection for improvement of animal well-being. In: Muir, W., Aggrey, S. (Eds.), Poultry Breeding and Biotechnology. CAB International, Wallingford, UK, pp. 221–245.

Faure, J.M., Jones, R.B., 2004. Genetic influences on resource use, fear and sociality. In: Perry, G.C. (Ed.), Welfare of the Laying Hen, Twenty-Seventh Poultry Science Symposium. CAB International, Wallingford, UK, pp. 99–108.

Fel, L.R., Colditz, I.G., Walker, K.H., Watson, D.L., 2003. Associations between temperament, performance and immune function in cattle entering a commercial feedlot. Aust. J. Exp. Agric. 39, 795–802.

Fisher, A., Matthews, L., 2001. The social behaviour of sheep. In: Keeling, L., Gonyou, H. (Eds.), Social Behaviour in Farm Animals. CAB International, Wallingford, UK, pp. 211–245.

Flaherty, C.F., 1996. Incentive Relativity. Cambridge University Press, Cambridge.

Forkman, B., Boissy, A., Meunier-Salaün, M., Canali, E., Jones, R.B., 2007. A critical review of fear tests used on cattle, pigs, sheep, poultry and horses. Physiol. Behav. 92, 340–374.

Fraser, D., Duncan, I.J.H., 1998. 'Pleasures', 'pains' and animal welfare: toward a natural history of affect. Anim. Welf. 7, 383–396.

von Frijtag, J.C., Reijmers, L.G.J.E., Van der Harst, J.E., Leus, I.E., Van den Bos, R., Spruijt, B. M., 2000. Defeat followed by individual housing results in long-term impaired reward- and cognition-related behaviours in rats. Behav. Brain Res. 117, 137–146.

Gentle, M.J., Waddington, D., Hunter, L.N., Jones, R.B., 1990. Behavioural evidence for persistent pain following partial beak amputation in chickens. Appl. Anim. Behav. Sci. 27, 149–157.

Gosling, S.D., 2001. From mice to men: what can we learn about personality from animal research? Psychol. Bull. 127, 45–86.

Gosling, S.D., John, O.P., 1999. Personality dimensions in nonhuman animals: a cross-species review. Curr. Dir. Psychol. Sci. 8, 69–75.

Gray, A.J., 1987. L'Anxiété comme cas-type d'émotion. Bull. de Psychol., 96–103.

Greiveldinger, L., Veissier, I., Boissy, A., 2007. Emotional experiences in sheep: predictability of a sudden event lowers subsequent emotional responses. Physiol. Behav. 92, 675–683.

Greiveldinger, L., Veissier, I., Boissy, A., 2009. Behavioural and physiological responses of lambs to controllable versus uncontrollable aversive events. Psychoneuroendocrinology 34, 805–814.

Greiveldinger, L., Veissier, I., Boissy, A., 2011. The ability of lambs to form expectations and the emotional consequences of a discrepancy from their expectations. Psychoneuroendocrinology 36, 806–815.

Greiveldinger, L., Boissy, A., Aubert, A., 2012. An ethological perspective of the relations between sociality and emotions in animals. In: Aubert, A. (Ed.), Social Interaction Evolution Psychology and Benefits. Nova Science Publishers, New York, in press.

Gutiérrez-Gil, B., Ball, N., Burton, D., Haskell, M., Williams, J.L., Wiener, P., 2008. Identification of quantitative trait loci affecting cattle temperament. J. Anim. Sci. 99, 629–638.

Hansen, I., Christiansen, F., Hansen, H., Braastad, B., Bakken, M., 2001. Variation in behavioural responses of ewes towards predator-related stimuli. Appl. Anim Behav. Sci. 70, 227–237.

Harding, E.J., Paul, E.S., Mendl, M., 2004. Animal behavior—cognitive bias and affective state. Nature 427, 312.

Hargreaves, A.L., Hutson, G.D., 1990. The stress response in sheep during routine handling procedures. Appl. Anim. Behav. Sci. 26, 83–90.

Hazard, D., Foulquié, D., Delval, E., François, D., Sallé, G., Moreno, C., and et al. (2012). Identification of QTL for behavioural reactivity in sheep using the ovine SNP50 beadchip.

In: Proceedings of the Annual Meeting of the European Federation of Animal Science, Bratislava (SL), p. 343.

Hemsworth, P.H., 2003. Human–animal interactions in livestock production. Appl. Anim. Behav. Sci. 81, 185–198.

Hemsworth, P.H., Barnett, J.L., Coleman, G.J., 1993. The human–animal relationship in agriculture and its consequences for the animal. Anim. Welf. 2, 33–51.

Hemsworth, P.H., Coleman, G.J. (Eds.), 1998. Human–Livestock Interactions: The Stockperson and the Productivity and Welfare of Intensively Farmed Animals. Londres.

Henriksen, R., Rettenbacher, S., Groothuis, T.G.G., 2011. Prenatal stress in birds: pathways, effects, function and perspectives. Neurosc. Biobehav. Rev. 35, 1484–1501.

Hernandez, C.E., Matthews, L.R., Oliver, M.H., Bloomfield, F.H., Harding, J.E., 2010. Effects of sex, litter size and periconceptional ewe nutrition on offspring behavioural and physiological response to isolation. Physiol. Behav. 101, 588–594.

Janczak, A.M., Torjesen, P., Palme, R., Bakken, M., 2007. Effects of stress in hens on the behaviour of their offspring. Appl. Anim Behav. Sci. 107, 66–77.

Jensen, P., Buitenhuis, B., Kjaer, J., Zanella, A., Mormède, P., Pizzari, T., 2008. Genetics and genomics of animal behaviour and welfare—challenges and possibilities. Appl. Anim Behav. Sci. 113, 383–403.

Jones, R.B., 2004. Environmental enrichment: the need for bird-based practical strategies to improve poultry welfare. In: Perry, G.C. (Ed.), Welfare of the Laying Hen Twenty-Seventh Poultry Science Symposium. CAB International, Wallingford, UK, pp. 215–225.

Jones, R.B., Boissy, A., 2010. Fear and other emotions, Chapter 6. In: Appleby, M.C., Mench, J.A., Olsson, I.A.S., Hughes, B.O. (Eds.), Animal Welfare, second ed. CAB International, Cambridge, pp. 78–97.

Jones, R.B., Manteca, X., 2009. Best of Breed. Public Sci. Rev. 18, 562–563.

Jones, R.B., Satterlee, D.G., 1996. Threat-induced behavioural inhibition in Japanese quail genetically selected for contrasting adrenocortical response to mechanical restraint. Brit. Poultry Sci. 37, 465–470.

Kalbe, C., Puppe, B., 2010. Long-term cognitive enrichment affects opioid receptor expression in the amygdala of domestic pigs. Genes. Brain Behav. 9, 75–83.

Kapoor, A., Dunn, E., Kostaki, A., Andrews, M.H., Matthews, S.G., 2006. Fetal programming of hypothalamo–pituitary–adrenal function: prenatal stress and glucocorticoids. J. Physiol. 572, 31–44.

Kappas, A., 2006. Appraisals are direct, immediate, intuitive, and unwitting... and some are reflective.... Cogn. Emot. 20, 952–975.

Keeling, L., Gonyou, H., 2001. Social Behaviour in Farm Animals. CAB International, Wallingford, UK.

Kilgour, R.J., Melville, G.J., Greenwood, P.L., 2006. Individual differences in the reaction of beef cattle to situations involving social isolation, close proximity of humans, restraint and novelty. Appl. Anim Behav. Sci. 99, 21–40.

Koolhaas, J.M., 2008. Coping style and immunity in animals: making sense of individual variation. Brain Behav. Immun. 22, 662–667.

Koolhaas, J.M., Korte, S.M., De Boer, S.F., Van der Vegt, B.J., Van Reenen, C.G., Hopster, H., et al., 1999. Coping styles in animals: current status in behavior and stress-physiology. Neurosci. Biobehav. Rev. 23, 925–935.

Lang, F.R., Staudinger, U.M., Carstensen, L.L., 1998. Perspectives on socioemotional selectivity in late life: how personality and social context do (and do not) make a difference. J. Gerontol. 53, P21–P30.

Lansade, L., Bertrand, M., Boivin, X., Bouissou, M.-F., 2004. Effects of handling at weaning on manageability and reactivity of foals. Appl. Anim. Behav. Sci. 87, 131–149.

Lansade, L., Bouissou, M.-F., 2008. Reactivity to humans: a temperament trait of horses which is stable across time and situations. Appl. Anim. Behav. Sci. 114, 492–508.

Lansade, L., Bouissou, M., Erhard, H.W., 2008a. Fearfulness in horses: a temperament trait stable across time and situations. Appl. Anim. Behav. Sci. 115, 182–200.

Lansade, L., Bouissou, M., Erhard, H.W., 2008b. Reactivity to isolation and association with conspecifics: a temperament trait stable across time and situations. Appl. Anim. Behav. Sci. 109, 355–373.

Laporte-Broux, B., Roussel, S., Ponter, A.A., Giger-Reverdin, S., Camous, S., Chavatte-Palmer, P., et al., 2012. Long-term consequences of feed restriction during late pregnancy in goats on feeding behavior and emotional reactivity of female offspring. Physiol. Behav. 106, 178–184.

Lazarus, R.S., 1991. Progress on a cognitive–motivational–relational theory of emotion. Am. Psychol. 46, 819–834.

Lensink, B.J., Veissier, I., Boissy, A., 2006. Enhancement of performances in a learning task in suckler calves after weaning and relocation: motivational versus cognitive control? a pilot study. Appl. Anim. Behav. Sci. 100, 171–181.

Lupien, S.J., McEwen, B.S., Gunnar, M.R., Heim, C., 2009. Effects of stress throughout the lifespan on the brain, behaviour and cognition. Nat. Rev. Neurosci. 10, 434–445.

Madec, I. 2008. Effets du semiochemique MHUSA (Mother Hens' Uropygial Secretion Analogue) sur le stress des poulets de chair. Approches zootechnique, physiologique et comportementale. PhD Thesis, University of Toulouse, France.

Mahony, D.L., Burroughs, W.J., Lippman, L.G., 2002. Perceived attributed of health promoting laughter: a cross-generational comparison. J. Psychol. 136, 171–181.

Malmkvist, J., Hansen, S.W., 2002. Generalization of fear in farm mink, *Mustela vison*, genetically selected for behaviour towards humans. Anim. Behav. 64, 487–501.

Martin, R.A., 2001. Humor, laughter, and physical health: methodological issues and research findings. Psychol. Bull. 127, 504–519.

Mason, J.W., 1971. A re-evaluation of the concept of 'non-specificity' in stress theory. J. Psychiatr. Res. 8, 323–333.

Matthews, S.G., 2002. Early programming of the hypothalamo–pituitary–adrenal axis. Trends Endocrinol. Metabol. 13, 373–380.

Meehan, C.L., Mench, J.A., 2007. The challenge of challenge: can problem solving opportunities enhance animal welfare? Appl. Anim. Behav. Sci. 102, 246–261.

Mehta, P.H., Gosling, S.D., 2006. How can animal studies contribute to research on the biological bases of personality? In: Canli, T. (Ed.), Biology of Personality and Individual Differences. Guilford, New York, pp. 427–448.

Mehta, P.H., Gosling, S.D., 2008. Bridging human and animal research: a comparative approach to studies of personality and health. Brain Behav. Immun. 22, 651–661.

Mellor, D.J., 2012. Animal emotions, behaviour and the promotion of positive welfare states. N-Z. Vet. J. 60, 1–8.

Mendl, M., Burman, O.H.P., Parker, R.M.A., Paul, E.S., 2009. Cognitive bias as an indicator of animal emotion and welfare: Emerging evidence and underlying mechanisms. Appl. Anim Behav. Sci. 18, 161–181.

Mendl, M., Paul, E.S., 2004. Consciousness, emotion and animal welfare: insights from cognitive science. Anim. Welf. 13, S17–S25.

Michelena, P., Sibbald, A.M., Erhard, H.W., Mcleod, J.E., 2009. Effects of group size and personality on social foraging: the distribution of sheep across patches. Behav. Ecol. 20, 145–152.

Mills, A.D., Faure, J.M., 1991. Divergent selection for duration of tonic immobility and social reinstatement behavior in Japanese quail (Coturnix coturnix japonica) chicks. J. Comp. Psychol. 105, 25–38.

Moe, R.O., Nordgreen, J., Janczak, A.M., Spruijt, B.M., Zanella, A.J., Bakken, M., 2009. Trace classical conditioning as an approach to the study of reward-related behaviour in laying hens: a methodological study. Appl. Anim Behav. Sci. 121, 171–178.

Mormède, P., Andanson, S., Auperin, B., Beerda, B., Guemene, D., Malmkvist, J., et al., 2007. Exploration of the hypothalamic–pituitary–adrenal function as a tool to evaluate animal welfare. Physiol. Behav. 92, 317–339.

Murphey, R.M., Moura Durate, F.A., Torres Penedo, M.C., 1981. Responses of cattle to humans in open spaces: breed comparisons and approach-avoidance relationships. Behav. Genet. 11, 37–48.

Müller, R., Von Keyserlingk, M.A.G., 2006. Consistency of flight speed and its correlation to productivity and to personality in Bos taurus beef cattle. Appl. Anim Behav. Sci. 99, 193–204.

Panksepp, J., 1994. Evolution constructed the potential for subjective experience within the neurodynamics of the mammalian brain. In: Eckman, P., Davidson, R.J. (Eds.), The Nature of Emotion: Fundamental Questions. Oxford University Press, Oxford, pp. 336–399.

Panksepp, J., 1998. Affective Neuroscience: The Foundations of Human and Animal Emotions. Oxford University Press, New York.

Panksepp, J., Burgdorf, J., 2003. "Laughing" rats and the evolutionary antecedents of human joy? Physiol. Behav. 79, 533–547.

Pardon, M.C., Perez-Diaz, F., Joubert, C., Cohen-Salmon, C., 2000. Influence of a chronic ultramild stress procedure on decision-making in mice. J. Psych. Neurosci. 25, 167–177.

Paul, E.S., Harding, E.J., Mendl, M., 2005. Measuring emotional processes in animals: the utility of a cognitive approach. Neurosc. Biobehav. Rev. 29, 469–491.

Penson, R.T., Partridge, R.A., Rudd, P., Seiden, M.V., Nelson, J.E., Chabner, B.A., et al., 2005. Laughter: the best medicine? Oncologist 10, 651–660.

Phillips, M.L., Drevets, W.C., Rauch, S.L., Lane, R., 2003. Neurobiology of emotion perception I: the neural basis of normal emotion perception. Biol. Psychiat. 54, 504–514.

Poindron, P., Raksanyi, I., Orgeur, P., Le Neindre, P., 1984. Comparaison du comportement maternel en bergerie à la parturition chez des brebis primipares ou multipares de race Romanov, Préalpes du Sud et Ile de. France. Genet. Sel. Evol. 16, 503–522.

Price, E.O., 1984. Behavioral aspects of animal domestication. Quart. Rev. Biol. 59, 1–32.

Ramos, A., Berton, O., Mormède, P., Chaouloff, F., 1997. A multiple-test study of anxiety-related behaviours in six inbred rat strains. Behav. Brain Res. 85, 57–69.

van Reenen, C.G., O'Connell, N.E., Van der Werf, J.T.N., Korte, S.M., Hopster, H., Jones, R.B., et al., 2005. Responses of calves to acute stress: individual consistency and relations between behavioral and physiological measures. Physiol. Behav. 85, 557–570.

van Reenen, C.G., Hopster, H., van der Werf, J.T.N., Engel, B., Buist, W.G., Jones, R.B., et al., 2009. The benzodiazpeine brotizolam reduces fear in calves exposed to a novel object test. Physiol. Behav. 96, 307–314.

Reisberg, D., Heuer, F., 1995. Emotion's multiple effects on memory. In: Mc Gaugh, J.L., Weiberger, N.M., Lynch, G. (Eds.), Brain and Memory: Modulation and Mediation of Neuroplasticity. Oxford University Press, Oxford, pp. 84–92.

Richard-Yris, M.-A., Michel, N., Bertin, A., 2005. Nongenomic inheritance of emotional reactivity in Japanese quail. Devel. Psychobiol. 46, 1–12.

Rodenburg, T.B., Turner, S.O., 2012. The role of breeding and genetics in the welfare of farm animals. Anim. Front. 2, 16–21.

van Rooijen, J., 1984. Impoverished environments and welfare. Appl. Anim Behav. Sci. 12, 3–13.

Rosen, J.B., 2004. The neurobiology of conditioned and unconditioned fear: a neurobehavioral analysis of the amygdale. Behav. Cog. Neurosci. Rev. 3, 23–41.

Roussel, S., Merlot, E., Boissy, A., Duvaux-Ponter, C., 2007. Prenatal stress: state of the art and possible consequences for farm animal husbandry. Prod. Anim. 20, 81–85.

Rutherford, K.M.D., Donald, R.D., Arnott, G., Rooke, J.A., Dixon, L., Mehers, J.J.M., et al., 2012. Farm animal welfare: assessing risks attributable to the prenatal environment. Anim. Welf. 21, 419–429.

Sander, D., Grandjean, D., Scherer, K.R., 2005. A systems approach to appraisal mechanisms in emotion. Neural Netw. 18, 317–352.

Sandi, C., Loscertales, M., Guaza, C., 1997. Experience-dependent facilitating effect of corticosterone on spatial memory formation in the water maze. Europ. J. Neurosci. 9, 637–642.

Scherer, K.R., 1999. Appraisal theory. In: Dalgleish, T., Power, M. (Eds.), Handbook of Cognition and Emotion, Vol. 30. John Wiley and Sons Ltd, pp. 637–663.

Scherer, K.R. (Ed.), 2001. Introduction to Social Psychology. Blackwell, Oxford.

Schmied, C., Boivin, X., Waiblinger, S., 2008. Stroking different body regions of dairy cows: effects on avoidance and approach behavior toward humans. J. Dairy Sci. 91, 596–605.

Schmutz, S.M., Stookey, J.M., Winkelman-Sim, D.C., Waltz, C.S., Plante, Y., Buchanan, F.C., 2001. A QTL study of cattle behavioral traits in embryo transfer families. J. Hered. 92, 290–292.

Selye, H., 1936. A syndrome produced by diverse nocuous agents. Nature, 32.

Shelton, M., Wade, D., 1979. Predatory losses: a serious livestock problem. Ann. Industr. Tod 2, 4–9.

Sibbald, A.M., Erhard, H.W., Mcleod, J.E., Hooper, R.J., 2009. Individual personality and the spatial distribution of groups of grazing animals: an example with sheep. Behav. Process. 82, 319–326.

Spinka, M., Newberry, R.C., Bekoff, M., 2001. Mammalia, play: training for the unexpected. Quart. Rev. Biol. 76, 141–168.

Spruijt, B.M., 2001. How the hierarchical organization of the brain and increasing cognitive abilities may result in consciousness. Anim. Welf. 10, 77–87.

Spruijt, B.M., van den Bos, R., Pijlman, F.T., 2001. A concept of welfare based on reward evaluating mechanisms in the brain: anticipatory behavior as an indicator for the state of reward systems. Appl. Anim. Behav. Sci. 72, 145–171.

Tallet, C., Veissier, I., Boivin, X., 2008. Temporal association between food distribution and human caregiver presence and the development of affinity to humans in lambs. Dev. Psychobiol. 50, 147–159.

Turner, S.P., Lawrence, A.B., 2007. Relationship between maternal defensive aggression, fear of handling and other maternal care traits in beef cows. Livest. Sci. 106, 182–188.

Valance, D., Boissy, A., Després, G., Arnould, C., Galand, C., Favreau, A., et al., 2008. Changes in social environment induce higher emotional disturbances than changes in physical environment in quail. Appl. Anim. Behav. Sci. 112, 307–320.

van de Weerd, H.A., Docking, C.M., Day, J.E.L., Avery, P.J., Edwards, S.A., 2003. A systematic approach towards developing environmental enrichment for pigs. Appl. Anim. Behav. Sci. 84, 101–118.

van der Harst, J.E., Baars, A.M., Spruijt, B.M., 2003. Standard housed rats are more sensitive to rewards than enriched housed rats as reflected by their anticipatory behavior. Behav. Brain Res. 142, 151–156.

van der Harst, J.E., Baars, A.M., Spruijt, B.M., 2005. Announced rewards counteract the impairment of anticipatory behaviour in socially stressed rats. Behav. Brain Res. 161, 183–189.

Veenema, A.H., Meijer, O.C., de Kloet, E.R., Koolhaas, J.M., 2003. Genetic selection for coping style predicts stressor susceptibility. J. Neuroendocrinol. 15, 256–267.

Veissier, I., Boissy, A., 2007. Stress and welfare: two complementary concepts that are intrinsically related to the animal's point of view. Physiol. Behav. 92, 429–433.

Veissier, I., Boissy, A., Désiré, L., Greiveldinger, L., 2009. Animals' emotions: studies in sheep using appraisal theories. Anim. Welf. 18, 347–354.

Veissier, I., Aubert, A., Boissy, A., 2012. Animal welfare: a result of animal background and perception of its environment. Anim. Front. 2, 7–15.

Vinke, C.M., van Den, R.B., Spruijt, B.M., 2004. Anticipatory activity and stereotypical behavior in American mink (*Mustela vison*) in three housing systems differing in the amount of enrichments. Appl. Anim. Behav. Sci. 89, 145–161.

Waiblinger, S., Boivin, X., Pedersen, V., Tosi, M.-V., Janczak, A.M., Visser, E.K., et al., 2006. Assessing the human–animal relationship in farmed species: a critical review. Appl. Anim. Behav. Sci. 101, 185–242.

Webster, J., 2005. Animal Welfare: Limping Towards Eden. Blackwell Publishing, Oxford.

Weiss, J.M., 1972. Psychological factors in stress and disease. Sci. Am. 226, 104–113.

Wiepkema, P.R., 1987. Behavioural aspects of stress. In: Wiepkema, P.R., Van Adrichem, P.W.M. (Eds.), Biology of Stress in Farm Animals: An Integrative Approach. Martinus Nijhoff Publishers, Dordrecht, Boston, Lancaster, pp. 113–133.

Wilson, S.C., Fell, L.R., Colditz, I.G., Collins, D.P., 2002. An examination of some physiological variables for assessing the welfare of beef cattle in feedlots. Anim. Welf. 11, 305–316.

Wohlt, J.E., Allyn, M.E., Zajac, P.K., Katz, L.S., 1994. Cortisol increases in plasma of holstein heifer calves from handling and method of electrical dehorning. J. Dairy Sci. 77, 3725–3729.

Wright, W.F., Bower, G.H., 1992. Mood effects on subjective probability assessment. Org. Behav. Hum. Decis. Proc. 52, 276–291.

Zebunke, M., Langbein, J., Manteuffel, G., Puppe, B., 2011. Autonomic reactions indicating positive affect during acoustic reward learning in domestic pigs. Anim. Behav. 81, 481–489.

FURTHER READING

Hall, C.S., 1934. Emotional behavior in the rat. I. Defaecation and urination as measures of individual differences in emotionality. *J. Comp. Psychol.* 18, 385–403.

Chapter 4

Genetics and Behavior During Handling, Restraint, and Herding

Temple Grandin* and Mark J. Deesing†
*Department of Animal Science, Colorado State University, Fort Collins, Colorado, USA;
†Grandin Livestock Handling Systems, Inc., Fort Collins, Colorado, USA

INTRODUCTION

The primary objective of this chapter is to discuss the effects of genetics on the behavior of grazing animals during handling and restraint. Some of the species that will be covered are cattle, sheep, horses, and goats. Since the publication of the first edition in 1998, there are numerous new research studies on how genetic factors and early experience affect the behavior of herding animals. In the first edition, the primary emotional variable discussed was fear. Many scientists studying animal behavior assert that emotional experiences of animals are questions science cannot answer since there is no direct evidence of subjective states—i.e., experiencing what they are experiencing. Recently, neurobiological evidence shows that mammals have additional emotional systems that drive behavior (Morris *et al.*, 2011; Panksepp, 1998). The brain mechanisms for psychological experiences exist in some of the most ancient regions of the brain and are similar in mice and men. Panksepp (2011) defined the basic emotional circuits of mammalian brains:

FEAR—Emotion caused by a perceived threat. Sometimes referred to as behavioral reactivity, behavioral agitation, or a highly reactive temperament.
PANIC—Separation distress. Sometimes referred to as social isolation stress or high social reinstatement behavior.
RAGE—A feeling of intense anger.
LUST—Sexual desire or appetite.
CARE—Maternal nurturing
PLAY—A system to help young animals acquire social knowledge

In this edition, research studies will be viewed through the lens of Panksepp's core emotional systems. Although Panksepp's use of the word

Genetics and the Behavior of Domestic Animals. DOI: http://dx.doi.org/10.1016/B978-0-12-394586-0.00004-4
© 2014 Elsevier Inc. All rights reserved.

PANIC to describe separation distress may be confusing, excessive fear or anxiety when an animal is separated from family members or its home environment can be considered panic.

Highly variable results from many studies may be due to confusing fear with separation distress and seeking behavior. These behaviors are controlled by two separate emotional networks in the brain. Research by Faure and Mills (1998; Chapter 8) on quail show that both fear and separation distress are separate systems, and the emotional responses produced by these systems are influenced by genetics. See Chapter 1 for a further discussion of the emotional systems and how they motivate behavior. During our careers we have observed thousands of animals under many different conditions. In this chapter we discuss our observations under field conditions and review research findings which affect behavior during handling, herding, or restraint.

Writing on the behavioral aspects of animal domestication for the *Quarterly Review of Biology*, Price (1984) stated:

It is difficult to generalize about the effects of domestication on either genetic or phenotypic variability because of different selection pressures on different traits and species. However, it is apparent that with respect to animal behavior, domestication has influenced the quantitative rather than the qualitative nature of the response.

To put it in simpler words, domestication changes the intensity of behavioral response. For example, domesticated Norway rats are less cautious than their wild counterparts in a wide variety of situations (Price, 1998). Parsons (1988) further wrote that domestic animals are more stress resistant because they have been selected for a calm attitude toward man. In either event, a genetic predisposition to be fearful or calm interacts with early experience and learning in very complex ways. We use the term temperament in this chapter to refer to an animal's level of fearfulness, which is determined by both genetic and environmental factors. Genetic influences on temperament interact with early experience and learning to shape adult patterns of behavior. During handling for veterinary or other husbandry procedures, fearfulness is a major determinant of an animal's behavior. In the following sections, we will discuss the interacting forces of genetics and experience. Some of the topics that will be covered are flocking behavior, herding behavior, flight zone, social behavior, attraction and reaction to novelty, effect of experiences on ease of handling, heritability of behavior, and sire effect. Some general principles for handling domestic animals will also be discussed. There will also be a discussion of the methods used to evaluate behavioral differences between individual animals of the same breed and differences between breeds.

PRINCIPLES OF HERD BEHAVIOR

Herding behaviors evolved as a defense against predators. A group of animals fleeing a predator will demonstrate herd behavior for protection. Herd animals find safety in numbers. One aspect of herd behavior often noted is that the herd is not completely interested in protection of the group. Instead self-interest is a primary motivator of the individual (Couzin and Krause, 2003; Hamilton, 1971). The selfish herd theory differed from the popular idea that evolution of such social behavior was based on mutual benefits to the population. Herd animals, when they fear a predator work to get into the center of the herd so they are less vulnerable. Genetic factors are responsible for both species differences in herd behavior, and individual differences in how the animals react to the presence of predators or people. Domestic herd animal behavior may be quite different from that of wild herds of the same or related species. Interestingly, people who work with wildlife have observed that African herding animals bunch together more tightly than similar animals in North America. This may be due to more predators such as lions, hyenas, and cheetahs.

Visual and Auditory Senses

With eyes positioned on the sides of the head, herd animals have a wide visual field (Hutson, 1993; Matthews, 1993; Prince, 1977). Panoramic vision enables animals to move together as a herd and constantly scan the surroundings for predators or other dangers. Grazing animals are dichromate's and do not see the color red (Carroll *et al.*, 2007; Jacobs *et al.*, 1998). Dichromatic vision may also provide a greater sensitivity to visual contrast. Research on sheep and horses demonstrate that their visual system is designed for horizontal scanning while grazing (Saslow, 1999; Shinozaki *et al.*, 2010). As herd animals graze on grass with their heads down, elongated horizontal pupils allow them to scan the horizon for movement (possible predators) much better than round pupils. In addition, eye preferences for viewing certain stimuli are common in a wide range of species. The left and right sides of the brain are specialized to process information in different ways and to control different types of behavior (Rogers and Andrew, 2002). The left hemisphere (right eye) plays a role in assessing novelty and the role of the right hemisphere (left eye) in processing negative emotional responses (De Boyer Des Roches *et al.* 2008; Farmer *et al.*, 2010; Robins and Phillips, 2009). Sheep that were separated from a group of sheep had more ear posture changes and a higher percentage of animals with forward ears than feeding sheep (Reefman *et al.*, 2009). Basile *et al.* (2009) found a right-ear, left-hemisphere preference of the auditory response to neighbor versus stranger calls in horses.

When in flight, cattle, deer, and bison all raise their tails. Kiley-Worthington (1976) suggests that the raised tail serves as a signal of danger to the rest of the herd. Social signals that trigger fear represent a particular type of fear-producing stimuli (Boissy, 1995). Research across species highlights the critical role of social fear learning, and the neural mechanisms are beginning to be understood (Olsson and Phelps, 2007).

Herding animals orient and point both their eyes and ears in the direction of any novel sight or sound. Orienting responses alert the animal to possible danger, are accompanied by an internal state of nervous system arousal (Boissy, 1995; Davis, 1992; Rogan and LeDoux, 1996), and elicit responses modulated by the nucleus accumbens in the brain. The nucleus accumbens mediates both fearful (FEAR) motivation toward threats and appetitive (SEEKING) motivation for rewards (Faure *et al.*, 2008; Reynolds and Berridge, 2008). When an animal orients towards a novel stimulus, it may freeze, stand completely still and watch (SEEKING) mode, or go into (FEAR) motivated anti-predator mode. The authors speculate that during the orienting response when animals are watching, the nucleus accumbens remains in SEEKING mode. The nucleus accumbens in the brain acts as a biochemical switch and contains neural circuits that can turn on fearful behaviors or approach-seeking behaviors. Typical fear-motivated behaviors are fleeing, bunching closely together, or fighting when the animal is trapped or defending calves. Both genetic factors and experience affect the response in the nucleus accumbens to either fearful stimuli or the motivation for rewards. For instance, rodents subjected to loud rock music and bright lights were more fearful because the fear-generating zones in the nucleus accumbens have expanded (Reynolds and Berridge, 2008). Conversely, a preferred home environment (familiar, dark, quiet) caused appetitive-generating zones to expand.

Differences in the Strength of Herding Behavior of Cattle and Sheep

The first author has observed that wild ungulates such as the American bison have stronger herding behavior compared to domestic cattle. Separation of a single animal from the herd will cause it to make an intense effort to rejoin its herdmates (Grandin, 1993). Possibly this behavior is driven by the emotion of PANIC (separation distress). Domestic cattle become more difficult to handle and sort when they become agitated and engage in bunching, or milling. Each animal will attempt to push itself into the middle of the group where it will be safe from predators. The strongest animals end up in the middle of the milling herd. In their natural environments, prey species animals such as elk and deer spread apart when grazing a hillside, but at the first sign of danger group closer together and flee as a herd. Domestic cattle living in areas with a large number of predators, such as wolves, graze in tighter groups than cattle reared in areas free of predators (Joe Stookey, personal communication, 2012). Small domestic herding animals such as sheep display similar behavior. When herding animals sense danger, they tend to flock together. Sheep tend to be more reactive and herd together more tightly compared to cattle. Sheep are more vulnerable to predators than cattle, therefore bunch together very tightly to find safety in numbers. Bunching is the only defense sheep have against predators. Cattle often turn and fight. Small, nimble animals such as antelope rely on the ability to quickly flee. Following the reintroduction of wolves in the Yellowstone Park area, ranchers are reporting that Angus cattle are becoming extremely aggressive towards both domestic dogs and wild canines.

Syme and Elphick (1982) found vocal and jumper sheep were seldom the first animals to move through a handling facility. Grandin (1980b) made similar observations in cattle. The wildest and most difficult to handle cattle move through a handling race at the end of the group. This was demonstrated by Orihuela and Solano (1994) with cattle in slaughter plants. Animals at the end of the group took longer to move through the race. Tulloh (1961) reported that Brahman cattle flock more tightly together. Whateley *et al.* (1974) and Schupe (1978) also showed breed differences in flocking behavior of sheep. Rambouillet flock tightly and Chevoits are more independent.

Social Behavior and Handling

When a herd is moving, social rank can determine an animal's position within the herd. Animals with high social rank seldom lead (Hedigar, 1955, 1968). Being a leader is a more dangerous position, so animals with high social rank lag behind or remain in the safety of the middle of the herd. Social behavior also has an influence on learning. Boissy and Le Neindre (1990) found that in tasks where cattle had to press a plate to obtain a food reward, the presence of a companion facilitated learning. Cattle trained with companions learned quickly compared to isolated animals. This was probably due to less separation distress (PANIC). Kilgour and de Langen (1970) further showed that presence of companions lowers secretion of stress hormones in sheep. The ability to detect and respond to signs of fear and pain in other members of the herd has probably conferred a strong selective advantage during evolution. Animals as diverse as primates, birds, mice, cats, and cows learn fear by observing other members of their species (Curio, 1988; Olsson *et al.*, 2007). Munksgaard *et al.* (2001) found the behavior of dairy cows was influenced by observing other members of the herd being handled gently or aversively. In a test of recognition, Elliker (2007) trained sheep to approach photographs of sheep with a calm expression rather than those of sheep with a startled expression. Ear position was the main feature used by the sheep to make this distinction, rather than eye features.

Social learning, social fear learning, and social recognition are familiar to most people that handle or raise domestic cattle and horses. Horse trainers use calm trained horses to help ease the fear of novelty in young inexperienced horses. For example, loading in a trailer for the first time, being ridden for the first time, or even the stress of weaning can be reduced by having a calm companion horse nearby. In Australia, tame "coacher" cattle are used to assist in gathering wild feral cattle (Roche 1988). In the U.S., tame "Judas" horses are used to lead wild (feral) horses into traps or corrals. After the horses are herded by helicopter into an area near the trap corrals, the "Judas" horse is then released leading the wild horses into the corrals. Fordyce (1987) used a few tame steers to facilitate training Brahman calves to novel handling procedures. Petherick *et al.* (2009) found that both good handling practices and increased exposure to people during feeding in a feedlot reduced fear of people. On "Old West" cattle drives, lead steers with a calm temperament were kept around and excitable leaders were destroyed (Harger, 1928). Excitable leaders caused stampedes.

Flight Zone

The flight zone is the distance within which a person can approach an animal before it moves away. Herd animals usually turn and face a potential threat when it is outside of their flight zone, but when it enters the flight zone, the animal turns and moves away. Hedigar (1968) first described the flight zone principle in wild animals. Kilgour (1971) and Grandin (1980a, b) further described flight zone principles for domestic sheep and cattle and showed how working along on the edge of the animal's flight zone, both cattle and sheep can be moved easily (Figure 4.1). Small, flighty herd animals such as deer and antelope have larger flight zones than domestic herding animals (Figure 4.2). They also have higher stress hormone levels than large, heavy animals such as cattle or Cape buffalo (Morton *et al.*, 1995).

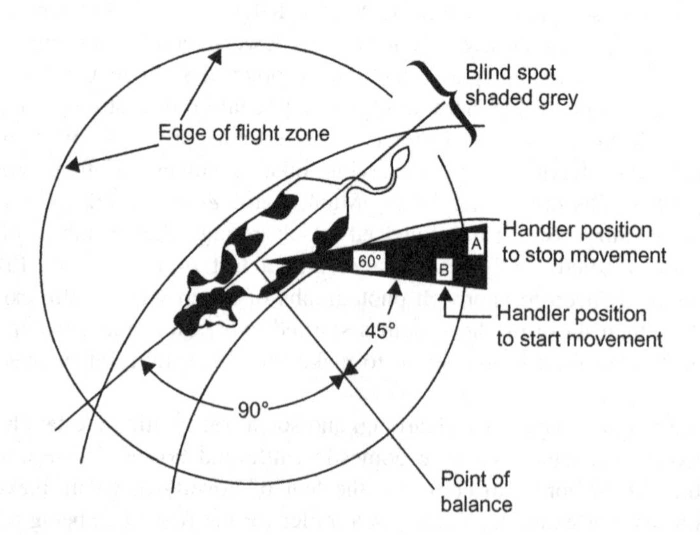

FIGURE 4.1 Flight zone diagram showing the correct positon for a handler to move an animal.

FIGURE 4.2 Flight zone of a large flock of extensively raised sheep. The size of the flight zone varies depending on both genetic factors and previous experiences.

Genetic factors influence individual flight zone sizes between animals of the same species or breed, and between individuals within a species or breed. Flight zone distance is also determined by the risk an animal perceives in a given circumstance. Threats associated with greater risk (speed, size, directness of approach) influence flight initiation distance. In a large group of animals with similar genetics and previous handling experiences, most members of the group have similar flight distances, but a few individuals have either a very small or a very large flight zone. Flight zone size is also strongly affected by experience and learning. An animal that survives a close call with danger learns to be more wary of similar dangers in the future and increases its flight zone size accordingly. Furthermore, a completely tame or trained animal has no flight zone and will allow people to approach and touch it (Price, 1984). Cattle that have people moving around them everyday have small flight zones compared to genetically similar cattle raised on ranches in the mountains who seldom see people. Thus, interactions between genetics and experience in shaping an animal's flight zone size can be complex.

Working the Flight Zone

Flight zone principles can be used by handlers to move single animals or large groups of animals. To move a single animal, the handler positions himself in relation to the animal's point of balance at the shoulder (Kilgour and Dalton, 1984). To move an animal forward, the handler is positioned behind the point of balance. To move the animal backward, the handler moves in front of the point of balance. When the handler is close to the animal, the point of balance is at the shoulder and tends to move towards the eye when the handler is further away (Grandin, 2007; Grandin and Deesing, 2008). When the handler moves rapidly the flight zone will get larger (Fukasawa, 2008). Figure 4.1 illustrates the correct handler positions for moving an animal forward. Controlling the movements of a large herd while on foot, on horseback, or in a vehicle is done by alternately entering and withdrawing from the edge of the herd's collective flight zone. An entire group of animals can be induced to move forward by the handler moving inside the collective flight zone in the opposite direction of desired movement. These positions work on all herd animals.

Dogs trained for herding use the principles of the flight zone and the same positions described above to control livestock movement. Coppinger and Coppinger (1993) suggest that dogs that herd livestock in this manner are displaying a natural hunting strategy used by wolves. Fortunately for the livestock, though, domestic dogs have been bred that do not attack or kill.

MEASURING TEMPERAMENT IN LIVESTOCK

Since the first edition of this book was written in 1998, several hundred papers have been published on temperament measurements in cattle, horses, pigs, sheep, and other domestic animals. Extensive reviews have been written by Burdick *et al.* (2011), Dodd and Pitchford (2012), Forkman *et al.* (2007), Sebastian *et al.* (2011), Turner *et al.* (2011), and Benhajali *et al.* (2010). There is extensive evidence that behavioral variation in temperament may reflect underlying hormonal and neuro-endocrine variation among individuals (Boissy, 1995). A variety of measures are used and there is a need to correctly associate each method with the correct core emotional system. Learning to assess the emotional system being tested may reduce many conflicting findings between studies.

Tests Designed to Measure Fearfulness

Restraint Tests

It is likely that restraint tests mainly measure fearfulness. Behavior is observed when the animal is held in either a single animal scale (Benhajali *et al.*, 2010; Holl *et al.*, 2010), a head gate (head stanchion) (Hoppe *et al.*, 2010), or tightly held in a squeeze chute (Voisinet *et al.*, 1997 and Baszczak *et al.*, 2006). One of the first chute scoring methods used a seven-point scale ranging from calm to highly agitated (Fordyce *et al.*, 1988). Grandin (1994) used a five-point scale. A four-point scale is recommended because it eliminates a middle score. The scores are: (1) calm; no movement; (2) intermittent movement; (3) continuous movement; and (4) violently struggling and attempting to escape (Grandin and Deesing, 2008). Hall *et al.* (2011) successfully used a five-point scale for assessing the temperament of pigs held loosely in a single animal scale. Restlessness and struggling is scored. Individual differences in fearfulness and reactivity may be easier to differentiate when the animal is held loosely in a single animal scale or head gate. Sebastian *et al.* (2011) suggests replacing these subjective measurements with objective strain gauges which measure the force exerted against the head gate by the animal. Another possible objective measurement method would be a computer program used to analyze shifts in weight to measure sheep movement in a single animal scale (Horton and Miller, 2011). Many smart mobile telephones are equipped with accelerometers which can also be used to measure how much an animal shakes a restraint device (John Church, personal communication, 2012). Aps originally designed for monitoring earthquakes can be easily downloaded from the internet. A phone with this application could be used to measure shaking by taping it to the framework of the chute. Tests that measure movement in a squeeze chute or scale may work poorly on purebred *Bos indicus* cattle such as Nelore if they have been subjected to a painful procedure, such as dehorning. *Bos indicus* may exhibit tonic immobility and freeze when they become extremely fearful. Tonic immobility is not likely to occur in crosses between *Bos indicus* and *Bos Taurus*, and almost never occurs in the *Bos Taurus* (English or European) breeders of cattle.

Exit Speed Tests

Exit speed tests are also called flight speed tests. It is likely that these tests mainly measure fear. The speed an animal exits a squeeze chute or single animal scale is measured with either an electronic device (Burrow and Dillon, 1997), or by gaits such as walk, trot, run (Baszczak *et al.*, 2006; Lanier and Grandin, 2002). In some studies, exit speed tests have been shown to be more accurate compared to restraint tests (Café *et al.*, 2010). This may be caused by cattle being too tightly restrained. In other studies where chute scoring was able to discriminate individual differences, the animals were restrained loosely. Hydraulic squeeze chutes that hold animals very tightly and restrict movement reduce the effectiveness of subjective scoring during restraint, making it difficult to differentiate between differences in behavior. In studies where restraint tests were effective for differentiating differences in behavior, researchers usually used looser restraint methods such as a single animal scale, a head gate or a manually (hand) operated squeeze chute (Benhajali *et al.*, 2010; Holl *et al.*, 2010; Voisinet *et al.*, 1997). Studies done by Baszczak *et al.* (2006) and Vetters *et al.* (2013) were conducted using hydraulic squeeze chutes that held cattle very tightly and exit speed scoring was shown to be more effective. Brazilian studies by Sant-Anna *et al.* (2012) report that subjective behavior scores were effective. It may be that the Brazilian restraint device held cattle more loosely than a hydraulic squeeze.

Exit speed scoring may not be accurate in Nelore or other purebred *Bos indicus* cattle if the animals freeze in response to handling stress. This is most likely to occur in Nelore which have been subjected to rough handling.

Startle Tests

Startle tests mainly measure fear. In a startle test, an animal's reaction to a sudden novel stimulus is used. Some examples of a sudden novel stimulus are sudden opening of an umbrella, dropping a heavy chain, or a person stamping his feet. The reaction of pigs to a person suddenly stamping his foot was very effective for differentiating the behavior of different genetic lines of pigs (Lawrence *et al.*, 1991). Startle tests are most appropriate for "one-time" temperament tests. Animals are likely to habituate quickly to repeated exposure to sudden novel stimulus.

Tests that Measure Separation Distress, Fear, and Seeking

Isolation Tests

Isolation tests measure PANIC (separation distress). In a typical isolation study, animals are separated from herd mates and placed alone in a pen. Vocalization and locomotor agitation is scored. In sheep, Basile *et al.* (2009); Lansade *et al.* (2008); Ligour *et al.* (2011); found that vocalization (bleats) was not correlated with a tendency to avoid or approach people. Isolating single sheep in an arena may be measuring two different emotional systems. The vocalization was probably measuring PANIC (separation distress) and the tendency to either approach or avoid people may be due to FEAR. In a heritability study of emotional reactivity, Boissy *et al.* (2005) exposed lambs to three challenging tests involving novelty, human contact, and social isolation. The highest heritability was for high bleats ($h2 = 0.48$) during social isolation. In horses, the frequency of vocalization (neighing) during isolation was stable over time (Lansade *et al.*, 2008). The research also measured attraction towards other horses (gregariousness) and concluded that the trait of gregariousness is stable over a series of tests from 8 months to 2.5 years (Lansade *et al.*, 2008). Possibly gregariousness and PANIC (separation distress) are separate but linked traits.

Open Field Tests

Open field tests may elicit a mix of emotional system responses: FEAR (a strong emotion caused by anticipation or awareness or danger), but may also measure SEEK (intensively energized to explore) and PANIC (separation distress). In an open-field test, an animal is placed in an arena and its activity or the distance it walks is measured. Beausoleil *et al.* (2008) tested Merino sheep bred for either low or high activity in an open field. With a stationary person standing inside the arena, the more active genetic line walked around more, vocalized (bleated) more, and spent more time closely investigating the person in the arena. Analysis showed two emotional systems affecting behavior of the more active genetic line. The active genetic line was less fearful and more exploratory compared to the less active line.

In a similar study using crossbred sheep which were not selected for high or low activity, Wolf *et al.* (2008) report the existence of considerable genetic variation in behavior. In addition, behavioral measures of locomotion/vocalization and proximity to the human are genetically and environmentally uncorrelated and represent independent measures of emotional state. The authors hypothesize that locomotion/vocalization in an open field is probably due to separation distress (PANIC).

Approach a Novel Objective or Novel Person Test

The tendency to approach or avoid a novel object or a person may be motivated by the emotion systems FEAR, SEEK, or PANIC. Previous experience and genetics strongly influence an animal's response to this type of test. In a typical novel object test, an object the animal has never encountered is placed in an arena and the time it takes to approach and/or touch the object is measured. Kilgour *et al.* (2006) conclude that in order to assess individual differences in beef cattle to conditions of social isolation, human proximity, novelty, and restraint, the tests that best do this are the restraint test, the open-field test, the following test, flight time, the fear of humans test and flight distance.

Pen Testing

Pen testing measures FEAR, PANIC (separation distress) and Aggression. In this test, cattle are placed in a small pen such as an auction arena and actively approached by a person. In France, this test is used by Limousin breeders to cull animals with a bad temperament. Cattle that charge people are not kept for breeding. Variations of the test are also used to determine size of the animal's flight zone.

Other Behavioral Indicators Which May Indicate Genetic Differences in Fear

Eye White

The percentage of eye white showing in cattle is highly correlated with exit speed score (Core *et al.*, 2009). Eye white percentage likely measures fear because anxiety-reducing drugs, such as diapazen, decrease the percentage of eye white that shows (Sandem *et al.*, 2006).

Physiological Measures

Animals highly agitated during restraint or with high exit speeds often have higher levels of the stress hormone cortisol in blood, and/or increased heart rate (Curley *et al.*, 2004; King *et al.*, 2006). Genetic factors influence physiological measures of handling stress. Zavy *et al.* (1992) found breed differences in cortisol (stress hormone) levels between *Bos taurus* and *Bos indicus* cattle. Furthermore, Zavy *et al.* (1992) found higher cortisol levels Brahman cross calves compared to crosses of Angus and Hereford. Stricklin *et al.* (1980) report Angus cattle having higher cortisol levels than Herefords. In Merino sheep, isolation stress increased plasma glucose 27% (Lennon *et al.*, 2009). Pigs that squeal, jam in a chute, or were electric prodded had significantly elevated blood lactate (Edwards *et al.*, 2010). Blood glucose and lactate are simple, economical tests but must be used IMMEDIATELY after an animal is subjected to restraint, isolation or handling.

Syme and Elphick (1982) used a commercial heart-rate monitor to study heart rate and handling stress in merino sheep. The sheep were all neutered males raised on the same farm. Temperament classifications were (1) quiet, no response to being separated from the group, (2) jumpers, which attempted to jump out of the pen after being separated, and (3) vocalizers. Quiet sheep had lower heart rates compared to jumpers and vocalizers. Syme and Elphick (1982) may have been measuring separation distress (PANIC) instead of fear. McCann *et al.* (1988a,b) found a similar correlation between emotionality scores and heart rate in horses. Tests of emotionality included response to being herded, isolation, and being approached by people. Higher heart rates were recorded in horses having strong reactions to isolation and being approached by people.

Production Traits

Restraint and exit speed tests in cattle, sheep, and other animals have been used to study temperament correlations with production traits. Animals with fearful, reactive temperaments gain less weight, have more dark cutters, and other meat quality problems (Café *et al.*, 2010; Hall *et al.*, 2011; Holl *et al.*, 2010; Pajor *et al.*, 2008; Voisinet *et al.*, 1997).

Relationship between Fear and Attraction to Novelty

Highly reactive wild herd animals and highly reactive farm animals are both attracted to novelty and fearful of it (see Chapter 1). Lawrence *et al.* (1991) found that reactive (fearful) pigs are most attracted to novelty (seeking). Pigs most likely to approach a novel bucket placed in their pen also showed the most intense startle responses to a person stamping his foot. In an open field test, highly reactive pigs also display behavior differences compared to non-reactive pigs. Dantzer and Mormede (1978) found that highly reactive Pectrain pigs walk around more in a novel arena compared to placid Large White Pigs. Pictrain pigs also learn an active avoidance task quicker than Large White pigs. William *et al.* (1964) also found breed differences in learning active avoidance tasks between excitable Hampshire piglets and placid Durocs. Hampshire piglets learn to avoid shocks quickly compared to Durocs. The breed effect was greater than the effect of individual differences between litters of the same breeds. Highly reactive pigs are vigilant and intensely aware of changes in their environment. Vigilance and reactivity may facilitate certain types of learning. The first author has observed that by confronting an animal with sudden novelty, the animal is more likely to show its true reactivity or level of fearfulness. Flight zone is possibly more affected by learning than reaction to a sudden novelty. For example, Hereford cattle, a generally calm, placid breed, may have large flight zones when raised where they seldom see people. But even extensively raised Herefords seldom become highly agitated and struggle violently in a squeeze chute. On the other hand, tame halter broke individuals of Saler or Limousin cattle in the U.S. are much more likely to go berserk when suddenly confronted with the novelty of an auction ring, compared to extensively raised Herefords.

Animals with a Fearful Temperament Have More Violent Reactions to a Sudden Novelty

Both authors have observed that the most reactive and fearful animals of a group are more sensitive to changes in the environment, and usually first to orient toward novel sights or sounds. Highly reactive herd animals constantly rotate their ear positions, raise their head quickly from the ground when grazing, excessively flick their tails, flinch when touched, or move away when approached by people. These behaviors are all signs of a highly vigilant reactive nervous system. Research studies and anecdotal observations suggest that responses to sudden placement in a novel environment, or reactions to sudden novel stimulus are very effective for identifying highly fearful animals. Cattle entering a novel auction ring and suddenly being confronted with people and noise, is analogous to being placed in a novel open field test, or suddenly stamping your foot in a group of pigs. Animals with genetically higher levels of nervous system reactivity and fearfulness are more likely to have violent reactions to a sudden novel stimulus (Grandin, 1997). In sheep, there is a relationship between high cortisol levels after slaughter and increased vocalization and vigilance during on-farm behavior tests (Diess *et al.*, 2009). It is likely the sheep in this study were reacting to the sudden novelty of the slaughter plant.

Animal Reaction to Novelty is Very Specific

Animals do not generalize well between different objects or stimuli. Leiner and Fendt (2011) find that horses habituated to a stationary blue and white umbrella show fear to an unfamiliar orange tarp. After habituation-training to the object, the fear-response to the object is specifically attenuated whereas the fear response to another object remains. Cattle with a very small flight zone when handled by a man on horseback have a much larger flight zone when moved the first time by a man on foot (Grandin and Deesing, 2008). Christensen *et al.* (2008) found that habituation to a novel object is stimulus specific. Horses tested with balls, boards, cones, or boxes were less fearful when all the objects were the same color. The first author hypothesizes that animals are sensory-based thinkers and remember objects as specific pictures (Grandin and Johnson, 2005).

Temperament and Strength of Emotional Responses are Highly Heritable

Temperament in cattle is highly heritable. The heritability estimate of temperament in dairy cattle is 0.40 (O'Blesness *et al.*, 1960), 0.53 (Dickson *et al.*, 1970), and 0.45 (Sato, 1978). In beef cattle, the estimates are 0.40 (Shrode and Hammack, 1971) and 0.48–0.44 (Stricklin *et al.*, 1980). The numbers of vocalizations in sheep isolated from flockmates during testing is highly heritable (Boissy *et al.*, 2005). The trait of fear of humans is also highly heritable (0.38) (Hemsworth *et al.*, 1990). Despite the abundance of evidence indicating that individual differences in temperament are genetically influenced, behavioral genetics research has more to offer the study of temperament than estimates of genetic influence. Environmental contributions to temperament are equally important, namely, does temperament change across age and through experience? Intuitively, one might guess that as an animal matures and becomes more interactive with increasingly diverse environments, the role of genetic factors on temperament might diminish. Although limited research has focused on whether temperament can be modified, Petherick *et al.* (2009) found that fear of people, flight speed, and the amount the cattle moved around a test arena was reduced by good handling.

Field Observations on Breed Differences in Cattle and Sheep Behavior

Cattle with Brahman genetics are generally considered "flighty" compared to breeds with British genetics such as Angus and Hereford. However, Brahman cattle are more inquisitive and investigate or follow a person or a dog. The tendency to follow a person is much greater in Brahman compared to British or European Continental breeds. Hohenboken (1987) found Brahmans can become extremely docile if handled gently. The first author has observed that Brahman cattle seek stroking from people more than Herefords. In Australia, a common practice used to move groups of Brahmans is allowing them to follow a person.

The author's hypothesize that *Bos Indicus* breeds may have stronger SEEKING and possibly stronger PANIC (separation distress) compared to *Bos Taurus* cattle. Murphey *et al.* (1980, 1981) discussed behavioral differences between several breeds in flight zone size, and the tendency to approach novel objects or a man lying on the ground. Breeds with the largest flight zones had the strongest tendency to approach novel objects. This observation is only true if animals are allowed to voluntarily approach novel objects. During forced movements when animals are being driven toward a novel object, just the opposite is true. The excitable individuals are more fearful and more likely to "spook" or balk. The second author has observed similar breed differences in the temperament of horses. High-strung breeds such as Arabians or Thoroughbreds take longer to habituate and stop fearing a novel situation such as new riding arenas, transport vehicles, or a strange person. Calmer, more placid animals habituate more easily to novel stimulus.

Many well-documented examples report differences in temperament between different breeds of cattle. Stricklin *et al.* (1980) report that Herefords are the most docile British breed and Galloways the most excitable. Early studies show that Angus and Angus cows are more temperamental than Herefords (Stricklin *et al.*, 1980; Tulloh, 1961; Wagnon *et al.*, 1966). Angus cattle dominated Hereford cattle in social rank even though they weighed less. Fordyce *et al.* (1988) found that Brahman cross cattle become more agitated during restraint than Shorthorns. Research by Voisinet *et al.* (1997) had similar results. Braford, SimmentalRed Angus, Red Brangus, and Simmental × Brahman cattle became more agitated during restraint than Angus, Simmental × Red Angus and Tarentais × Angus. Numerous studies show that English breeds such as Hereford or Angus are calmer than continental breeds such as Limousin or Brahma- cross cattle (Baszczak *et al.*, 2006; Hoppe *et al.*, 2010). Sheep breeds bred for meat such as Suffolk or Texel, became more agitated in an auction ring compared to Merino and Corriedale breeds bred for wool (Dias Barbosa Silveira *et al.*, 2010).

Problems with Crossbred Cattle

In 1993, the first author began to hear reports from cattle feeders in the United States about certain cattle unable to adapt to a feedlot. Most newly arrived cattle in feedlots adapt to vehicles and people within a few days. However, feedlot operators began to notice some crossbreeds with European Continental genetics Limousin or Saler were unable to adapt to the feedlot. Some animals had to be removed from the feedlot and returned to pasture. The breed associations responded to these problems by introducing to ranchers programs which used restraint tests to evaluate their cattle. In 2012, Limousin cattle are on average are much calmer compared to herds in the early 1990s. Further observations by the first author suggest that behavior during handling or restraint of highly reactive European Continental crosses is different from the behavior of Brahman or Brahman-cross cattle. For example, Brahman cattle are seldom self-destructive when they become agitated. However, the first author has observed European Continental crosses more likely to injure themselves and panic during handling in squeeze chutes or truck-loading ramps. When Brahman cattle become extremely agitated they are more likely to lie down and become immobile. People in Brazil report similar tonic immobility behavior in *Bos indicus* Nelore cattle. Certain genetic lines of European Continental cattle are more excitable than British breeds (Grandin, 1994; Stricklin *et al.*, 1980).

Sire Effect on Behavior

Studies of sire effect on behavior clearly show in differences in behavior of the offspring of different sires which cannot be attributed to learning the behavior from the mother. Dickson *et al.* (1970) found that sire had a significant effect on milking parlor behavior and temperament of cows. Arave *et al.* (1974) showed that Holstein heifers from certain sires had higher activity levels and greater learning ability. Both ranchers and dairymen learn from practical experience that calves from certain sires are more nervous and excitable. Torres-Hernandez and Hohenboken (1979) found a similar relationship in sheep. Ewes by Romney sires are more agitated when placed in a stressful test situation compared to ewes sired by Suffolk or Columbia ewes. Beef cows sired by cattle developed for use in the mountains graze more frequently on hilly, rugged terrain compared to cows sired by breeds such as Angus not selected to live in rugged terrain (Van Wagoner *et al.*, 2006).

INDIVIDUAL DIFFERENCES WITHIN A BREED

In "Behavior and Evolution" Mayr (1958) wrote,

The time has come to stress the existence of genetic differences in behavior, in view of the enormous amount of material the students of various forms of learning have accumulated on genetic variation in behavior. Striking individual differences have been described for predator–prey relationships, for the reactions of birds, or mimicking, or to warning colorations for child care among primates, and for maternal behavior in rats. It is generally agreed by observers that much of the individual difference is not affected by experience but remains essentially constant throughout the entire lifetime of the individual.

Individual differences in animal behavior have been widely discussed by Hirsh (1963), Scott and Fuller (1965), Fuller and Thompson (1978), and Plomin (1990). Livestock animals show individual differences in behaviors similar to those found in laboratory animals and dogs. Kerr and Wood-Gush (1985) investigated a relationship between individual behavioral differences and production measures in dairy cows. They propose that behavior of dairy heifers in one context can be used to predict behavior in other situations. Furthermore, Kerr and Wood-Gush (1987) found that an individual dairy heifer's reaction to being touched by a human observer showed a consistent individual pattern that remained consistent for a two-year period. In young horses from 9 to 22 months of age, Visser *et al.* (2001) measured individual variation and consistency of behavioral variables related to temperament. One out of four temperament traits, "flightiness" (novel object test), proved to be consistent over time.

Kilgour (1975) used the open-field test as an assessment of temperament in dairy cows. Le Neindre *et al.* (1995) documented individual differences in docility of Limousin cattle. In this study, cattle were separated from herd mates by a person working their flight zone. Individuals were considered aggressive if they lowered their heads or charged the handler. This test may have been measuring separation distress (PANIC) instead of aggression. Murphey *et al.* (1980) found he could approach dairy breeds more closely than beef breeds. Individual differences also affect grazing behavior. In cattle with similar genetics, some individuals prefer to graze on steeper slopes compared to other individuals that prefer flatter ground (Bailey *et al.*, 2004).

Grandin (1992) assessed the temperament of bulls during restraint in a squeeze chute four times at 30-day intervals, and concluded that in order to obtain accurate temperament evaluations on individual animals, each animal should be scored more than once. Temperament scores were stable over time for the calmest individuals and stable over time for the most agitated individuals. A test conducted four times easily identified the extremely calm individuals versus the extremely agitated. Wild Bighorn sheep show a similar stability of temperament over time. Reale *et al.* (2000) reported that behavior during capture was similar during repeated captures.

Effect of Pig Genetics on Behavior During Handling

Using several different handling tests, Lawrence *et al.* (1991) compared two separate groups of modern hybrid Cotswold Landrace × Large White pigs. Both groups were reared under the same conditions, but differed in their behavior. The first author also observed differences in behavior of the same genetic line of hybrid pigs on different farms. Pigs from one farm were calmer than pigs from another farm. The first author speculates that differences are due to differences in how the managers selected breeding stock and differences in handling. One farm had a scale that was difficult to read if a pig became highly agitated in it. This may have caused the manager to select calmer breeding stock.

Handling tests used by Lawrence *et al.* (1991) were: (1) willingness to leave the home pens; (2) movement ease through a hallway; (3) response to a suddenly approaching human; (4) restraint-resistance using a snout snare; and (5) vocalization during restraint. With the exception of vocalization, all tests had a strong tendency to correlate. The single-best predictor of temperament was a suddenly approaching human who stamped his foot. The response was scored on a five-point scale: (1) no response; (2) mild flinch; (3) animal backed away from the person; (4) animal backed away and squealed; and (5) animal attempted to escape. Lawrence *et al.* (1991) concluded that the temperament differences between individuals were stable over time. Yoder *et al.* (2010) divided pigs into vocal, or not vocal categories while loading on a single animal scale. Landrace pigs vocalized compared to Chester Whites, Durocs, or Yorkshires. Yorkshire pigs were also more difficult to load onto the scale.

Lean, Rapid-Growing Pigs are More Excitable

Ultra-lean hybrid pigs with very little back fat became popular in the United States during the mid-1990s. The first author has observed some genetic lines of lean hybrid pigs are much more excitable during handling compared to fatter breeds of pigs. She speculates this effect cannot be attributed entirely to early rearing environments. Striking differences in pig behavior occur in lean hybrids of different genetic backgrounds raised in the same building operated by the same people. Many pork producers have commented that some genetic lines of lean hybrids are more excitable than older crossbred pigs with large amounts of back fat. The differences in behavior were first noticed when producers changed breeding stocks. Lean hybrids are more reactive and more likely to squeal when touched. Research by Holl *et al.* (2010) helps confirm these observations. They found pigs with a docile calm temperament and more backfat grew more quickly. Leaner pigs in a composite population of Duroc, Large White, and Landrace became more agitated on a single animal scale. Pigs with a very excitable temperament are much more difficult to handle. Grandin (1994) observed that excitable pigs can be moved easily if the handler moves slowly and avoids sudden movements. Sudden movements cause excitable pigs to bunch together and pile up.

Lean hybrid pigs cause serious handling problems in slaughter plants which process 1000 pigs per hour on a single line. The reactive pigs balk more and are difficult to drive. Most large slaughter plants were forced to install a second complete handling and stunning systems in order to handle the lean hybrids at a rate of 500 pigs per hour in each system.

Behavior Problems in Crossbred Beef Cattle During the 1990s

Behavior problems with crossbred cattle have been observed by breeders. Bonsma (1975a, b) stated that Brahman crossbreds have more temperament problems than purebred Brahman. The first author has made similar observations in crossbreeds of Brahman and Hereford cattle. Compared to the purebred parents, crossbred calves are more difficult to halter break. The authors speculate that when Brahman are crossed with Hereford or Angus, emotional traits of either separation distress (PANIC) or CARING is reduced. In the early 1990s, there were problems with very excitable beef cattle in the U.S. that were likely to become extremely agitated when suddenly confronted with the novel surroundings of an auction market or slaughter plant (Grandin, 1994). Problems with highly reactive animals are most evident in cattle breeds imported from Europe and crosses of these breeds. Observations in feedlots and slaughter plants by the first author indicate that crossbred animals are more likely to become agitated compared to purebred European Continental cattle. These cattle were introduced to extensive ranching operations in the United States during the late 1960s and early 1970s, and came into favor when the drive for lean beef motivated producers to select for lean rapidly growing cattle.

An example of genetic selection for leanness which causes handling problems in a slaughter plant was observed by both authors. Three hundred highly excitable fine-boned heifers of Saler and British genetics were impossible to handle calmly. Some engaged in self-destructive behavior. One heifer tore off her front leg, and two others had ripped off pieces of a hoof. This occurred after they stepped in a crack between the loading ramp and the truck which caused them to react with instant panic. Cattle with a calmer temperament are more likely to stop and withdraw their foot from the crack. One heifer fell down and became so fearful that she thrashed around on the floor and was unable to make coordinated movements to get up. Another pounded her head against the side of the chute after she fell down. The heifers constantly bellowed and if touched they kicked with both back feet. When a plant employee leaned over the side of the race to encourage one to move, she responded by kicking with both back feet to a height of about 7 feet (2 m), narrowly missing his head. One hind foot caught the brim of his hard hat with such force that the hat flew high in the air. This type of extremely agitated behavior has a strong genetic basis. Cattle with calmer temperaments do not display such frenzied behavior, even when they originate from a remote ranch and have minimal previous contact with people. The first author has observed thousands of cattle move through this particular facility. In several years she had never observed such berserk behavior. This type of behavior is similar to that of a high-strung horse. Compared to cattle, horses are more likely to became extremely fearful and injure themselves.

ANIMAL SIZE, BODY SHAPE, COAT COLOR, AND TEMPERAMENT

A phenotype is the composite of an animal's observable characteristics or traits: such as its morphology (body shape and size), biochemical or physiological properties, coat color, phenology (biological phenomenon correlated with climatic conditions), behavior, and temperament. Phenotypes result from the expression of an organism's genes as well as the influence of environmental factors and the interactions between the two. Phenotypic variation (due to underlying heritable genetic variation) is a fundamental prerequisite for evolution by natural selection. Within any given population, there are multiple potential sources of phenotypic variation, and each of these sources reflects a different underlying cause.

One source of variation is found in Bergmann's rule. This principle that states that within a broadly distributed genus, species of larger size are found in colder environments, and species of smaller size are found in warmer regions. These differences are described by Goodwin (2002), and Moen (1973) identified a geographic cline (gradual change) that affects morphology and behavior. Individuals from the northern extent of the range tend to be stockier—with adaptations for energy conservation in cold climate—and to have less reactive temperament. Individuals from the southern extent tend to be more gracile (slender) in build—with adaptations to heat dissipation in hot climate and a tendency to be more reactive. A clear example of this principle can be seen in differences between *Bos Tauras* (English and Continental breeds) and *Bos Indicus* (Brahman) cattle. The phenotypic variation is obvious in body shape and size, coat colors, and temperament.

Considerably more phenotypic variation exists in domesticated species like dogs than in non-domesticated species. Domesticated species differ from their wild ancestors in a number of physical characteristics and traits, generally referred to as the domesticated phenotype. Domestication produces simultaneous changes in a number of both physiological and behavioral traits (Kukova, *et al.*, Chapter 10). It has been shown that selection for one trait associated with the domestic phenotype can lead to simultaneous changes in others.

Body Type and Temperament in Domestic Animals

Body shape in animals is related to temperament. Krushinski (1961) reported that slender, narrow-bodied dogs had increased excitability compared to "athletic wide-bodied" dogs. The narrow-bodied dogs were more fearful. Martins *et al.* (2009) used 11 body measurements in Braford cows and classified temperament using both a restraint test and an exit speed test. Animals of big body size performed better in the temperament tests compared to the groups with medium and small body size. The first author has observed that selecting pigs and cattle for lean meat and thin back fat has resulted in animals that are more reactive to sudden novelty. This idea is supported by Holl *et al.* (2010) who found that leaner pigs were more agitated on a weigh scale. When animals are selected for leanness and bulging muscles they tend to be less reactive than lean, smooth-bodied slender animals. However, both types of lean animals are probably more reactive than animals with more body fat. Holmes *et al.* (1972) found that double-muscled Angus cattle had more excitable temperament scores. The first author observed that Charolais cattle bred in Quebec, Canada, with heavy bones and muscular bodies are very calm at a noisy livestock auction compared to slender lean Charolais cattle bred in the U.S. When we looked at the different body shapes and sizes of different breeds of pigs, cattle, and horses, the physique–temperament relationship was obvious. Lanier and Grandin (2002) found that beef steers with thicker, wider cannon bones exited a squeeze chute more slowly than steers with thinner bones. Lighter breeds of sheep have stronger flocking behavior and larger flight zones when confronted with threatening stimuli (Hansen *et al.*, 2001). In this study a stuffed lynx or bear on a trolley caused a bigger reaction than a rain poncho on a trolley.

Another example of the physique–temperament relationship is seen in highly prolific Chinese pigs. The sows have an abundance of fat and bear many offspring. These animals are very placid and non-reactive, whereas pigs carrying the porcine stress gene are slender, have low amounts of fat, and are more reactive. Pictrain pigs and the Chinese breeds are extreme examples of the relationship between temperament and body morphology.

Type and Temperament in Wild Herd Animals

Body size and shape of wild herd animals is also related to the temperament. Highly reactive, delicate, lightweight antelope and deer survive in the wild by fleeing quickly from predators. Heavy animals like cattle and bison survive by either attacking or fleeing. However, one of the heaviest herding animals, the Cape buffalo, survives by attacking because it is too heavy to flee. Fine-boned, defenseless animals such as antelope have the most excitable temperaments. This is especially true if they live in environments with many predators. Small species of antelope are hypervigilant and equipped with large ears. This enables them to orient and localize faint sounds of danger. Any change in the environment results in instant orientation and possible flight. Larger, heavier animals are more placid but can be dangerous to handle in a confined space. They are more likely to attack if handlers enter their flight zone. In handling facilities, the American bison is one of the most dangerous animals. The first author has observed bison butt a fence with their heads up to 20 times. Bison have also been known to eviscerate a horse with their upward curving horns. They appear to be intermediate in the flight/fight survival strategy. Bison will flee from danger, but being fairly heavy, they are known to attack when unable to flee. In a variety of hoofed animals, Morton *et al.* (1995) measured the stress hormone cortisol. This research showed that large animals such as the Cape buffalo are less stressed by capture and restraint compared to flighty antelope species such as the nyala. When 18 different species of African animals were captured, the Cape buffalo was the only species in which cortisol levels did not rise.

Coat Color and Temperament

Agouti is the natural color of wild rats. Instead of being solid, each individual hair has three bands of color. The typical Agouti color is a rich chestnut with dark slate at the base of the hair. The genetics of Agouti are very simple. The Agouti has two alleles "A" Agouti, and "a" non-Agouti. Dominant "A" produces the banding of hair we recognize as Agouti. Recessive "a" produces solid colored hairs. Many coat color mutations appear in Agouti rats bred in captivity: albino, black, hooded, and others. In a population of rats bred in captivity for 35 generations, Keeler (1947) noted that some of these mutations were associated with varying degrees of tameness. Black rats were much tamer than the agoutis. They described the black rats as "Already tame by nature. If very excited they may click their teeth, but are not apt to bite." They conclude that the strain of tame albino rat (homozygous for black and hooded), "was not domesticated by painstaking selection over long periods of time, but was modified in morphology principally by the introduction of three coat color genes [albino, hooded, and black], and in behavior principally by the (non-agouti) black gene." Breeding for the black non-agouti coat color in rats resulted in tamer, easier to handle rats (Keeler, 1942). In foxes, graded differences in temperament were found between foxes selected for different coat colors (Keeler, 1975; Keeler *et al.*, 1968). Foxes that were all raised in the same wooded compound had flight zone differences that ranged from silvers (200 yards), to platinums (100–200 yards), to ambers (3–100 yards). It is interesting to note that the foxes became progressively tamer as more and more mutant coat color genes were added. Tameness was also highly correlated with adrenal gland weights. The size of the adrenal gland decreased and the body weight increased as the foxes became tamer. The animals with the most mutant coat-color genes were the tamest.

Trut *et al.* (1997) selected wild agouti rats for tameness for over 30 generations. In the beginning of the experiment, white patches of hair began to appear on the rats and increased until about 73% of the rats had white bellies. Over time, the white patches on the piebald rats grew larger and larger. In sum, selection for tameness correlated with the frequency and extent of depigmentation.

Genetic selection for body traits has an effect on behavioral traits, and selection for behavioral traits leads to changes in body traits (Belyaev, 1979). Coat color and behavior are connected via a common biochemical synthesis pathway of the pigments determining color—the melanins—and the catecholamide group of neurotransmitters forming the basis of the information processing system (Hemmer, 1990). In artificial selection studies conducted for the U.S. Department of agriculture on over 34,000 guinea pigs, Wright (1978) showed that selecting for different hair colors changed many other characteristics. This early research revealed patterns of associated changes in body shape, relative size of certain internal organs, and changes in temperament. In many cases, coat color can be used as a genetic marker for such traits (Keeler, 1975). Coat color is associated with differences in physiology, morphology, and behavior in foxes (Belyaev, 1979; Keeler *et al.*, 1968; Trut, 1999), guinea pigs (Wright, 1978), Cocker Spaniel dogs

(Perez-Guisado, *et al.*, 2006; Podberscek and Serpell, 1996), and cats (Kendall and Ley, 2006).

Both authors have heard reports from dairymen that white Holsteins are flighty and often more dangerous to handle than more pigmented cows. These casual observations made by dairymen may be explained partially by findings from studies on coat color–temperament relationships in other animals. Tozser *et al.* (2003) used a scale test and exit-speed scores to assess temperament in a group of Red and Black Angus bull calves. In this sample the red bulls were calmer than the black bulls in the temperament score, and were found easier to handle by the Stockmen. Experiments on horses also demonstrate that chestnuts are more reactive than bays and dark-colored sheep and goats were more motile than white ones (in Hemmer, 1990).

Hair Whorls

In humans, it is well known that hair whorl patterns form in the fetus at the same time the brain is forming (Smith and Gong, 1974). Hair whorl patterns are associated with handedness (Beaton and Mellor, 2007; Hatfield, 2006; Klar, 2003; Schmidt, *et al.* 2008; Zarate and Zarate, 1991), hemispheric language dominance (Jansen *et al.*, 2007; Weber *et al.*, 2006), sexual preference (Hatfield, 2006; Klar, 2004), and schizophrenia (Alexander *et al.*, 1992; Yousefi-Nooraie and Moratz-Hedjri, 2008). Scalp hair patterning is determined at 10–16 weeks of fetal life and is secondary to the growth and shape of tissues which underlie the fetal skin, especially the brain. Thus aberrant scalp hair patterning may be utilized as a clinical indicator of aberrant growth and/or shape of the early fetal brain prior to 16 weeks' gestation (Scott *et al.*, 2005; Smith and Gong, 1974; Tirosh et al., 1987). Children with developmental disorders, such as Down's syndrome and Prader-Willi syndrome have a high incidence of abnormal scalp hair whorl patterns (Smith and Gong, 1973, 1974). Many body traits form very early in development at the same time the brain is forming.

Hair whorl patterns and temperament in horses was first observed hundreds of years ago in Arabia. In his work as a horse trainer, the second author observed that a "high" hair whorl on a horses' forehead, or those with two spiral hair whorls were more reactive and "high strung." Other horse trainers and veterinarians have observed a casual relationship between hair whorl patterns and temperament (Barker, 1990; Friedly, 1990; Tellington-Jones and Bruns, 1985). Many scientists believed these observations were rubbish and that horsemen were wrong. In the early 1990's we set out to find if scientific evidence support these observations. We decided to study hair whorls and temperament in cattle. Cattle and horses have similar hair whorl patterns on their foreheads (Figures 4.3 and 4.4). It was easy to observe large numbers of cattle with similar genetics and handling experience. Cattle are not handled in the same way as horses, and their behavior is

FIGURE 4.3 Hair whorl height is related to temperament in cattle. This animal has a hair whorl in the middle position between the eyes. Cattle with a spiral hair whorl below the eyes tend to be calmer.

FIGURE 4.4 This steer has a hair whorl above the eyes and this is related to a more reactive temperament. Since the late 1990s, American beef producers have selected cattle for a calmer temperament. Today it is harder to find beef cattle with a hair whorl high above the eyes.

not confounded by the experience of being trained for riding. The survey was conducted at a commercial feedlot in Colorado. A total of 1500 cattle were temperament-ranked (with the rating scale described previously) while restrained in a hydraulic squeeze chute for vaccinations. The cattle originated from many different ranches. Seventy-two per cent were European × British crosses and the rest were zebu crosses from Mexico. To prevent bias, temperament rankings were collected by one person standing to the side and behind the squeeze chute. From this position he was unable to see the hair whorl. A second observer recorded hair whorl location from a position at the back of the squeeze chute. The hair whorl location was recorded before each animal entered the squeeze chute. We found that cattle with hair whorls above the eyes became significantly ($p = 0.001$) more agitated during restraint than cattle with hair whorls below the eyes (Grandin *et al.*, 1995a). To our knowledge, this study was the first published evidence of a relationship between behavior and hair whorls.

Randle (1998) assessed the relationship between hair whorl position and temperament in 57 *Bos taurus* cattle in a post hoc investigation. Individuals were scored on 19 measures of personality, 14 of which related to temperament. Individuals with mid-whorls exhibited significantly greater flight distances to an unfamiliar human than did individuals with low whorls ($P < 0.01$). Cattle with high hair whorls above the eyes are more vigilant, excitable, and reactive (Bueno *et al.*, 2012; Florcke *et al.*, 2012; Martins, *et al.*, 2009). Lanier *et al.* (2000) report that extensively raised beef cattle with high hair whorls are more likely to become agitated due to loud intermittent noise in an auction ring. Florcke *et al.* (2012) found mother cows with high hair whorls became aware of a strange approaching vehicle at a greater distance. In semi feral Konik horses, the height of a single hair whorl had no effect on startle response, but horses with elongated double hair whorls took significantly longer to approach a novel object (Gorecka *et al.*, 2007). Possibly feral Konik horses have fewer variations in hair whorl height than more heterogeneous groups of domestic horses. Hair whorl patterns are highly heritable in horses (Gorecka *et al.*, 2006), and Holstein cattle (Shirley *et al.*, 2006).

The beef cattle in our first study had a higher percentage of hair whorls above the eyes compared to Holstein cows at a large commercial dairy (Grandin *et al.*, 1995a; Tanner *et al.*, 1994). Holsteins are much less reactive to novelty and aversive stimuli than beef cattle and there was no effect of hair whorl position on behavior (Shirley *et al.*, 2006). This is probably due to less variation in both genetics and hair whorl positions in Holsteins. In commercial feedlots Holsteins are less likely to become excited during handling compared to beef cattle. In 2011, the second author collected hair whorl data on over 600 beef cattle in Nebraska. A small percentage of these cattle had high whorls compared to the data collected by Grandin *et al.* (1995a). This may be due to 15 years of beef industry programs designed to select for calm temperaments. Significant improvements in levels of fearfulness have been achieved by selecting populations against these traits. Selective breeding makes a group of animals more homogenous. Hair whorl position is likely to be more variable and have a significant relationship with behavior in groups of cattle or horses with more heterogeneous diverse genetic backgrounds.

Hair Whorl Height and Vigilance

Animals with hair whorls located high on the forehead may have high nervous system reactivity, increased fearfulness and emotionality, highly acute senses, stronger orienting responses, and vigilance. A low hair whorl may be related to reduced vigilance and a reduced reactivity and fearfulness. Florcke *et al.* (2012) observed that Red Angus beef cows with high hair whorls were more vigilant when their newborn calf was threatened by an approaching vehicle. Cows looked up and oriented towards the vehicle when at a greater distance compared to cows with lower hair whorls. In Brazil, cattle with high hair whorls above the eye exited more rapidly from a restraining chute Bueno Ribeiro *et al.* (2012), and struggled more in a scale. Brazilian cattle have had much less intensive selection for temperament.

In cattle and horses, fine-boned, slender-bodied animals have a highly reactive temperament, a high hair whorl, and are more likely to have an explosive reaction when suddenly confronted with a novel stimulus that moves suddenly. Heavy-boned, muscular animals are more likely to have a calm temperament and a low hair whorl. Holstein dairy cows lack heavy muscling but fit the above criteria because they are heavy boned. We began to speculate about two separate genetic mechanisms influencing temperament after the first author visited the Lasater Beefmaster herd in 1996. Lasater's cattle been closed to new genetics for 60 years, and subjected to a unique set of selection pressures. Lasater (1972) selected his cattle using the natural principle of survival of the fittest. Heifers unable to give birth unassisted or to protect their calf from coyotes were culled. However, Lasater wanted commercially useful cattle so he also selected for temperament and carcass traits. The Beefmaster's breed is half Brahman, quarter Hereford, and quarter Shorthorn. Heifer's calves were selected as herd replacements if they willingly ate a food treat off a stick held by a seated person. Tom Lasater's son Dale explained how they selected for temperament. Instead of using sudden aversive novelty (such as restraint) as a temperament test, they assessed temperament of hungry, newly weaned calves by sitting in the pen with them. Calves that failed to eat from a person's hand after two days were culled. When the herd was first started, about a quarter of the calves were culled. Today only 1% are culled for temperament (Dale Lasater, personal communication, 1996).

Lasater's selection criteria resulted in cows that are very protective of their calves, but extremely tame and seek contact with people. It is unusual to observe range cows that will come up and lick people and will stand still while being scratched and petted. It is likely they are selected for low FEAR, high CARE (maternal nurturing), and PANIC (separation distress). The appearance of these reddish brown animals is striking. They are muscular and heavy boned with either high hair whorls on the forehead or no hair whorls.

Faure and Mills (Chapter 8) demonstrated that the traits of fearfulness and social reinstatement are genetically separate. Social reinstatement and PANIC (separation distress) are probably the same emotional system. In Japanese quail, social reinstatement is defined as the tendency of an isolated bird to rejoin flockmates. In a wild population it is likely that both fear and high social reinstatement would appear naturally in the same animal because this would improve survival. In domestic cattle, fearful animals bunch together tightly when excited. Tight bunching is probably motivated by high fear. Social reinstatement makes animals bond and is probably not motivated by fear. Faure and Mills showed that social reinstatement is separate from fear because they were able to select and breed both low-fear and high-social reinstatement birds and *vice versa*.

A second group of cattle similar to the Lasater Beefmasters's were observed by the first author during a trip to England in 2012. The cattle were Limousin × Devon cross cows that were very docile. They allowed strange people to approach them out on pasture, even though they had young calves. The cows were attentive to their calves and vocalized to call them as soon as they saw a strange tractor towing a trailer full of people entering the pasture. Most of the hair whorls on the cows were slightly above the eyes and no animals with extremely low whorls were observed. The cows were heavy boned yet extremely curious and quickly approached and touched the novel tractor. Even though heavy boned, their nervous system was vigilant. When I jumped off the trailer, most of the cows flinched but did not run away. The authors speculate that the combination of Devon genetics and Limousin had reduced FEAR, and produced animals that were high SEEKING, high PANIC (separation distress), and high CARE (maternal nurturing). Devons are bred to be docile and some farmers report they may push other cattle away from resources such as feed troughs. Possibly Limoisin genetics helped to maintain the vigilance trait.

THE EFFECTS OF EARLY EXPERIENCE ON HANDLING

Animals vary genetically in their susceptibility to environmental influences. Evidence indicates that some animals, such as those with emotional temperaments, are more vulnerable to the adverse effects of negative experience. In fact, they are actually more vulnerable to *both* positive and negative experiences. Early rearing effects in cattle have been studied by Dellmeier *et al.* (1985). Calves were reared in individual stalls, which restricted movement.

When later tested in an open-field pen, the stall-raised calves were more energetic, had a greater level of activity, and ran around excessively compared to calves reared in group pens. Fordyce (1987) showed that adult cattle handled quietly as calves are quieter during handling compared to non-handled controls. Calves reared in an intensive system can also be sorted quickly compared to calves reared in a more extensive system (Boivin *et al.*, 1992). Near weaning time, Boissy and Bouissou (1988) found that handling and contact with people improves the ease of handling in cattle. Early experiences will affect how animals react in the future to handling, restraint, or novel situations (Gonyou, 1993; Hemsworth *et al.*, 1986).

The effects of early experience on handling are complex. Lensink *et al.* (2001) found that reactions of calves to handling depend on housing condition and previous experience with humans. Calves were housed individually or in pairs. Half of the individually housed calves were stroked on the neck for 90 seconds each day, and half of the calves housed in pairs received stroking. Paired calves took more time to interact and interacted less with a person compared to calves housed individually. Calves that received additional contact with people interacted longer with a person than calves with minimal contact. It took more time and effort to load calves housed in pairs compared to individually housed calves, and less time to load calves that had additional contact with people. During loading additional-contact calves had lower heart rates than minimal contact calves, while during transport pair-housed calves had lower heart rates than calves housed individually.

Lyons (1987) found that goats hand-reared by people, react less strongly to changes in the environment compared to goats raised on their dams. Differences in early rearing methods resulted in long-term, stable differences in temperament. Several studies show previous experience has a significant effect on an animal's flight zone size the physiological responses to handling and restraint (Boissy and Bouissou, 1988; Ewbank, 1993; Gonyou, 1993; Hastings *et al.*, 1992; Ried and Mills, 1962). For example, both hand-reared cattle and hand-reared deer show lower cortisol levels during restraint compared to animals with less early contact with people (Boandle *et al.*, 1989; Hastings *et al.*, 1992). Hastings *et al.* (1992) also found "tame" hand-reared deer had lower cortisol stress hormone levels when restrained compared to deer raised by their mothers. Interestingly, both groups actively resisted restraint and vocalized. Compared to cattle with rough handling, Grandin (1984, 1987) concludes that cattle with gentle previous experience during handling will be calmer and easier to handle in the future. This principle applies to all domestic livestock, wildlife, even dogs.

Effects of Environment and Experience on Pigs

Early environment effects on behavior of pigs reared indoors has been studied by Grandin *et al.* (1987), Grandin (1989), and Beattie *et al.* (1995). Pigs reared with assorted objects to play with, straw bedding, and daily petting were more willing to approach a novel object or novel person compared to pigs reared in barren pens. Environmental enrichment made the pigs calmer and less reactive, and they startled less when a person suddenly jerked upon the gate to their pens. Environmental enrichment also makes pigs more willing to walk through a single-file chute (Grandin, 1989; Pederson, 1992). Less prodding was required for the pigs from the "enriched" environment. Dudink *et al.* (2006) investigated whether announcement of an environmental enrichment, more than enrichment alone could facilitate play behavior and reduce weaning stress. A sound cue was paired with environmental enrichment with a 30-second delay in which anticipatory behavior develops. Results indicate that announcement of enrichment and not enrichment alone increased play behavior, decreased aggression before and after weaning, and reduced the amount of injuries after weaning.

The first author has experience training market-weight pigs with excitable genetics to drive quietly at the slaughter plant. To facilitate handling and adaptability to commercial conditions, the training is done on the farm during the entire finishing time (fattening period). She found that by walking through the pens in a random pattern each day (10–15 seconds per pen) and training the pigs to move away in an orderly manner, the pigs learned to get up and flow around her as she walked through the pen. In this manner pigs are taught to be driven quietly by people instead of learning to follow them. However, Grandin cautions the person walking the pens not to frighten the pigs. The idea is to teach pigs to become accustomed to people, to get up and move when people enter the pen, and not get upset and pile up. Walking the pens is especially beneficial for lean hybrids with nervous temperaments. Pigs differentiate between a person walking through their pens and a person walking in the aisle. This is why it is essential for a person to enter the pen and walk among the pigs. Further studies have shown that getting pigs accustomed to walking in the aisles and other handling procedures improved ease of handling (Geverink *et al.*, 1998) and (Abbott *et al.*, 1997). Pigs with practice moving through alleys and a ramp were easier to move through them in the future (Lewis *et al.*, 2008).

Excellent Memory of Aversive Experiences

Cattle, sheep, and other herd animals have excellent memories and remember aversive handling experiences (Grandin *et al.*, 1986; Hutson, 1980, 1993; Pascoe, 1986). For example, Hutson (1980) found that sheep subjected to aversive procedures were reluctant to re-enter the test facility six months to a year later. Pascoe (1986) also found that dairy cows that had received shocks, remembered the shocks and showed higher heart rates when returned to the facility. Previous rough handling also makes dairy cows reluctant to enter a milking parlor (Seabrook, 1987). Cows previously handled in a gentle manner had smaller flight zones and entered a milking parlor more quickly than cows handled in an aversive manner. It is extremely important that an animal's initial experience with a new handling facility be as positive as possible. If the animal's initial experience is aversive, it may become difficult to induce the animal to re-enter the facility (Grandin, 1993). The second author had an experience with a horse that hit its head the first time it backed out of a horse trailer. This one incident caused pain and fright and the horse remembered it. Following the incident, the horse learned to enter the trailer quietly, but during unloading he continued to rush out of the trailer quickly. It was almost impossible to train the horse not to back out of the trailer suddenly.

Training Highly Reactive Animals by Introducing Novelty Gradually

Highly reactive animals can be conditioned to cooperate during veterinary procedures such as injections (Grandin *et al.*, 1995b). To summarize the results of this study (discussed in detail in Chapter 1), adult antelope at the Denver Zoo were trained to accept routine blood tests and their cortisol remained at baseline levels. This showed the antelope were not fearful (Grandin *et al*, 1996). The training was done in small, incremental steps over a period of weeks, and great care was taken to avoid triggering a massive flight reaction. For instance, when the antelope oriented and froze during any small change in the procedure, the procedure was stopped. Each day new procedures were gradually introduced until the antelope habituated and no longer oriented. However, it is very important not to confuse taming effects on adult animals with changes that take place in the nervous system of young animals after handling. When the trained nyala were suddenly confronted with something not encountered during training, their true nervous system reactivity was displayed and they reacted with explosive fear reaction. Bongo antelope trained at the Denver Zoo had low (almost baseline) levels of cortisol when voluntarily standing in a box to have blood drawn (Grandin, *et al.*, 1996). Training experiments with pronghorn antelope show high specificity of their perception. Pronghorns trained to stand and receive an IV injection in the neck would remain calm and quiet because they had been carefully habituated to this procedure. When a student attempted to give an injection in the shoulder, they went berserk. An injection in the shoulder was novel and scary and an injection in the neck was familiar and safe (Grandin and Johnson, 2010).

In contrast to the highly reactive nyala or other reactive herd species, training placid animals such as Hereford cattle or Suffolk sheep to accept novel procedures can be done quickly and habituation to non-painful procedures can occur even when the animals are pushed beyond the orienting response. However, if a highly reactive animal is trained too quickly and has a massive fear reaction during the training, it may be very slow to habituate afterwards. Highly reactive animals may become more fearful with each forced handling trial, whereas an animal with a more placid temperament may habituate and become less fearful. Grandin *et al.* (1994) found this in Angus × Hereford × Charolais × Simmental crossbred cattle that became increasingly agitated when they were repeatedly run through a squeeze chute on the same day. Animals need time to calm down between training trials and practical experience has shown that training trials should be spaced at least 24 hours apart.

During handling and driving in alleys, lean hybrid pigs have been observed to startle frequently and are more easily excited compared to fatter genetic lines of pigs (Grandin, 1994). The hybrids are highly reactive when a whip is cracked and raise their ears instantaneously in response to a sudden novel sound. Lanier *et al.* (1995) conducted experiments on genetically similar crossbred pigs with identical experiences and found individual differences in their adaptability to a forced novel swimming task. Some of the pigs readily adapted to a series of swimming tasks and others remained highly stressed. Some pigs had baseline epinephrine (adrenalin) levels after swimming, and other had high levels. In essence, the stressed pigs never adapted. To summarize, animals with a placid, calm temperament habituate more easily to forced non-painful procedures than animals with a flighty, excitable temperament.

Innate Nervous Fearfulness or Reaction to Separation Stress

A common question asked by cattle breeders in the United States is why cows with flighty European Continental genetics may be difficult to handle here in the U.S. and easy to handle in France. The answer: French calves are raised in close association with people and become tame. On ranches in the western U. S., cows are normally handled only two or three times a year. Early intensive contact of calves with people in France produces calm, easy to handle adult animals. The highly reactive nature of some genetic lines of European Continental cattle may be not be displayed when they are handled in a quiet manner in close contact with people. In familiar surroundings, they may appear calm, but when suddenly confronted with novelty, they are more likely to have a violent reaction. Boissy *et al.* (1996) report similar findings in sheep. Romanov ewes originally bred for intensive conditions are stressed under extensive conditions due to a tendency to avoid humans. Further research has shown that gentling procedures designed to get young lambs accustomed to human contact had the greatest beneficial effect on the Gentile di Publia breed of sheep compared to the Comisana breed (Caropresе *et al.*, 2012). This study may indicate that the breed with the high separation distress (PANIC) benefitted the most. When a person was present in the test arena, the trained lambs vocalized less and had less attempts at jumping out (Caroprese *et al.*, 2012). It is well documented that early handling will make cattle and horses easy to handle (Boivin *et al.*, 1992; Fordyce *et al.*, 1985).

The managers of a bull test station reported two accounts of bulls that went berserk when brought to the station. The bulls charged fences and attacked the handlers, but afterwards, the owner was able to walk them quietly into a trailer. They calmed down when they heard his familiar voice. In horses, both authors observed that the true temperament can be masked by learning. Some horses "blow up" when they are taken to a new place or exposed to a strange stimulus. For example, owners often teach their young horses to allow their feet to be handled by a horseshoer. The horse may learn to stand and accept this handling by the owner but when the farrier arrives for the first time, the horse may react violently. This may be due to the novelty of a new person. Some horses become fearful every time they are handled by a farrier. Owners often comment that this is "out of character" for the horse. When highly reactive animals are in familiar surroundings, they may be calm and their true temperament can be masked. They learn that certain sights and sounds are harmless. Recently, the owner of a high-strung horse commented, "He's 12 years old. You'd think he'd learn to get over the little things by now." Taming and training adult animals may reduce their reactions to specific handling procedures but it does not change their level of innate emotional reactivity. The level of emotional reactivity in an animal is like the level of water in a drinking glass. The level of water in the glass is influenced by both genetics and the environment during early life. Environmental factors can either raise or lower the initial level of water in the glass. Handling and contact with people early in life can reduce emotionality in an adult animal but it is almost impossible to manipulate the environment to such an extent that a highly reactive animal can become as calm as an animal which has calm genetics. For example, no amount of early handling can turn a high-strung Arab stallion into an "old plug" riding stable horse. You might make him calmer, but he will never be as calm as a horse that is genetically calm.

Facilitating Animal Movement in Handling Facilities

An experienced dairy cow will walk over a drain grate in the milking parlor but young heifers often balk (Grandin, 1980a,b, 1996; Kilgour, 1971; Lynch and Alexander, 1973). Dairy cows can learn that drain grates and other things are harmless. However, animals being handled in an unfamiliar facility will balk and refuse to walk over puddles or drain grates. Shining reflections, shadows, and changes in flooring type will cause livestock to balk and refuse to move through a handling facility. This is especially problematic in places such as auction yards or slaughter plants where the animals have no opportunity for learning. If leader animals are given a few minutes to investigate a shadow or puddle, they will usually step around or over it and other animals will follow (Hutson, 1980). Unfortunately, in a high-speed slaughter plant, there is not enough time to allow the leaders to investigate novel objects or areas of high contrast.

Emotionally reactive animals are often in a high state of nervous system arousal due to the novelty of the situation, which makes them more vigilant and likely to balk at small distractions in alleys and races that draw their attention. A jiggling chain, people up ahead, a sparkling reflection on a piece of metal, or air drafts and hissing exhausts blowing in their faces can all cause the animals to stop (Grandin, 1996; Grandin and Deesing, 2008; Tanida *et al.*, 1996). More placid animals will usually hesitate and look at small distractions, but are more likely to keep moving. However, an area of high contrast or something that looks out of place is more likely to cause balking in nervous animals. In five different slaughter plants, the removal of all distractions such as sparkling reflections, air drafts, and a presence of people up ahead made it possible to greatly reduce the use of electric prods (Grandin, 1996, 2007; Grandin and Deesing, 2008).

Genetic Effects on Handling Facility Design

The first author has observed that the temperamental differences between Brahman breeds (*Bos Indicus*) and European Continental breeds (*Bos Tauras*) has had a substantial influence on the design of cattle handling facilities in the United States. These differences in design evolved over many years because ranchers used what worked. For example, in the northern United States, when cattle are being sorted into groups by sex or weight, the mostly placid British cattle are sorted in 3.5- to 5-meter-wide alleys by people moving among them either on foot or on horseback. This method works because animals with a calm temperament can be easily separated from the group. Cattle with Brahman genetics are more excitable and have to be sorted in a 70-cm-wide, single-file race with gates at the end leading to separate pens. When Brahman cattle get excited, they engage in bunching behavior and herd tightly, which makes it difficult to sort out a single animal in a wide alley. Removing a single animal from the group is more difficult. It is much easier to sort cattle with an excitable temperament through a single-file chute. Hohenboken (1987) also commented that Brahmans behave differently in corrals and working facilities compared to *B. taurus* cattle.

CONCLUSION

Both genetic factors and the animal's rearing environment will determine how an animal will behave during handling. Animals with a highly reactive, excitable temperament will become more fearful and agitated when confronted by sudden novelty compared to animals with a calm, placid temperament. Animals with a calm temperament will habituate more easily to new handling procedures. Stress levels will remain low in flighty excitable animals if new procedures are introduced very slowly. An understanding of the influence of genetics on behavior will facilitate animal handling and training.

REFERENCES

Abbott, T.A., Hunter, E.J., Guise, J.H., Penny, R.H.C., 1997. The effect of experience on pig's willingness to move. Appl. Anim. Behav. Sci. 54, 371–375.

Alexander, C., Breslin, N., Molnar, C., Richter, J., Mukherjee, S., 1992. Counterclockwise scalp hair whorl in schizophrenia. Biol. Psychiatry 32, 842–845.

Arave, C.W., Albright, J.L., Sinclair, C.L., 1974. Behavior, milk yield, and leucocytes of dairy cows in reduced space and isolation. J. Dairy Sci. 59, 1497–1501.

Bailey, D.W., Keil, M.R., Rittenhouse, L.R., 2004. Research observation: daily movement patterns of hill climbing and bottom dwelling cows. Rangeland Ecol. Manage. 57, 20–28.

Barker, R., 1990. The mind behind the swirls, Rocky Mt. Quarter Horse 28 (March 26–27).

Basile, M., Boivin, S., Boutin, A., Blois-Heulin, C., Hausberger, M., Lemasson, A., 2009. Socially dependent auditory laterality in domestic horses (*Equus caballus*). Anim. Cogn. 12 (4), 611–619.

Baszczak, J.A., Grandin, T., Gruber, S.L., Engle, T.E., Platter, W.J., Schroeder, A.L., et al., 2006. Effects of ractopamine supplementation on behavior of British, Continental, and Brahman crossbred steers during routine handling. J. Anim. Sci. 84, 3410–3414.

Beaton, A.A., Mellor, G., 2007. Direction of hair whorl and handedness. Laterality: asymmetries of body. Brain Cogn. 12 (4), 295–301.

Beausoleil, N.J., Blache, D., Strafford, K.J., Mellor, D.J., Noble, A.D.L., 2008. Exploring the basis of divergent selection for 'temperament' in domestic sheep. Appl. Anim. Behav. Sci. 109 (2), 261–274.

Belyaev, D.K., 1979. Destabilizing selection as a factor in domestication. J. Heredity 70, 301–308.

Benhajali, N., Bovin, X., Sapa, J., Pellegrin, P., Boulesteix, P., LaJudie, P., LaJudie, P., Phocas, F., 2010. Assessment of different on-farm measures of beef cattle temperament for genetic evaluation. J. Anim. Sci. 88 (11), 3529–3537.

Boandle, K.E., Wohlt, J.E., Carsia, R.V., 1989. Effect of handling, administration of local anesthetic and electrical dehorning on plasma cortisol in Holstein calves. J. Dairy Sci. 72, 2193–2197.

Boissy, A., 1995. Fear and fearfulness in animals. Q. Rev. Biol. 70, 165–191.

Boissy, A., Bouissou, M., 1988. Effects of early handling on heifers' subsequent reactivity to humane and to unfamiliar situations. Appl. Anim. Behav. Sci. 20, 259–283.

Boissy, A., LeNeindre, P., 1990. Social Influences on the reactivity of heifers: implications for learning abilities in operant conditioning. Appl. Anim. Behav. Sci. 25, 149–165.

Boissy, A., Le Neindre, P., Orgeur, P., Bouix, J., 1996. Genetic variability psychobiological reactivity of lambs reared under open range management. Proceedings of the 30th International Congress of the International Society for Applied Ethology. University of Guelph, Guelph, Ontario, Canada, 1996, p. 59.

Boissy, A., Boouix, J., Orgeur, P., Poindron, P., Bibe, B., LeNeindre, P., 2005. Genetic analysis of emotional reactivity in sheep: effects of the genotype of the lambs and of their dams. Genet. Sel. Evol. 37, 381–401.

Boivin, X., Le Neindre, P., Chupin, J.M., Garel, J.P., Trillat, G., 1992. Influence of breed and early management on ease of handling and open-field behavior of cattle. Appl. Anim. Behav. Sci. 32, 313–323.

Bonsma, J., 1975a. Judging cattle for functional efficiency. Beef Cattle Sci. Handb. 12, 23–36, papers delivered at the International Stockmen's School, San Antonio, TX.

Bonsma, J., 1975b. Crossbreeding for increased cattle production. Beef Cattle Sci. Handb. 12, 37–49, papers delivered at the International Stockmen's School, San Antonio, TX.

Bueno Ribeiro, A.R., Mello du Alenear, M., Andrea de S. Bastos, P., Luis Paco A., M.G. Ibelli A., 2012. Relationships among temperaments, facial hair whorl position and productivity traits of beef heifers in three genetic groups. International Symposium on Beef Cattle Welfare. University of Saskatchewan, Saskatoon, Canada, June 5, 7, 2012, pp. 17–19.

Burdick, N.C., Randel, R.D., Carroll, J.A., Welsh, T.H., 2011. Interactions between temperament, stress, and immune function in cattle. Int. J. Zool. 2011 (2011), Article ID 373197, 9 pages.

Burrow, H.M., Dillon, R.D., 1997. Relationship between temperament and growth in feedlot and commercial carcass traits of *Bos indicus* crossbreds. Aust. J. Exp. Agric. 37, 407–411.

Café, L.M., Robinson, D.L., Ferguson, D.M., McIntyre, B.L., Gesink, G.H., Greenwood, P.L., 2010. Cattle temperament: persistence of assessments and associations with productivity, efficiency, carcass, and meat quality traits. J. Anim. Sci. 88 (9), 3059–3069.

Caroprese, M., Napolitano, F., bovin, X., Abenzio, M., Annicchiarico, G., Savi, A., 2012. Development of affinity to the stock person from two lamb breeds. Physiol. Behav. 105, 251–256.

Carroll, J.M., Murphey, C.J., Neitz, M., Hoeve, J.N., 2007. Photo pigment basis for dichromatic vision in the horse. J. Vis. 1, 80–87.

Christensen, J.M., Zharkikh, T., Ladewig, J., 2008. Do horses generalize between objects during habituation? Appl. Anim. Behav. Sci. 114, 509–520.

Coppinger, L., Coppinger, R., 1993. Dogs for herding livestock. In: Grandin, T. (Ed.), Livestock Handling and Transport. CABI, Wallingford, UK, pp. 179–196.

Core, S., Widowski, T., Mason, G., Miller, S., 2009. Eye white percentage as a predictor of temperament in beef cattle. J. Anim. Sci. 87 (6), 2168–2174.

Couzin, I.D., Krause, J., 2003. Self-organization and collective behavior in vertebrates. Adv. Study Behav. 32, 1–75.

Curio, E., 1988. Cultural transmission of enemy recognition. In: Zentall, R.E., Galef Jr, B.G. (Eds.), Social Learning: Psychological and Biological Perspectives. Erlbaum, Hillsdale, New Jersey, USA, pp. 75–97.

Curley, K.O., Neuendorf, D.A., Lewis, A.W., Cleere, J.J., Welsh, T.H., Randal, R.D., 2004. Effects of temperament on stress indicators in Brahman heifers. J. Anim. Sci. (Supl. 1), 459 (Abstract).

Dantzer, R., Mormede, P., 1978. Behavioral and pituitary–adrenal characteristics of pigs differing in their susceptibility to the malignant hyperthermia syndrome induced by halothane anesthesia. Ann. Vet. Res. 9 (3), 559–567.

Davis, M., 1992. The role of the amygdala in fear and anxiety. Annu. Rev. Neurscience 15, 353–375.

De Boyer Des Roches, A., Richard-Yris, M.A., Henry, S., Ezzaouia, M., Hausberger, M., 2008. Laterality and emotions: visual laterality in the domestic horse (*Equus caballus*) differs with objects' emotional value. Physiol. Behav. 9, 487–490.

Dellmeier, G.R., Friend, T.H., Gbur, E.E., 1985. Comparison of four methods of calf confinement, II. Behav. J. Anim. Sci. 60, 1102–1109.

Dias Barbosa Silveira, I., Fischer, V., deMendonca, G., 2010. Effects of genotype and age of sheep on reactivity measured at market auction. R. Bras. Zootec. 39 (10).

Dickson, D.P., Bart, G.R., Johnson, P., Wiekert, D.A., 1970. Social dominance and temperament in dairy cows. J. Dairy Sci. 53, 904.

Diess, V., Temple, D., Liyour, S., Pacine, C., Boux, J., Terlouw, et al., 2009. Can emotional reactivity predict stress response at slaughter in sheep. Appl. Anim. Behav. Sci. 119, 193–202.

Dodd, C.L., Pitchford, W.S., 2012. Measures of behavioral reactivity and production traits in sheep: a review. Appl. Anim. Behav. Sci. 140 (1–2), 1–15.

Dudink, S., Simonse, H., Marks, I., deJonge, F.H., Spruiit, B.M., 2006. Announcing the arrival of enrichment increases play behavior and reduces weaning-stress-induced behaviours of piglets directly after weaning. Appl. Anim. Behav. Sci. 101 (1), 86–101.

Edwards, L.N., Grandin, T., Engle, T.E., Porter, S.P., Ritter, M.J., Sosnicki, A.A., Anderson, D.B., 2010. Use of exsanguination bloodlactate to assess quality of pre-slaughter pig handling. Meat Sci. 86, 384–390.

Elliker, K.R., 2007. Social cognition and its implications for the welfare of sheep. Ph.D. Thesis. University of Cambridge.

Ewbank, R., 1993. Handling cattle in intensive systems. In: Grandin, T. (Ed.), Livestock Handling and Transport. CABI, Wallingford, UK, pp. 59–73.

Farmer, K., Krueger, K., Byrne, R.W., 2010. Visual laterality in the domestic horse (*Equus caballus*) interacting with humans. Anim. Cogn. 13, 229–238.

Faure, A.S.M., Reynolds, J.M., Richard, J.M., Berridge, K.C., 2008. Mesolimbic dopamine in desire and dread: enabling motivation to be generated by localized glutamate disruption in the nucleus accumbens. J. Neurosci. 28, 7184–7192.

Faure, J.M., Mills, A.D., 1998. Improving the adaptability of animals by selection. Genetics and the Behavior of Domestic Animals. Academic Press.

Florcke, C., Engle, T.E., Grandin, T., Deesing, M.J., 2012. Individual differences in calf defense patterns in Red Angus beef cows. Appl. Anim. Behav. Sci. 139 (3–4), 203–208.

Fordyce, G., 1987. Weaner training. Qld. Agric. J. 113, 323–324.

Fordyce, G., Goddard, M.E., Tyler, R., Williams, G., Toleman, M.A., 1985. Temperament and bruising of *Bos indicus* cross cattle. Aust. J. Exp. Agric. 25, 283–288.

Fordyce, G., Dodt, R.M., Wyther, J.R., 1988. Cattle temperaments in extensive herds in northern Queensland. Aust. J. Exp. Agric. 28, 683–687.

Forkman, B., Boissey, A., Meunier-Salaün, M.C., Cancli, E., Jones, R.B., 2007. A critical review of fear tests used in cattle, pigs, poultry, and horses. Physiol. Behav. 92, 340–374.

Friedly, J., 1990. Dang ... there's a tricoglyph on your horse. Rocky Mt. Quarter Horse 28 (July), 26–27.

Fukasawa, F., 2008. The effect of approach direction and pace on flight distance of beef breeding cows. Anim. Sci. J. 79, 722–726.

Fuller, J.L., Thompson, W.R., 1978. Foundations of Behavior Genetics. Mosby, St. Louis, MO.

Geverink, N.A., Kappers, A., von deBurgwal, E., Lambooji, E., Blokhuis, J.H., Wieyant, V.M., 1998. Effects of regular moving and handling on the behavioral and physiological responses of pigs to pre-slaughter treatment and consequences for meat quality. J. Anim. Sci. 76, 2080–2085.

Gonyou, H.W., 1993. Behavioral principles of animal handling and transport. In: Grandin, T. (Ed.), Livestock Handling and Transport. CABI, Wallingford UK.

Goodwin, D., 2002. Horse behavior: evolution, domestication, and feralization. In: Waren, N. (Ed.), The Welfare of Horses. Kluwer, Dordrecht. The Netherlands, pp. 1–18.

Gorecka, A., Slonicewski, K., Golonka, M., Joworski, Z., Jezierski, T., 2006. Heritability of the hair whorl position on the forehead in konik horses. J. Anim. Breed. Genet. 123, 396–398.

Gorecka, A., Golonka, M., Chruszczewski, M., Jezierski, T., 2007. A note on behavior and heart rate in horses differing in facial hair whorl. Appl. Anim. Behav. Sci. 105, 244–248.

Grandin, T., 1980a. Livestock behavior related to handling facilities design. Int. J. Study Anim. Probl. 1, 33–52.

Grandin, T., 1980b. Observations of cattle behavior applied to the design of handling facilities. Appl. Anim. Ethol. 6, 19–33.

Grandin, T., 1984. Reduce stress of handling to improve productivity of livestock. Vet. Med. 79, 267–831.

Grandin, T., 1987. Animal handling. Vet. Clin. North Am. 3, 323–338.

Grandin, T., 1989. Effects of rearing environment and environmental enrichment on behavior and development of young pigs. Doctoral Dissertation, University of Illinois, Urbana-Champaign.

Grandin, T., 1992. Behavioral agitation is persistent over time. Appl. Anim. Behav. Sci. 36, 1–9.

Grandin, T., 1993. Behavioral principles of cattle handling under extensive conditions. In: Grandin, T. (Ed.), Livestock Handling and Transport. CABI, Wallingford UK, pp. 43–57.

Grandin, T., 1994. Solving livestock handling problems. Vet. Med. 89, 989–998.

Grandin, T., 1996. Factors which impede animal movement in slaughter plants. J. Am. Vet. Med. Assoc. 209, 757–759.

Grandin, T., 1997. Assessment of stress during handling and transport. J. Anim. Sci. 75, 249–257.

Grandin, T., 2007. Behavioral principles of handling cattle and other grazing animals under extensive conditoins: In: Grandin T. (Ed.) Livestock Handling and Transport. CABI Publishing, Wallingford, Oxfordshire, UK, pp. 44–64.

Grandin, T., Deesing, M., 2008. Humane Livestock Handling, North Adams, Mass.

Grandin, T., Johnson, C., 2005. Animals in Translation. Scribner, New York, NY.

Grandin, T., Johnson, C., 2010. Animals Make us Human. Houghton Mifflin Harcourt, New York.

Grandin, T., Curtis, S.E., Widowski, T.M., Thurman, J.C., 1986. Electro-immobilization versus mechanical restraint in an avoidance choice test. J. Anim. Sci. 62, 1469–1480.

Grandin, T., Curtis, S.E., Taylor, I.A., 1987. Toys, mingling, and driving reduce excitability in pigs. J. Anim. Sci. 65 (Suppl. 1), 230 (Abstract).

Grandin, T., Odde, K.G., Schutz, D.N., Behrns, L.M., 1994. The reluctance of cattle to change a learned choice may confound preference tests. Appl. Anim. Behav. Sci. 39, 21–28.

Grandin, T., Deesing, M.J., Struthers, J.J., Swinker, A.M., 1995a. Cattle with hairwhorl patterns above the eyes are more behaviorally agitated during restraint. Appl. Anim. Behav. Sci. 46, 117–123.

Grandin, T., Rooney, M.B., Phillips, M., Canbre, R.C., Irlbeck, N.A., Graffam, W., 1995b. Conditioning of nyala (*Tragelaphus angasi*) to blood sampling in a crate with positive reinforcement. Zoo Biol. 14, 261–273.

Grandin, T., Irlbeck, N., Phillips, M., 1996. Training antelope to cooperate with veterinary procedures. Proceedings of the 30th International Congress of the International Society for Applied Animal Ethology. University of Guelph, Guelph, Ontario Canada.

Hall, N.L., Buchanan, D.S., Anderson, V.L., Iise, B.R., Carlin, K.R., Berg, E.P., 2011. Working chute behavior of feedlot cattle can be an indicator of cattle temperament and beef carcass composition and quality. Meat Sci. 89, 52–57.

Hamilton, W.D., 1971. Geometry for the selfish herd. J. Theor. Biol. 31, 295–311.

Hansen, I.I., Christiansen, F., Hansen, H.S., Braastad, B., Bakken, M., 2001. Variation behavioral responses of ewes towards predator related stimuli. Appl. Anim. Behav. Sci. 70, 227–237.

Harger, C.M., 1928. Frontier Days. Macrae Smith Company, Philadelphia.

Hastings, B.E., Abbot, D.E., George, L.M., Stradler, S.G., 1992. Stress factors influencing plasma cortisol levels and adrenal weights in Chinese Water deer. Reds. Vet. Sci. 53, 375–380.

Hatfield, J.S., 2006. The genetic basis of hair whorl, handedness, and other phenotypes. Med. Hypotheses 66, 708–714.

Hedigar, H., 1955. Studies of the Psychology and Behavior of Captive Animals in Zoos and Circuses. Criterion Books, New York.

Hedigar, H., 1968. The Psychology and Behavior of Animals in Zoo and Circuses. Dover, New York.

Hemmer, H., 1990. Domestication: The Decline of Environmental Appreciation. Cambridge University Press.

Hemsworth, P.H., Barnett, J.L., Hansen, C., Gonyou, H.W., 1986. The heritability of the trait fear of humans and the association between this trait and subsequent reproductive performance of gilts. Appl. Anim. Behav. Sci. 25, 85–95.

Hemsworth, P.H., Barnett, J.L., Treasy, D., Madgwick, P., 1990. The heritability of the trait fear of humans and the association between this trait and subsequent reproductive performance of gilts. Appl. Anim. Behav. Sci. 25, 85–95.

Hirsh, J., 1963. Behavior genetics and individuality understood. Science 142, 1436–1442.

Hohenboken, W.D., 1987. Behavioral genetics. Vet. Clin. North Am. Food Anim. Prac. 3, 217–229.

Holl, J.W., Rohrer, G.A., Brown-Brandi, T.M., 2010. Estimates of genetic parameters among scale activity scores in growth and fatness in pigs. J. Anim. Sci. 88, 455–459.

Holmes, J.H.G., Robinson, D.W., Ashmore, C.R., 1972. Blood lactic acid and behavior in cattle with hereditary muscular hypertrophy. J. Anim. Sci. 35, 1011–1013.

Hoppe, S., Brandt, H.R., Konig, S., Erhardt, G., Gauly, M., 2010. Temperament traits in beef calf measured under field conditions and relationships on performance. J. Anim. Sci. 88, 1982–1989.

Horton, B.J., Miller, D.L., 2011. Validation of an algorithm for real time measurement of sheep activity in confinement by recording movement within a commercial weighing crate. Appl. Anim. Behav. Sci. 129, 74–82.

Hutson, G.D., 1980. Visual field, restricted vision and sheep movement in laneways. Appl. Anim. Ethol. 6, 233–240.

Hutson, G.D., 1993. Behavioral principles of sheep handling. In: Grandin, T. (Ed.), Livestock Handling and Transport. CABI, Wallingford, UK, pp. 127–146.

Jacobs, G.H., Deegan II, J.S., Neitz, J., 1998. Photopigment basis for dichromatic color vision in cows, goats, and sheep. Vis. Neurosci. 15, 581–584.

Jansen, A., Lohmann, H., Scharfe, S., Sehlmeyer, C., Deppe, M., Knecht, S., 2007. The association between scalp hair whorl direction, handedness and hemispheric language dominance: is there a common genetic basis of lateralization? NeuroImage 35 (2), 853–861.

Keeler, C.E., 1942. The association of the black (non-agouti) gene with behavior in the Norway rat. J. Heredity 33, 371–384.

Keeler, C.E., 1975. Genetics of behavior variations in color phases of the red fox. In: Fox, M.W. (Ed.), The Wild Canida: Their Systematics, Behavioral Ecology and Evolution. Van Nostrand-Reinhold, New York, p. 1.

Keeler, C.E., 1947. Coat color, physique, and temperament. J. Heredity 38, 271–277.

Keeler, C.E., Ridgway, S., Lipscomb, L., Fromme, E., 1968. The genetics of adrenal size and lameness in color phase foxs. J. Heredity 59, 82–84.

Kendall, K., Ley, J., 2006. Cat ownership in Australia: barriers to ownership and behavior. J. Vet. Behav.: Clin. Appl. Res. 1 (1), 5–16.

Kerr, S.G.C., Wood-Gush, D.G.M., 1985. Investigation of relationships between behavioral and production measures of dairy heifers. Appl. Anim. Behav. Sci. 15, 181–182.

Kerr, S.G.C., Wood-Gush, D.G.M., 1987. The development of behavior patterns and temperament in dairy heifers. Behav. Processes 15, 1–16.

Kiley-Worthington, M., 1976. Tall movements of ungulates, canids and felids with particular reference to their causation and function as displays. Behaviour 56, 69–115.

Kilgour, R., 1971. Animal handling in words. Pertinent behavior studies. Proceedings of the 13th Meat Industry Research Conference. Hamilton, NZ pp. 9–12.

Kilgour, R., 1975. The open-field test as an assessment of the temperament of dairy cows. Anim. Behav. 23, 615–624.

Kilgour, R., Dalton, D.C., 1984. Livestock Behaviour. University of New South Wales Press, Sydney, Australia.

Kilgour, R., de Langen, H., 1970. Stress in sheep resulting from management practices. Proc. N. Z. Soc. Anim. Prod. 30, 65–76.

King, D.E., Scheuble-Pfeiffer, C.E., Randel, R.D., Welsh, T.H., Oliphant, R.A., Baird, B.E., 2006. Influence of animal temperament and stress responsiveness on carcass quality and beef tenderness of feedlot cattle. Meat Sci. 74, 546–556.

Klar, A.J.S., 2003. Human handedness and scalp hair-whorl direction develop from a common genetic mechanism. Genetics 165, 269–276.

Krushinski, L.V., 1961. Animal behavior. Its normal and abnormal development. In: Wortis, J. (Ed.), International Behavioral Sciences Service, Consultants Bureau, New York (Original Russian version published by Moscow University Press).

Lanier, E.K., Friend, T.H., Bushong, D.M., Knabe, D.A., Chamney, T.H., Lay, D.G., 1995. Swine habituation as a model for eustress and distress in the pig. J. Anim. Sci. 73 (Suppl. 1), 126 (Abstract).

Lanier, J., Grandin, T., Green, R.D., Avery, D., McGee, K., 2000. The relationship between sudden intermittent movements and sounds and temperament. J. Anim. Sci. 78, 1467–1474.

Lanier, J.L., Grandin, T., 2002. The relationship between *Bos taurus* feedlot cattle temperament, and cannon bone measurements. Proc. Western Section Am. Soc., Anim. Sci. 53.

Lansade, France Bouissou, M., Erhard, H.W., 2008. Reactivity to isolation and association with conspecifics: a temperament trait stable across time and situations. Appl. Anim. Behav. Sci. 105, 355–373.

Lasater, L.M., 1972. The Lasater Philosophy of Cattle Raising. University of Texas Press, El Paso.

Lawrence, A.B., Terlouw, E.M.C., Illius, A.W., 1991. Individual differences in behavioral responses of pigs exposed to non-social and social challenges. Appl. Anim. Behav. Sci. 30, 73–78.

Le Neindre, P., Trillat, G., Sapa, F., Menissier, F., Bonnet, J.N., Chupin, J.M., 1995. Individual differences in docility of LImousin cattle. J. Anim. Sci. 72, 2249–2253.

Leiner, L., Fendt, M., 2011. Behavioral fear and heart rate response of horses after exposure to novel objects: effects of habituation. Appl. Anim. Behav. Sci. 131, 104–109.

Lennon, K.L., Hebart, M.L., Hyund, P.I., 2009. Physiological response to isolation in Merino ewes of differing temperament. Proceedings of the 43rd Congress of International Society Applied Ethology Cairns. Queensland <http://hdl.handle.net>.

Lensink, B.J., Raussi, S., Bovin, X., Veissier, I., 2001. Reactions of calves to handling depend on housing condition and previous experience with humans. Appl. Anim. Behav. Sci. 70 (3), 187–199.

Lewis, G.R.G., Hulbert, L.E., McGlone, J.J., 2008. Novelty causes elevated heart rate and immune changes in pigs exposed to a handling alley and ramps. Livestock Sci. 116, 338–341.

Ligour, S., Fouiquie, D., Sebe, F., Bouix, J., Boissy, B., 2011. Assessment of sociability in farm animals: the use of the arena test for lambs. Appl. Anim. Behav. Sci. 135, 57–62.

Lynch, J.J., Alexander, G., 1973. The Pastoral Industries of Australia. University Press, Sydney, Australia.

Lyons, D.M., 1987. Individual differences in temperament of dairy goats and the inhibition of milk ejection. Appl. Anim. Behav. Sci. 22, 269–282.

Martins, C.E.N., Quadros, S.A.F., Trindade, J.P.P., Quadros, F.L.F., Costa, J.H.C., Raduenz, G., 2009. Shape and function in Braford cows: the body shape as an indicative of performance and temperament. Arch. Zootec. 58 (223), 425–433.

Matthews, L.R., 1993. Deer handling and transport. In: Grandin, T. (Ed.), Livestock Handling and Transport. CABI, Wallingford, UK, pp. 253–272.

Mayr, E., 1958. Behavior and systematics. In: Roe, A., Simpson, G.G. (Eds.), Behavior and Evolution. Yale University Press, New Haven, pp. 341–362.

McCann, J.S., Heird, J.C., Bell, R.W., Lutherer, L.O., 1988a. Normal and more highly reactive horses, I. Heart rate, respiration rate and behavioral observations. Appl. Anim. Behav. Sci. 19, 201–214.

McCann, J.S., Bell, R.W., Lutherer, L.O., 1988b. Normal and more highly reactive horses. II. The effects of handing and reserpine on the cardiac response to stimuli. Appl. Anim. Behav. Sci. 192, 215–226.

Moen, A.A., 1973. Wildlife Ecology. W.H. Freeman & Co., San Francisco.

Morris, C.L., Grandin, T., Irlbeck, N.A., 2011. Companion animals symposium: environmental enrichment for companion, exotic, and laboratory animals. J. Anim. Sci. 89 (12), 4227–4238.

Morton, D.J., Anderson, E., Foggin, C.M., Kock, M.D., Tran, E.P., 1995. Plasma cortisol as an indicator of stress due to capture and translation in wildlife species. Vet. Rec. 136, 60–63.

Munksgaard, L., DePassille, A.M., Rushen, B., Herskin, M.S., Kristensen, A.M., 2001. Dairy cows' fear of people: social learning, milk yield and behavior at milking. Appl. Anim. Behav. Sci. 73, 15–26.

Murphey, R.M., Moura Duarte, F.A., Torres Penedo, M.C., 1980. Approachability of bovine cattle in pasture: breed comparison and a breed x treatment analysis. Behav. Genet. 10, 171–181.

Murphey, R.M., Moura Duarte, F.A., Torres Penedo, M.C., 1981. Responses of cattle to humans in open spaces: breed comparison and approach–avoidance relationships. Behav. Genet. 11, 37–48.

O'Blesness, G.V., Van Vleck, L.D., Henderson, C.R., 1960. Heritabilities of some type appraisal traits and their genetic and phenotypic correlations with production. J. Dairy Sci. 42, 1490–1498.

Olsson, A., Nearing, K.I., Phelps, E.A., 2007. Learning fears by observing others: the neural systems of social fear transmission. Soc. Cog. Affect. Neurosci. 2, 3–11.

Olsson, A., Phelps, E.A., 2007. Social learning of fear. Nat. Neurosci. 10, 1095–1102.

Orihuela, J.A., Solano, J.J., 1994. Relationship between order of entry in slaughterhouse raceway and time to traverse raceway. Appl. Anim. Behav. Sci. 40, 313–317.

Pajor, F., Szentlek, A., Laczo, E., Tozer, J., Poti, P., 2008. The effect of temperament on weight gain on Hungarian Merino, German Merino and German Blackface lambs. Arch. Tierz, Dummerstorf 5, 247–254.

Panksepp, J., 1998. Affective Neuroscience: The Foundations of Animal and Human Emotions. Oxford University Press, NY.

Panksepp, J., 2011. The basic emotional circuits of mammalian brains. Do animals have affective lives? Neurosci. Biobehav. Rev. 35, 1791–1804.

Parsons, P.A., 1988. Behavior, stress, and variability. Behav. Genet. 18, 293–3008.

Pascoe, P.J., 1986. Humaneness of an electroimmobilization unit for cattle. Am. J. Vet. Res. 10, 2252–2256.

Pederson, B.K., 1992. Comprehensive evaluation of well being in pigs: environmental enrichment and pen space allowance, PhD. Thesis, University of Illinois, Urbana.

Perez-Guisado, J., Lopez-Rodriguez, Munoz-Serrano, A., 2006. Heritability of dominant–aggressive behaviour in English Cocker Spaniels. Appl. Anim. Behav. Sci. 100 (3–4), 219–227.

Petherick, J.C., Doogan, V.J., Holroyd, R.C., Olsson, P., Venus, B.K., 2009. Quality of handling and yarding environment and beef cattle temperament: relationship between flight speed and fear of humans. Appl. Anim. Behav. Sci. 120, 18–27.

Plomin, R., 1990. The role of inheritance in behavior. Science 248, 183–188.

Podberscek, A.L., Serpell, J.A., 1996. The English cocker spaniel: preliminary findings on aggressive behaviour. Appl. Anim. Behav. Sci. 47 (1–2), 75–89.

Price, E.O., 1984. Behavioral aspects of animal domestication. Q. Rev. Biol. 59, 1–32.

Price, E.O., 1998. Behavioral genetics and the process of domestication. Genetics and the Behavior of Domestic Animals. Academic Press.

Prince, J.H., 1977. The eye and vision. In: Swenson, M.J. (Ed.), Duke's Physiology of Domestic Animals. Cornell University Press, Ithaca, NY.

Randle, H.D., 1998. Facial hair whorl position and temperament in cattle. Appl. Anim. Behav. Sci. 56, 139–147.

Reale, D., Gallent, B.Y., LeBlanc, M., Festa-Biarchez, M., 2000. Consistency of temperament in bighorn ewes and correlates with behavior and its history. Anim. Behav. 30, 589–597.

Reefman, N., Kaszas, F.B., Wechler, B., Gygax, L., 2009. Ear and tail positions as indicators of emotional valance in sheep. Appl. Anim. Behav. Sci. 118, 199–207.

Reynolds, S.M., Berridge, K.C., 2008. Emotional environmental retune of the balance of appetitive versus fearful functions in the nucleus accumbens. *Nat. Neurosci.* 11, 423–425 and 589–597.

Ried, R.L., Mills, S.C., 1962. Studies of Carbohydrate metabolism of sheep, XVI, The adrenal response to physiological stress. Aust, J. Agric. Res. 13, 282–294.

Robins, A., Phillips, C., 2009. Lateralised visual processing in domestic cattle herds responding to novel and familiar stimuli. Laterality: Asymmetries Body, Brain Cogn. 15 (5).

Roche, B.W., 1988. Coacher mustering. Queensl. Agric. J. 114, 215–216.

Rogan, M.T., LeDoux, J.E., 1996. Emotion: systems, cells, and synaptic plasticity. *Cell* (Cambridge, Mass.) 85, 369–475.

Rogers, L., Andrew, R.J., 2002. Comparative Vertebrate Lateralization. Cambridge University Press.

Sandem, A.I., Janczak, A.M., Salle, R., Braastad, B.O., 2006. The use of diazepam as a pharmacological validation of eye white as an indicator of emotional state in dairy cows. Appl. Anim. Behav. Sci. 96, 177–183.

Sant-Anna, A., Paranhos de Costa, M.J.R., Valente, T.S., Baldi, F., Golvao de Albuquerque, L., 2012. Genetic and phenotypic association of flight speed and temperament, visual scores in Nellore beef cattle. International Symposium on Beef cattle Welfare. University of Saskatchewan, Saskatoon, Canada, p. 21–22.

Saslow, C.A., 1999. Factors affecting stimulus visibility for horses. Appl. Anim. Behav. Sci. 61 (4), 273–284.

Sato, S., 1978. Factors associated with temperament in beef cattle. Jpn, J. Zootech. Sci. 52, 595–605.

Schmidt, H., Depner, M., Kabesch, M., 2008. Medial Position and Counterclockwise Rotation of the Parietal Scalp Hair-Whorl as a Possible Indicator for Non-Right Handedness. Sci. World J. 8, 848–854.

Schupe, W.L., 1978. Transporting sheep to pastures and market. Technical Paper No. 78–6008, American Society of Agricultural Engineers, St. Joseph, MI.

Scott, J.P., Fuller, J.L., 1965. Genetics and the Social Behavior of the Dog. University of Chicago Press, Chicago.

Scott, N.M., Weinberg, S.M., Neiswanger, K., Brandon, C.A., Marazita, M.L., 2005. Hair whorls and handedness: informative phenotypic markers in nonsyndromic cleft lip with or without cleft palate (NS CL/P) and their unaffected relatives. Am. J. Med. Genet. 136A (2), 158–161.

Seabrook, M. (Ed.), 1987. The role of the stockman in livestock productivity and management. Proceedings of a Seminar Report EUR 10982 EN. Commission of the European Communities, Luxenburg.

Sebastian, T., Watts, J., Stookey, J., Buchanan, F., Waldner, C., 2011. Temperament in beef cattle: methods of measurement and their relationship to production. Can. *J. Anim. Sci.* 91 (4), 557–565.

Shinozaki, A.Y., Hosaka, et al., 2010. Topography of ganglian cells and photo receptors in the sheep retina. J. Comp. Neurol. 518, 2305–2315.

Shirley, K.L., Garrick, D.J., Grandin, T., Deesing, M., 2006. Inheritance of facial hair whorls attributes in Holstein cattle. Proc. Western Section, Am. Soc. Anim. Sci. 57, 62–65.

Shrode, R.R., Hammack, S.P., 1971. Chute behavior of yearling beef cattle. J. Anim. Sci. 33, 193 (Abstract).

Smith, D.W., Gong, B.T., 1973. Scalp hair patterning as a clue to early fetal development. J. Pediatr. 83, 374–380.

Smith, D.W., Gong, B.T., 1974. Scalp hair patterning: its origin and significance relative to early fetal brain development. Teratology 9, 17–34.

Stricklin, W.R., Heislet, C.E., Wilson, L.L., 1980. Heritaiblity of temperament in beef cattle. J. Anim. Sci. 5 (1. Suppl. 1), 109 (Abstract).

Syme, L.A., Elphick, G.R., 1982. Heart-rate and the behavior of sheep in yards. Appl. Anim. Ethol. 9, 31–35.

Tanida, H., Miura, A., Tanaka, T., Yosimoto, T., 1996. Behavioral responses of piglets to darkness and shadows. Appl. Anim. Behav. Sci. 49, 173–183.

Tanner, M., Grandin, T., Cattell, M., Deesing, M., 1994. The relationship between facial hair whorls and milking parlor side preference. J. Anim. Sci. 72 (Suppl. 1), 207 (Abstract).

Tellington-Jones, L., Bruns, V., 1985. The Tellington-Jones Equine Awareness Method. Breakthrough Publications, Millwood, NY.

Tirosh, E., Jaffe, M., Dar, H., 1987. The clinical significance of multiple hair whorls and their association with unusual dermatoglyphics and dysmorphic features in mentally retarded Israeli children. Eur. J. Pediatr. 146 (6), 568–570.

Torres-Hernandez, G., Hohenboken, W., 1979. An attempt to assess rates of emotionality in crossbred ewes. Appl. Anim. Ethol. 5, 71–83.

Tozser, J., Maros, K., Szentleleki, A., Zandoki, R., Nikodemusz, E., Balazs, F., Bailo, A., Alfoldi, L., 2003. Evaluation of temperament in cows of different age and bulls of different colour variety. Czech J. Anim. Sci. 48 (8), 344–348.

Trut, L.N., 1999. Early canid domestication: the farm-fox experiment: foxes bred for tamability in a 40-year experiment exhibit remarkable transformations that suggest an interplay between behavioral genetics and development. Am. Sci. 87 (2), 160–169.

Trut, L.N., Iliushina, I.Z., Prasolova, L.A., Kim., A.A., 1997. The hooded allele and selection of wild Norway rats Rattus norvegicus for behavior. Genetika 33 (8), 1156–1161 (Translation in English available in: *Russion Journal of Genetics.* V. 33 (8) pp. 983–989.

Tulloh, N.M., 1961. Behavior of cattle in yards. II. A study of temperament. Anim. Behav. 9, 25–30.

Turner, S.P., Gibbons, J.M., Haskell, M.J., 2011. Developing and validating measures of temperament in livestock. Primatol. Monogr. 201–224, Part 1.

VanWagoner, H.C., Bailey, D.W., Kress, O.D., Anderson, D.C., Davis, K.C., 2006. Differences among beef sire breeds and relationships between terrain use and performance when daughters graze foothills range as cows. Appl. Anim. Behav. Sci. 97, 105–121.

Vetters, M.D., Engle, T.E., Ahola, J.K., Grandin, T., 2013. Comparison of flight speed and exit score as measurements of temperament in beef cattle. J. Anim. Sci. 91, 374–381.

Visser, E.K., van Reenan, C.G., Hopster, H., Schindler, M.B.H., Knapp, J.H., 2001. Quantifying aspects of young horses' temperament: consistency of variables. Appl. Anim. Behav. Sci. 74, 241–258.

Voisinet, B.D., Grandin, T., Tatum, J.D., O'Conner, S.F., Struthers, J., 1997. Feedlot cattle with calm temperaments have higher average daily gains than cattle with excitable temperaments. J. Anim. Sci. 75, 892–896.

Wagnon, K.A., Loy, R.G., Rollins, W.C., Carroll, F.D., 1966. Social dominance in a herd of Angus, Hereford and Shorthorn cows. Anim. Behav. 14, 474–479.

Weber, B., Hoppe, C., Faber, J., Axmacher, N., FlieBbach, K., Mormann, F., et al., 2006. Association between scalp hair whorl direction and hemispheric language dominance. NeuroImage 30 (2), 539–543.

Whateley, J., Kilgore, R., Dalton, D.C., 1974. Behavior of hill country sheep breeds during farming routines. Prac. NZ, Soc. Anim. Prod. 34, 28–36.

William, R.L., Karas, G.G., Henderson, D.C., 1964. Partial acquisition and extinction of an avoidance response in two breeds of swine. J. Comp. Physiol. Psychol. 57, 117–122.

Wolf, B.T., McBride, S.D., Lewis, R.M., Davies, M.H., Haresign, W., 2008. Estimates of the genetic parameters and repeatability of behavioural traits of sheep in an arena test. Appl. Anim. Behav. Sci. 112, 68–80.

Wright, S., 1978. The relation of livestock breeding to theories of evolution. J. Anim. Sci. 46, 1192–1200.

Yoder, C.L., Matecca, C., Casssady, J.P., Fowers, W.L., Price, S., See, M.T., 2010. Breed differences in pig temperament scores during a performance test and their phenotypic relationship with performance. Livestock Prod. Sci. 138, 93–101.

Yousefi-Nooraie, R., Moratz-Hedjri, S., 2008. Dermatoglyphic asymmetry and hair whorl patterns in schizophrenic and bipolar patients. Psychiatry Res. 157 (1–3), 247–250.

Zavy, M.T., Juniewicz, P.E., Williams, A.P., von Tungeln, D.L., 1992. Effects of initial restraint, weaning, and transport stress on baseline and ACTH stimulated cortisol responses in beef calves of different genotypes. Am. J. Vet. Res. 53, 352–357.

Zarate, J.C.O., Zarate, C.O.O., 1991. Hairwhorl and handedness. Brain and Cognition 16, 228–230.

FURTHER READING

Amaral, A., 1977. Mustang Life and Legends of Nevada's Wild Horses. University of Nevada Press, Reno.

Beattie, V.E., Walker, N., Sneddon, I.A., 1995. Effect of rearing environment and change of environment on the behavior of gilts. *Appl. Anim. Behav. Sci.* 46, 57–65.

Broadhurst, P.L., 1960. Analysis of maternal effects in the inheritance of behavior. *Anim. Behav.* 9, 129–141.

Fowler, M.E., 1978. Restraint and Handling of Wild and Domestic Animals. Iowa University Press, Ames, IA.

Keeler, C.E., King, H.D., 1947. Multiple effects of coat color genes in the Norway rat, with special reference to temperament and domestication. *J. Comp. Psychol.* 34, 241–250.

Kilgour, R.J., Melville, G.J., Greenwood, P.L., 2006. Individual differences in the reaction of beef cattle to situations involving social isolation, close proximity of humans, restraint and novelty. Appl. Anim. Behav. Sci. 99, 21–40.

Lay, D.C., FGriend, T.H., Grissom, K.K., Hale, R.L., Bowers, C.C., 1992. Novel breeding box has variable effect on heartrate and cortisol response of cattle. J. Anim. Sci. 35, 1–10.

Chapter 5

Reproductive and Maternal Behavior of Livestock

Peter J. Chenoweth*, Antonio J. Landaeta-Hernández[†], and Cornelia Flöercke[‡]

**Charles Sturt University, Wagga Wagga, Australia; [†]Universidad del Zulia, Facultad de Ciencias Veterinarias, Unidad de Investigaciones Zootécnicas, Maracaibo, Venezuela; [‡]Department of Animal Sciences, Colorado State University, Fort Collins, Colorado, USA*

GENERAL INTRODUCTION

Species survival is ensured by the regular production of viable offspring. This in turn depends upon the successful execution of a number of processes which are behavior driven. These processes include copulation and fertilization, pregnancy maintenance, and the subsequent successful delivery and raising of the young. In pursuit of these outcomes, mammals have evolved widely differing mating and birthing strategies in response to a number of influences which include susceptibility to predation, geographical dispersion and social contexts. Additional demands on sexual and reproductive behavior, social tolerance and adaptive ability have been imposed by domestication (Jensen and Andersson, 2005; McPherson and Chenoweth, 2012). The fact that only 6% of current species of ungulates and elephants are domesticated (Tennessen and Hudson, 1981) indicates that those species which did become domesticated had predisposing behaviors and traits, many of which are reproductive or maternal (Chenoweth and Landaeta-Hernández, 1998). In the process of domestication, natural selection, which occurs during the natural breeding process, is replaced by artificial selection, which occurs prior to breeding (Katz, 2008). Artificial selection accentuates certain traits while relaxing selection pressure on others, with the latter causing increased genetic variation. In addition, some behaviors which are important for reproduction and/or survival lose adaptive significance (Price, 1984), although this can also occur in the absence of domestication (Gorelick and Heng, 2011).

The seasonal nature of the breeding cycle in many species has evolved to ensure that offspring are born at the most opportune time of the year for survival and growth as well as for re-mating of the dam, with this usually being in spring. However, even in those species which have strongly seasonal

Genetics and the Behavior of Domestic Animals. DOI: http://dx.doi.org/10.1016/B978-0-12-394586-0.00005-6
© 2014 Elsevier Inc. All rights reserved.

breeding patterns, male reproductive capability is maintained throughout the year, even though libido, spermatogenesis, and male fertility may be lowered during the non-breeding period. Although artificial selection has changed many characteristics of domestic animals, it has had relatively little effect in altering the breeding patterns of seasonal breeders despite economic pressures to do so, for example in horses (Chemineau *et al.*, 2008).

Environmental change is also relevant to this discussion on genetic aspects of reproductive and maternal behavior, with the greatest effects being likely in those species which rely upon seasonal changes for breeding, hibernation, or migration (McPherson and Chenoweth, 2012). For example, the time of mating in many mammalian species has evolved within their environmental context in which seasonal differences occur in nutritional resources. This is most evident in seasonal breeders which rely on photoperiod as well as other cues such as ambient temperature and food availability to initiate breeding behavior (Chemineau *et al.*, 2008; Scaramuzzi and Martin, 2008). In addition, adverse effects may derive from direct anthropogenic influences on mating systems and efficiency (Lane *et al.*, 2011) such as occurs with the changing of habitats and phonologies (either by changing the local environment or relocations) and increasing the presence of environmental factors such as endocrine disruptors.

In summary, each livestock species has developed a distinct repertoire of courtship, copulatory, and maternal behaviors to ensure optimal reproductive success within its native environment. Those features of domestication which reinforce these behaviors should lead to greater reproductive "success" than those which are in conflict. Thus a better understanding of these behaviors has practical considerations for reproductive management of captive and domestic animals. The ensuing discussion will provide an overview of the reproductive and maternal behaviors among different domestic animals, emphasizing those aspects which are of evolutionary and genetic interest. As there is a significant amount of relevant information available on domestic cattle, this will be used to illustrate a number of the phenomena and principles involved.

LIVESTOCK REPRODUCTIVE BEHAVIOR

In general, livestock species are both polygynous and promiscuous. Most evolved as prey animals (although some, such as swine, are both prey and predator) and generally exist in social groups, which may be matriarchal or patriarchal in nature. Females undergo estrous cycles in which there are defined periods of mating. This combination of prey status, female estrus and grouping has led to a variety of mating strategies.

Female sexual behavior in domestic animals has been classified into categories of attractivity, proceptivity, and receptivity (Beach, 1976). Here, attractivity represents the female's passive ability to stimulate male interest, usually measured by male approach responses such as numbers of flehmens. Proceptive behaviors are those female behaviors that initiate or maintain male sexual interest or activity. Representative behaviors here include male-seeking behavior and female–female mounting. Receptivity represents those female actions or postures that enable successful copulation by the male. Such behaviors include immobility (Becke and Serrano, 2002) and tail diversion (Katz and McDonald, 1992). Estrus behavior has also been described in terms of both duration and intensity, with the latter including such observations as mounting activity and mobility (Landaeta-Hernández *et al.*, 2004b).

Although the estrous cycle is governed by hormonal events, which in turn may be seasonally influenced, the expression, or signaling, of estrus is behavioral and can be influenced by a number of factors including stress and genetics (Landaeta-Hernández *et al.*, 2002a). Examples of the former include suppressive effects of extreme weather or social stress. Breed, line, and individual differences in the intensity of estrus behavior have been reported in livestock (Landaeta-Hernández *et al.*, 2004b; Løvendahl and Chagunda, 2009; Price, 1985; Rydhmer *et al.*, 1994). Other factors influencing estrus expression include nutritional (such as estrogenic feedstuffs), pathological (such as granulose cell tumors in horses), pain (e.g., foot abscess in dairy cattle), different forms of intersex (such as male pseudohermaphrodites), and stability and type of flooring in cattle.

Male livestock, in general, share a number of commonalities in their sexual behavior. Libido, or sex drive, has a strong genetic basis, tends to be active year-round and can often be revived by exposure to new stimuli. The copulatory process is usually rapid, particularly in prey species such as most food animals. A behavioral progression can often be observed in which males tend to be first attracted to individual females in late pro-estrus and where the ultimate cue for attempted copulation is immobility, or standing behavior of the female. Male courtship behaviors vary, but all seek to elicit or detect standing behavior in the female. In turn, females play a major role in mate selection and often determine the timing of mating and the identity of the successful male(s). Males tend to be hierarchical and compete for eligible females, with successful males generally being those higher in social rank. In general males of high status have greater access to females than those of lower rank (Price, 1987). This may be advantageous if dominance is associated with other favorable reproduction and production traits. Unfortunately, however, social dominance in male livestock does not appear to be synonymous with either fertility or mating ability (Price, 1985), leading to the possibility that dominant males may, on occasion, act to depress herd or flock fertility (Chenoweth, 1994a). Competition can increase male sexual response (Mader and Price, 1984), although audience effects may also act to inhibit male sexual activity (Lindsay and Ellsmore, 1968).

Maternal behavior includes those behaviors of the dam which occur around the time of birth and which are associated with the responsiveness, delivery, attentiveness, and concomitant care to guarantee greatest possible survival of the young (Buddenberg *et al.*, 1986). Included here are separation behavior, shelter seeking, nest building, parturition, cleaning and stimulating the neonate to suckle, and establishment of the maternal–offspring bond. Those unique species-specific behaviors which have evolved in relation to birth and care of the neonate are essential for species survival (Gonyou and Stookey, 1987). In many cases these same behaviors also represent valued production traits in domestic livestock. Proper expression of such behavior is, however, under threat in modern livestock systems where genetic selection for extensive conditions is favored (Simm *et al.*, 1996). Here, attempts to increase productivity place unprecedented demands upon mothering ability while procedures which congregate dams within unnatural environments can lead to undue stress and failure of bonding between mother and young (Edwards, 1983).

In ungulates, the mother–young relationship was described by Walther (1965) who distinguished between "follower"- and "hider"-types. Aggregation behavior and breeding synchrony are often associated with follower-types (Estes, 1976; Singh *et al.*, 2010), allowing for greatest offspring survival in combination with maternal and group defense. Separation behavior is frequently observed in hider-types, where females disperse from the herd to give birth unaccompanied (Bowyer *et al.*, 1999; Langbein and Raasch, 2000; Rettie and Messier, 2001) to reduce predation and intraspecific aggression (Lent, 1974). Additionally, it has been suggested that in followers, mother–offspring acoustic recognition is mutual (Torriani *et al.*, 2006) whereas it is unidirectional in hiders (Marchant-Forde *et al.*, 2002; Torriani *et al.*, 2006). None the less, follower–hider categorization should be regarded as flexible (Ralls *et al.*, 1986) as some species employ either type depending upon environmental cues.

REPRODUCTIVE BEHAVIOR IN CATTLE

Cattle were domesticated relatively recently in terms of human prehistory. Prior to this, they were mainly located within groups, or herds, on open savannah and grassland range country, although some antecedents, such as the Gaur, might also be found in woodland fringes. Studies with feral cattle and conserved "wild" herds indicate that these groups were essentially matriarchal, comprising adult females and young offspring. Breeding tended to be dictated by seasons, with calving and breeding occurring during more favorable nutritional periods, usually in spring. Adult males would spend much of the year either solitary or in bachelor groups, joining the female herd when females were sexually active. Domesticated cattle breed all year round, although some seasonal influences are detectable in the cyclicity of *Bos indicus* cattle. For example, females of *Bos indicus* breeds show a preference for long-day breeding (Randel, 1984) with lowered cyclicity being evident during the late fall and winter months, even in subtropical and tropical environments (Chenoweth, 1994b; Randel, 1984; Taylor *et al.*, 1995).

Although mounting behavior is an important component of the cattle reproductive behavioral repertoire, it is not restricted to reproductive purposes only. Young calves show mounting behavior as early as 1 week of age, although this becomes more common as puberty approaches. In prepubertal animals, mounting is probably associated with play behavior as well as mimicry of adult behavior. In addition, mounting behavior plays a strong social role in cattle in which dominant animals tend to mount those which are more subordinate.

Male Reproductive Behavior

Bulls are sexually attracted to females primarily by the sight of females standing to be mounted. In fact, this attraction can be elicited by any inverted U structure which resembles the rear end of the female (Buechlmann, 1950). This is useful in semen collection centers which often employ steers as mount animals to reduce the risk of venereal disease transmission (Chenoweth, 1981). Pheromonal action is also useful in helping bulls to detect estrous females, although close physical proximity is probably necessary for this to occur in cattle (Jacobs *et al.*, 1980). Flehmen behavior, which is often displayed during courting, is considered to primarily represent an investigatory, rather than a cognitive process (Chenoweth and Landaeta-Hernández, 1998).

The most important special sense that range bulls employ to detect estrous females is vision (Chenoweth 1986; Chenoweth and Landaeta-Hernández, 1998). This is facilitated by formation of a sexually active group (SAG) comprising females in late pro-estrus and estrus which, although constantly in motion, usually stays within visual contact of bulls (Chenoweth, 1981; Williamson *et al.*, 1972). Females newly in estrus tend to be most attractive to bulls, and sexual interest can be restored in satiated males by the presentation of new stimuli. Bulls will form hierarchical groups in which social status influences relative reproductive success (Chenoweth and Landaeta-Hernández, 1998). For young bulls, a learning process may be necessary before they achieve competence and confidence in their mating ability (Boyd *et al.*, 1989; Landaeta-Hernández *et al.*, 2001). Once a bull achieves successful mounting, copulation usually follows rapidly with ejaculation being completed within 1–2 seconds of intromission (Seidel and Foote, 1969).

Bull Libido/Sex-Drive

In this discussion, the terms sex-drive and libido are used synonymously and refer to the eagerness or willingness of the bull to mount and service a female (Chenoweth, 1981). Bull libido can be measured in a relatively objective manner because: (a) bulls are polygynous and tend to distribute their services among receptive females; (b) the greatest single stimulus for bulls to mount and service is an immobile object that resembles the rear end of a female; (c) prestimulation of bulls increases their sexual response (and semen harvest); and (d) competition among bulls can increase their sexual response (Chenoweth, 1997). Despite these underlying principles, however, one study showed that six to eight tests were required of 2-year-old Angus and Hereford bulls to achieve reasonably repeatable results for both libido score and serving capacity, although fewer tests were required to identify the "best" bulls (Landaeta-Hernández *et al.*, 2001). For young bulls, a learning process is often necessary before competent mating is achieved (Boyd *et al.*, 1989; Chenoweth, 1994a; Landaeta-Hernández *et al.*, 2001; Lunstra, 1984). Such considerations probably contribute to study differences in which bull sex-drive was either favorably associated with aspects of herd fertility (Blockey, 1989; Blockey *et al.*, 1978, Godfrey and Lunstra, 1988) or not (Boyd *et al.*, 1991; Farin *et al.*, 1989).

Although there are anecdotal reports of bull breeds differing in sex-drive, relatively few definitive studies have been conducted on this question. Bulls of dairy breeds may generally be more sexually active than those of beef breeds (Amann and Almquist, 1976; Chenoweth, 1997; Hafez 1987) and *Bos indicus* bulls are generally considered to be more "sexually sluggish" than *Bos taurus* bulls (Chenoweth and Landaeta-Hernández, 1998) with one study from an A.I. center in India reporting that 22.6% of Sahiwal bulls were culled due to inadequate sex drive (Mukhopadhyay *et al.*, 2010). This is in contrast to the findings of a Brazilian survey of beef bulls of largely *Bos taurus* extraction (n = 30,700) in which 3.6–5.2% were rejected for sexual behavior problems (Menegassi *et al.*, 2011). In both Australian and U.S. studies, *Bos taurus* bulls performed better in sex-drive assessments than did *Bos indicus* bulls (Chenoweth and Landaeta-Hernández, 1998) even though the latter can be equally effective in detecting, serving and impregnating estrous females (Chenoweth, 1994a). This discrepancy may be explained in part by observations that *Bos indicus* bulls tend to be selective and shy breeders, and do not perform well in pen tests to assess sex drive (Chenoweth *et al.*, 1996). Zebu bulls are reputedly reluctant to breed with females of other breeds (Chenoweth and Landaeta-Hernández, 1998). Selectivity in breeding partners has also been reported among *Bos taurus* breeds in bull mating activity (Trautwein *et al.*, 1958). Although these studies have generally implicated the bull as being the partner that initiates and controls sexual contact, increasing evidence suggests that female choice might play a major role in such outcomes. Although it as yet unclear as to which gender plays the major role in such selectivity, observations on interactions between *Bos indicus* (Gyr and Brahman) and tropicalized *Bos taurus* (Criollo-Limonero) cattle in Venezuela indicate that female rejection of males of different breed-types does occur (Landaeta-Hernández, unpublished).

The role that genetic influences play on the expression of bull sex-drive has been confirmed in a number of studies. An early classic study was that of Bane (1954) who, despite raising six pairs of monozygous twin bulls on differing regimes, observed greater similarity within than among pairs in mating behavior and temperament. Subsequent studies have established that sires and lines are significant sources of bull sex-drive variation (see review by Chenoweth, 1981). A heritability estimate of 0.59 ± 0.16 was obtained for serving capacity in a study of 157 paternal half-sib bull groups in Australia (Blockey *et al.*, 1978). In addition, high sex-drive in bulls is not necessarily in those with either superior production traits (average daily gain or final test weight) or high social ranking (Ologun *et al.*, 1981). It is apparent that conventional breeding soundness criteria, such as scrotal circumference and semen traits, are not predictive of bull sex-drive (Chenoweth, 2000) indicating that these are separate traits. This raises the possibility that bulls of high sex-drive may not be either dominant or fertile, which has important implications for the management of cattle breeding programs.

Female Reproductive Behavior

The duration of observable estrus in *Bos indicus* females is usually shorter and its external signs less intense than in *Bos taurus* cattle (Chenoweth, 1994b; Galina *et al.*, 1996). Part of this difference may be due to difficulties encountered with estrus detection in *Bos indicus* females, especially as they tend to display behavioral estrus during the evening hours (Chenoweth, 1994b). In *Bos taurus* cattle, the mounted female is more likely to be in estrus than the mountee, particularly if she exhibits immobility (Baker and Seidel, 1985), whereas social status of *Bos indicus* females can influence an individual's chances of being ridden by another female (Mukasa-Mugerwa, 1989; Orihuela *et al.*, 1988). Also, some high-ranking *Bos indicus* females appear to not participate in mounting behavior at all, either actively or passively. If so, this differs from *Bos taurus* cattle in which cows higher in the social order are usually those which initiate mounting activity (Mylrae and Beilharz, 1964).

Females of dairy breeds reputedly exhibit more mounting behavior than those of beef breeds (Baker and Seidel, 1985), possibly because of greater selection for this trait in systems where males are largely or completely absent. In this respect, the widespread use of artificial insemination in dairy cattle has probably encouraged selection for overt estrous behavior as females showing weak estrus signs would tend not be inseminated (Hohenboken, 1987). Despite this, intense genetic selection towards highly efficient dairy cows has occurred over the past 50 years with probable adverse effects on reproductive efficiency and fertility (Lucy, 2001). This reinforces observations that high milk production in dairy cows may be associated with weak or irregular estrus (Morrow, 1966; Walsh *et al.*, 2011), with poor estrus expression being a major contributor to reduced fertility

(Walsh *et al.*, 2011). Despite this, the expression of standing heat and its visual detection by trained observers is still regarded as a reliable management tool (Lyimo *et al.*, 2000). In turn, the intensity of estrus expression has been associated with conception rates in both *Bos taurus* (Bonfert, 1955; Evans and Walsh, 2012) and *Bos indicus* derived females (Morales *et al.*, 1983). An heritability estimate of 0.21 was obtained for estrous intensity by Rottenstein and Touchberry (1957), who also obtained a within-year repeatability for estrus behavior score of 0.29.

In many of the studies quoted above, human observation was employed to detect and quantify estrus behavior. This has limitations due to human inability to provide both 24 h quality observation as well as detect subtle behavioral cues. In addition, factors such as climate, confinement, type of housing and flooring can influence estrus expression in female cattle (Landaeta-Hernández *et al.*, 2002a; Rodtian *et al.*, 1996). The relatively recent advent of electronic devices to monitor mounting behavior in cattle allows more accurate monitoring of such behaviors (At-Taras and Spahr, 2001), particularly with *Bos indicus* cattle (Cavalieri and Fitzpatrick, 1995). One study using this method found that measures of estrus activity were more heritable than fertility traits in several dairy breeds (Løvendahl and Chagunda, 2009).

Cattle Maternal Behavior

Cows in free-ranging herds are reported to leave the herd for calving, although this is understandably less common in more intensive pasture systems. Red Angus cattle appear to show a considerable degree of behavioral plasticity in calving behavior and calf defense patterns when approached by a strange object (e.g., a vehicle that differed from the familiar trucks on the ranch) (Floercke *et al.*, 2012). In post-parturient cows, individual differences in protection-, aggression-, and vocalization-behaviors towards the newborn calf exist when cows are approached by a vehicle. In the former study, 99% of cows were protective, 13% showed signs of aggression by lowering the head or pawing the ground and 78% vocalized towards the calf. The expression of these behaviors reflects individual differences in temperament and cows also differed in the level of "vigilance" towards the surroundings. Important criteria for selection of a birthing site include dry, soft bedding with provision of cover. Further influencing factors are the availability of nutrition, type of terrain, threat of predation, and the need for bonding with the calf.

The time of day at which most births occur varies in different reports. Edwards (1979) found no bias towards day or night calving whereas Keyserlingk and Weary (2007) reported increased calving in the late afternoon and evening; nevertheless both findings may simply reflect routine management practices. Females bond with their newborn very early in the postpartum period and this sensitive period between mother and young is triggered by amniotic fluids (Gonyou and Stookey, 1987; Lévy and Keller, 2009) and hormonal changes in the brain of the dam (Nowak *et al.*, 2000). Cattle have been described as a "hider" species as there is a preference for secretion of the young. This time of separation from the herd varies from a few hours to up to several days after birth. During this time the mother grazes within hearing distance and returns regularly to the calf (Langbein and Raasch, 2000). Older calves may be left under the watchful eye of a "nanny" cow under extensive conditions, with this behavior being particularly observed in *Bos indicus* breeds.

Large individual differences have been observed to occur in calving behavior during the perinatal period (Kunowska-Slósarz and Różańska, 2009). However, breed differences were also reported in a study by Le Neindre (1989) in which maternal behavior exhibited by a beef breed (Salers) was more intense than that shown by a dairy breed (Friesian). It has also been observed that beef females tend to leave the herd at calving more readily than dairy females (Lidfors *et al.*, 1994). Dairy cattle have been selected for less intense maternal behavior than beef breeds, in which strong maternal behavior is valued (Le Neindre, 1989).

Recently, extensive beef systems in North America have been facing the challenge of increased predation loss by wolves (Bangs and Fritts, 1996; Clark and Johnson, 2009), black vultures, and golden eagles (Avery and Cummings, 2004). Selection since the late 1990s towards calmer temperament cattle (Hyde, 2010) may have reduced maternal protectiveness. Ranchers have reported mothering problems and maternal neglect in very calm females resulting in weak calves and calf starvation (Sime and Bangs, 2010). Females of *Bos indicus* breeds are generally regarded as being strongly protective mothers, an observation supported by Williams *et al.* (1991), who reported a direct positive additive genetic influence for weaning rate of calves in Brahmans. In Red Angus cows, individual differences in protectiveness towards the calf have been found (Floercke *et al.*, 2012). The hair whorl pattern located high on the forehead may be used as an indicator for selection towards more reactive cattle (Grandin *et al.*, 1995). Selection for superior maternal ability in regions with high predation pressure could have advantages for ranchers. Temperament score at calving, a supposed indicator of maternal ability, was shown to differ among beef breeds in New Zealand. Heritabilities for behavioral traits were generally low (Morris *et al.*, 1994). However, a German study showed that Angus cows were more protective than Simmentals (Hoppe *et al.*, 2008) although maternal protective behavior was not associated with weight gain of the offspring. From a production standpoint, strong maternal protective behavior may have less importance in highly intensive systems with few predators, such as Germany, than in more extensive areas in which predation is common. Last, an important contributor to calf losses in *Bos indicus* (e.g., Guzerat) cattle is the development of oversized or "boule" teats in dams which hinder suckling (Frisch, 1982; Holroyd, 1987; Schmidek *et al.*, 2008).

Another important cause of *Bos indicus* calf losses is the neonatal weakness syndrome (also known as "weak calf" and "dummy calf" syndrome) which is associated with high morbidity and mortality in young *Bos indicus* calves (DeRouen *et al.*, 1967; Franke *et al.*, 1975; Landaeta-Hernández *et al.*, 2004c; Radostits *et al.*, 1994). Affected calves show clinical signs which include poor cognitive and sensory responses, poor or absent suckling ability, difficulty in standing and movement and marked intolerance to cold weather (see review by Landaeta-Hernández *et al.*, 2002b). Although similar symptoms occur in *Bos taurus* calves, the relatively high occurrence of this syndrome in *Bos indicus* cattle in general, as well as its association with certain sires and breed types, suggests a genetic association (Landaeta-Hernández *et al.*, 2004b; Rowan, 1992). More recently, a congenital myasthenic syndrome caused by homozigosity for 20 base pair deletion in the CHRNE gene (CHRNE 470del20) was identified in Brahman cattle in South Africa. The CHRNE 470del20 leads to a non-functional acetylcholine receptor causing progressive muscle weakness and mortality in young calves (Thompson *et al.*, 2003, 2007). Rapid progress in defining cattle genetic traits can be expected in association with the complete sequencing of the bovine genome (Elsick *et al.*, 2009) which should ultimately permit genetic modification and enhanced selection for desired traits, such as improved maternal ability, while ensuring other favorable traits are not compromised.

Biostimulation in Cattle

The effect of the male, or male-like factors, on the physiological and reproductive status of exposed females is part of a phenomenon termed biostimulation (Chenoweth, 1983; Chenoweth and Landaeta-Hernández, 1998). Although biostimulatory effects do occur in cattle (Chenoweth and Spitzer, 1995), they tend to be less evident than in other livestock species such as sheep or swine (Rekwot *et al.*, 2001). Biostimulatory effects can be induced via a number of mechanisms including pheromonal, visual, auditory, and mechanical as well as others which are as yet unknown. Genetic influences are suggested by observations that individuals, breeds, and lines differ in their ability to either cause or respond to biostimulation.

Early speculation on possible biostimulatory effects in cattle came from breeding programs where natural breeding showed advantages over artificial insemination. Subsequently, a number of studies confirmed a positive effect on the resumption of ovarian activity in both *Bos taurus* and *Bos indicus* beef breeds (Chenoweth and Spitzer, 1995; Landaeta-Hernández *et al.*, 2004a, 2008; Pérez-Hernández *et al.*, 2002; Rekwot *et al.*, 2000; Soto-Belloso *et al.*, 1997). Positive effects of biostimulation on other reproductive parameters such as estrous expression and length of first postpartum estrous cycle have also been reported (Berardinelli and Joshi, 2005; Landaeta-Hernández *et al.*, 2004a, 2006). Here it is relevant that a number of studies comparing estrus detection effectiveness between bulls and human sourced systems have not ruled out the possibility that biostimulation could influence female sexual activity (Rekwot *et al.*, 2001).

A positive effect of biostimulation on age of puberty in females, has been reported in *Bos taurus* (Chenoweth and Landaeta-Hernández, 1998) and *Bos indicus* heifers (Oliveira *et al.*, 2009; Rekwot *et al.*, 2000) although other studies could not demonstrate such an effect (Berardinelli *et al.*, 1979; Roberson *et al.*, 1987; Wehrman *et al.*, 1996). It is possible that confounding factors include social, managerial, and nutritional effects as well as genetic influences.

A number of factors can influence the effectiveness of biostimulation in different species and we could assume that this is also true of cattle. These include social effects, stress, novelty of male introduction, and effectiveness of the stimulating male. The latter can embrace age, dominance, experience, sex-drive, and relative size. An example is illustrated by ongoing work in Venezuela (Landaeta-Hernández, unpublished data) in which differences occurred in the responses of *Bos indicus* cross females served by bulls of differing relative size and experience. Here, the number of served and/or pregnant cows after 90 days of exposure to males above 400 kg was higher ($P < 0.005$) than in cows exposed to young males below 400 kg. (Landaeta-Hernández, unpublished data) and *Bos indicus* females appeared to be more receptive to males of *Bos indicus* type than were *Bos taurus* females.

Biostimulatory influences are not limited to the direct presence of the male, as both the presence of estrous females as well as those which have been testosterone-primed may also affect the resumption of ovarian activity after calving, the intensity of estrus expression, ovulation rates, and fertility under conditions of natural mating to a synchronized estrus (Burns and Spitzer, 1992; Chenoweth and Lennon, 1984).

In contrast with the relatively large amount of evidence supporting the effects of biostimulation in beef cattle, corresponding studies in dairy cattle are relatively scarce and biostimulatory effects on either the resumption of ovarian activity postpartum and estrus expression have not been conclusively demonstrated (Roelofs *et al.*, 2008; Shipka and Ellis, 1998, 1999). For example, in a study with Brown Swiss cows in Mexico (Izaguirre-Flores *et al.*, 2007), biostimulation was reported to reduce the interval from calving to conception, although this finding may have been compromised by factors such as low milk production, nutrition and study design as suggested by an earlier New Zealand study (Macmillan *et al.*, 1979).

REPRODUCTIVE BEHAVIOR IN SHEEP AND GOATS

Sheep and goats are seasonally polyestrus, with breeding season length being influenced by both genotype and environment (Hulet *et al.*, 1975). For example, female Alpine and Saanen dairy goats cease ovulating for much of the year (summer and early fall) even when in excellent body condition. Males generally show less evidence of seasonality than females, although sperm production, libido, and breeding competence may all be lowered during the non-breeding season (Hulet *et al.*, 1975).

Male Reproductive Behavior

As much of the information available on the mating behavior of rams has arisen from pen trials in which rams were used to breed synchronized females, differences may occur under more extensive, natural breeding conditions. However, in common with bulls and boars, immobility of the female appears to be the greatest single stimulus for male copulatory behavior (Signoret, 1975) and this is often tested by the male with actions such as nudging, butting, and leg striking. Pheromonal influences appear to be of lesser importance as anosmic rams could detect estrous ewes as well as intact rams, although their precopulatory behavior was modified (Lindsay, 1965).

The ram has a fibro-elastic penis characterized by a filiform extension of the urethra. As with bulls, once mounting and orientation is achieved, copulation is rapid and accompanied by an ejaculatory thrust. Rams also form social hierarchies in which subordinate rams may be sexually inhibited (Chenoweth, 1981; Hulet *et al.*, 1962). Proceptive and receptive females tend to form a female "harem" in proximity to the dominant ram who attempts to monopolize them (Shreffler and Hohenboken, 1974; Wodzicka-Tomaszewska *et al.*, 1981). As with bulls, females newly in estrus are most attractive to rams and introduction of a novel female stimulus will restore sexual interest in satiated rams (Beamer *et al.*, 1969).

Both rams and bucks are capable of great sexual activity within limited periods, and relatively high female to male ratios are often employed. For example rams may breed eight to 10 ewes daily or well over 100 ewes over a single estrous cycle. However, not all rams are capable of such feats and considerable individual variation occurs in ram reproductive capacity (Chenoweth, 1981). Here a number of factors, including managerial and genetic, may play a role. With the former, sexual inhibition was associated with raising young rams in all-male groups (Hulet *et al.*, 1964). However, the case for significant genetic influences on ram sexual activity is strong (Chenoweth and Landaeta-Hernández, 1998) with an estimated heritability of 0.30 ± 0.62 being obtained for serving capacity in one study (Kilgour, 1981). As with bulls, high sex-drive does not imply superiority in other reproductive traits such as scrotal circumference (Tulley and Burfening (1983). Also, as with bull studies, some trials have shown that libido or serving capacity tests of rams were predictive of subsequent breeding performance, whereas others were not. Interestingly, several studies have indicated that rams of higher sex-drive will sire females which achieve earlier puberty and higher fecundity than do rams of lower sex-drive (Chenoweth and Landaeta-Hernández, 1998).

Despite numerous anecdotal reports, there is relatively little scientific evidence for breed differences in ram or buck sexual activity. For example, no significant differences in sexual activity were observed between Ramboulliet, Targhee, and Colombia breed rams in one study (Hulet *et al.*, 1962), nor between Dorset Horn, Merino, and Border Leicester rams in another (Lindsay and Ellsmore, 1968). However, breeds do differ in their seasonal expression of male reproductive activity. For example, Alpine bucks displayed dramatic variations in sexual behavior between the spring–summer and autumn–winter periods (Delgadillo and Chemineau, 1992). Schanbacher and Lunstra (1976) observed seasonal variations in the sexual activity of both Finnish Landrace and Suffolk rams, with the former generally showing higher levels of sexual activity for a more extended period than the latter.

Female Reproductive Behavior

Female sheep and goats exhibit relatively few overt behavioral indications of estrus. Although does may show some female–female mounting, this is unusual for ewes. As with cattle, female–female mounting activity in goats stimulates male sexual behavior (Shearer and Katz, 2006). Tail wagging is a behavioral characteristic of estrous does and may play a role in attracting males and maintaining their sexual interest (Haulenbeek and Katz, 2011). Males of both species often undertake considerable investigatory activity, including sniffing and exhibiting the flehmen response (Hurnik, 1987) although, as mentioned above, pheromones appear to play a minor role in male detection of receptive females (Signoret, 1975).

If there are sufficient females concurrently in late proestrus and estrus, a sexually active group will often become established (Lindsay and Fletcher, 1972) which, despite high mobility, remains within visual contact of the ram or ram group. Such ram–seeking behavior was demonstrated in a study by Lindsay and Robinson (1961) where tethered rams serviced a large proportion of estrous ewes. When few, if any, ewes are in estrus, rams tend to expend more investigatory effort than they do when many are in estrus (Wodzicka-Tomaszewska *et al.*, 1981), with the result that individual females may receive less attention (Hulet *et al.*, 1962).

In studies attempting to define ewe proceptivity, it was shown that ewes preferred larger rams and those which scored higher on serving capacity tests (Estep *et al.*, 1989). Rams were shown to prefer "woolly" ewes over recently shorn ewes (Tillbrook and Cameron, 1989). In terms of the ram's ability to stimulate ewe cyclicity, low-libido rams were less effective than high libido rams (Perkins and Fitzgerald, 1994; Signoret *et al.*, 1984). In goats, it appears that photoperiod effects on males can affect the onset of estrous behavior in females. In one study (Rivas-Muñoz *et al.*, 2007), does joined with males that were exposed to artificially long days were more likely to exhibit estrous behavior than those joined with males which had received shorter light exposure.

Maternal Behavior in Sheep

In sheep, as in cattle, the selection of a proper birth site is crucial for the survival of the offspring. At the time of birth, the ewe and her lambs are in a weakened state and reducing susceptibility to predators is of major importance in certain areas. In wild Bighorn Sheep (*Ovis canadensis*), strong separation behavior for parturition has been observed (Bangs *et al.*, 2005), whereas Merino sheep tend to give birth wherever they are. With the initiation of labor, progressive restlessness, pawing and stamping, and cessation of grazing may occur (Hulet *et al.*, 1975). Much of the birth process occurs with the ewe in recumbency, especially toward the end of the delivery process, even though she may stand intermittently. The duration of labor is relatively short (often less than 1 hour) and it has been shown that ewes respond to amniotic fluid via licking and sniffing. Maternal responsiveness seems to be induced partly by the release of progesterone and estradiol (Dwyer, 2008; Lévy and Keller, 2009) although it appears that these steroids may have a priming role, allowing the expulsion of the fetus (Nowak *et al.*, 2000). This mechanical action seems to trigger maternal care in ewes. Once the first lamb is born, maternal behavior develops rapidly and can be described in two phases (Lindsay, 1996). During the first phase, which may last for 1 h or less, the ewe will respond to any newborn. This has been termed the responsiveness phase. Normal behavior during this phase includes cleaning, licking and nuzzling the newborn. Ewes will start at the head and nose of the lamb, licking away any fluids that may impede breathing and this action also stimulates peripheral blood flow, promoting homeothermy within the new environment (Darwish and Ashmawy, 2011). In the second phase, the ewe develops a bond with a specific lamb or lambs; a period termed the selectivity phase. The onset of the selectivity phase varies from 4–12 hours after birth and it is influenced by genotype, age, experience, and environmental factors. If, however, this second phase is not achieved, a lasting bond is not developed between the mother and young (Poindron and Le Neindre, 1980) and lamb survival is compromised. Therefore, both periods are often referred to as a "critical period" because survival is compromised if a good mother–young bond is not established.

Sheep are classified as a follower species. Olfactory cues play a major role in the development of the maternal bond (Nowak *et al.*, 2000) and it has been shown that the odor of monozygotic twins is more similar than that of dizygotic twins, thereby complicating discrimination between monozygotic twins for ewes (Lévy and Keller, 2009; Romeyer *et al.*, 1993). Losses in twins are usually greater than those of singletons, especially in breeds such as the Merino, which is reputedly defective in its ability to care for twins (Alexander *et al.*, 1984). Mismothering can also occur in the form of lamb "stealing" by ewes that have lost their lamb. Overall, pneumonia and mismothering account for more than 30% of lamb losses (Mandal *et al.*, 2005). Whately *et al.* (1974) observed that Merinos and Romneys in New Zealand exhibited the poorest mothering ability of those breeds compared, whereas crossbred sheep were best. Merino dams were the most apt to abandon lambs, although Romneys showed the poorest ability to seek shelter when lambing in poor weather. Alexander *et al.* (1983) reported poor maternal behavior in twin-bearing Merino, Dorset Horn and Border Leicester ewes in Australia. Romanov ewes, on the other hand, are known for their shy behavior as reflected by the weak level of exploration and superior maternal abilities (Boissy *et al.*, 2005). A recent multibreed sheep study in Brazil found that polymorphism (genotype AA or AB) in the aromatase gene (*Cyp 19*) affects growth, reproduction and maternal ability. Genotype frequencies are 0.64 AB and 0.36 BB and ewes with genotype AB exhibited higher maternal abilities, defined by lamb birth weight, weight gain and weaning weight (Lôbo *et al.*, 2009). Furthermore, breed differences in the expression of maternal care at parturition persist throughout the lactation period with Blackface sheep being more vigilant and Suffolk sheep allowing more sucking bouts than the former (Pickup and Dwyer, 2011).

Significant differences in maternal behavior have been observed between two sheep breeds, i.e. Scottish Blackface and Suffolk. Here, Scottish Blackface ewes were more vigilant, groomed their lambs more, stayed in closer proximity to them and underwent longer sucking bouts than did Suffolk ewes (Pickup and Dyer, 2011). Nordic ewes remained closer to their lamb when it was tagged compared with Karakas or Ile de France × Ahkarinen ewes (Yilmaz *et al.*, 2011). A recent study in Scottish Suffolk sheep found that birth difficulty and lamb vigor are moderately heritable, a finding which might help increase lamb health and welfare if included in breeding programs (Macfarlane *et al.*, 2010). However, the heritability of rearing ability in ewes is rather low and has not been confirmed (Hatcher *et al.*, 2010; Haughey, 1984; Piper *et al.*, 1982) perhaps because the lamb also plays an active role in establishing the maternal bond (Nowak, 1989). Also, the quantification of maternal behavior in sheep presents particular logistic problems as expression of this trait may be influenced by parity, number of lambs, experience, nutrition, and environmental factors (Poindron and Le Neindre, 1980). Temperament of both lamb and ewe can also play a role. For example, Plush *et al.* (2011) found a genetic relationship between agitation scores of lambs separated from their mothers and lamb survival, and Rech *et al.* (2008) observed that more reactive or nervous ewes exhibited lower maternal care and weaned less heavy lambs than did less reactive ewes.

Maternal Behavior in Goats

Gestation length in goats is variable and depends on the breed of the dam, parity and season, with shortest length in summer and longest gestations in winter (Hoque *et al.*, 2002). Granadina goats tend to have shorter gestation than those of the Toggenburg and Alpine breeds. In general, prolonged gestation length is considered as beneficial because it is associated with increased viability of neonatal kids (Mellado *et al.*, 2000) although this does not apply to pathological causes of prolonged gestation. Depending upon nutritional opportunity and the risk of predation, feral goats may exhibit either following or lying-out behavior at birth (O'Brien, 1984). However, domestic goats mostly isolate themselves from conspecifics at birth, with this trait being more evident in multiparous dams than in those kidding for the first time.

Amniotic fluids are considered to be the main trigger for the onset of maternal responsiveness and development of the doe-kid bond (Poindron *et al.*, 2010). In this study, kids generally first suckled within the first hour after birth, with the mother–young pair remaining together at the birth site for an average of 18 hours during which they were in constant contact. Vocal cues were exchanged in the period immediately following birth and the doe engaged in a considerable amount of licking and grooming of kids during the first 2 hours. In an investigation of vocal recognition, does were able to discriminate their own kids' vocal cues from other non-related kids (Terrazas *et al.*, 2003). In a comparative study of intrinsic maternal factors of Saanen, German Fawn and Damascus goats, Saanen had an increased likelihood of requiring birth assistance and Damascus goats exhibited the longest duration of licking and grooming (Ocak and Onder, 2011). It is possible that intensive genetic improvement programs for increased production have reduced maternal abilities as less intensively selected Norduz breeds had higher maternal abilities than more intensively selected Karakas breeds (Yilmaz *et al.*, 2011). If so, this needs to be considered in future selection programs.

Biostimulation in Sheep and Goats

It has been long recognized that rams, under certain circumstances, can stimulate estrus and ovulation in females, with this phenomenon being termed "the ram effect". This effect is exploited in both sheep and goat management to stimulate the onset of both puberty and estrus, and when employed at the beginning of the breeding season, to facilitate synchronization of both estrus and ovulation. Although the degree of response obtained varies with genotype and latitude, it represents a valuable management tool under most circumstances (Delgadillo *et al.*, 2009).

The mechanism by which this occurs is considered to be largely pheromonal and mediated via the female vomero-nasal organ (VNO) as reviewed by Gelez and Fabre-Nys (2004). This is because direct physical or visual contact between rams and ewes is not necessary to induce stimulation; urine, wax and wool from rams are as effective as is the presence of the ram (Izard, 1983) and this appears to be the same in goats (Iwata *et al.*, 2000) in which a priming pheromone has been identified (Iwata *et al.*, 2003). Many male ungulates apply urine to their coats during the breeding season; a practice which probably augments their biostimulatory ability (Izard, 1983). Despite this, it is probable that non-olfactory cues, such as auditory or visual, play either a complementary or synergistic role (Delgadillo *et al.*, 2009).

The enabling circumstances for this to occur include the novel introduction of the male(s) to transitional females, as is commonly exploited as a management tool at the beginning of the breeding season. A similar phenomenon occurs with goats and it has been suggested that the "ram effect" evolved as a strategy in wild ungulates to ensure synchrony of mating (Delgadillo *et al.*, 2009). It is now apparent that male "novelty" is more important than prior isolation, and also that this effect is not restricted to anovulatory females (Delgadillo *et al.*, 2009). Further, the effectiveness of biostimulation is related to the sexual activity of the males involved (Delgadillo *et al.*, 2009; Perkins and Fitzgerald, 1994; Signoret *et al.*, 1984).

REPRODUCTIVE BEHAVIOR IN SWINE

Wild pigs generally form small matriarchal herds containing one or more sows and their young, with males living elsewhere and joining the herd for breeding (Signoret *et al.*, 1975). Although feral swine reportedly show a preference for breeding in the autumn, or fall, this is not generally evident in domesticated animals (Signoret *et al.*, 1975). Sows will actively seek boars, with both proestrous and estrous females being strongly attracted to the male (Signoret, 1970).

Males, in turn, show little discrimination between estrus and nonestrous females, often identifying receptive females by a process of trial and error (Chenoweth, 1981). Despite this, mounting attempts are rarely made with anoestrous sows (Tanida *et al.*, 1989). Flehmen is generally not observed in boars. In free-range systems, courtship is often lengthy, with the sow showing relatively few overt signs of impending estrus. Immobility of the female is the most important indication that she is in estrus, although she may show special interest in the male, including attempts to mount him (Signoret *et al.*, 1975). Stimuli from the male are, however, important in eliciting standing behavior in estrous females, with both olfactory and vocal stimuli being important (Chenoweth, 1981; Signoret *et al.*, 1975). The mating reaction of boars is stimulated by both visual and tactile factors, and they will readily mount immobile objects resembling the rear end of a female. Full erection of the fibroelastic penis generally occurs after mounting is achieved and copulation time is often prolonged compared with cattle and sheep (Chenoweth, 1981; Signoret *et al.*, 1975).

Male Reproductive Behavior

A female in late pro-estrus receives increased attention from the boar, who will attempt to stay in close attendance while attempting nosings and nudgings, and emitting a series of soft, guttural grunts. Categories of boar courtship and service behavior noted by Tanida *et al.* (1989) included sniffing, head to head, nosing, following, chin resting, mounting, and copulation. When the estrous female assumes the mating stance, more vigorous muzzling of the genital region occurs accompanied by continuous grunts, teeth grinding, foaming from the mouth and rhythmic urination. Experienced boars will generally mount rapidly at this juncture, although some will mount and dismount a number of times prior to copulation. Ejaculation is prolonged (3–20 minutes) compared with other livestock species, with the female remaining immobile throughout. The mating reaction of boars is stimulated by both visual and tactile factors, and can easily be elicited by immobile objects resembling the rear end of a female. Full erection of the fibroelastic penis generally occurs once mounting is achieved (Signoret *et al.*, 1975).

Significant breed and strain differences in both swine libido and mating behavior have been reported (Chenoweth and Landaeta-Hernández, 1998) as well as differences in the duration of ejaculation (Signoret *et al.*, 1975). When comparing successful mating attempts in young boars, Yorkshires were more successful than Durocs and crossbreds were generally superior to purebreds (Neely and Robison, 1983).

When boars were selected for 10 generations on the basis of testosterone (T) response (high or low) to GnRH challenge, the heritability was moderate. In addition, litter sizes in females of the high-T response line were significantly larger than those in females of the low T-response line (Robison *et al.*, 1994). Rothschild (1996), quoting a number of studies on the heritability of boar libido, provided a general estimate of 20% (range 10–50%).

Female Reproductive Behavior

Under open range conditions, pre-copulatory behavior can last for some time, with the female becoming more restless and nervous several days before estrus. During this phase she will resist the boar's advances although staying in his proximity. When she reaches the stage of receptivity, she will show more direct interest in the boar via nuzzling, head to head contact, and occasional mounting attempts of the boar. When willing to mate, she adopts the mating stance allowing the boar to mount and copulate. This stance is characterized by rigidity, arching of the back, and cocking of the ears.

The presence of the boar facilitates estrus detection. For example, in one study, only 45% of estrous gilts exhibited the standing reaction in the absence of the male. However, 90% of these displayed the standing reaction when the boar was proximate and the addition of visual and tactile stimuli resulted in all responding.

Breed differences are evident in the age at which females attain puberty (Hurnick, 1987). Rydhmer *et al.* (1994) reported heritability estimates for estrous traits, such as length of proestrus and ability to show the standing reflex, of approximately 20 and 30% respectively; estimates supported by Knauer *et al.* (2010). Thus, selection for estrous traits in swine could be economically attractive in terms of reduced age at puberty, reduced seasonal infertility, and increased sow reproductive lifetime (Rydhmer, 2000). However, selection for reproduction traits is not always favorably reflected in production traits, so these relationships, as well as environmental factors such as housing and handling, should be considered in multi-trait breeding programs (Rydhmer, 2000).

Maternal Behavior in Swine

In modern swine production systems, sows are often kept in crates which limit maternal behaviors due to space confinements. However, when straw and branches are provided to provide a semi-natural environment, sows readily engage in nest building behavior prior to parturition (1–7 h) and this can increase time spent in lateral recumbency during the process (Damm *et al.*, 2000). Both farrowing and nest building behaviors appear to be associated with stress reaction patterns, especially in gilts. Gilts that reacted calmly to stress exhibited better nest building and farrowing behaviors than did more behaviorally active individuals (Thodberg *et al.*, 2002).

Further, personality traits, as reflected by fear and anxiety expressed in the presence of humans, have been linked with maternal ability. Higher levels of fear and anxiety were associated with a longer duration of farrowing and more stillborns (Janczak *et al.*, 2003) whereas sows that were more responsive to piglet demands and not afraid of people had better piglet survival (Grandinson *et al.*, 2003). Losses could possibly be reduced by genetically selecting for the maternal component of farrowing survival which would reduce stillbirths while not negatively affecting total piglets born (Leenhouwers *et al.*, 2003). Further, genetic selection targeting piglet survival by reducing crushing of piglets could be considered in breeding programs (Baxter *et al.*, 2011). The pre-lying behavior of sows tends to be crucial for piglet survival. Sows that perform "sniffing", "looking around", and "nosing" before lying down tend to be less likely to crush piglets than sows that don't perform these behaviors (Wischner *et al.*, 2010). Additionally, sows that did not crush piglets were more restless pre-partum and engaged more frequently in nest building than more relaxed sows. There may be differences between the best genetic selection program for intensively housed sows and extensively raised pastured sows.

Poor maternal behaviors, such as savaging and eating piglets, have been linked to certain (QTL) quantitative trait loci (Chen *et al.*, 2009). Vangen *et al.* (2005) reported that possible behaviors that could be influenced by genetic selection are a sow's reaction to people, including fear and aggression, especially when her piglets are handled. Selection of sows for calm behavior during farrowing may help make intensively raised sows more productive (Snieder *et al.*, 2011). In addition, behavior of gilts assessed at five months of age associated with subsequent economically important traits such as piglets born alive and wean to estrus interval (Snieder *et al.*, 2011).

Olfactorial, thermal, and tactile cues are used by piglets to locate the teats (Rohde Parfet *et al.*, 1990). Newborn pigs engage in teat seeking very rapidly after birth, with an established and relatively consistent teat order being evident after about 4 days (de Passillé *et al.*, 1988). Although larger, faster-growing piglets tend to suckle the more anterior teats, exceptions regularly occur (Winfield *et al.*, 1974), with less stable teat orders being more evident in larger litters than in smaller ones. However, survival of piglets appears to be sex biased. Even though maternal investment in male piglets is higher, male-mortalities exceed female ones (Baxter *et al.*, 2012). It is speculated that piglet survival represents a combination of the maternal genetic component (genotype of the sow) and the direct genetic component (genotype of the piglet) (Leenhouwers *et al.*, 2001). In contrast to other species, sows are tolerant of foreign young, particularly in the first few days postpartum, and fostering can be achieved relatively easily at this time (Signoret *et al.*, 1975).

Biostimulation in Swine

Biostimulation is used as a management tool in swine which can accelerate puberty in gilts and reduce the postpartum period in sows (Brooks and Cole, 1970; Kirkwood *et al.*, 1981). The effects on puberty are most pronounced in gilts raised in confinement, where the introduction of a mature boar can lower the age at puberty by approximately 30 days in concert with a marked degree of synchrony when the gilts are approximately 190 days of age (Brooks and Cole, 1970). The ability of boars to act as biostimulators is related to their age with more mature boars (2 years old) being more effective than young (6.5 months old) boars in inducing puberty in gilts (Kirkwood *et al.*, 1981).

The major stimulus associated with pubertal advancement is probably olfactory, as exposing gilts to a pen which previously contained a boar is also effective. Here, the mechanism appears to be the release of boar pheromone(s) secreted in saliva (Kirkwood *et al.*, 1981). In addition, biostimulation significantly enhances the standing reaction of sows in estrus (Rekwot *et al.*, 2001). These phenomena are widely utilized in confinement swine management systems where, for example, re-breeding of sows at the lactational estrus and thus reducing time to re-breeding, is an important economic consideration.

REPRODUCTIVE BEHAVIOR IN HORSES

Horses have several reproductive behavioral aspects which distinguish them from many other livestock species. In females, these include a prolonged period of estrus (5–7 days), variable estrus manifestation during seasonal transitions, a fertile estrus in the early post-partum period ("foal heat") and, not uncommonly, display of estrous signs during pregnancy (Crowell-Davis, 2007). The stallion has a vascular-muscular type penis and, under wild conditions, is associated with his "harem" year-round (McDonnel, 2000).

Horses are seasonal breeders with increasing daylight hours being the trigger for increased ovulatory activity (Palmer and Guillaume, 1992). However, although feral horses may breed between May and October in the Northern Hemisphere, domesticated horses have been selected to breed as early as possible such that it is not uncommon for mares to cycle during winter (Aurich, 2011).

In the wild, horses form patriarchal family groups, usually containing a stallion plus adult mares (the "harem") and their young offspring, which are maintained year-round and are non-territorial. The dominant stallion protects the group while maintaining a strong social bond and discouraging internecine disputes among the mares. This organizational structure enhances estrus detection as well as reproductive success. Once individuals of either sex reach sexual maturity, they are actively encouraged to leave their natal band, probably to minimize the risk of inbreeding (Heitor *et al.*, 2006). Here, distinction is made between horses and some other members of the equid family, such as donkeys, wild asses, and certain Zebras, which are territorial breeders (McDonnell, 2000) and exhibit behaviors typical of this category such as the formation of sexually active groups of females and a relative lack of herding behavior by the male (Henry *et al.*, 1991).

Colts will often exhibit mounting behavior towards their mothers or other horses in the group during their first weeks of life, and sexual behavior may be observed in 2–3-month-old colts (Tyler, 1972). Young fillies will also show similar behavior although to a lesser degree. The age at puberty in fillies varies between 12 and 18 months, varying with the season of birth, and mares will continue cycling until old age.

Male Reproductive Behavior

The social status of the mature stallion with respect to his ability to gain access to a harem or not has important implications for reproductive physiology and behavior (McDonnell 2000), with these being significantly lower in non-breeding "bachelor" groups. Here it is possible that the management practices on multi-stallion breeding farms in which stallions are grouped away from mares for extended periods may suppress reproductive capabilities (McDonnell, 2000).

Breeding stallions can show herding or driving behavior of mares, as well as prancing behavior (Houpt, 1998). When a stallion attends an estrous mare, he "nickers" while smelling her external genitalia and groin. This is usually followed by flehmen behavior as the penis becomes erect; a state which must be fully achieved before intromission can occur. Erection is relatively slow due to the vascular–muscular mechanisms involved, and copulation depends upon successful foreplay which is of greater importance in horses than in most livestock species (Houpt, 1998). Visual and sensory stimuli are essential for successful foreplay by the stallion, with an important aspect being the posture of the mare in concert with other signs of receptivity as discussed below. However, in common with other livestock species, males can be trained to accept a surrogate, for example a dummy, which has some similarities to the rear end of an immobile mare. Here, experienced stallions will not show inhibitions in mounting an appropriate dummy, although inexperienced males often require application with urine from an estrous female before they will attempt to mount (Houpt, 1998).

Experienced stallions will mount readily with an average of 1.4 mounts (range 1–4) per copulation. Ejaculation, accompanied by rhythmical contractions of urethral muscle and characteristic flagging tail movements occur an average 13 seconds after intromission. Stallions may breed several times in a day with satiation being achieved on average after 2.9 ejaculations (range 1–10) in succession (Bielanski and Wierzboweski, 1961). Young stallions can be very aggressive in approaching an estrous mare and may show inexperience in mounting. Masturbation is common in confined stallions, although rarely accompanied by ejaculation (Houpt, 1998). Stallions show peak sexual behavior in spring, with this being lowest in winter and highest in spring–summer (Pickett *et al.*, 1975).

Female Reproductive Behavior

Under natural conditions, mares play the major role in the mating process through their proceptive behaviors by stimulating the stallion and timing copulation (McDonnell, 2000), with older mares being more proactive than younger ones (Heitor *et al.*, 2006). Proceptive females will approach and actively solicit the stallion. Once receptive, they will present their hindquarters to the stallion and lower the pelvis while deviating the tail and exhibiting clitoral "winking", often while voiding small amounts of urine. Under conditions in the wild, this will usually involve the "harem" male, for which the mare has established strong social bonds. Under domesticated conditions, the mare and stallion are often not familiar (Crowell-Davis, 2007), and mares can show preferences for, and dislikes of, particular stallions to the extent that estrus signs may be modified accordingly. Sexual attraction of mares to a particular stallion is influenced by coat color, age, dominance rank, and days from ovulation (Heitor *et al.*, 2006; McDonnell, 2000).

Hand-breeding of horses, where the mare and stallion are brought together at the time of breeding only, is often preceded by the process of "teasing" whereby a mare is exposed to a stallion which can interact with her without fear of injury to either animal. The behavioral responses which the stallion elicits from the mare are indicators of her degree of receptivity. Estrous behavioral responses have also been obtained from mares when they were exposed to simulations of stallion courting behaviors such as playing recordings of stallion vocalizations and manually manipulating their external genitalia (Crowell-Davis, 2007) with the latter possibly also improving passive sperm transport.

Maternal Behavior of Horses

Gestation length in horses is approximately 11 months in duration and is influenced by the genotype of the fetus and the uterine environment (Rossdale and Short, 1967) with heritability, in an early study, being estimated at 0.36 (Rollins and Howell, 1951). During the last month of gestation, the mare may undergo false labors on several occasions. These usually last no longer than 10 minutes, are not rhythmic and will slow down after a short period of time. Signs of approaching parturition are: the distention of the udder (2–6 weeks before parturition (b. p.), relaxation of the muscles in the croup area (7–10 days b. p.), teat enlargement (4–6 days b. p.), and a waxy secretion oozing out of the nipples (2–4 days b. p.) (Evans, 2000). The day before parturition, a drop in body temperature is observed, and during the night of parturition (about 80% of mares foal at night) mares show increased walking and lying in recumbency with decreased standing (Shaw *et al.*, 1988). Once the fetal membranes rupture, normal labor takes usually not longer than 20–30 minutes.

Zúrek and Danek (2011) described the following as normal maternal behaviors occurring immediately after parturition: nuzzling, licking, grooming, avoiding stepping or laying on the newborn, enabling and giving assistance with suckling, protecting the foal from potential danger by positioning itself between the infant and the hazard, and even attacking and repelling the intruder. Whether in the wild or in domesticity, maternal care is crucial for the survival of the foal (Heitor and Vicente, 2008) and the strength of the maternal attachment can be measured by the duration of nursing bouts, mutual grooming, and general proximity between mare and foal (Waran *et al.*, 2008). Horses are classified as "followers" and the foal must be ready to follow the mare within hours after birth. Mare–foal bonding occurs immediately after birth, triggered by maternal hormones and the mare's view of the wet, uncoordinated foal (Houpt, 2000, 2002). The mare's attachment to the foal is almost immediate whereas foals need some days to be truly linked to their mothers. In Arab horses, a relationship between temperament, measured as the behavioral reactivity (fearfulness) and heart rate in response to rotating black–white squares (Fearfulness-test), and maternal ability has been identified. Fearless dams vocalized more and displayed greater maternal behavior than fearful mares which is vitally important for survival of the foal (Budzynska and Krupa, 2011). In general, poor maternal abilities have been reported in Arabian mares (Houpt, 2002).

In feral horses stable social relationships are maintained with the stallion defending the harem band against other males. However, sneak copulations have been observed. Mares that raise a foal that is not sired by the dominant stallion have been observed to be more protective of their foal due to increased infanticide risk for the offspring (Gray *et al.*, 2012). The rate of aggression between the stallion and the foal can be used as a predictor of maternal protectiveness (Cameron *et al.*, 2003) with mare protectiveness being negatively correlated with reproductive success in the subsequent year.

REPRODUCTIVE BEHAVIOR IN WATER BUFFALO

Water buffalo, which are among the most important sources of food and draft power in developing countries, are often differentiated into swamp (chromosome number 48) and river (chromosome number 50) types. Analyses of maternal lineages and nuclear DNA indicate that river and swamp buffalo originated from different wild populations (Yindee *et al.*, 2010). These types can interbreed, resulting in individuals with 49 chromosomes which are apparently fertile. Water buffalo females are polyestrus, with less obvious overt signs of estrus being shown than in cattle (Perera, 2011). The most reliable sign is acceptance by the female of the male, who often employs visual observation of mounting behavior to detect estrous females. The intensity of sexual activity (both male and female) is reduced during daylight hours; although mating can occur during the day, it is suppressed during the hotter periods. The water buffalo, particularly the swamp breeds, will breed throughout the year although nutritional, and possibly other constraints, may influence this (Jainudeen, 1983). In general, information on genetic influences on water buffalo reproductive behavior is lacking (Jainudeen, 1983).

Male Reproductive Behavior

Buffalo bulls are capable of breeding throughout the year, although some seasonal fluctuations occur in reproductive functions. For example, libido score was significantly higher and reaction time less in river buffalo bulls during the peak breeding season (autumn, early winter) than in the low breeding season, or summer (Younis *et al.*, 2003). Male behavior is similar to that of *Bos taurus* bulls, although less intense. A reputation for being a sluggish breeder is however, considered to be unfounded (Cockrill, 1981). Sniffing of the vulva and of urine often precedes flehmen and mounting with the occurrence of flehmen behavior being most evident during estrus (Rajanarayanan and Archunan, 2004) where several estrous-specific urinary compounds have been identified as putative sexual pheromones (Rajanarayanan and Archunan, 2011). Mating is brief, lasting a few seconds only, and the ejaculatory thrust is less marked than it is in *Bos taurus* bulls.

Following ejaculation, buffalo bulls generally dismount slowly and undergo a sexual refractory period (Jainudeen and Hafez, 2000).

Maternal Behavior in Water Buffalo

Water buffalo tend to calve within the herd (Tulloch, 1979), although they prefer sites where there is cover and soft bedding. However, no consistent patterns in calving behavior were discerned in domesticated buffalo when different pasture and housing systems were compared. In addition, relevant information comparing maternal behavior between the swamp and river types appears to be lacking.

CONCLUSIONS

Although each livestock species has developed a distinct repertoire of courtship, copulatory, and maternal behaviors, significant differences occur within species and between individuals in the expression of all of these traits. This permits selection for positive reproductive and maternal behaviors, both of which are linked with livestock productivity and profitability. Despite this, indiscriminate artificial selection for maximum productivity has been linked with adverse reproductive and maternal outcomes. It is considered that informed selection and management based on good behavioral principles should lead to greater reproductive "success" and improved animal welfare within the context of profitable production.

REFERENCES

Alexander, G.A., Kilgour, R., Stevens, D., Bradley, L.R., 1984. The effect of experience on twin care in New Zealand Romney sheep. Appl. Anim. Behav. Sci. 12, 363–372.

Alexander, G.D., Stevens, D., Kilgour, R., de Langen, H., Mouershead, B.E., Lynch, J.J., 1983. Separation of ewes from lambs: incidence in several sheep breeds. Appl. Anim. Ethol. 10, 301–317.

Amann, R.P., Almquist, J.O., 1976. Bull management to optimize sperm output. Proceedings of the Sixth NAAB Technical Conference on Artificial Insemination and Reproduction. pp. 1–10.

At-Taras, E.E., Spahr, S.L., 2001. Detection and characterization of estrus in dairy cattle with an electronic heatmount detector and an electronic activity tag. J. Dairy Sci. 84 (4), 792–798.

Aurich, C., 2011. Reproductive cycles of horses. Anim. Repro. Sci. 124, 220–238.

Avery, M.L., Cummings, J.L., 2004. Livestock depredation by black vultures and golden eagles. Sheep Goat Res. J. 19, 58–63.

Baker, A.E., Seidel, G.E., 1985. Why do cows mount other cows? Appl. Anim. Behav. Sci. 13, 237–241.

Bane, A., 1954. Studies on monozygous cattle twins. XV. Sexual functions of bulls in relation to rearing intensity and somatic conditions. Acta. Agric. Scand. 4, 95–208.

Bangs, E.E., Fritts, S.H., 1996. Reintroducing the gray wolf to Central Idaho and Yellowstone National Park. Wildl. Soc. Bull. 24 (3), 402–413.

Bangs, P.D., Krausman, P.R., Kunkel, K.E., Parson, Z.D., 2005. Habitat use by desert bighorn sheep during lambing. Eur. J. Wildl. Res. 51, 178–184.

Baxter, E.M., Jarvis, S., Sherwood, L., Farish, M., Roehe, R., Lawrence, A.B., Edwards, S.A., 2011. Genetic and environmental effects on piglet survival and maternal behaviour of farrowing sows. Appl. Anim. Behav. Sci. 130, 28–41.

Baxter, E.M., Jarvis, S., Palarea-Albaladejo, J., Edwards, S.A., 2012. The weaker sex? The propensity for male-biased piglet mortality. Plos One 7 (1), DOI: 0.1371/journal.pone.0030318.

Beach, F.A., 1976. Sexual attractivity, proceptivity, and receptivity in female mammals. Horm. Behav. 7, 105–138.

Beamer, W., Bermant, G., Clegg, M.T., 1969. Copulatory behavior of the ram, *Ovis aries*. II. Factors affecting copulatory satiation. Anim. Behav. 17, 706–711.

Becke, N.G.S., Serrano, E.R., 2002. Evaluation of oestrous detection in dairy farms. Rev. Cient-Fac. Vet. 12, 169–174.

Berardinelli, J.G., Dailey, R.A., Butcher, R.L., Inskeep, E.K., 1979. Source of progesterone prior to puberty in beef heifers. J. Anim. Sci. 49, 1276–1280.

Berardinelli, J.G., Joshi, P.S., 2005. Introduction of bulls at different days postpartum on resumption of ovarian cycling activity in primiparous beef cows. J. Anim. Sci. 83, 2106–2110.

Bielanski, W., Wierzboweski, S., 1961. Depletion test in stallions. Proceedings of the Fourth International Congress on Animal Reproduction. pp. 279–282.

Blockey, M.A.deB., 1989. Relationship between serving capacity of beef bulls as predicted by the yard test and their fertility during paddock mating. Aust. Vet. J. 66, 348–351.

Blockey, M.A.deB., Straw, W.M., Jones, L.P., 1978. Heritability of serving capacity and scrotal circumference in beef bulls. Proceedings of the American Society of Animal Science AGM, Abstr. No. 253.

Boissy, A., Bouix, J., Orgeur, P., Poindron, P., Bibé, B., Le Neindre, P., 2005. Genetic analysis of emotional reactivity in sheep: effects of the genotypes of the lambs and their dams. Genet. Sel. Evol. 37, 381–401.

Bonfert, A., 1955. Beziehungen zwischen brunstgeschen und fruchtbarkeit. Fortpflanz. Zuchthyg. Haustierbesamung 5, 125–128.

Bowyer, R.T., van Ballenberghe, V., Kie, J.G., Maier, A.K., 1999. Birth-site selection by Alaskan moose: maternal strategy for coping with a risky environment. J. Mammal. 80, 1070–1083.

Boyd, G.W., Lunstra, D.D., Corah, L.R., 1989. Serving capacity of crossbred yearling bulls. 1. Single-sire mating behavior and fertility during average and heavy mating loads at pasture. J. Anim. Sci. 67, 60–71.

Boyd, G.W., Healy, V.M., Mortimer, R.G., Piotrowski, J.R., 1991. Serving capacity tests are unable to predict the fertility of yearling bulls. Theriogenology 36, 1015–1025.

Brooks, P.H., Cole, D.J., 1970. The effect of the presence of a boar on the attainment of puberty in gilts. J. Reprod. Fertil. 23, 435–440.

Buddenberg, B.J., Brown, C.J., Johnson, Z.B., Honea, R.S., 1986. Maternal behavior of beef cows at parturition. J. Anim. Sci. 62, 42–46.

Budzynska, M., Krupa, W., 2011. Relation between fearfulness level and maternal behaviour in Arab mares. Anim. Sci. Pap. Rep. 29, 119–129.

Buechlmann, E., 1950. Das sexuelle verhahen des rindes. Wien. Tierauztl. Monatsschr. 37, 225–230.

Burns, P.D., Spitzer, J.C., 1992. Influence of biostimulation on reproduction in postpartum beef cows. J.Anim. Sci. 70, 358–362.

Cameron, E.Z., Linklater, W.L., Stafford, K.J., Minot, E.O., 2003. Social grouping and maternal behaviour in feral horses (*Equus caballus*): the influence of males on maternal protectiveness. Behav. Ecol. Sociobiol. 53, 92–101.

Cavalieri, J., Fitzpatrick, L.A., 1995. Oestrus detection techniques and insemination strategies in *Bos indicus* heifers synchronised with norgestomet-oestradiol. Aust. Vet. J. 72 (5), 177–182.

Chemineau, P., Guillaume, D., Migaud, M., Thiéry, J.C., Pellicer-Rubio, M.T., Malpaux, B., 2008. Seasonality of reproduction in mammals: intimate regulatory mechanisms and practical implications. Reprod. Domest. Anim. 43 (2), 40–47.

Chen, C., Guo, Y., Yang, G., Tang, Z., Zhang, Z., Yang, B., et al., 2009. The genome wide determination of quantitative trait loci on pig maternal infanticide behavior in a large scale white Duroc x Erhualian resource population. Behav. Genet. 2, 213–219.

Chenoweth, P.J., 1981. Libido and mating behavior in bulls, boars and rams. A review. Theriogenology 16, 155–177.

Chenoweth, P.J., 1983. Reproductive management procedures in control of breeding. Anim. Prod. Aust. 15, 24–38.

Chenoweth, P.J., 1986. Reproductive behavior of bulls. In: Morrow, D.A. (Ed.), Current Therapy in Theriogenology, second ed. Saunders, Philadelphia.

Chenoweth, P.J., 1994a. Bull behavior, sex-drive, and management. In: Fields, M.A., Sand, R.S. (Eds.), Raising Calf Crops. CRC Press, Boca Raton, Fl.

Chenoweth, P.J., 1994b. Aspects of reproduction in female *Bos indicus* cattle: a review. Aust. Vet. J. 71, 422–426.

Chenoweth, P.J., 1997. Bull Libido/Serving capacity. In: Van Camp, S.D. (Ed.), Veterinary Clinics of N.America: Food Animal Practice. W.B. Saunders Co., Philadelphia, 13(2), pp. 331–344.

Chenoweth, P.J., 2000. Bull sex-drive and reproductive behavior. In: Chenoweth, P.J. (Ed.), Topics in Bull Fertility. International Veterinary Information Service, Ithaca, NY.

Chenoweth, P.J., Landaeta-Hernández, A.J., 1998. Genetic influences on reproductive and maternal behavior in livestock. In: Grandin, T. (Ed.), Genetics of Animal Behavior. Academic Press, New York, pp. 145–165.

Chenoweth, P.J., Lennon, P.E., 1984. Natural breeding trials in beef cattle employing oestrus synchronization and biostimulation. Anim. Prod. Aust. 15, 293–296.

Chenoweth, P.J., Chase Jr., C.C., Larsen, R.E., Thatcher, M.J.D., Bivens, J.F., Wilcox, C.J., 1996. The assessment of sexual performance in young *Bos taurus* and *Bos indicus* beef bulls. Appl. Anim. Behav. Sci. 48, 225–236.

Chenoweth, P.J., Spitzer, J.C., 1995. Biostimulation in livestock with particular reference to cattle. Assist. Reprod. Technol./Andro. (ARTA) 7, 271–278.

Clark, P.E., Johnson, D.E., 2009. Wolf–cattle interactions in the northern rocky mountains. In: Range Field Data 2009 Progress Report. Special Report 1092, June 2009. Agricultural Experiment Station, 1–7. Oregon State University, Corvallis, OR.

Cockrill, W.R. (Ed.), 1981. The water buffalo: new prospects for an underutilized animal. Report of Ad Hoc Panel of the advisory committee on technology innovation board on science and technology for international development. NAP (National Academy Press).

Crowell-Davis, S.L., 2007. Sexual behavior of mares. Horm. Behav. 52, 12–17.

Damm, B.I., Vestergaard, K.S., Schrøder-Petersen, D.L., Ladewig, J., 2000. The effect of branches on prepartum nest building in gilts with access to straw. Appl. Anim. Behav. Sci. 69, 113–124.

Darwish, R.A., Ashmawy, T.A.M., 2011. The impact of lambing stress on post-parturient bahe-viour of sheep with consequences on neonatal homeothermy and survival. Theriogenology 76 (6), 999–1005.

de Passillé, A.M., Rushen, J., Hartsock, G., 1988. Ontogeny of teat fidelity in pigs and its relation to competition at suckling. Can. J. Anim. Sci. 68, 325–338.

Delgadillo, J.A., Chemineau, P., 1992. Abolition of the seasonal release of luteinizing hormone and testosterone in Alpine male goats (*Capra hircus*) by short photoperiodic cycles. J. Reprod. Fert. 94 (1), 45–55.

Delgadillo, J.A., Gelez, H., Ungerfield, R., Hawken, P.A.R., Martin, G.B., 2009. The "male effect" in sheep and goats – Revisiting the dogmas. Behav. Brain Res. 200, 304–314.

DeRouen, T.M., Reynolds, W.L., Meyerhoeffer, D.C., 1967. Mortality of beef calves in the Gulf coast Area. J. Anim. Sci. (abst.) 26, 202.

Dwyer, C.M., 2008. Individual variation in the expression of maternal behaviour: a review of the neuroendocrine mechanism in the sheep. J. Neuroendocrinol. 20, 526–534.

Edwards, S.A., 1979. The timing of parturition in dairy cattle. J. Agri. Sci. 93, 359–363.

Edwards, S.A., 1983. The behavior of dairy-cows and their newborn calves in individual or group housing. Appl. Anim. Ethol. 10, 191–198.

Elsik, R., Tellam, L., Worley, K.C., 2009. The genome sequence of taurine cattle: a window to ruminant biology and evolution. Sci. 324 (5926), 522–528.

Estep, D.Q., Price, E.O., Dally, M.R., 1989. Social preference of domestic ewes for rams (*Ovis aries*). Appl. Anim. Prod. Sci. 24, 287–300.

Estes, R.D., 1976. The significance of breeding synchrony in the wildebeest. E. Afr. Wildl. J. 14, 135–152.

Evans, A.C.O., Walsh, S.W., 2012. The physiology of multifactorial problems limiting the establishment of pregnancy in dairy cattle (Review). Reprod. Fert. Dev. 24 (1), 233–237.

Evans, J.W., 2000. Horses, a Guide to Selection, Care and Enjoyment, third ed. W.H. Freeman and Company, NY.

Farin, P.W., Chenoweth, P.J., Tomky, D.F., Ball, L., Pexton, J.E., 1989. Breeding soundness, libido and performance of beef bulls mated to estrus synchronized females. Theriogenology 32 (5), 717–725.

Floercke, C., Engle, T.E., Grandin, T., Deesing, M.J., 2012. Individual differences in calf defence patterns in Red Angus beef cows. Appl. Anim. Behav. Sci. 139, 203–208.

Franke, D.E., Combs Jr., G.E., Burns, W.C., Thatcher, W.W., 1975. Neonatal health status in Brahman calves and blood components. J. Anim. Sci. 40, 193 (abst.).

Frisch, J.E., 1982. The use of teat-size measurements or calf weaning weight as an aid to selection against teat defects in cattle. Anim. Prod. 32, 127–133.

Galina, C.S., Orihuela, A., Rubio, I., 1996. Behavioral trends affecting oestrus detection in Zebu cattle. Anim. Reprod. Sci. 42, 465–470.

Gelez, H., Fabre-Nys, C., 2004. The "male effect" in sheep and goats: a review of the respective roles of the two olfactory systems. Horm. Behav. 46, 257–271.

Godfrey, R.W., Lunstra, D.D., 1988. Influence of serving capacity and single or multiple sires on pasture mating of beef bulls. J Anim. Sci. (abst.) 66, 235.

Gonyou, H.W., Stookey, J.M., 1987. Maternal and neonatal behavior. Vet. Clin. North Am.: Food Anim. Pract. 3, 231–250.

Gorelick, R., Heng, H.H.Q., 2011. Sex reduces genetic variation: a multidisciplinary review. Evolution 65 (4), 1088–1098.

Grandin, T., Deesing, M.J., Struthers, J.J., Swinker, A.M., 1995. Cattle with hair whorl patterns above the eyes are more behaviorally agitated during restraint. Appl. Anim. Behav. Sci. 46, 117–123.

Grandinson, K., Rydhmer, L., Strandberg, E., Thodberg, K., 2003. Genetic analysis of on-farm tests of maternal behaviour in sows. Livest. Prod. Sci. 83, 141–151.

Gray, M.E., Cameron, E.Z., Peacock, M.M., Thain, D.S., Kirchoff, V.S., 2012. Are low infidelity rates in feral horses due to infanticide? Behav. Ecol. Sociobiol. 66 (4), 529–537.

Hafez, E.S.E., 1987. Reproductive behavior. In: Hafez, E.S.E. (Ed.), Reproduction in Farm Amrnals, sixth ed. Lea & Febiger, Malver, PA.

Hatcher, S., Atkins, K.D., Safari, E., 2010. Lamb survival in Australian Merino sheep: a genetic analysis. J. Anim. Sci. 88 (10), 3198–3205.

Haughey, K.G., 1984. Can rearing ability be improved by selection? In: Lindsay, D.R., Pierce, D.T. (Eds.), Reproduction in Sheep. Australian Wool Corporation and Australian Academy of Science, Canberra.

Haulenbeek, A.M., Katz, L.S., 2011. Female tail wagging enhances sexual performance in male goats. Horm. Behav. 60, 244–247.

Heitor, F., Oom, M.M, Vicente, L., 2006. Social relationships in a herd of Sorraia horses: Part II, factors affecting affiliative relationships and sexual behaviors. Behav. Processes 73, 231–239.

Heitor, F., Vicente, L., 2008. Maternal care and foal social relationship in a herd of Sorraia horses: influence of maternal rank and experience. Appl. Anim. Behav. Sci. 113, 189–205.

Henry, M., McDonnell, S.M., Lodi, L.D., Gastal, E.L., 1991. Pasture mating behavior of donkeys (*Equus asinus*) under natural and induced oestrus. J. Reprod. Fert.(Suppl. 44), 77–86.

Hohenboken, W.D., 1987. Behavioral genetics. Vet. Clin. North Am.: Food Anim. Pract. 3, 217–229.

Holroyd, R.G., 1987. Foetal and calf wastage in *Bos indicus* beef genotypes. Aust. Vet. J. 64, 133–137.

Hoppe, S., Brandt, H.R., Erhardt, G., Gauly, M., 2008. Maternal protective behavior of German Angus and Simmental beef cattle after partuition and its relation to production traits. Appl. Anim. Behav. Sci. 114, 297–306.

Hoque, M.A., Amin, M.R., Baik, D.H., 2002. Genetic and non-genetic causes of variation in gestation length, litter size and litter weight in goats. Asian–Australian J. Anim. Sci. 15, 772–776.

Houpt, K.A., 1998. Sexual behavior. In: Houpt, K.A. (Ed.), Domestic Animal Behavior for Veterinarians and Animal Scientists, third ed. Iowa State Press, Ames, Iowa, pp. 111–168.

Houpt. K.A., 2000. Equine maternal behavior and its aberrations. International Veterinary Information Service. <www.ivis.com> Document No. A0802.0800.

Houpt, K.A., 2002. Formation and dissolution of the mare-foal bond. Appl. Anim. Behav. Sci. 78, 319–328.

Hulet, C.V., Ercanbrack, S.K., Price, D.A., Blackwell, R.L., Wilson, L.O., 1962. Mating behavior of the ram in the one-sire pen. J. Anim. Sci. 21, 857–864.

Hulet, C.V., Blackwell, R.L., Ercanbrack, S.K., 1964. Observations on sexually inhibited rams. J. Anim. Sci. 23, 1095–1097.

Hulet, C.V., Alexander, G., Hafez, E.S.E., 1975. The behavior of sheep. In: Hafez, E.S.E. (Ed.), The Behaviour of Domestic Animals, third ed. William Clowes, London.

Hurnik, F., 1987. Sexual behavior of female domestic animals. Vet. Clin. North Am: Food Anim. Pract. 3, 423–461.

Hyde, L., 2010. Limousine Breeders tackle temperament—genetic trend shows power of selection. <http://www.nalf.org/pdf/2010/aug19/tackletemperament> (accessed 29.02.2012.).

Iwata, E., Wakabayashi, Y.K., Kakuma, Y., 2000. Testosterone dependent primer pheromone production in the sebaceous gland of male goat. Biol. Repro. 62, 806–810.

Iwata, E., Kikusui, T., Takeuchi, Y., Mori, Y., 2003. Substances derived from 4-ethyl octanoic acid account for primer pheromone activity for the "mate effect" in goats. J. Vet. Med. Sci. 65, 1019–1027.

Izaguirre-Flores, F., Martinez-Tinajero, J., Sánchez-Orozco, L., Ramón-Castro, M.A., Pérez-Hernández, P., Martínez-Priego, G., 2007. Influence of suckling and sire presence on the productive and reproductive performance of brown swiss cows in the humid tropics. Rev. Científica FCV-LUZ XVII, 614–620.

Jacobs, V.L., Sis, R.F., Chenoweth, P.J., Klemm, W.R., Sherry, C.J., Coppock, C.E., 1980. Tongue manipulation of the palate assists estrus detection in the bovine. Theriogenology 13, 353–356.

Jainudeen, M.R., 1983. The water buffalo. Pertanika 6 (Rev. Suppl.), 133–151.

Jainudeen, M.R., Hafez, E.S.E., 2000. Cattle and buffalo. In: Hafez, E.S.E., Hafez, B. (Eds.), Reproduction in Farm Animals. Lippincott, Williams and Wilkins, Phil. (Hafez, E. S. E. ed.).

Janczak, A.M., Pedersen, L.J., Rydhmer, L., Bakken, M., 2003. Relation between early fear- and anxiety-related behaviour and maternal ability in sows. Appl. Anim. Behav. Sci. 88, 121–135.

Katz, L.S., 2008. Variation in male sexual behavior. Anim. Reprod. Sci. 105, 64–71.

Katz, L.S., McDonald, T.J., 1992. Sexual behavior of farm animals. Theriogenology 38, 239–253.

Keyserlingk, M.A.G., Weary, D.M., 2007. Maternal behaviour in cattle. Horm. Behav. 52, 106–113.

Kilgour, R.J., 1981. The mating performance of rams in pens and its usefulness in predicting flock mating performance. MSc. Thesis, University of Sydney, Sydney, Australia.

Kirkwood, R.N., Forbes, J.M., Hughes, P.E., 1981. Influence of boar contact on attainment of puberty in gilts after removal of the olfactory bulbs. J. Reprod. Fertil. 61, 193–198.

Knauer, M.T., Cassady, J.P., Newcom, D.W., See, M.T., 2010. Estimates of variance components for genetic correlations among swine estrus traits. J. Anim. Sci. 88, 2913–2919.

Kunowska-Slósarz, M., Różańska, J., 2009. High yielding cows and their calves' behaviour in the perinatal period. Ann. Pol. Zootec. Soc. 5 (2), 191–199.

Landaeta-Hernández, A.J., Chenoweth, P.J., Berndtson, W.E., 2001. Assessing sex-drive in young *Bos taurus* bulls. Anim. Repro. Sci. 66, 151–160.

Landaeta-Hernández, A.J., Yelich, J.V., Tran, T., Lemaster, J.W., Fields, M.J., Chase Jr., C.C., et al., 2002a. Environmental, genetics, and social factors affecting the expression of estrus in beef cows. Theriogenology 57, 1357–1370.

Landaeta-Hernández, A.J., Rae, D.O., Olson, T.A., Archbald, L.F., 2002b. Aspectos genéticos de la debilidad al nacer en becerros Brahman. Proceedings of the 11th Congreso Venezolano de Producción e Industria Animal. Valera, Oct 22–26. Trujillo, Venezuela, pp. 1–13.

Landaeta-Hernández, A.J., Giangreco, M.A., Meléndez, P., Bartolomé, J., Bennet, F., Rae, D.O., et al., 2004a. Effect of biostimulation on uterine involution, early ovarian activity and first postpartum estrous cycle in beef cows. Theriogenology 61, 1521–1532.

Landaeta-Hernández, A.J., Palomares-Naveda, R., Soto-Castillo, G., Atencio, A., Chase Jr., C. C., Chenoweth, P.J., 2004b. Social and breed effects on the expression of a PGF2a-induced estrus in beef cows. Reprod. Dom. Anim. 39, 315–320.

Landaeta-Hernández, A.J., Rae, D.O., Olson, T.A., Ferrer, J.M., Barboza, M., Archbald, L.F., 2004c. Preweaning traits of Brahman calves under a dual-purpose management system in the tropics. Rev. Científica FCV-LUZ XIV, 344–353.

Landaeta-Hernández, A.J., Meléndez, P., Bartolomé, J., Rae, D.O., Archbald, L.F., 2006. Effect of biostimulation on expression of estrus in postpartum angus cows. Theriogenology 66, 710–716.

Landaeta-Hernández, A.J., Meléndez, P., Bartolomé, J., Rae, D.O., Archbald, L.F., 2008. The effect of bull exposure on the early postpartum reproductive performance of suckling angus cows. Rev. Científica FCV-LUZ IXI (6), 682–691.

Lane, J.E., Forrest, M.N.K., Willis, C.K.R., 2011. Anthropogenic influences on natural mating systems. Anim. Behav. 81, 909–917.

Langbein, J., Raasch, M.L., 2000. Investigations on the hiding behaviour of calves at pasture. Archiv. fur. Tierzucht–Arch. Anim. Breed. 43, 203–210.

Le Neindre, P., 1989. Influence of cattle rearing conditions and breed on social relationships of mother and young. Appl. Anim. Behav. Sci. 23, 117–127.

Leenhouwers, J.I., de Almeida, C.A., Knol, E.F., van der Lende, T., 2001. Progress of farrowing and early postnatal pig behaviour in relation to genetic merit for pig survival. J. Anim. Behav. 79, 1416–1422.

Leenhouwers, J.I., Wissink, P., van der Lende, T., Paridaans, H., Knol, E.F., 2003. Stillbirth in the pig in relation to genetic merit for farrowing survival. J. Anim. Sci. 81, 2419–2424.

Lent, P.C., 1974. Mother–infant relationships in ungulates. In: The Behavior of Ungulates and its Relation to Management. Geist, V., Walther, F. (Eds.), IUCN Publ. N. S., pp. 14–55.

Lévy, F., Keller, M., 2009. Olfactory mediation of maternal behavior in selected mammalian species. Behav. Br. Res. 200, 336–345.

Lidfors, L.M., Moran, D., Jung, J., Jensen, P., Castren, H., 1994. Behavior at calving and choice of calving place in cattle kept in different environments. Appl. Anim. Behav. Sci. 42, 11–28.

Lindsay, D.R., 1965. The importance of olfactory stimuli in the mating behavior of the ram. Anim. Behav. 13, 75–78.

Lindsay, D.R., 1996. Environment and reproductive behavior. Anim. Reprod. Sci. 42, 1–12.

Lindsay, D.R., Ellsmore, J., 1968. The effect of breed, season, and competition on mating behavior of rams. Aust. J. Exp. Agric. Anim. Husb. 8, 649–652.

Lindsay, D.R., Fletcher, I.C., 1972. Ram seeking activity associated with oestrous behavior in ewes. Anim. Behav. 20, 452–456.

Lindsay, D.R., Robinson, T., 1961. Studies on the efficiency of mating of sheep. II. The effect of freedom of rams, paddock size and age of ewes. J. Agric. Sci. 57, 141–145.

Lôbo, A.M.B.O., Lôbo, R.N.B., Paiva, S.R., 2009. Aromatase gene and its effects on growth, reproduction, and maternal ability traits in a multibreed sheep population from Brazil. Genet. Mol. Biol. 32 (3), 484–490.

Løvendahl, P., Chagunda, M.G.G., 2009. Short communication: genetic variation in estrus activity traits. J. Dairy Sci. 92 (9), 4683–4688.

Lucy, M.C., 2001. Reproductive loss in high-productive dairy cattle: where will it end?. J. Dairy Sci. 84, 1277–1293.

Lunstra, D.D., 1984. Changes in libido–fertility relationships as bulls mature. J. Anim. Sci. 59 (Suppl. 1), 351.

Lyimo, Z.C., Nielen, M., Ouweltjes, W., Kruip, T.A.M., van Eerdenburg, F.J.C.M., 2000. Relationship among estradiol, cortisol, and intensity of estrous behavior in dairy cattle. Theriogenology 53, 1783–1795.

Macfarlane, J.M., Matheson, S.M., Dwyer, C.M., 2010. Genetic parameters for the birth difficulty, lamb vigor and lamb suckling ability in Suffolk sheep. Anim. Welf. 19, 99–105.

Mader, D.R., Price, E.O., 1984. The effects of sexual stimulation on the sexual performance of Hereford bulls. J. Anlm. Sci. 59, 294–300.

Mandal, A., Barua, S., Rout, P.K., Roy, R., Prasad, H., Sinha, N.K., et al., 2005. Factors affecting the prevalence of mortality associated with pneumonia in a flock of Muzaffarnagari sheep. Indian J. Anim. Sci. 75 (4), 407–410.

McDonnell, S.M., 2000. Reproductive behavior of stallions and mares: comparison of free-running and domesticated in-hand breeding. An. Reprod. Sci. 60–61, 211–219.

McPherson, F.J., Chenoweth, P.J., 2012. Mammalian sexual dimorphism. Anim. Reprod. Sci. 131 (2012), 109–122.

Mellado, M., Amaro, J.L., Garcia, J.E., Lara, L.M., 2000. Factors affecting gestation length in goats and the effect of gestation period on kid survival. J. Agric. Sci. 135, 85–89.

Menegassi, S.R.O., Barcellos, J.O.J., Peripolli, V., Camargo, C.M., 2011. Behavioral assessment during breeding soundness evaluation of beef bulls in Rio Grande do Sul. Anim. Reprod. 8 (3–4), 77–80.

Morales, R., Mika, J., Holy, L., 1983. Reproductive performance of 3/4 Holstein–Friesian x 1/4 Zebu cows. 4. Clinical checks of postpartum ovarian activity and oestrus. Rev. Cubana Reprod. Anim. 3, 62–71.

Morris, C.A., Cullen, N.G., Kilgour, R., Bremner, K.G., 1994. Some genetic factors affecting temperament in *Bos taurus* cattle. N.Z. J. Agric. Res. 37, 167–175.

Morrow, D.A., 1966. Postpartum ovarian activity and uterine involution in dairy cattle. J. Am. Vet. Med. Assoc. 149, 1596–1609.

Mukasa-Mugerwa, E., 1989. A review of reproductive performance of female *Bos indicus* (Zebu) cattle. ILCA Monogr., 6.

Mukhopadhyay, C.S., Gupta, A.K., Yadav, B.R., Khate, K., Raina, V.S., Mohanty, T.K., et al., 2010. Subfertility in males: an important cause of bull disposal in bovines. Asian–Australas. J. Anim. Sci. 23 (4), 450–455.

Mylrae, P.J., Beilharz, R.G., 1964. The manifestation and detection of oestrus in heifers. Anim. Behav. 12, 25–30.

Neely, J.D., Robison, O.W., 1983. Estimates of heterosis for sexual activity in boars. J. Anim. Sci. 56, 1033–1038.

Nowak, R., 1989. Early recognition of the mother by the new-born lamb. Ph.D Thesis, University of Western Australia, Perth.

Nowak, R., Porter, R.H., Lévy, F., Orgeur, P., Schaal, B., 2000. Role of mother–young interactions in the survival of offspring in domestic mammals. Rev. Reprod. 5, 153–163.

O'Brien, P.H., 1984. Leavers and stayers: maternal post-partum strategies in feral goats. Appl. Anim. Behav. Sci. 12, 233–243.

Ocak, S., Onder, H., 2011. Placental traits and maternal intrinsic factors affected by parity and breed in goats. Anim. Reprod. Sci. 128, 45–51.

Oliveira, C.M., Oliveira-Filho, B.D., Gambarini, M.L., Viu, M.A., Lopes, D.T., Sousa, A.P., 2009. Effects of biostimulation and nutritional supplementation on pubertal age and pregnancy rates of Nelore heifers (*Bos indicus*) in a tropical environment. Anim. Reprod. Sci. 113, 38–43.

Ologun, A.G., Chenoweth, P.J., Brinks, S., 1981. Relationships among production trails and estimates of sex drive and dominance value in yearling beef bulls. Theriogenology 15, 379–388.

Orihuela, A., Galina, C.S., Duchateau, A., 1988. Behavioral patterns of Zebu bulls towards cows previously synchronized with prostaglandin F2a. Appl. Anim. Behav. Sci. 21, 267–276.

Palmer, E., Guillaume, D., 1992. Photoperiodism in the equine species—what is a long night? Anim. Reprod. Sci. 28, 21–30.

Perera, B.M.A.O., 2011. Reproductive cycles of buffalo. Anim. Reprod. Sci. 124 (3–4), 194–199.

Pérez-Hernández, P., García-Winder, M., Gallegos-Sánchez, J., 2002. Bull exposure and an increased within-day milking to suckling interval reduced postpartum anoestrus in dual purpose cows. Anim. Reprod. Sci. 74, 111–119.

Perkins, A., Fitzgerald, J.A., 1994. The behavioral component of the ram effect: the influence of ram sexual behavior on the induction of estrus in anovulatory ewes. Anim. Sci. 72, 51–55.

Pickett, B.W., Faulkner, L.C., Voss, J.L., 1975. Effect of season on some characteristics of stallion semen. J. Reprod. Fert. (Suppl. 23), 25–28.

Pickup, H.E., Dyer, C.M., 2011. Breed differences in the expression of maternal care at parturition persist throughout the lactation period in sheep. Appl. Anim. Behav. Sci. 132, 33–41.

Piper, L.R., Hanrahan, J.P., Evans, R., Bindon, B.M., 1982. Genetic variation in individual and maternal components of lamb survival in Merinos. Proc. Aust. Soc. Anim. Prod. 14, 29–30.

Plush, K.J., Hebart, M.L., Brien, F.D., Hynd, P.I., 2011. The genetics of temperament in Merino sheep and relationships with lamb survival. Appl. Anim. Behav. Sci. 134, 130–135.

Poindron, P., Le Neindre, P., 1980. Endocrine and sensory regulation of maternal behavior in the ewe. Adv. Study Behav. 11, 75–119.

Poindron, P., Otal, J., Ferreira, G., Keller, M., Guesdon, V., Nowak, R., Lévy, F., 2010. Amniotic fluid is important for the maintenance of maternal responsiveness and the establishment of maternal selectivity in sheep. Anim. 4, 2057–2064.

Price, E.O., 1984. Behavioral aspects of animal domestication. Q. Rev. Biol. 59 (1), 1–32.

Price, E.O., 1985. Sexual behavior of large domestic farm animals: an overview. Anim. Sci. 61, 62–74.

Price, E.O., 1987. Male sexual behavior. Vet. Clin. North Am.: Food Anim. Pract. 3, 405–422.

Radostits, O.M., Blood, D.C., Gay, C.C., 1994. Specific diseases of uncertain etiology, Veterinary Medicine, eighth ed. Bailliere-Tindall, London. UK.

Rajanarayanan, S., Archunan, G., 2004. Occurrence of flehmen in male buffaloes (*Bubalus bubalis*) with special reference to estrus. Theriogenology 61 (5), 61–866.

Rajanarayanan, S., Archunan, G., 2011. Identification of urinary sex pheromones in female buffaloes and their influence on bull reproductive behaviour. Res. Vet. Sci. 91 (2), 301–305.

Ralls, K., Kranz, K., Lundrigan, B., 1986. Mother–young relationships in captive ungulates: variability and clustering. Anim. Behav. 34, 134–145.

Randel, R.D., 1984. Seasonal effects on female reproductive functions in the bovine (Indian breeds). Theriogenology 21 (1), 170–185.

Rech, C.L.S., Rech, J.L., Fischer, V., Osório, M.T.M., Silveira, I.D.B., Tarouco, A.K., 2008. Temperament and maternal behavior in Corriedale and Pollwarth sheep and its relation with lamb's survival. Ciência Rural Santa Maria 38 (5), 1388–1393.

Rekwot, P.I., Ogwu, D., Oyedipe, E.O., 2000. Influence of bull biostimulation, season and parity on resumption of ovarian activity of Zebu (*Bos indicus*) cattle following parturition. Anim. Reprod. Sci. 63, 1–11.

Rekwot, P.I., Ogwu, D., Oyedipe, E.O., Sekoni, V.O., 2001. The role of pheromones and biostimulation in animal reproduction. Anim. Reprod. Sci. 65, 157–170.

Rettie, W.J., Messier, F., 2001. Range use and movement rates of woodland caribou in Saskatchewan. Canad. J. Zool. 79, 1933–1940.

Rivas-Muñoz, R., Fitz-Rodríguez, G., Poindron, P., Malpaux, B., Delgadillo, J.A., 2007. Stimulation of estrous behavior in grazing goats by continuous or discontinuous exposure to males. J. Anim. Sci. 85, 1257–1263.

Roberson, M.S., Ansortegui, R.P., Berardinelli, J.G., Whirman, R.W., McInerney, M.J., 1987. Influence of biostimulation by mature bulls on occurrence of puberty in beef heifers. J. Anim. Sci. 64, 1601–1605.

Robison, O.W., Lubritz, D., Johnson, B., 1994. Realized heritability estimates in boars divergently selected for testosterone levels. J. Anim. Breed. Genet. 111, 35–42.

Rodtian, P., King, G.J., Subrod, S., Pongpiachan, P., 1996. Oestrus behavior of Holstein cows during cooler and hotter tropical seasons. Anim. Reprod. Sci. 45, 47–58.

Roelofs, J.B., Soede, N.M., Voskamp-Harkema, W., Kemp, B., 2008. The effect of fenceline bull exposure on expression of oestrus in dairy cows. Anim. Reprod. Sci. 108, 226–235.

Rohde Parfet, K.A., Gonyou, H.W., 1990. Attraction of newborn piglets to auditory, visual, olfactory and tactile stimuli. J. Anim. Sci. 69, 125–133.

Rollins, W.C., Howell, C.E., 1951. Genetic sources of variation in the gestation length of the horse. J. Anim. Sci. 10, 797–806.

Romeyer, A., Porter, R.H., Poindron, P., Orgeur, P., Chesné, P., Poulain, N., 1993. Recognition of dizygotic and monozygotic twin lambs by ewes. Behav. 127, 119–139.

Rossdale, P.D., Short, R.V., 1967. The time of foaling in thoroughbred mares. J. Reprod. Fert. 13, 341–343.

Rothschild, M., 1996. Genetics and reproduction in the pig. Anim. Reprod. Sci. 42, 143–151.

Rottenstein, K., Touchberry, R.W., 1957. Observations on the degree of expression of estrus in cattle. J. Dairy Sci. 40, 1457–1465.

Rowan, K.J., 1992. Foetal and calf wastage in *Bos indicus*, *Bos taurus* and crossbred beef genotypes. Proc. Australian Assoc. Anim. Breed. Genet. 10, 370–375.

Rydhmer, L., 2000. Genetics of sow reproduction, including puberty, oestrus, pregnancy, farrowing, and lactation (Review). Livest. Prod. Sci. 66 (1), 1–12.

Rydhmer, L., Eliasson-Selling, K., Johansson, S., Stern, K., Andersson, K., 1994. A genetic study of estrus symptoms at puberty and their relationship to growth and leanness in gilts. J. Anim. Sci. 72, 1964–1970.

Scaramuzzi, R.J., Martin, G.B., 2008. The importance of interactions among nutrition, seasonality and socio-sexual factors in the development of hormone-free methods for controlling fertility. Reprod. Domest. Anim. 43 (2), 129–136.

Schanbacher, B.D., Lunstra, D.D., 1976. Seasonal changes in sexual activity and serum levels of LH and testosterone in finnish landrace and suffolk rams. J Anim. Sci. 43, 644–650.

Schmidek, A., Mercadante, M.E.Z., da Costa, M.J.R.P., Razook, A.G., de Figueiredo, L.A., 2008. Failure to suckle in Guzera calves: underlying factors and genetic parameters. Rev. Bras. Zootec-Braz. J. Anim. Sci. 37, 988–1004.

Seidel Jr., G.E., Foote, R.H., 1969. Motion picture analysis of ejaculation in the bull. J. Reprod. Fertil. 20, 313–317.

Shaw, E.B., Houpt, K.A., Holmes, D.F., 1988. Body temperature and behaviour of mares during the last two weeks of pregnancy. Equine Vet. J. 20 (3), 199–202.

Shearer, M.K., Katz, L.S., 2006. Female–female mounting among goats stimulates sexual performance in males. Horm. Behav. 50, 33–37.

Shipka, M.P., Ellis, L.C., 1998. No effects of bull exposure on expression of estrous behavior in high-producing dairy cows. Appl. Anim. Behav. Sci. 57, 1–7.

Shipka, M.P., Ellis, L.C., 1999. Effects of bull exposure on postpartum ovarian activity of dairy cows. Anim. Reprod. Sci. 54, 237–244.

Shreffler, C., Hohenboken, W.D., 1974. Dominance and mating behavior in ram lambs. J. Anim, Sci. 39, 725–731.

Signoret, J.-P., 1970. Swine behavior in reproduction. Proceedings of the Symposium on the Effects of Disease and Stress on Reproductive Efficiency of Swine. Univ. Nebraska and USDA, pp. 28–46.

Signoret, J.-P., 1975. Influence of the sexual receptivity of a teaser ewe on the mating preference in the ram. Appl. Anim. Ethol. 1, 229–232.

Signoret, J.-P., Baldwin, B.A., Fraser, D., Hafez, E.S.E., 1975. The behaviour of swine. In: Hafez, E.S.E. (Ed.), The Behaviour of Domestic Animals, third ed. Williams and Wilkins, Baltimore.

Signoret, J.-P., Cognie, Y., Martin, G.B., 1984. The effect of males upon female reproductive physiology. In: Courot, M. (Ed.), The Male in Farm Animal Reproduction. Martinus Nijhof, Dordrecht, The Netherlands.

Sime, C.A., Bangs, E.E. (Eds.), 2010. Rocky Mountain Wolf Recovery 2010 Interagency Annual Report. U.S. Fish and Wildlife Service, Ecological Services. 585 Shepard Way, Helena, MT, 59601.

Simm, G., Conington, J., Bishop, S.C., Dwyer, C.M., Pattinson, S., 1996. Genetic selection for extensive conditions. Appl. Anim. Behav. Sci. 49, 47–59.

Singh, N.J., Grachev, I.A., Bekenov, A.B., Milner-Gulland, E.J., 2010. Saiga antelope calving site selection is increasingly driven by human disturbance. Biol. Convers. 143, 1770–1779.

Snieder, J.F., Rempel, L.A., Rohrer, C.A., Brown-Brandt, T.M., 2011. Genetic parameter estimate among scale activity score and farrowing disposition with reproductive traits in swine. J. Anim. Sci. 89, 3514–3521.

Soto-Belloso, E., Ramírez-Iglesias, L., Guevara, L., Soto-Castillo, G., 1997. Bull effect on the reproductive performance of mature and first calf-suckled Zebu cows in the tropics. Theriogenology 48, 1185–1190.

Tanida, H., Murata, Y., Tanaka, T., Yoshimoto, T., 1989. Mounting efficiencies, courtship behavior and mating preferences of boars under multi-sire mating. Appl. Anim. Behav. 22, 245–253.

Taylor, R., Gamboa, F., Ramirez, N., 1995. Aspectos reproductivos en el nanejo del ganada doble proposito. In: Madrid-Bury, N., Soto-Belloso, E. (Eds.), Nanejo de Ia Ganaderia Metiza de Doble Proposito. Universidad del Zulia, Maracaibo, Venezuela.

Tennessen, T., Hudson, R.J., 1981. Traits relevant to the domestication of herbivores. Appl. Anim. Ethol. 7, 87–102.

Terrazas, A., Serafin, N., Hernández, H., Nowak, R., Poindron, P., 2003. Early recognition of newborn goat kids by their mother: II. Auditory recognition and evidence of an individual acoustic signature in the neonate. Develop. Psychol. 43, 311–320.

Thodberg, K., Jensen, K.H., Herskin, M.S., 2002. Nest building and farrowing in sows: relation to the reaction pattern during stress, farrowing environment and experience. Appl. Anim. Behav. Sci. 77, 21–42.

Thompson, P.N., Steinlein, O.K., Harper, C.K., Kraner, S., Sieb, J.P., Guthrie, A.J., 2003. Congenital myasthenic syndrome of Brahman cattle in South Africa. Vet. Rec. 153, 779–781.

Thompson, P.N., Van der Werf, J.H.J., Heesterbeek, J.A.P., Van Arendonk, J.A.M., 2007. The CHRNE 470del20 mutation causing congenital myasthenic syndrome in South African Brahman cattle: prevalence, origin, and association with performance traits. J. Anim. Sci. 85, 604–609.

Tillbrook, A.J., Cameron, A.W.N., 1989. The contribution of the sexual behaviour of rams to successful mating of ewes under field conditions. In: Oldham, C.M., Martin, G.B., Purvis, L.V. (Eds.), Reproductive Physiology of Merino Sheep: Concepts and Consequences. University of Western Australia, Perth.

Torriani, M.V.G., Vannoni, E., McElligott, A.G., 2006. Mother–young recognition is a unidirectional process. Americ. Nat. 168 (3), 412–420.

Trautwein, K., Bauer, H., Fluhr, F., 1958. Beobachrungen zur psychologie der bullen speziell zum deckverhalten. Zuchthyg 2, 217–234.

Tulley, D., Burfening, P.J., 1983. Libido and scrotal circumference of rams as affected by season of the year and altered photoperiod. Theriogenology 20, 435–448.

Tulloch, D.G., 1979. The water buffalo (*Bubalus bubalis*) in Australia: reproductive and parent offspring behavior. Aust. Wildl. Res. 6, 265–287.

Tyler, S.J., 1972. The behavior and social organization of the new forest ponies. Anim. Behav. Monogr. 5, 85–196.

Vangen, C., Holm, B., Valros, A., Lund, L., Rydhmer, L., 2005. Genetic variation in sow's maternal behavior, recorded under field conditions. Livest. Prod. Sci. 93, 63–71.

Walsh, S.W., Williams, E.J., Evans, A.C.O., 2011. A review of the causes of poor fertility in high producing dairy cows (Review). Anim. Reprod. Sci. 123 (3–4), 127–138.

Walther, F., 1965. Verhahensstudien an der Grangazelle in Ngorongoro Krater. Z. Tierpsycltol. 22, 167–208.

Waran, N.K., Clarke, N., Farnworth, M.L., 2008. The effect of weaning on the domestic horse (*Equus callabus*). Appl. Anim. Behav. Sci. 110, 42–57.

Wehrman, M.E., Kojima, F.N., Sanchez, T., Mariscal, D.V., Kinder, J.E., 1996. Incidence of precocious puberty in developing beef heifers. J. Anim. Sci. 74, 2462–2467.

Whately, J., Kilgour, R., Dalton, D.C., 1974. Behavior of hill country sheep breeds during farming routines. Proc. N.Z. Soc. Anim. Prod. 34, 28–36.

Williams, A.R., Franke, D.E., Saxton, A.M., 1991. Genetic effects for reproductive traits in beef cattle and predicted performance. J. Anim. Sci. 69, 531–542.

Williamson, N.B., Morris, R.S., Blood, D.C., Cannon, C.M., Wright, P.J., 1972. A study of oestrus behaviour and oestrus detection methods in a large commercial dairy herd. Vet. Rec. 91, 58–62.

Winfield, C.G., Hemsworth, P.M., Taverner, M.R., Mullaney, P.D., 1974. Observations on the suckling behavior of piglets in litters of varying sizes. Proc. Aust. Soc. Anim. Prod. 10, 307–310.

Wischner, D., Kemper, N., Stamer, E., Hellbrügge, B., Presuhn, U., Krieter, J., 2010. Pre-lying behaviour patterns in confined sows and their effects on crushing of piglets. Appl. Anim. Behav. Sci. 122, 21–27.

Wodzicka-Tomaszewska, M., Kilgour, R., Ryan, M., 1981. "Libido" in the larger farm animals: a review. Appl. Anim. Ethol. 7, 203–238.

Yilmaz, A., Karaca, S., Bingal, M., Kor, A., Koke, B., 2011. Effect of maternal behavior score (MBS) on weaning weight and litter survival in sheep. Afric. J.Agric. Res. 6, 1393–1397.

Yindee, M., Vlamings, B.H., Wajjwalku, W., Techakumphu, M., Lohachit, C., Sirivaidyapong, S., et al., 2010. Y-chromosomal variation confirms independent domestications of swamp and river buffalo. Anim. Gen. 41, 433–435.

Younis, M., Samad, H.A., Ahmad, N., Ahmad, I., 2003. Effects of age and season on the body weight, scrotal circumference and libido in Nili-Ravi buffalo bulls maintained at the semen production unit, Qadirabad Pakistan. Vet. J. 23, 59–65.

Zúrek, U., Danek, J., 2011. Maternal behaviour of mares and the condition of foals after parturition. Bull. Vet. Inst. Pulawy. 55, 451–456.

FURTHER READING

Godfrey, R.W., Gray, M.L., Collins, J.R., 1998. The effect of ram exposure on uterine involution and luteal function during the postpartum period of hair sheep ewes in the tropics. J. Anim. Sci. 76, 3090–3094.

Izard, M.K., Vandenbergh, J.G., 1982. The effects of bull urine on puberty and calving date in crossbred beef heifers. J. Anim. Sci. 55, 1160–1169.

Chapter 6

Differences in the Behavior of Landraces and Breeds of Dogs

Kathryn Lord, Lorna Coppinger, and Raymond Coppinger
School of Cognitive Science, Hampshire College, Amherst, MA, USA

In 1758 Linnaeus classified the domestic dog as *Canis familiaris,* a species designation. Subspecies of dogs are called landraces or breeds, which are thought to have breed-specific behaviors that predispose its members to perform some task in a manner superior to any other landrace or breed. Since there are more than 300 breeds of dogs, it might be said that behaviorally (ethologically) dogs are the most varied species of mammals. However, many of the modern breeds are developed by breed clubs, which often have little interest in working ability.

Indeed many of the variations in dog sizes and shapes are simply the preservation of mutations (Kemper *et al.*, 2012). Various forms of dwarfism, gigantisms, and skull shapes are reasonably common in many mammals including humans (Kemper *et al.*, 2012; Millien, 2006; Perry and Dominy, 2009; Zabek and Slota, 2007), but these are bred for capriciously in dogs (Asher *et al.*, 2009; Bond, 2007; Coppinger and Coppinger, 2001; Serpell, 1995). For example, modern breeds such as dachshunds or bassett hounds are achondroplastic dwarfs, the result of a single gene mutation and not the result of people artificially selecting for short legs.

Within the subspecies "dog", landraces and breeds vary widely and are phenotypically distinct from each other and from other members of the genus *Canis*. At the same time, skeletal characteristics that clearly discriminate dogs as a whole from other species and sub-species of the genus *Canis* do not exist.

Taxonomists often characterize (normal) mesaticephalic dogs as having robust frontal bones, overlapping dentiton or a shorter facial length, but these characteristics are not well differentiated, statistically (Coppinger *et al.*, 2010). So-called diagnostic features of dogs can be found in ontogenetic stages of other members of the genus. For example, sub-adult wolves have short, wide palate lengths with overlapping dentition. Descriptive differentiation of dogs is always a matter of how the sample was chosen.

Genetics and the Behavior of Domestic Animals. DOI: http://dx.doi.org/10.1016/B978-0-12-394586-0.00006-8
© 2014 Elsevier Inc. All rights reserved.

Characters we diagnose as strictly "dog," such as hypertrophied barking (Lord *et al.*, 2009), or diestrous cycling (rare or nonexistent in wild canids) (Lord *et al.*, 2013), or coat color and pattern, do not fossilize. Dog-like skeletons found in proximity to humans are difficult to distinguish from fossil remains of dog-like animals not found near humans. Further, it is difficult to distinguish wolf skulls from those of dogs if the animals being compared are the same size. One distinguishing characteristic is that the brain case of "big" dogs is smaller in proportion to the brain case of big wolves. However, dogs in the 14-kg range do not have small brains, but rather have genus-typical brain size.

Honacki *et al.* (1982) suggested a reclassification of the dog as a subspecies of the wolf, *Canis lupus familiaris.* From a biological standpoint we agree with that and only object that it doesn't go far enough. *C. lupus*, *C. familiaris*, *C. aureus*, *C. dingo*, and *C. simensis* should all be classified as incipient species or subspecies of one another (Coppinger *et al.*, 2010). Of course, from a political standpoint, renaming the members of this genus as one species would be problematic for endangered species policy, given the multitude of domestic dogs.

Selection for breed-specific behavior is often interpreted as an analog to Darwin's theory of transmutation of species by natural selection. Darwin (1858) himself equated breed selection with natural selection:

It is wonderful what the principle of selection by man, that is the picking out of individuals with any desired quality, and breeding from them, and again picking out, can do."

Darwin, 1899

Darwin was fascinated with domestic dogs and pigeons not only because of the morphological varieties, but because roller and tumbler varieties of pigeons, for example, displayed breed-specific behaviors. Domestication was for Darwin a way to picture the natural selection argument, and not as a phenomenon that tested his hypothesis. Using selection by humans as the model to explain breed diversification, he then equated it to wild species diversification:

Now suppose there were a being who did not judge by mere external appearances, but who could study the whole internal organization, who was never capricious and should go on selecting for one object during millions of generations . . .

Darwin, 1858

Many scientists have applied the tenets of Darwin's theory back onto the domestication process, following the rules of natural selection. Researchers of domestication assign selective advantages for each observable trait, in what Maderson *et al.* (1982) describe as "*a posteriori* rationalizations of presumed selective advantage which cannot by definition, be experimentally verified." For example, Price (1984) wrote that ". . . characteristics of the juvenile . . . are retained into adulthood, perhaps as a result of selection to preserve the greater aesthetic appeal of the young animal." The point is that domestication is an analog to natural selection; it is not the experimental test of the hypothesis. The argument becomes circular.

There is ample evidence that domestication happened very quickly (perhaps 9000 years ago), and that landrace forms of domestic breeds appeared as the result of natural selection. Breeds, however, are not products of natural selection but the result of founder effects, mutational anomalies, and hybridization, causing saltations (phylogenetic leaps, or novelties in form, as distinct from Darwin's gradualism). In other words, there were sudden changes in both physical characteristics and behavior. In this paper we explore the processes of breed divergence, especially exploring breed-specific behaviors.

In the last 20 years researchers have attempted to determine the phylogenetic origins of dogs and the location of those origins by means of neutral mutation theory. So far, such techniques cannot account for effects of population size or genetic drift (Coppinger *et al.*, 2010) or the effects of frequent hybridization (admixture) (Larson *et al.*, 2012).

Rindos (1980) described domestication as a developing symbiotic relationship between humans and the domestic species. Symbiosis is often used as a synonym for mutualism, defined as two species living together for mutual benefit. Ecologists, however, reserve the word mutualism for a permanent and obligatory relationship that benefits both species. Other forms of symbiotic relationships are (obligatory) commensalism, where the benefit is entirely for one species while the other neither benefits nor is harmed, and (obligatory) parasitism, which benefits the one species to the detriment of the other. As we discuss the various breeds of dogs, we will explore their ecology and suggest how their behavior evolved and what kind of symbiotic relationship they have developed with humans.

FIVE KINDS OF DOGS

Five types of dogs are described, each representing a behavioral "type form" (a taxonomic standard for comparison) of domestic dogs. Descriptions of morphological and behavioral uniqueness of each of these types provide a background for discussion of the genetic bases for their breed-specific behaviors.

The five types are based on contemporary observations and studies. There are perhaps 700,000,000 to a billion dogs in the world, with humans having reproductive control over a minority of them (Stafford, 2007; Lord *et al.*, in review). Probably 75% or more of the dogs in the world reproductively behave as any wild species and are subject to natural selection. These are commonly the landrace dogs, those local types of domestic canids that evolve mainly on their own by, as it were, natural selection. All over the world, their adaptation to their immediate environment has resulted in a huge population of dogs that are usually short-haired, weigh about 14 kg, stand about 45 cm at the shoulder, and can be any color, with ash-yellow being common. The constraints on reproduction of these animals is entirely post-zygotic (selective advantage is conferred to individuals after they are born). The fewer number of working dogs and pet-class dogs are the result of pre-zygotic selection (bred by selecting for certain parental traits). We would like to imagine the five types as a phylogenetic sequence. The five types are presented in an order that suggests both the origin of the dog as a species and its subsequent diversification into subspecific and racial breeds. The five types are:

1. Village dogs. The vast majority of dogs are natural scavengers, and they obey the rules of a niche-adapted species.
2. Livestock-guarding dogs. These are mostly landrace animals, rather than breeds; they are village dogs that ontogenetically/developmentally acquire working behaviors.
3. Sled dogs. The unique morphology and behavior of sled-pullers are the result of choosing individuals from a background population, and hybridization.
4. Herding dogs, gun dogs, and hounds. These breeds result from pre-zygotic selection and have unique sets of innate motor patterns necessary for their specific task.
5. Household dogs. This very large population of dogs is adopted for pet or companion qualities.

Village Dogs

Ecology

Neolithic humans have lived in permanent settlements for approximately 10,000 years. These settlements now range in size from small seasonal shelters to cities, but in this chapter we call them all villages. Coexisting in and around these villages are many animal species. A short list includes cockroaches, rats, house mice, pigeons, cats, and a variety of canids—foxes, jackals, wolves, and, of interest here, dogs. It is rare to find some of these species (especially the dog) anywhere else but in the village environment. Thus the environment in and around a village can be viewed as a niche(s). The species living exclusively in this niche have adapted to forage, reproduce, and avoid hazards in the presence of humans. In some cases these adaptations (process) are developed ontogenetically, while others appear to be evolutionary adaptations (results of natural selection).

Dogs are well-known village occupants the world over, existing sometimes in great numbers—estimates range to ±1 billion. In some places there may be 100–1000 dogs/square kilometer. For example, (the village of) Baltimore, Maryland, has supported an estimated 40,000 free-ranging dogs (Beck, 1973). Boitani and Fabbri (1983) estimated 800,000 free-ranging dogs in Italy. In January 1989, in the mountain town of Mucuchíes, Venezuela (human population, 2000), we estimated a population of 800 dogs (1 dog/2.5 people). In 1994 it was estimated that 1,360,000 free-ranging dogs lived in the communal lands of Zimbabwe (Butler and Bingham, 2000). Human population there was estimated at 6,190,000, or 22 dogs/100 people, or 1 dog per 4.5 people. The current data on dog populations in developing countries shows a range of 8.1 dogs/100 people to 35 dogs/100 people (Jackman and Rowan, 2007). In resource-rich locations such as the Mexico City dump, we estimated 700 dogs/square kilometer and that is by no means a record.

The United Nations World Health Organization classifies dogs in four categories (WHO, 2004), and in the town of Mucuchíes all four were represented: feral dogs, which are independent and unrestricted and live outside the village (although individuals often came in during twilight hours); neighborhood dogs living in town and observable at any time of day independent of humans even though they may recognize and beg from individuals; family dogs that are more or less managed by humans but semi-restricted; and the fourth category, the fully restricted dog that is totally dependent on humans. This type of habitat partitioning by dogs is familiar (e.g., Boitani *et al.*, 1995; Macdonald and Carr, 1995; Macdonald *et al.*, 2004).

People often refer to the semi- or unrestricted free-ranging village dogs as strays, mongrels, pavement specials, or curs. The assumption is that these are dogs which are the results of irresponsible ownership, or have severed their relationship with humans, and live by foraging on garbage within or on the margins of the village. But worldwide these dogs are phenotypically remarkably similar in shape and size, suggesting to us that they are the result of natural selection and adaptation to a niche. The uniformity of size, shape, and behavior also suggests a lack of artificial selection. There is historical evidence that smallish canids have continuously occupied villages since the beginning of writing (Coppinger and Coppinger, 2001).

Researchers have studied various ecotypes and varieties of village dogs, using names that reflect their assumed ontogenetic or phylogenetic history, or their foraging strategy: they are stray dogs, pariahs, sylvatic, feral, or wild dogs. Within the village environment are probably several ecozones which elicit different foraging strategies, and over time and geography, differences in the riches of the niches account for a variety of the observed adaptive strategies. Ecologists interested in the behavior of carnivores make predictions about these social arrangements based on the "resource dispersion hypothesis" (Carr and Macdonald, 1986; Macdonald *et al.*, 2004).

Three of our study sites are Zanzibar (Coppinger and Coppinger, 2001), Ethiopia (Ortolani *et al.*, 2009), and South Africa (Gallant, 2002). The dogs of Zanzibar are presented here as the type form of village dog. The study site in Zanzibar, in 1996, was primarily rural Pemba (island) where people live in villages off the road or in small towns along connecting roads. The people regard dogs as unclean animals, based on religious predilection. They feel that dogs have diseases and have parasitic organisms living in their mouths and nasal passageways. The wet noses of dogs are indications of these infectious agents and should not be touched. In East Africa and many other places in the world, rabies inflicted by dog bites is common. People are also repulsed by dogs because they eat human feces and corpses; it is a common notion to bury bodies deeply and cover them with stones so dogs cannot get to a body. In places like Zulu-Natal, dogs digging up shallow graves and eating bodies is a common problem in shallow rocky soil. In countries where bodies are sometimes rafted away down rivers, when they strand up on a beach dogs often eat the remains. Dogs are regarded the way we regard rats: an animal ubiquitously present, a potential vector of disease, a scavenger, and occasionally a thief, whose population needs to be culled from time to time. In most areas of Africa the dog is regarded as both the reservoir and the vector of rabies whether true or not.

In our interviews, the cultural dislike for dogs was invariably presented first, followed by various individual modifications. These ranged from people who were disgusted by the thought of touching a dog, to others who thought dogs had some value as alarms or hunters of pests.

Ortolani *et al.*'s (2009) study in Ethiopia suggested there was a difference in behavior of village dogs depending on the predominant religion of the village. In Muslim villages the dogs tended to disperse from approaching strangers, whereas in Christian towns they tended to ignore strangers. However, in Muslim mountain towns, a slightly higher percentage of dogs were found in houses, a consequence of keeping sheep and goats in a section of the house.

Phenotype

The dogs of Zanzibar are phenotypically similar to other village dogs we have studied. As has been noted, village dogs around the world appear remarkably similar. They are medium to small (12–16 kg), with solid color or piebald coats, in any mammalian color possible and in any combination. Conformation is not without variation; ears, for example, range from pricked to pendulous.

The small but numerous variations in size and shape are most likely the results of local adaptation (dogs at the equator tend to be smaller than those at increasing latitude and altitude), local founder effects (such as a high frequency of some color pattern), or genes introduced by household dogs or local working breeds that stray into the village dog population. The dogs of Mucuchíes are larger than 10 kg, and are referred to locally as if they were a breed descended from Simon Bolivar's dog, Nevado. Occasionally, Mucuchíans introduce/release European dogs (Pyrenean mountain dogs or Saint Bernards) to "improve" the qualities of this free-ranging "breed."

FIGURE 6.1 Mucuchies Venezuela village dog with pups "improved" by the introduction of St Bernards. *Photo credit: Gail Langeloh.*

Boitani *et al.* (1995) noted that the stray dogs in Italy often resemble the Italian maremma shepherd dog (20–30 kg), and in Patagonia, village dogs often carry some aspect of the locally popular greyhounds. The Zulus in South Africa will frequently introduce greyhounds into their local dogs (Gallant, 2002).

The New Guinea singing dog is an example of a village dog that was island-isolated for so long that it developed unique morphological and behavioral characteristic. Larson *et al.* (2012) think that it now may be extinct on the island as a breed due to the admixture with introduced European breeds. The current breed of New Guinea singing dogs descended from fewer than 14 individuals.

Behavior

Foraging

In Zanzibar, village dogs foraged in and around villages, in trash heaps, and latrines. Dogs were common in and around marketplaces, fishing ports, slaughterhouses—wherever human food was being processed. Killing domestic animals for food was not observed, as was also noted in Italy by Boitani *et al.* (2006). From our own personal observations in South America, Africa, and Europe, it was rare for village dogs to kill anything on its feet. However, sick animals on the ground, including humans, could be killed and eaten.

Several people in Zanzibar reported that they fed dogs, but not routinely. Most acted confused when asked what they fed to dogs. Food, especially protein, was not abundant in human diets. Whether people liked dogs or not, foraging behavior for the most part was independent of humans, except that the source was usually human waste. Populations of dogs in places like the

Mexico City dump are well fed, suggesting that something other than food was limiting the population. The niche of the village dog, like the niche of any species, has a lush center which declines in abundance toward the margins.

Reproduction

There is little to no manipulation of dog breeding claimed by Zanzibaris, except for one hunter who had a favorite female. Like foraging, reproduction is the responsibility of the dog. Female and male dogs are promiscuous (Daniels, 1983; Ghosh *et al.*, 1984/1985). In Zulu-Natal, the Mexico City dump, and Mucuchíes, males were observed to line up and breed a female sequentially, with little aggression between them. This kind of promiscuity is common in our studies. Indeed when one sees a "pack" of dogs it is almost always a bitch in heat followed by any number of males. Litters of pups can have multiple fathers.

Females raise their own puppies independently of humans, and the literature suggests that no help is provided by male dogs or other members of a group (Lord *et al.*, 2013). Pups begin to follow their mother to food sources at about 8 weeks. Pups often beg for food from people but for all practical purposes pups are competing (poorly) with adults from the post-nursing stage (Lord *et al.*, 2013).

The reproductive success of village dogs was discussed by Beck (1973), Boitani *et al.* (1995, 2006), Macdonald and Carr (1995), and Pal (2001). These studies note little "successful" reproduction. Mortality of juveniles was high for a variety of reasons, but partly because these populations are pests and harassed by humans. Reece *et al.* (2008) estimated that in spite of the harassment, village dogs in Jaipur sustained an average fecundity of 1.33 female pups per year, which suggests growing populations. Chawla and Reece (2002) found that the dog population in this area was growing with the human population (about 4% per year). This increase is likely lower than normal as both Reece *et al.* (2008) and Chawla and Reece (2002) were performing these studies to assess an attempt to control this city's dog population through surgical sterilization. Similarly, on Pemba, it was reported that when the population of dogs gets too high it has to be culled. However, since there are no household dogs to restock this population, we must conclude that reproduction among village dogs is self-sustaining no matter how high puppy mortality is in any given year (Lord *et al.*, 2013).

Hazard Avoidance

A striking difference between village dogs and wild canids lies in the avoidance of people. That avoidance is called "flight distance." The onset of flight in village dogs is initiated only at short distances and the flight is just adequate to prevent capture; it ceases with the achievement of the minimum distance. In contrast, wild canids, which scavenge dumps at night, usually initiate flight before observational contact. The flight is rapid, increasing the interspacing many times and perhaps terminating foraging for long periods of time.

Village dogs in Zanzibar were shy of people although they often rested in proximity to them or begged for food. Village dogs withdrew only when approached too closely. Many showed little interest in people unless the people were eating. They paid scarce attention to us crawling around taking their pictures. Most often they "foraged" in a way that increased the distance between them and the observer. To lay a hand on one was difficult and attempts were sometimes greeted with threats (snarl/growl or barking) and withdrawal. (Rats displayed similar behaviors on Pemba, often approaching our picnics and begging for food, often successfully, but scampering off, squealing, if we tried to catch one.)

In Ethiopia, Ortolani *et al.* (2009) reported differences in flight distance behaviors among dogs in different villages, suggesting developmental differences. The attitudes toward and treatment of dogs by local people affected their adult hazard-avoidance behaviors.

In some villages of the world, dogs are preyed upon by people. One common report on Pemba was that North Korean construction workers bought or captured dogs to eat. Local people viewed this behavior with a combination of mirth and revulsion. However, the eating of village dogs (often puppies) is reported virtually worldwide (Corbett, 1985; Lantis, 1980; Podberscek, 2009; Serpell, 1995; Titcomb, 1969).

Dogs that intrude in some noxious way are killed by people. On Pemba, when the dog population is high, the army can be called out to reduce the number of dogs in the face of obnoxious behavior or the presence of disease.

Evolution

One of the requirements for evolution and divergence of species is large populations of animals, over time. Village dogs, with loose or no attachments to people, living within or on the outskirts of villages, are a good model for understanding the origins of the dog and perhaps other domestic animals. The people on Pemba, generally ambivalent to dogs, are likely to echo the attitude of late Mesolithic village people who would not have realized that a transformation of wild canids to domestic dogs was possible.

The hypothesis presented here is that the growth of villages in the early Neolithic period created a new niche, which attracted wild species. The initiator of the adaptation to village life is the wild species itself (Coppinger and Smith, 1983). Those species which could invade successfully, adapted quickly. For some large portion of the dog population this niche has remained open and growing to the present, and people still have little influence on the direction of morphological and behavioral adaptation. This is true of other species such as the house mouse, pigeon, rat, or the particular bird species that use bird feeders.

One might hypothesize that the apparent similarity of size and shape worldwide suggests selective pressures for energetic efficiency for the resources, climate, and other environmental factors involved in scavenging the village environment. Selection favored those individuals which could feed close to human habitation. There may not have been any net benefit for people, just as there may or may not be selective advantages to having house mice or rats foraging in the human environment. The symbiotic relationship is therefore commensalism—and for dogs it is an obligatory commensalism.

It could be that people derived sanitary benefits from having wastes removed from the living area, and that dogs provided food, and that barking at impending dooms was valuable. But (and this is a big "but") there is no more implication that selection occurred on the basis of these benefits than there is for rats that lived on garbage, that were eaten on occasion, and that warned sailors about their sinking ship by running up the mast. For the most part, even in modern times, dogs are seen as noisy pests from which people need to protect themselves (Lord *et al.*, 2009).

The diagnostic characteristics of domestic dogs, i.e. floppy ears, multicolored piebald coats, and diestrous cycling, have also been subjected to the *a posteriori* argument. One must keep in mind the results of Belyaev and Trut (1975). In an experiment to produce manageable and tame behaviors in farmed fur foxes, Belyaev and Trut chose from large populations only those animals with *shorter flight distances*. Breeding these animals together and choosing the progeny with shorter flight distances, in a mere 20 generations they essentially produced a domesticated (genetically tame) form. And, in addition to the tame behavior, a number of unexpected anatomical changes occurred, including piebald coats, diestrous cycles, floppy ears, and bizarre vocalizations; "... they even sound like dogs" (Belyaev, 1979). "Natural" selection for flight distance produced morphological and behavioral saltations which were not selected for.

The transformation from wild fox into dog-like creature simply by choosing tame behavior has to mirror in some way the original transformation of wild-type canid into dog. Animals that can forage on village wastes (a new niche) in the presence of humans would have a selective advantage over their siblings which were limited to scavenging at night on the outskirts of town.

This hypothesis is based on an observable population of animals in which selection for reduced flight distance accounts for a differential mortality. One advantage of this hypothesis is that there is a substantial population in which a differential mortality can occur, which the Darwinian Theory requires. The common belief that humans consciously chose those wild animals that were tamable, and entered into a mutualistic working relationship with them (hunting), lacks evidence of a substantial population. None of the characteristic dog traits are the result of simple Mendelian gene segregation, nor did Belyaev select for them, nor did he desire them as outcomes, nor are there known phylogenetic homologs. As Bemis (1984) noted, "... the observation of evolutionary change by itself is insufficient evidence that adaptation actually occurred." For "adaptation," one could also read "selection." The fact that dogs have a diestrous cycle is not evidence that people selected for it.

In a sense, then, it is reasonable to argue that the domestic dog is adapted in the evolutionary sense to the energetics of a village scavenger and, along with the genetic changes for reduced flight distance, acquired several observable diagnostic features, saltations with no known adaptive value.

Livestock-Guarding Dogs

Ecology

Livestock-guarding dogs are the most common working dog in the world. It is hard to find a pastoral community without them. The reason is simple—they are village dogs that are associated with pastoral cultures. Dogs living in pastoral communities socially bond with the livestock during early development and display dog-specific behaviors to that stock. Technically they are not breeds but rather landraces, since no pre-zygotic selection has taken place. However, in the last 75 years small samples taken from several pastoral regions (e.g., in France, Hungary, and Turkey), have been registered with national kennel clubs as breeds in the U.S. and Europe.

In Turkey we observed dogs in remote villages in which the people are pastoralists. They depend on sheep (and other livestock) for a large portion of their food and fiber. Dogs accompany transhumance migrations and live among the sheep continuously. Dogs which attend to livestock in this way are called livestock-guarding dogs (Coppinger *et al.*, 1983; Dawydiak and Sims, 2004; Landry, 2001; Rigg, 2001).

FIGURE 6.2 Zulu village dog, which has developed a social bond with goats. *Photo credit: Daniel Stewart*

Archaeologists discovered evidence in China of dogs and pigs together by about 4000 BC, and with cattle and sheep after about 3200 BC (Olsen, 1985). In the 2nd century BC, treatises by Cato the Elder and Marcus Terentius Varro dealt extensively with the use of dogs to protect livestock (Fairfax, 1913). In the Bible, Job (30:1), mentions "... the dogs of my flock."

Behaviorally, there are two distinct kinds of sheepdog. The behavioral individuality of the two types makes them among the most interesting dogs in the world. Here are distinct dog populations that work in identical grassland habitats, directing their behaviors to identical environmental stimuli (livestock), but in strikingly different ways. The duty of livestock-guarding dogs is to be attentive to sheep, to not disrupt their behavior, and to protect them from predators. Herding dogs are expected to disrupt livestock and to conduct them from place to place (Coppinger and Coppinger, 1993, 2001; Spady and Ostrander, 2008). Neither dog breed can be trained to do the other's job.

Phenotype

Livestock-guarding dogs vary enormously in body size. The dogs involved in Eurasian transhumance migrations tend to be large (25 kg), and in African mountain communities such as Lesotho they are similar in size and shape to the Eurasian dogs, while the dogs of desert or semi-arid scrub country such as those used by the Navajo peoples of the American southwest, and the Damara of Angola, or the Masai cattle dogs, tend to be small (12–16 kg). Size appears to be an adaptation to the particular climate or elevation, and to the task the dogs are required to do—for example walk on a 700-km migration twice a year.

Characters that are detectable at birth are sometimes chosen as breed markers, such as rough or smooth coats, distinctive coloring, and markings. Pups without the favored regional marker are often culled at birth. What people notice is that the regional characteristics of these dogs give the animals a breed-like phenotype. But because the selection is post-zygotic, they are more formally (and properly) referred to as a landrace.

Behavior

Foraging

Livestock-guarding dogs in pastoral villages forage much the same way as village dogs, scavenging in refuse and latrines. There they seem to have at least two foraging strategies, being attentive either to the village or to a flock. Pastoralists are often seasonally nomadic, and on trips away from the villages the dogs that accompany the flock are routinely allowed to scavenge sheep by-products such as afterbirths, food scraps, flock carcasses, and by-products of milk, yogurt, and cheese-making.

Dogs perceived to be useful in protecting livestock receive more attention than those that are "just" village dogs—in two ways. They gain preferential access to waste animal products simply because of their proximity to livestock activity. They also elicit feeding because humans notice their behavior. The feed from livestock activities (including milk, cheese, whey, dead animals, and grains) is perhaps higher quality than what dogs can find in village dumps and latrines.

Some of the newly founded "breeds" of these animals hunt and kill wildlife (Coppinger *et al.*, 1988). Many, however, seem incapable of predatory behaviors. Stillborn calves fed to a pen with 10 livestock-guarding dogs imported from Italy, Yugoslavia, and Turkey had to be cut open before the dogs would feed, indicating that they did not have dissecting motor patterns, whereas the herding dogs (border collies) had no difficulty dissecting a carcass (Coppinger and Coppinger, 2001).

Reproduction

Reproductive survival of food-enriched dogs may be better than that of their village counterparts. As in village dogs, there is little or no imposed reproductive isolation of individuals by people. However, there can be reproductive isolating mechanisms. A group of dogs attentive to a flock of sheep will have an edge over outside males to breeding their flock-guarding females in heat. A breakdown in this dog/sheep group plus a similar social disruption in a diminishing wolf population, Boitani (1983) argues, are the causes of hybridizations between the landrace dogs and free-ranging dogs and between dogs and wolves in Italy. Wayne *et al.* (1995) state that wolf/coyote hybridization occurs with population (social) and habitat disruptions, and further suggest that this hybridization resulted in the present red wolf (*C. rufus*). Although no geographic isolation occurs over the generations, some temporal isolation does, resulting in temporal homozygosities of racial formation (landrace).

Additionally, humans practice culling for the purpose of limiting litter sizes, often to two pups, for a variety of management reasons. Shepherds frequently cull pups that do not display the guarding dog characteristics. The survivors are often chosen simply for the preferred local color, perhaps a color that is reminiscent of a previous good animal. Often the color becomes the distinguishing characteristic of the "breed." An Italian shepherd showing us a litter with a large color and pattern variation identified the two white pups as the "pure" maremmas. Another shepherd in Puglia when asked why maremmas are white just laughed and said only the white ones were allowed to live.

Hazard Avoidance

Dogs that kill, injure, or are disruptive to the management of stock are killed. The display of predatory motor patterns, such as eye-stalk or chase and bite, is considered a fault and may be the cause of culling adults, thereby eliminating animals with these behaviors from the gene pool. There is little discussion of corrective methods for such animals, and there is a matter-of-factness worldwide about what to do if animals misbehave around stock.

Mortality specifically related to livestock-guarding dogs is discussed by Lorenz *et al.* (1986) and Marker *et al.* (2005). Most deaths were those that could happen to any dog anywhere. One category in Lorenz *et al.* (1986) included "getting lost" (40% of accidental deaths). In this case dogs are lost to the system but not necessarily dead. Within migratory sheep flocks this represents a differential mortality.

Evolution

Livestock-guarding dogs are local adaptations to pastoral communities, and hence they are known as landrace dogs. Because livestock-guarding dog behavior is perceived by pastoralists as beneficial to human endeavors, the dogs receive more feed and care from humans than do local village dogs. The resulting interactions can lead to local diversification (larger size and distinctive color patterns) and a differential mortality. The symbiotic relationship could be described as mutualism.

The question is, what leads to their distinctive guarding behavior? The behavior that distinguishes good (i.e. attentive, trustworthy, and protective) livestock-guarding dogs results primarily from two processes: raising pups during their critical (sensitive) socialization period in proximity to sheep or other livestock, and removing animals that are inattentive to the stock and/or are disruptive of livestock. Because of the pastoral village economy, livestock-guarding dogs are likely to be born and raised close to or in with the livestock. Livestock-guarding dogs not raised with livestock during the post-neonatal period rarely can become attentive to them (Coppinger *et al.*, 1983).

FIGURE 6.3 Navajo village dog, which has developmentally acquired livestock-guarding behavior towards sheep. *Photo credit: Ray Coppinger*

Dogs and other canids develop their social repertoire during a critical period (Fox and Bekoff, 1975; Scott and Fuller, 1965; Scott and Marston, 1950). This period (sometimes called the sensitive period; Bateson, 1979; Estep, 1996) for social development in dogs is thought to occur between approximately 4 and 14 weeks of age. Dogs that do not have close interspecific interaction with people or livestock during this time develop shyness, a behavior similar to if not identical to village dog behavior. Shyness is a heritable characteristic in dogs (Willis, 1995).

Developmentally, individuals pass through several sensitive periods with the onset of genetically programmed motor patterns. Animals develop their species (breed)-typical behavior repertoire during these periods of receptivity and "learn" where and when to display the motor pattern. As they pass out of a stage, their chances of learning the associative behaviors become increasingly diminished, if not impossible. If the proper environment is not present during that period, then the resulting behavior cannot become functional.

Thus, village dogs, on their way to specialization, can display three distinct (innate?) behavioral repertoires depending on their developmental environment. (1) Dogs born in and around villages, with social contact limited almost exclusively to other dogs, display shy behavior with humans and livestock. (2) Dogs raised in proximity to humans are not shy with them, but are attentive, perhaps following them on hunting and gathering forays. (3) Animals raised in proximity to livestock are attentive to them, following them on their daily foraging activities.

In each case dogs will display their "dog" social motor patterns to the species of primary contact during the sensitive period and to a lesser extent to the other two, which are also present in the pastoral environment. There is no genetic difference among the three behaviorally distinct types (village dogs, livestock-guarding dogs, or pets) and indeed in many areas of the world they are phenotypically indistinguishable (Coppinger and Coppinger, 2001).

Good livestock-guarding dogs do not display predatory motor patterns to sheep, because of their early social bonding. However, although some do show chase and bite behaviors even if they are raised with livestock, most do not display these patterns even if they are not raised with livestock, suggesting that these motor patterns are heritable and responsive to selective forces.

Unfortunately, our knowledge of the details of critical-period development is incomplete for the dog, and whether specific breeds such as livestock-guarding dogs have lost predatory inclinations is suspected but not proven (Coppinger *et al.*, 1987). Not only do they not display predatory behaviors even when not raised with livestock, but even those raised as pets commonly do not chase a ball. Biological mechanisms for the diversification of village dogs into breeds, whether they be hunting dogs or livestock-guarding dogs or pets, first of all require little or no genetic change from the village dog, but rather are results of their developmental environment. The developmental environment may be consciously or unconsciously provided by humans. Second, selection by culling puppies can produce a marker for dogs which are managed for specific tasks. Third, differences in behavior of emerging breeds can be created through culling animals that misbehave or through reproductive isolation of superior animals. Fourth, variation in foraging energetics results in local adaptation.

Specific aspects of culling and reproductive isolation may or may not be manipulated by humans. The developing symbiotic mutualism (reciprocal benefits to dogs and people) need not have been the result of any genetic divergence of the village dogs, and our observations of Damara tribe livestock-guarding dogs in Namibia reinforce Black and Green's (1985) contention that any dog, especially "mongrels" (Coppinger *et al.*, 1985), can be a livestock guardian if the proper developmental recipe is followed.

Small samples of dogs from various pastoralist societies have been collected and registered with kennel clubs primarily in the U.S. and Europe. These dogs are described by a breed standard by the people who create and manage the kennel club. Thus we have breeds of livestock-guarding dogs such as Great Pyrenees, Anatolian Shepherds, Sarplaninac, and so on, which reflect the area of origin. They are historical representations of the dogs that breeders envisioned were the "original" landrace dog. The idealized phenotypic standard tends to change in their "restricted environment" because of capricious breeder preference for bigger dogs or uniform coat color (Drake and Klingenberg, 2007, 2010) or by genetic drift, and founder effects.

FIGURE 6.4 Landrace breed from the Maluti Mountains in Lesotho, Africa, used as livestock-guarding dogs. *Photo credit: Daniel Stewart*

Sled Dogs

Ecology

Our discussion now jumps to the modern era, and traces the development of a highly specialized type of dog, the sled dog. While village and livestock-guarding dogs may date from early Neolithic, sled dogs are relatively recent (historic times), and tandem-harnessed teams are perhaps only 150 years old (Lantis, 1980). The dogs of the Eskimos have little to do with modern freighting and racing sled dogs. The specialized variety of these dogs originated in the U.S., Canada, and Asia/Siberia, at the end of the 19th century. Their function was moving freight on packed trails, in tandem harnesses, with a lead dog. Both the initial equipment and the vocabulary (e.g., "gee," "haw," "marche") of early dog driving was adopted from French fur trappers, and from horse and ox driving.

Pulling behavior is technically "play" since there is no immediate reward for the behavior (Coppinger and Schneider, 1995), and it provides the social facility to perform as a team. Coincidentally, play behavior can reduce intergroup aggression (Pellis and Pellis, 1996), so it is very important in a dog team. Lead dogs are not socially defined, but rather are spirited individuals that can set the pace and take directional commands. They are often females, or pairs of females, and a team may consist of several lead dogs that can be rotated into the position during a race.

The behavior of sled dogs is *not* homologous to adult wolf pack hierarchy behaviors (Coppinger and Schneider, 1995).

Phenotype

The phenotypic value of sled dogs came in the early 20th century with the sport of sled dog racing. Since then the shape has steadily changed (evolved). In the early 20th century dogs were capable of 5-minute miles for marathon distances. By mid-century, 4-minute miles was racing speed. By the end of the century sled dogs were pulling sled and driver in 3-minute miles.

The phenotype of the sled dogs is distinct. In order for the dog to work (behave) well, the dog must have the proper size and shape (Coppinger and Coppinger, 2001; Coppinger and Schneider, 1995). Racing dogs weigh slightly more than 20 kg but less than 27 kg. Maximum size is limited by surface-to-volume ratio, which determines the heat dispersal capacity during performance (Phillips *et al.*, 1981). Larger dogs have longer strides. Dogs with more mass and greater stride can pull more and faster than smaller dogs, but the detrimental side-effect to greater mass is the inability to radiate excess heat. This can result very quickly in heat stress problems.

Freight dogs and long-distance racing dogs trot, while racing sled dogs, in a 20- to 30-mile race, lope. These gaits allow sled dogs to always keep at least one foot in contact with the ground. Dogs with gaits containing single or double flights, such as greyhounds, would be pulled off balance by their harnesses because all four feet leave the ground during the flight. The particular gait that a dog performs is predisposed to and limited by skeletal and muscular morphology.

All dogs on a sled dog team should have identical gaits and strides. Ideally, they should move together, each pair a mirror image. Dogs differing in gait and not synchronized in their motions introduce inefficiencies in the forward movement of the sled because the angular vectors of the harnesses have energetic costs. In order for the dogs to run synchronously they should have identical morphologies.

Behavior

Foraging

Modern sled dogs are captive animals. They are not allowed to forage on their own; they are fed by their owners. Dog drivers recognize that their dogs must be fed well in order for them to perform well. Distance racing dogs can burn 10,000 calories a day and the trick is how to get that much food into a relatively small stomach. Heating food and water to body temperature can reduce the amounts to be ingested. Foods high in calories (fats and oil) are preferred. The peristaltic action of a running dog moves food through the intestine rapidly, decreasing the times for digestion. Grinding foods is helpful in increasing the digestive surface area, and has the positive effect of increasing the speed of digestion.

Reproduction

Modern racing sled dogs are not free to roam and have no choices in mates. The founding stocks (original) were dogs collected from populated areas in the U.S. and Canada and, later, Siberia (Coppinger, 1977). They tended to be large pet and hunting/working breeds. They were trained to pull sleds. Dogs that were not able to perform were destroyed. Often in the Canadian and American gold rush periods, freight dogs were bought from ships, imported at the beginning of winter and then destroyed at the end of the winter season (Walden, 1928).

Humans chose from the remaining dogs, the better ones to be bred together. Occasionally, new or different breeds are still introduced for a variety of reasons, such as increasing the variability and hybrid vigor. However, mates are usually chosen from dogs that display the correct functional behaviors. Dogs showing particular conformation and behavioral talents are picked more frequently for breeding.

Hazard Avoidance

In order to successfully reproduce, sled dogs need to willingly pull the sled continuously with other dogs. They therefore must not disrupt other dogs pulling sleds by fighting or entering into other social displays. Any dog exhibiting dominance hierarchy behavior which escalates repeatedly to aggression is culled. Other causes for not making the team include problems defecating while running or the tendency to collect snow and ice on their feet.

Evolution

Sled dogs were hybridized from a non-local population of dogs. The ancestors of sled dogs were individuals of hunting, sheep, and household "breeds" of dogs. In the beginning, size was the major criterion for choosing an animal. Differential mortality resulted from culling animals that did not perform well, and from choosing superior performers for breeding. The resulting divergent form has a specialized and uniform morphology which is an adaptation to the specific behavioral performance. "Choosing" is used here to avoid confusion with the word "selecting"; chosen animals are already "fit." The dogs of early freight handlers tended to have little value. It wasn't until dog racing became lucrative that individual dogs became valuable for their abilities to win.

There was no recognition cognizant by humans about details of the adaptive morphology; for example, they did not know the thermodynamics of heat load (Phillips *et al.*, 1981) or the histology of foot pad sweat glands (Sands *et al.*, 1977). Early attempts to breed greyhounds to racing sled dogs belied recognition that the 32-second sprint races of greyhounds are run on liver glycogen and the leaping gait is inappropriate for a harnessed animal. Very quickly, the human selectors of good running sled dogs abandoned such Darwinian (1858) "capriciousness," and selected for performance. The resulting divergence toward sled dog morphology, although orchestrated by humans, was an adaptation "of the whole internal organization" based on the sorting of varieties.

Hybridization provides the genetic diversity from which the new form was chosen. With sled dogs, the divergent morphology and behaviors were created in perhaps 20 years. This rapid evolution can be accounted for by the tremendous variation among the founding stock. There is not a gradual shifting of any particular character. Hybridization is rather a recombining and re-sorting of characters plus the creation of novel combinations of characters. After the initial hybridizations, choosing only those dogs that perform a task, restricting breeding to the best performers, and choosing among the offspring over generations, sounds like natural selection, but it should not be confused with gradualism.

Hybrids are often thought to be a blending or averaging of characters. However, mongrelization also creates unpredictable outcomes and saltatory events where there are sudden changes in the dog's morphology (Arons and Shoemaker, 1992). Hybridization is a common methodology of breeding for new and novel working dogs. Cattle dogs and truffle hounds are recombinations of other breeds. Competitors breed a little of this to a little of that in an effort to produce a superior animal.

It has long been realized by biologists that saltatory changes can be the products of hybridization (Stebbins, 1959). Haldane (1930) wrote that there is "... every reason to believe that new species may arise quite suddenly, sometimes by hybridization, sometimes perhaps by other means. Such species do not arise, as Darwin thought, by natural selection. When they have arisen they must justify their existence before the tribunal of natural selection." If we substitute "type, race or breed" for "species" in that statement, we think we have described the evolution of sled dogs. A look at histories of working dogs suggests that this is the most accurate and frequent methodology for the creation of types. Hybridization of breeds perhaps is the most frequent cause of morphological diversification and "structural disharmonies" (Stockard, 1941). Thus there is some natural selection (artificial selection), but most of the adaption is sorting through hybrids for superior animals. Many now use the agricultural techniques developed in breeding hybrid corn, chickens, and hogs.

Modern racing sled dogs are a good illustration of how working-dog behavior is derived. The important point to remember is that the genetic bases of behavior are based on the shape and structure of the animal. Technically, all behavior is epigenetic (above the genes). Even as professionals we often refer (incorrectly) to behavior as a product of genes or to genetically programmed behaviors (Coppinger and Coppinger, 2001; Estep, 1996). Behavior is defined as an organism moving through space and time. The form of that movement is a consequence of its shape. The sled dogs move through space pulling a sled more efficiently than other breeds because the genes and gene products have created (interactively with the environment) a shape that predisposes a 20-kg animal to keep cool while running, with at least one non-sweaty foot on the ground at all times. Sweaty feet are detrimental because they collect ice crystals.

As in the livestock-guarding dogs, "breeders" will collect dogs of some origin and sexually isolate them into a breed. A breed standard is developed of what the ideal sled dog should look like and the resulting animals are registered with a kennel club. Often the "breeds" will develop from a founding stock of a handful of animals. Because of selective breeding they can be traced back to fewer than a handful (Larson *et al.*, 2012). The Siberian husky is a good example of the process.

Herding Dogs and Gun Dogs

Ecology

Herding dogs conduct livestock from one place to another. Their behavior elicits fear-flocking and/or flight behaviors in livestock. These "chase and bite" dogs are represented by many breeds, often bred specifically to work with a particular class of livestock, e.g., with sheep (border collies) or cattle (Queensland blue heelers). There are also subcategories based on whether the livestock are to be gathered and penned (the dogs are "headers"), or driven ("heelers"), or captured ("catch dogs"). Within each of these categories, some breeds operate close to the handler, others at great distances without much contact (e.g., New Zealand huntaways). Some breeds vocalize while working (Australian shepherds); others are silent (border collies). Each behavior is specific to the breed, and it is inappropriate, counterproductive, and perhaps impossible, for example, to use cattle heeling dogs to pen sheep.

Working performance is based on the display of one or more specific motor patterns, which the handler directs by commanding the direction of display with whistles, vocalizations, or hand signals (McConnell and Baylis, 1985). The "chase" motor pattern is universal, but some dogs are expected to precede the chase with an eye-stalk behavior. Cattle heelers are expected to chase and grab-bite (nip at the heels), but the grab-bite is a fault in sheep dogs.

The presence, absence, frequency, or sequencing of the display of each of these motor patterns differentiates the breeds of herding dogs. Each herding breed shows one or more of these motor patterns and the form is usually unique. For example, the orientation of the outrun can be a direct pursuit, as in the heelers, driving the animals away from the handler, or a circling orientation, as in the headers, where the dog works 180° from the handler and moves the sheep toward the handler. Different strains of border collies display different outruns, which are locally popular because of management practices.

Like herding dogs, gun dogs have breed-specific motor patterns. Retrievers have the spontaneous onset of "chase and retrieve." Early in development a pup appears with an object in its mouth, and to throw it guarantees the owner a lifetime of repetitive performance. Grab-bites are part of the retrieving behavior, but crush or kill-bites are faults known as "hard mouth." Dissect and eat motor patterns are, obviously, disqualifications.

Pointers silently hold the eye-stalk behavior until commanded otherwise. Foxhounds and coonhounds are expected to vocalize while chasing (called "voice") and are expected to grab and kill-bite.

Phenotype

Some herding dogs have unique morphological traits. Cattle dogs such as blue heelers and corgis have shorter legs, ostensibly so a kicking cow's hoof passes over their head. Since sheep do not kick, sheep dogs are not short-legged. In fact, herding dog handlers rarely talk about morphology, in contrast to sled dog drivers, who never give up on the subject. However, complaints about herding dogs not having endurance, or being uncontrollable on hot days, suggest that size and heat storage capacities are relevant subjects for herding and hunting dog handlers. The careful observer notes that water retrievers in colder climates tend to be bigger-volumed animals than the upland field dogs, suggesting that the sportsman also understands that morphology limits performance. Larger dogs have a better ability to retain body heat when they swim in cold water. Herding dogs have to be big enough to frighten livestock and retrievers have to be big enough to carry the game.

Breed markers are prevalent among gun dogs and herding dogs. There are no studies that compare animals with different variations in phenotype, but these are claimed anecdotally. "Sheep move away from a black collie better than a white one." "Livestock-guarding dogs should be white so they blend in with sheep or can be distinguished from attacking wolves." These are examples of *a posteriori* rationalizations, and it is not clear even that the observations are correct.

Behavior

Foraging

Herding and gun dogs are confined so they will not perform without supervision, requiring them to be fed by their keepers. Like the sled dogs, performance is exhausting and the handler's great fear is that overworked dogs, or dogs in poor condition, will quit. Thus every effort is made to keep the dog in good physical condition.

Feeding these dogs, valuable because of their specialized motor patterns, is thought of as increasing the bond with the dog, and becomes part of a daily ritual, with anticipatory reaction on the part of the dog.

Unlike working sled dogs, which tend to be continuously constrained, herding dogs and gun dogs are free-ranging during performance. Most exciting in this regard is that although their working behavior is thought to be based on predatory motor patterns, the actual working performance is not motivated by food. They do not hunt food on their own, nor does satiation inhibit performance.

Reproduction

Herding dogs and gun dogs which display the proper innate stereotyped motor patterns are field-tested and judged on their abilities to take directions and produce a desired result. Breed creation is a modern affair. The original creation of retrievers, pointers, and herding dogs, as with the sled dogs, was achieved by hybridization. Selection was for the behavioral phenotype. Almost unlike the racing sled dogs, herding dog and gun dog breed definition changed in the late 19th and early 20th centuries from a phenotypic description to genealogy. Thus all border collies can be traced to a single individual born in 1893 (Larson *et al.*, 2012) and all golden retrievers back to two brothers (Coppinger and Coppinger, 2001).

Whichever the definition, good dogs are bred to good dogs. Breeding is restricted not only within the breed but often to superior performers (further decreasing heterozygosity). Crossbreeding is thought to disrupt the motor pattern-specific behavior in the offspring (Coppinger *et al.*, 1985). Except for accidents, modern herding and hunting breeds are sexually isolated from the larger population of dogs.

Hazard Avoidance

In the motor pattern-specific breeds, the hazards to be avoided are similar to those for sled dogs. Both individual and reproductive survival depend on the quality of the performance. Humans provide the care for and protection of these animals. Hazards are capricious, and survival rests with the handler. In a sense this is a true mutualism: one species is responsible for the foraging, reproductive, and hazard-avoidance functions in return for the proper display of a behavior which immediately and materially benefits the human, but not the performing animal. Dogs are physically or reproductively culled on the basis of their relationship with the human.

Evolution

Existing breeds of herding dogs and gun dogs are all very recent. Breed books often state that a particular breed is "ancient," but was improved by crossbreeding "recently." Creation of new breeds is still active in the livestock industry, especially in places of intense livestock development such as New Zealand and Australia (huntaways, kelpies, Queensland blue heelers). Creation of new breeds of gun dogs slowed in the U.S. with the banning of commercial hunting, but recently the interest in dogs used in finding or assessing endangered species has stimulated new types. There was considerable breed creation activity in the market-gunning era of the 19th century (e.g., Chesapeake Bay retriever).

The point is that these motor pattern-specific breeds are in a sense species-specific in that an inventory of the quality, frequency, and sequencing of behavior (an ethogram) will identify the animal taxonomically. They are also locally created, often by a single breeder or group. What this means is that a particular population of animals was isolated from the rest of the gene pool in some discrete locality for some short evolutionary period. Historically they are often traced to founder animals.

How is it possible in a few generations to create a population of animals which displays breed-specific motor patterns, displays to birds and not to rabbits (or *vice versa*), and, just as important, does not display other motor patterns specific to other breeds? Certainly, environmental conditioning plays a part in these matters, but our attempts to raise one breed in the situation of another (e.g., a Chesapeake Bay retriever as a livestock-guarding dog, livestock-guarding dogs as border collies, and *vice versa*) not only failed to produce good working dogs but essentially hindered the individual for retraining in its natural work.

It is hypothesized that herding-dog and gun-dog behaviors are analogous to the individual motor units of the ancestral predatory sequence (eye-stalk > chase > grab-bite > kill-bite > dissect) that characterize the functional carnivore foraging behavior (Coppinger and Coppinger, 2001).

Two differences are evident in these motor units in dogs, compared with their ancestors. First, in several wild species these behaviors are often linked together in the adult. If the performance is interrupted, the animal is inhibited from proceeding. Some species of cats are classic in this regard and cannot perform a motor pattern unless the prerequisite behavior is performed. Thus, they cannot eat carrion because they have to perform the entire eye-stalk > chase > bite > kill sequence in order to dissect and feed (Leyhausen, 1973).

Second, in dogs, these motor patterns are not only individually displayed but are often exaggerated (hypertrophied) or ritualized. Dogs get stuck in the performance and cannot proceed. Pointers and border collies get stuck on a point and have to be rescued. The performance in these cases is not functional in moving an animal toward a food reward, but the display appears to be its own reward (autotelic), as well as the end of the sequence.

Coppinger and Smith (1989) argue that there are three stages in the development of a mammal: the neonatal period of innate stereotyped motor patterns, a transition stage where the neonatal motor patterns cease and adult behaviors onset (metamorphosis), and an adult stage with stereotypical innate and functional motor patterns. During the transitional or metamorphic period, motor patterns individually onset and offset, and can be displayed individually, in nonfunctional sequences, and against a backdrop of behaviors contextually inappropriate to either neonatal or adult functional sequences. This same theory was reiterated by Burghardt (2005).

Changes in the timing (heterochrony of a breed in a juvenile or metamorphic stage provide an event for creating an hypothesis to explain the display of individual predatory motor patterns throughout the adult life of a dog, independently of the ancestral functional sequence. Differential retardation or acceleration would then change each breed's development in a narrow ontogenic (developmental) stage where some but not other predatory motor patterns had emerged.

Lord (2010) found that the onset of adult foraging motor patterns in border collies and shutzhund German shepherds did vary heterochronically as compared with wolves. These variations also translated into differences in the frequency at which motor patterns were incorporated into play. However, behavioral differences between wolves and dogs and breeds thereof were not entirely explained by these timing differences, suggesting the importance of early environment in the production of these breed-typical behaviors.

Fox's (1978) experiments have provided an alternate hypothesis. Hybrids between dogs and coyotes performed some predatory motor patterns but could not organize them in functional sequences. Coppinger *et al.* (1985) argue that mongrelization of dogs is a method of disrupting functional sequences in a way that mimics neotenization. Since we have good records of hybridization between breeds in the creation of new breeds, this hypothesis is very plausible. For these reasons many true breeds have evolved where animals get sexually isolated into small populations. And genealogy becomes a frequent topic of breeding discussions.

Household Dogs

Ecology

Household dogs are a very large population in western countries but a small proportion of the total world population. In the U.S. and Europe there are about 150 million household dogs (AVMA, 2007; European Pet Food Industry Federation, 2010). In the U.S., about 72 million dogs live in 37% of the households (AVMA, 2007).

The literature on the function of pets and companions deals with physical and mental health benefits for humans. The health and well-being of the assistance and therapy dogs, also, is highlighted in the literature (e.g., Wells, 2007, 2009). However, current findings of this research are mixed (Esposito *et al.*, 2011; Herzog, 2011).

The literature generates the impression that "feeling better" is a biological benefit. Yet, a "sense of calm" in the presence of a dog does not demonstrate increased reproductive or economic capacities, which are the true test of mutualism. Any true biological benefits would also have to be measured against the cost of the dogs in terms of food, health, environment, and wider issues about their presence in human society (Serpell *et al.*, 2000).

Authors concluding that dog is "man's best friend" do not usually mention the estimated 4,500,000 dog bites/year in the U.S. (Sacks *et al.*, 1996), an average of 323,085 per year of which lead to emergency room visits (Quirk, 2012). Nor do they include the 85,000 emergency room visits in the U.S. as a result of falls caused by pet dogs (Herzog, 2011), or the enormous nuisance problems of barking dogs and dog feces.

Phenotype

This group is phenotypically the most variable, from very, very large to very, very tiny. The claim is that the domestic dog is the most variable species in Mammalia. However, the mutations for gargantuanism or dwarfism are found in "all" species but tend not to become fixed unless selected for by humans, or appearing on isolated islands.

Many household dogs are the approximate size, color, and shape of their working ancestors. Many of these are registered by national or international kennel clubs under the breed name of their phylogenetic working ancestors. They tend to be historical representations of an ancestral breed without the breed-specific behaviors. Working breed-specific behaviors often turn out to be obnoxious to pet owners.

Phenotypically this is a difficult group to generalize. Choice is often for novel shapes and sizes for their own sake. Morphological exaggerations exist for gargantuanism or miniaturization of an ancestral breed. Tendencies toward brachycephaly (short heads) or dolicephaly (long heads) are enhanced, producing deformations such as occluded jaws, dental anomalies, and nasal restrictions. These often have adverse behavioral and physiological consequences, i.e. lowered oxygen tensions, improper thermoregulation, atypical ocular overlap (Evans and Christensen, 1979), or inability to feed properly. Some household breeds cannot walk without pain or damage to themselves, or reproduce without assistance or caesarian section (Wolfensohn, 1981). The English bulldog, no longer an historical representation of its infamous ancestor (the "catch" dog), is perhaps an extreme case of these selective processes (Thomson, 1996).

There has been growing awareness regarding the health and welfare of pedigree dogs, particularly in the U.K. Darwin mentioned the connection of extreme breed size and breed-typical health problems as early as 1858 (Darwin, 1858). But concern for the welfare of pedigree dogs has come to the forefront with the 2008 BBC documentary, "Pedigree Dogs Exposed", which spurred several independent reports on dog breeding (APGAW, 2009; Bateson, 2010; Gallant and Gallant, 2008; Rooney and Sargan, 2009) and the formation of the Advisory Council on the Welfare Issues of Dog Breeding. The reports by the Associate Parliamentary Group for Animal Welfare (2009), Rooney and Sargan (2009), and Bateson (2010) all determined that the breeding for extreme morphological characteristics and the practice of inbreeding to reach breed standards, were a cause for concern in regards to the welfare of the dogs created through these practices.

Character selection is frequently capricious. There is no real functional purpose for the morphology. For those of us who think of behavior as the selective agent for morphological diversification, the term "unnatural selection" should apply to household dogs (Coppinger, 1996). As we have seen, sled dogs are not chosen for their legs, but for their running abilities. In household dogs, a morphology (e.g., double dew-claws; long, pointed nose; short, flat nose; achondroplasia) which is in some way pleasing to humans, is chosen regardless of the animal's ability to cope with the form.

Behavior

Foraging

Direct human feeding of household dogs has replaced foraging behavior. Foraging efforts are discouraged by the separation of waste from dogs. People buy special food formulations, processed from grains and domestic animal products. These agricultural crops are produced for human consumption and should not be thought of as waste or by-products.

Therefore, household dogs are ecologically competitive with humans for food and the land to produce it. Agricultural land is the prime cause of wildlife displacement (e.g., Warner *et al.*, 1996), and thus the added cropland necessary to grow dog food is in competition with those species. An animal that depends on another species to expend energy for its feed and does not return that energy in some form is biologically defined as an obligate parasite.

Pound for pound, dogs are metabolically about twice as expensive as humans. Humans require 41 kcal/day/kg of lean body mass (Schoeller *et al.*, 1986), while a small dog requires 80.2 kcal/day/kg of lean body mass (Speakman *et al.*, 2003). Rough calculations compare calorie consumption of the estimated 72,000,000 household dogs in the U.S. to that of the human residents of New York, Chicago, Los Angeles, and Dallas, combined.

Feed is not a limiting factor for these dogs and has no effect on energetics of size and shape. The lack of behavioral limitations on size and shape (as there is in our other dog categories) and the minimal need for behavioral output often result in obesity and other nutritional diseases.

Reproduction

The human intention is to control all reproductive activity of household dogs, which Bradshaw and Nott (1995) claim is "likely to have a major effect on any inherited aspects of social behavior". There are three major ways in which household dogs are bred, each with different behavioral consequences: (1) dog owners arrange matings, (2) specialized breeders maintain and publicly display breeding stock, and (3) breeding farms purchase and/or raise pups to be distributed to pet stores. Lack of knowledge or precaution sometimes allows dogs to "choose" for themselves. Neutering animals (reproductive culling) is common. However, New *et al.* (2000) found that 56% of households which responded in the affirmative to having had one or more litters, also reported that litters were not planned. This suggests that a large proportion of American people do not have, or do not maintain, reproductive control over their household dogs.

Household dogs can become free-living. They can interbreed (hybridize) with wild canids (Boitani *et al.*, 1995) or endemic village dogs (Brisbin *et al.*, 1994), contaminating gene pools to the point of extinction of these adapted wild types.

Many household dogs have breed-specific diseases (Asher *et al.*, 2009; Willis, 1992). Culling animals that have symptoms of these diseases increases the rate at which homozygosity occurs within a breed. It is doubtful, with what we know about inbreeding (isolating a small population of animals for the purpose of decreasing heterozygosity), that the genetic health of a breed can be maintained for very long. Breed-specific diseases are symptomatic of loss of genetic health, which cannot be restored with continued genetic isolation (APGAW, 2009; Bateson, 2010; Rooney and Sargan, 2009).

Hazard Avoidance

The most severe hazard these animals face is culling by humans. Shelters take in 3–4 million dogs each year in the U.S. alone (Luescher and Medlock, 2009). It is estimated that 60% of these animals are euthanized (Clevenger and Kass, 2003). Some of the remaining 40% that are adopted are returned to the shelter due to behavioral problems (Mondelli *et al.*, 2004; Wells and Hepper, 2000) and must go through the whole process again. Many dogs are relinquished for behavioral reasons in the first place—behaviors that are often worsened by the kennel environment (New *et al*, 2000; Tuber *et al.*, 1999). If dogs do not have behavioral problems upon relinquishment, they will often develop problems in the shelter (Tuber *et al.*, 1999). If they can be housebroken and are reasonably quiet, submissive, and nondestructive, their survival opportunities increase.

The nuisance costs of household dogs are high. Culling of unwanted animals is only partly reported by agencies that provide this service. Charitable humane societies and municipalities throughout the U.S. maintain dog officers and kennels to collect lost, stray, and behaviorally obnoxious dogs.

An ancestral uniqueness in working behavior is often unacceptable in pets (Houpt *et al.*, 2007). For example, the displays of specialized motor patterns such as eye-stalk, chase, grab-bite, and kill-bite are essential in some working dogs, but not when directed at joggers, cars, or bicycle riders. Also anathema to the household environment are dogs showing dominance or aggressiveness, or simply scavenging trash barrels, wandering and hunting widely, and emitting rally calls continuously.

Many household dogs are abandoned daily when their owners leave for work. Neurotic separation anxieties are clinically common (Sherman and Mills, 2008). Owners seek psychiatric care for them. The fatal flaw for many household dogs is that the human companion has little understanding of dog behavior and welfare needs (Hubrecht, 1995; Sherman and Mills, 2008). Animal behavior therapists try to educate owners about the nature of dogs (Voith, 2009). But too often, one finds an animal genetically unfit to be a pet, in an environment not fit for a pet, and exhibiting behaviors that the human companion does not understand or desire.

Household dogs are dangerous to humans. According to Langley's (2009) study, dog bites exist at epidemic proportions with an average of 19 human fatalities per year in the U.S. alone. The number of fatal dog bites has been increasing with the growth of the human population and concurrent growth of the dog population (Langley, 2009). One of the most common questions asked of behavior therapists is how to treat aggression in dogs (Bamberger and Houpt, 2006; Dodman, 1996; Mugford, 1995; Voith, 2009). Certain people (most notably, postal carriers on foot, and also running children) are frequently threatened by dogs.

Uncontrolled barking by dogs is also more than a nuisance. Beck (1973) posits that barking may reduce human productivity. An article by Lord *et al.* (2009) generated numerous responses from people who were tormented by barking dogs.

Seldom mentioned by humans is that depriving the animals of their reproductive potential might be the biggest hazard faced by the pet dog population.

Evolution

For household dogs, being a good companion should be the object of selection and evolution. Even though 5 million are culled annually, this may not be a high percentage compared with other working breeds. Information on culling rates across the various classes of dogs is not known, but theory predicts that genetic divergence (in this case selection for good companion behavior) is dependent on over-reproduction and culling those animals that perform unacceptably.

The question has to be asked, is good behavior being selected for? Behavior should be more important than unique colors or morphologies.

Some studies have focused on what makes a good pet or companion. The work of King *et al.* (2009) investigates what traits people think they would like in a pet dog. The most popular characteristics listed for an ideal dog were obedient, friendly, affectionate, healthy, and loyal. The authors note that while obedience was important to many (40.7%), trainability was not (only 3.6%). Many people also desired to get their dog as a puppy (75.8%). This is interesting because adult behavior is greatly influenced by early environment. Therefore, it would be difficult to identify the adult behavior of the pup they are adopting. Furthermore, the owner would need to pay attention to the early rearing of their new pup in order to develop the dog they desire.

Geneticists estimate the heritability of behavioral characteristics of temperament, nervousness, shyness, or aggressiveness (see Houpt, 2007, for a review). The assistance dog industry funds heritability studies as well as temperament testing, attempting to predict adult behavior from juvenile behavior (Beaudet *et al.*, 1994; Goddard and Beilharz, 1986; Serpell and Hsu, 2001). It is difficult to find behaviors that are the results of simple Mendelian ratios, or to control for developmental events, and in essence to understand the synergistic (interactive) effects of environment on developmental processes (Serpell and Jagoe, 1995). Indeed the criticism of heritability studies is their inability to test genetic and environmental interactions (Hirsch, 1990; Wahlsten, 1990).

Most working dog breeders solve these problems by breeding the best to the best, independently of other morphological or behavioral considerations. Perhaps the lesson for the household dog industry is that until we understand behavior genetics it would be advisable to breed the best companion dogs together irrespective of other considerations.

"Now suppose there were a being who did not judge by mere external appearances, but who could study the whole internal organization, who was never capricious . . ."

Darwin, 1858

DISCUSSION

Dog and dog breed behavioral and genetic divergence has been illustrated with five types of dogs. For each type we hypothesized mechanism(s) of behavioral divergence.

First, we postulate that the domestic dog that we know today is an adaptation to permanent human settlement. The existence of the niche is evident from the number of species that are "exclusively" associated with permanent human settlement. That background population of village scavengers has a long and unbroken history and is still growing as the human population grows (Coppinger and Coppinger, 2001). This hypothesis differs considerably from the popular, widely held view of many, many authors, that divergence was the result of people consciously removing animals from the wild, and selectively breeding them to produce domestic phenotypes (for a recent review of this hypothesis, see Shipman, 2009).

Second, the alternative view is that the wild species' self-initiated reduction of flight distance from humans is what led to domestication. This reduced flight distance, and interspecific food-begging behaviors, are also common behavioral traits of rats, pigeons, and other species that are self-adapted to human settlement. Natural selection for reduced flight distance could be the result of the behavioral energetics of a scavenger associated with human settlements. Diagnostic characters of dogs, including coat color and patterning, ear carriage, and aberrant vocalizations, need not be viewed as adaptations, but simply as saltations incurred during the processes of rapid selection for reduction of flight distance. Non-seasonal and diestrous cyling are thought to be adaptations to a stable food supply of human waste (Lord *et al.*, 2013).

Third, in both models, domestication precedes taming. Taming is a result of variation in the developmental environment. Breed divergence and symbiotic behaviors with humans are results of developmental proximity. The crucial ingredients for breed divergence are not genetic but rather are found in the recipe for manipulating the developmental environment (e.g., Black and Green, 1985). Taming can be defined in terms of the quality, frequency, and sequencing of interspecifically directed social motor patterns. The quality of tamed behaviors will be different in the village scavenger than in the wild type, given similar developmental environments (Fentress, 1967; Klinghammer and Goodman, 1987; Zimen, 1987).

Fourth, specialized breeds are developed by hybridization of size, shape, and behavior, chosen from a background of existing dogs. A new breed is the result of choosing from animals that already perform the behavior. Performing animals are bred together, which might perfect the form in a manner similar to natural selection. Superficial characteristics, i.e. breed markers, are the result of sorting between puppies rather than selection for the character in the Darwinian sense.

Working breeds are divided into two groups: those that are the result of a specific developmental environment (pastoralist), and those that have a specialized phenotype (including behavior). The latter breeds of dogs are highly refined specialists for specific tasks. In the Darwinian sense they are more complex, more specialized than the ancestor. They outperform the ancestor for the specific task. They are not degenerated forms, as is often claimed (see Price, 1984, for a review).

Fifth, each of the five types of dogs and their ancestors are not genetically discrete. Individuals can, and do, genetically invade any of the categories, both locally and temporally. Boitani *et al.* (1995) and Gottelli *et al.* (1994) write about the consequences of hybridization with wolves. Wild types and village dogs are niche-adapted, and although in a sense humans provide the environment for village dog evolution, their symbiotic relationship is commensal and not mutualistic. It is neither conscious human behavior nor mutual benefit that leads to domestication (see Rindos, 1980, for a counter-argument). We believe that the obligatory commensal behavior of village dogs is ancestral to that of the various working dogs which *are* mutualistic with humans. In some cases, this is simply the result of the specific developmental environment, as with the livestock guardians, while other cases involve genetic selection for breed-specific motor patterns, as in the herding and gun dogs. Modern western companion animals are often capricious exaggerations of the working and hunting breeds that are dependent on human care. It is difficult to measure the benefit of conferring that care, although many authors try. Consideration of an obligatory parasitism should be discussed for understanding the welfare of humans, and breeding capriciousness should be analyzed in terms of the dogs' welfare. The clichés like "man's best friend" seem hypocritical, given the advent of debilitating structural anomalies which shorten life spans and lead to the inability to perform any movement without causing pain.

Mechanisms for the evolutionary divergence of dogs are reasonably clear. All the species of the genus *Canis* are allelomorphic. This simply means that wolves, coyotes, the jackal (*C. aureus*), and dogs are karyotypically identical; that is, they have the same number of chromosomes, the same shape of chromosomes, and the same mapping of genes on those chromosomes. Allelic differences are minor differences in the base pairs of individual genes. Morphological and behavioral differences in allelomorphs must be the result of heterochrony—changes in the timing of onsets, rates, and offsets of gene products. Heterochronic shifting of gene products in such a way as to change development during the juvenile stage (heterochrony), is an explanation that biologists have hypothesized to account for the tremendous phenotypic divergence of breeds of dog. However, many divergent morphologies (such as achondroplasia in dachshunds and basset hounds) are simply point mutations that show up periodically in any species.

Coppinger and Schneider (1995) have found that various aspects of behavior appear homologous to juvenile canid behavior. Coppinger and Coppinger (1982) predicted that breed-specific behaviors would be displayed against a concordant set of morphological traits such as face length. And they were wrong.

There is nothing wrong with the neoteny theory in theory, but while the "snapshot" hypothesis may have conceptual value for the kind of processes that occur, the predictions of youthful characteristics (either morphological or behavioral) being ontologically preserved in the descendent adult form have so far escaped verification. In fact, the reverse is true. One purpose in proposing such an idea is to attract the attention of other scientists who may generate data to confirm or refute the idea.

Research on the subject since 1982 has yielded important information. Global retardation in a developmental stage provides no predictions as to the form of the character of the descendant. Dogs are not short-faced wolves, as is commonly thought (Morey, 1994; Olsen, 1985). Differences in skull morphologies are not the cessation of growth at some intermediate ontogenetic juncture of the ancestor (Drake, 2011). Dogs such as the borzoi have heads that are longer and narrower than the wild types, which also can be described as neotenic. Arresting development in a growth stage would mean that the animal would grow and grow, creating skull shapes that were phylogenetically bizarre. The brachycephalic (short-faced) dogs don't have heads that are puppy-proportioned but rather have normal adult wild-type palate length proportions (Wayne, 1986). Many breeds of dogs have head shapes that are virtually indistinguishable from the adult allometries of the rest of the genus. Dog head shape is not a reflection of an ontogenetic stage (paedomorphosis), but rather a result of the developmental processes. All members of the genus start with neonatal heads, which are virtually identical in shape and size. The borzoi is an exception; the characteristic lengthening of the skull starts before birth (Drake, 2011). Allometric remodeling of the neonatal head shape starts with the palate. Rostral changes start later and end later than palate allometries, followed by a period of isometric growth. At about 16 weeks, all members of the genus have essentially adult-shaped skulls, including the borzois, in their palate length proportions, but not the adult rostral proportions. Sub-adult wolves have short faces and overlapping dentition once thought to be diagnostic of dogs.

Growth, both isometric and allometric, continues through the acquisition of the permanent teeth, and accommodation of the various parts to one another modify the resulting shape. Animals neutered early in ontogeny end up without the secondary sexual characteristics of adults. Environmental over- or under-stimulation can affect the formation of the brain which in a sense builds its own brain case, and affects the final outcome of both size and shape. The point is that the resulting head shape in any of the species is not strictly genetic, but rather epigenetic. Brachycephalic dogs have adult palates but neonatal rostrums only in the sense that they do not grow very much. Examination of that growth indicates that it is the nasal bones that don't grow and the rest of the face is an accommodation of the maxillary and pre-maxillaries to the genus-specfic palate proportions, adult teeth, and the short adult nasals. Describing such skulls as paedomorphic is incorrect (Drake, 2011).

Coppinger and Schneider (1995) described mesaticephalic dogs as having normally shaped canid skulls but tending to be small proportionately to body size (more pronounced in large breeds). The claim was that dogs have reduced brain sizes. Brains of big dogs are equivalent to 4-month-old wolves (approx. 100 cc in big dogs). Arresting brain development early in ontogeny would support the behavioral observations that these animals lack adult foraging and hazard avoidance behaviors and are more puppy-like in behavior. But what is true of big dogs is not true of dogs in the 12- to 16-kg size range, which have genus-typical head sizes, brain sizes and tooth sizes. Brain sizes grow with a negative allometry, so indeed big dogs have small proportions compared to small dogs. The same is true of big wolves which have proportionately smaller heads and brains than smaller wolves. Certainly one could not make the argument that big wolves were ontogenetic snapshots of smaller wolves. Goodwin *et al.* (1997) take the nonsensical argument one step further, arguing that breeds of dogs that least look like wolves are the most paedomorphic

What is true of morphology is also true of behavior. Just as the elongated rostrum of the borzoi is a result of changing juvenile allometries, so is the eye-stalk behavior of the border collie. These herding dog behaviors may be homologous to the ancestors' predatory behaviors just as dogs' nasal bones are homologous to wolves' nasal bones. They are not juvenile characters carried into the adult stage of the dog. During play, border collies display many of the component parts of the ancestral hunting behavior significantly earlier than wolves (Lord, 2010).

Because so many different systems are developing simultaneously early in life, any heterochronic changes occurring during the juvenile stage can create these novel breed morphologies and behaviors. The results are phylogenetically original (Lord, 2010).

For example, the reduced flight distance of dogs as compared to wolves has been hypothesized to be the result of neoteny (Coppinger and Smith, 1989; see Price, 1998, for a review), since wolf pups are less fearful of novelty than adults. However, recent research suggests that this may only be a superficial relationship. The curiosity of young wolves (and mammals in general) is the result of the critical period of socialization. There is a period in early development when animals acquire species recognition and the related ability to discriminate between what is familiar and safe as opposed to what is novel and dangerous. This period begins with the ability to explore, and ends with the avoidance of novelty. Dogs begin this period at 4 weeks of age (Scott and Fuller, 1965), and 2 weeks after wolves (Frank and Frank, 1982). However, dogs and wolves develop their sensory systems (an integral part of exploration) at the same rate (Lord, 2013). At 4 weeks, when dogs begin to explore, all of their sensory systems are developed (Scott and Fuller, 1965). At 2 weeks, when wolves begin to explore, they are still deaf and blind (Lord, 2013). Thus, during the critical period of socialization, dogs can generalize their early experiences of the world and have a very rich picture of what is familiar. For wolves, familiarity with the world is largely based on their sense of smell. This difference in the interaction between sensory and motor development could explain the difference in flight distance between dogs and wolves (see Lord, 2013, for a full discussion).

This change in flight distance can affect the frequency of display of other ancestral behaviors in turn. It has been suggested that increased barking in dogs is paedomorphic. The logic is that wolf pups bark more than adults, and dogs bark more than adult wolves. However, new literature suggests that the bark is present in adult wolves and a variety of other species in conflict situations, and thus increased barking is likely the result of dogs being placed in these conflict situations as a result of the captive environment, and a reduced flight distance as compared with wolves (see Lord *et al.*, 2009, for a full discussion).

Many argue (correctly or incorrectly) that the dog is a descendant of the wolf. But then we look to wolf behavior to explain dog behavior, which is somewhat bizarre. Comparing closely related species might be useful in a phylogenetic sense but *Canis lupus familiaris* is an adaptation to a new niche, a specialist whose morphology and behavior are adaptations to living in that niche. Wolves cannot survive in the dog's environment because they have inappropriate behavior. The dog is behaviorally unique.

ACKNOWLEDGMENTS

We thank Josephine Coppinger for editorial assistance.

REFERENCES

American Veterinary Medical Association. 2007. "U.S. Pet Ownership & Demographics Sourcebook." Washington DC.

Arons, C.D., Shoemaker, W.J., 1992. The distributions of catecholamines and beta-endorphin in the brains of three behaviorally distinct breeds of dogs and their F_1 hybrids. Brain Res. 594, 31–39.

Asher, L., Diesel, G., Summers, J.F., McGreevy, P.D., Collins, L.M., 2009. Inherited defects in pedigree dogs. Part 1: disorders related to breed standards. Vet. J. 182, 402–411.

Associate Parliamentary Group for Animal Welfare.,2009. A healthier future for pedigree dogs. The Report of the APGAW Inquiry into the Health and Welfare Issues Surrounding the Breeding of Pedigree Dogs. House of Commons, London, pp. 54.

Bamberger, M., Houpt, K.A., 2006. Signalment factors, comorbidity, and trends in behavior diagnoses in dogs: 1644 cases (1991–2001). JAVMA 229, 1591–1598.

Bateson, P., 1979. How do sensitive periods arise and what are they for? Anim. Behav. 27, 470–486.

Bateson, P., 2010. Independent Inquiry into Dog Breeding. Micropress, Halesworth, Suffolk (64 pp).

Beaudet, R., Chalifoux, A., Dallaire, A., 1994. Predictive value of activity level and behavioral evaluation on future dominance in puppies. Appl. Anim. Behav. Sci. 40, 273–284.

Beck, A.M., 1973. The Ecology of Stray Dogs: A Study of Free-Ranging Urban Animals. York Press, Baltimore, MD.

Belyaev, D.K., 1979. Destabilizing selection as a factor in domestication. J. Hered. 70, 301–308.

Belyaev, D.K., Trut, L.N., 1975. Some genetic and endocrine effects of selection for domestication in silver foxes. In: Fox, M.W. (Ed.), The Wild Canids: Their Systematics, Behavioral Ecology and Evolution. Van Nostrand-Reinhold, NY, pp. 416–426.

Bemis, W.E., 1984. Paedomorphosis and the evolution of the Dipnoi. Paleobiol. 10, 293–307.

Black, H.L., Green, J.S., 1985. Navajo use of mixed-breed dogs for management of predators. J. Range Manage. 38, 11–15.

Boitani, L., 1983. Wolf and dog competition in Italy. Acta Zool Fenn. 174, 259–264.

Boitani, L., Fabbri, M.L., 1983. Censimento dei cani in Italia con particulari reguardo alfenomeno del randagismo. Ric. Bio. Selvaggina (INBS, Bologna) 73, 1–51.

Boitani, L., Francisci, F., Ciucci, P., Andreoli, G., 1995. Population biology and ecology of feral dogs in central Italy. In: Serpell, J. (Ed.), The Domestic Dog: Its Evolution, Behaviour, and Interactions with People. Cambridge University Press, Cambridge, UK, pp. 217–244.

Boitani, L., Ciucci, P., Ortolani, A., 2006. Behaviour and social ecology of free-ranging dogs. In: Jensen, P. (Ed.), The Behavioural Biology of Dogs. Cromwell Press, Trowbridge, pp. 147–165.

Bond, C., 2007. Human influence on physical and temperamental traits of the dog. J. Appl. Companion Anim. Behav. 1, 39–45.

Bradshaw, J.W.S., Nott, H.M.R., 1995. Social and communication behaviour of companion dogs. In: Serpell, J. (Ed.), The Domestic Dog: Its Evolution, Behaviour, and Interactions with People. Cambridge University Press, Cambridge, UK, pp. 115–130.

Brisbin, I.L., Coppinger, R.P., Feinstein, M.H., Austad, S.N., Mayer, J.J., 1994. The new guinea singing dog: taxonomy, captive studies and conservation priorities. Sci. New Guinea 20, 27–38.

Burghardt, G.M., 2005. The Genesis of Animal Play. The MIT Press, Cambridge MA.

Butler, J.R.A., Bingham, J., 2000. Demography and dog–human relationships of the dog population in Zimbabwean communal lands. Vet. Rec. 147, 442–446.

Carr, G.M., Macdonald, D.W., 1986. The sociality of solitary foragers: a model based on resource dispersion. Anim. Behav. 34, 1540–1549.

Chawla, S.K., Reece, J.F., 2002. Timing of oestrus and reproductive behaviour in Indian street dogs. Vet. Rec. 150, 450–451.

Clevenger, J., Kass, P.H., 2003. Determinants of adoption and euthansia of shelter dogs spayed or neutered in the University of California veterinary student surgery program compared to other shelter dogs. JVME 30, 372–378.

Coppinger, L., 1977. The World of Sled Dogs. Howell Book House, NY.

Coppinger, L., Coppinger, R., 1982. Livestock-guarding dogs that wear sheep's clothing. Smithsonian April, 64–73.

Coppinger, L., Coppinger, R., 1993. Dogs for herding and guarding livestock. In: Grandin, T. (Ed.), Livestock Handling and Transport. CAB International, Wallingford, UK, pp. 179–196.

Coppinger, R., 1996. Fishing Dogs. Ten Speed Press, Berkeley, CA.

Coppinger, R., Coppinger, L., 2001. Dogs: A New Understanding of Canine Origin, Behavior and Evolution. Scribner, NY.

Coppinger, R., Schneider, R., 1995. The evolution of working dog behavior. In: Serpell, J. (Ed.), The Domestic Dog: Its Evolution, Behaviour, and Interactions with People. Cambridge University Press, Cambridge, UK, pp. 21–47.

Coppinger, R., Lorenz, J., Glendinning, J., Pinardi, P., 1983. Attentiveness of guarding dogs for reducing predation on domestic sheep. J. Range Manage. 36, 275–279.

Coppinger, R., Coppinger, L., Langeloh, G., Gettler, L., Lorenz, J., 1988. A decade of use of livestock guarding dogs. Proc.:Vertebr. Pest. Conf. 13, 209–214.

Coppinger, R., Spector, L., Miller, L., 2010. What, if anything, is a Wolf? In: Musiani, M., Boitani, L., Paquet, P.C. (Eds.), The World of Wolves: New Perspectives on Ecology, Behaviour, and Management. University of Calgary Press, Calgary, Ontario, pp. 41–67.

Coppinger, R.P., Glendinning, J., Torop, E., Matthay, C., Sutherland, M., Smith, C., 1987. Degree of behavioral neoteny differentiates canid polymorphs. Ethol. 75, 89–108.

Coppinger, R.P., Smith, C.K., 1983. The domestication of evolution. Environ. Conserv. 10, 283–292.

Coppinger, R.P., Smith, C.K., 1989. A model for understanding the evolution of mammalian behavior. Curr. Mammal. 2, 335–374.

Coppinger, R.P., Smith, C.K., Miller, L., 1985. Observations on why mongrels may make effective livestock protecting dogs. J. Range Manage. 38, 560–561.

Corbett, L.K., 1985. Morphological comparisons of Australian and Thai dingoes: a reappraisal of dingo status, distribution and ancestry. Proc. Ecol. Soc. Aust. 13, 277–291.

Daniels, T.J., 1983. The social organization of free-ranging urban dogs. II. Estrous group and the mating system. Appl. Anim. Ethol. 10, 365–373.

Darwin, C., 1858. On the tendency of species to form varieties and on the perpetuation of varieties of species by natural selection. J. Linn. Soc. London, Zool. 3, 45–62.

Darwin, C., 1899. The Variation of Animals and Plants under Domestication, vol. 1. Appleton, NY.

Dawydiak, O., Sims, D.E., 2004. Livestock Protection Dogs: Selection, Care and Training, second ed. Alpine, Loveland, CO.

Dodman, N.H., 1996. The Dog Who Loved Too Much. Bantam, NY.

Drake, A.G., 2011. Dispelling dog dogma: an investigation of heterochrony in dogs using 3D geometric morphometric analysis of skull shape. Evol. Dev. 13, 204–213. 10.1111/j.1525-142X.2011.00470.

Drake, A.G., Klingenberg, C.P., 2007. The pace of morphological change: historical transformation of skull shape in St Bernard dogs. Proceedings of the Royal Society B: Biological Sciences, First Cite 23 October 2007, doi: 10.1098/rspb.2007.1169.

Drake, A.G., Klingenberg, C.P., 2010. Large-scale diversification of skull shape in domestic dogs: disparity and modularity. Amer. Nat. 175, 289–301.

Esposito, L., McCune, S., Griffin, J.A., Maholmes, V., 2011. Directions in human–animal interaction research: child development, health, and therapeutic interventions. Child Dev. Perspectives 5, 205–211.

Estep, D.Q., 1996. The ontogeny of behavior. In: Voith, V.L., Borchelt, P.L. (Eds.), Readings in Companion Animal Behavior. Veterinary Learning Systems, Trenton, NJ, pp. 19–31.

European Pet Food Industry Federation, 2010. Facts and Figures. Bruxelles, Belgium.

Evans, H.E., Christensen, G.C., 1979. Miller's Anatomy of the Dog, second ed. Saunders, Philadelphia.

Fairfax, H., 1913. Roman Farm Management: The Treatises of Cato and Varro Done into English, with Notes of Modern Instances by a Virginia Farmer. Macmillan, NY.

Fentress, J.C., 1967. Observations on the behavioral development of a hand-reared male timber wolf. Amer. Zool. 7, 339–351.

Fox, M.W., 1978. The Dog: Its Domestication and Behaviour. Garland STPM Press, NY.

Fox, M.W., Bekoff, M., 1975. The behaviour of dogs. In: Hafez, E.S.E. (Ed.), The Behaviour of Domestic Animals, third ed. Baillière Tindall & Cassell, London, pp. 370–409.

Frank, H., Frank, M.G., 1982. On the effects of domestication on canine social development and behavior. Appl. Anim. Ethol. 8, 507–525.

Gallant, J., 2002. The Story of the African Dog. University of KwaZulu-Natal Press, South Africa.

Gallant, J., Gallant, E., 2008. SOS DOG: The Purebred Dog Hobby Reexamed. Crawford CO, Alpine Publications.

Ghosh, B., Choudhuri, D.K., Pal, B., 1984. Some aspects of the sexual behaviour of stray dogs, *Canis familiaris*. Appl. Anim. Behav. Sci. 13, 113–127.

Goddard, M.E., Beilharz, R.G., 1986. Early predictions of adult behaviour in potential guide dogs. Appl. Anim. Behav. Sci. 15, 247–260.

Goodwin, D., Bradshaw, J.W.S., Wickens, S.M., 1997. Paedomorphosis affects agonistic visual signals of domestic dogs. Anim. Behav. 53, 297–304.

Gottelli, D., Sillero-Zubiri, C., Applebaum, G.D., Roy, M.S., Girman, D.J., Garcia-Moreno, J., et al., 1994. Molecular genetics of the most endangered canid: the Ethiopian wolf, *Canis simensis*. Mol. Ecol. 3, 301–331.

Haldane, J.B.S., 1930. The Causes of Evolution. Longmans, Green, London.

Herzog, H., 2011. The impact of pets on human health and psychological well-being: fact, fiction, or hypothesis? Curr. Dir. Psychol. Sci. 20, 236–239. 10.1177/0963721411415220.

Hirsch, J., 1990. A nemesis for heritability estimation. Behav. Brain Sci. 13, 137–138.

Honacki, J.H., Kinman, K.E., Koeppl, J.W., 1982. Mammal Species of the World: A Taxonomic and Geographic Reference. Allen Press and Association of Systematic Collections, Lawrence, KS.

Houpt, K.A., 2007. Genetics of canine behavior. Acta Vet. Brno. 76, 431–444.

Houpt, K.A., Goodwin, D., Uchida, Y., Baranyiová, E., Fatjó, J., Kakuma, Y., 2007. Proceedings of a workshop to identify dog welfare issues in the US, Japan, Czech Republic, Spain and the UK. Appl. Anim. Behav. Sci. 106, 221–233.

Hubrecht, R., 1995. The welfare of dogs in human care. In: Serpell, J. (Ed.), The Domestic Dog: Its Evolution, Behaviour, and Interactions with People. Cambridge University Press, Cambridge, UK, pp. 179–198.

Jackman, J., Rowan, A., 2007. Free-roaming dogs in developing countries: the public health and animal welfare benefits of capture, neuter, and return programs. In: Salem, D., Rowan, A. (Eds.), State of the Animals 2007. Humane Society Press, Washington, D.C..

Kemper, K.E., Visscher, P.M., Goddard, M.E., 2012. Genetic architecture of body size in mammals. Genome Biol. 13, 244.

King, T., Marston, L.C., Bennett, P.C., 2009. Describing the ideal Australian companion dog. Appl. Anim. Behav. Sci. 120, 84–93.

Klinghammer, E., Goodman, P.A., 1987. Socialization and management of wolves in captivity. In: Frank, H. (Ed.), Man and Wolf. Dr. W. Junk Publishers, Dordrecht, pp. 31–59.

Landry, J.-M., 2001. The guard dog: protecting livestock and large carnivores. In: Field, R., Warren, R.J., Okarma, H., Sievert, P.R. (Eds.), Wildlife, Land, and People: Priorities for the 21st Century. The Wildlife Society, Bethesda MD, pp. 209–212.

Langley, R.L., 2009. Human fatalities resulting from dog attacks in the United States, 1979–2005. Wilderness Environ. Med. 20, 19–25.

Lantis, M., 1980. Changes in the alaskan eskimo relation of man to dog and their effect on two human diseases. Arctic. Anthropol. 17, 2–24.

Larson, G., Karlsson, E.K., Perri, A., Webster, M.T., Ho, S.Y.W., Peters, J., et al., 2012. Rethinking dog domestication by integrating genetics, archeology, and biogeography. PNAS 2012, 1203005109v1-201203005

Leyhausen, P., 1973. On the function of the relative hierarchy of moods: as exemplified by the phylogenetic and ontogenetic development of prey-catching in carnivores. In: Lorenz, K., Leyhausen, P. (Eds.), Motivation of Humans and Animal Behavior: An Ethological View. Van Nostrand, NY, pp. 144–247.

Lord, K., 2010. A heterochronic explanation for the behaviorally polymorphic genus Canis: a study of the development of behavioral difference in dogs (Canis lupus familiaris) and wolves (Canis lupus lupus), (Ph.D. Dissertation). University of Massachusetts, Amherst.

Lord, K., 2013. A comparison of the sensory development of wolves (Canis lupus lupus) and dogs (Canis lupus familiaris). Ethology. 119, 110–120.

Lord, K., Feinstein, M., Coppinger, R., 2009. Barking and mobbing. Behav. Process. 81, 358–368.

Lord, K., Feinstein, M., Smith, B., and Coppinger, R., 2013. Variation in reproductive traits of members of the genus *Canis* with special attention to the domestic dog (*Canis familiaris*). Behav Process. 92, 131–142.

Lorenz, J., Coppinger, R., Sutherland, M., 1986. Causes and economic effects of mortality in livestock guarding dogs. J. Range Manage. 39, 293–295.

Luescher, A.U., Medlock, R.T., 2009. The effects of training and environmental alterations on adoption success of shelter dogs. Appl. Anim. Behav. Sci. 117, 63–68.

Macdonald, D.W., Carr, G.M., 1995. Variation in dog society: between resource dispersion and social flux. In: Serpell, J. (Ed.), The Domestic Dog: Its Evolution, Behaviour, and Interactions with People. Cambridge University Press, Cambridge, UK, pp. 199–216.

Macdonald, D.W., Creel, S., Millis, M.G.L., 2004. Canid society. In: Macdonald, D.W., Sillero-Zuberi, C. (Eds.), Biology and Conservation of Wild Animals. Oxford University Press, Oxford, Ch. 4.

Maderson, P.F.A., Alberch, P., Goodwin, B.C., Gould, S.J., Hoffman, A., Murray, J.D., et al., 1982. The role of development in macroevolutionary change: group report. In: Bonner, J.T. (Ed.), Evolution and Development. Springer-Verlag, Berlin, pp. 279–312.

Marker, L., Dickman, A.J., Macdonald, D.W., 2005. Survivorship and causes of mortality for livestock-guarding dogs on Namibian rangeland. Rangeland Ecol. Manage. 58, 337–343.

McConnell, P.B., Baylis, J.R., 1985. Interspecific communication in cooperative herding: acoustic and visual signals from human shepherds and herding dogs. Z Tierpsychol. 67, 302–328.

Millien, V., 2006. Morphological evolution is accelerated among island mammals. PLoS Biol. 4 (10), e321.

Mondelli, F., Prato Previde, P., Verga, M., Levi, D., Magistrelli, S., Valsecchi, P., 2004. The bond that never developed: adoption and relinquishment of dogs in a rescue shelter. Appl. Anim. Welf. Sci. 7, 253–266.

Morey, D.E., 1994. The early evolution of the domestic dog. Amer. Sci. 82, 336–347.

Mugford, R.A., 1995. Canine behavioural therapy. In: Serpell, J. (Ed.), The Domestic Dog: Its Evolution, Behaviour, and Interactions with People. Cambridge University Press, Cambridge, UK, pp. 139–152.

New, J.C.J., Salman, M.D., King, M., Scarlett, J.M., Kass, P.H., Hutchinson, J., 2000. Characteristics of shelter-relinquished animals and their owners compared with animals and their owners in U.S. pet-owning households. Appl. Anim. Welf. Sci. 3, 179–201.

Olsen, J.W., 1985. Prehistoric dogs in mainland East Asia. In: Olsen, S.J. (Ed.), Origins of the Domestic Dog: The Fossil Record. University of Arizona Press, Tucson, pp. 47–70.

Ortolani, A., Vernooij, H., Coppinger, R., 2009. Ethiopian village dogs: behavioural responses to a stranger's approach. Appl. Anim. Behav. Sci. 119, 210–218.

Pal, S.K., 2001. Population ecology of free-ranging urban dogs in West Bengal, India. Acta Theriol. 69–78.

Pellis, S.M., Pellis, V.C., 1996. On knowing it's only play: the role of play signals in play fighting. *Aggress. Violent Behav.* 1, 249–268.

Perry, G.H., Dominy, N.J., 2009. Evolution of the human pygmy phenotype. Trends Ecol. Evolut. 24, 218–225.

Phillips, C.J., Coppinger, R.P., Schimel, D.S., 1981. Hyperthermia in running sled dogs. J. Appl. Physiol. 51, 135–142.

Podberscek, A.L., 2009. Good to pet and eat: the keeping and consuming of dogs and cats in South Korea. J. Soc. Issues 65, 615–632.

Price, E.O., 1984. Behavioral aspects of animal domestication. Q. Rev. Biol. 59, 1–32.

Price, E.O., 1998. Behavioral genetics and the process of animal domestication. In: Grandin, T. (Ed.), Genetics and the Behavior of Domestic Animals. Academic Press, San Diego (Chapter 2).

Quirk, J.T., 2012. Non-fatal dog bite injuries in the USA, 2005–2009. Public Health 126, 300–302.

Reece, J.F., Chawla, S.K., Hiby, E.F., Hiby, L.R., 2008. Fecundity and longevity of roaming dogs in Jaipyr, India. BMC Vet. Res. 4, 6.

Rigg, R., 2001. Livestock guarding dogs: their current use world wide. IUCN/SSC Canid Specialist Group Occasional Paper No 1 [online]. <http://www.canids.org/occasionalpapers/>.

Rindos, D., 1980. Symbiosis, instability and the origins and spread of agriculture: a new model. Curr. Anthropol. 21, 751–772.

Rooney, N., Sargan, D., 2009. Pedigree Dog-breeding in the UK: A Major Welfare Concern?. RSPCA, London.

Sacks, J.J., Kresnow, M., Houston, B., 1996. Dog bites: how big a problem? Injury Prevention 2, 52–54 (doi:10.1136/ip.2.1.52.).

Sands, M.W., Coppinger, R.P., Phillips, C.J., 1977. Comparisons of thermal sweating and histology of sweat glands of selected canids. J. Mammal. 58, 74–78.

Schoeller, D.A., Ravussin, E., Schutz, Y., Acheson, K.J., Baertschi, P., Jequier, E., 1986. Energy expenditure by doubly labeled water: validation in humans and proposed calculations. Am. J. Physiol. 250 (5 Pt 2), R823–R830.

Scott, J.P., Fuller, J.L., 1965. Genetics and the Social Behavior of the Dog. University of Chicago Press, Chicago.

Scott, J.P., Marston, M., 1950. Critical periods affecting the development of normal and maladjustive social behavior of puppies. J. Genet Psychol. 77, 25–60.

Serpell, J., 1995. From paragon to pariah: some reflections on human attitudes to dogs. In: Serpell, J. (Ed.), The Domestic Dog: Its Evolution, Behaviour, and Interactions with People. Cambridge University Press, Cambridge, UK, pp. 245–256.

Serpell, J., Coppinger, R., Fine, A.H., 2000. The welfare of assistance and therapy animals: an ethical comment. In: Fine, A.H. (Ed.), Handbook of Animal-Assisted Therapy: Theoretical Foundations and Guidelines for Practice. Academic Press, San Diego, pp. 415–431. (Chapter 18).

Serpell, J., Jagoe, J.A., 1995. Early experience and the development of behaviour. In: Serpell, J. (Ed.), The Domestic Dog: Its Evolution, Behaviour, and Interactions with People. Cambridge University Press, Cambridge, UK, pp. 79–102.

Serpell, J.A., Hsu, Y., 2001. Development and validation of a novel method for evaluating behavior and temperament in guide dogs. Appl. Anim. Behav. Sci. 72, 347–364.

Sherman, B.L., Mills, D.S., 2008. Canine anxieties and phobias: an update on separation anxiety and noise aversions. Vet. Clin. North Amer. Small Anim. Pract. 38, 1081–1106.

Shipman, P., 2009. The wolf at the door. Amer. Sci. 97, 286.

Spady, T.C., Ostrander, E.A., 2008. Canine behavioral genetics: pointing out the phenotypes and herding up the genes. Am. J. Hum. Genet 82, 10–18. 10.1016/j.ajhg.2007.12.001 (Published online 2008 January 4.).

Speakman, J.R., van Acker, A., Harper, E.J., 2003. Age-related changes in the metabolism and body composition of three dog breeds and their relationship to life expectancy. Aging cell 2, 265–275.

Stafford, K., 2007. Animal Welfare Volume 4: The Welfare of Dogs. Springer Publ.

Stebbins, G.L., 1959. The role of hybridization in evolution. Proc. Am. Philos Soc. 103, 231–251.

Stockard, C.R., 1941. The genetic and endocrinic basis for differences in form and behavior. Am. Anat. Mem. (No. 19. Wistar Institute of Anatomy and Biology, Philadelphia.).

Thomson, K.S., 1996. The fall and rise of the English bulldog. Amer. Sci. 84, 220–223.

Titcomb, M., 1969. Dog and Man in the Ancient Pacific with Special Attention to Hawaii, Spec. Pub. (No. 59. Bernice P. Bishop Museum, Honolulu.)

Tuber, D.S., Miller, D.D., Caris, K.A., Haler, R., Linden, F., Hennessy, M.B., 1999. Dogs in animal shelters: problems, suggestions and needed expertise. Psychol. Sci. 10, 379–386.

Voith, V.L., 2009. The impact of companion animal problems on society and the role of veterinarians. Vet. Clin. North Am. Small Anim. Pract. 39, 327–345.

Walden, A.T., 1928. A Dog Puncher on the Yukon. Houghton Mifflin, NY.

Wahlsten, D., 1990. Insensitivity of the analysis of variance to heredity–environment interaction. Behav. Brain Sci. 13, 109–120.

Warner, S., Feinstein, M., Coppinger, R., Clemence, E., 1996. Global population growth and the demise of nature. Environ. Values 5, 285–301.

Wayne, R.K., 1986. Cranial morphology of domestic and wild canids: the influence of development on morphological change. *Evolution* (*Lawrence, Kans.*) 40, 243–261.

Wayne, R.K., Lehman, N., Fuller, T.K., 1995. Conservation genetics of the gray wolf. In: Carbyn, L.N., Fritts, S.H., Siep, D.R. (Eds.), Ecology and Conservation of Wolves in a Changing World. Canadian Circumpolar Institute, Edmonton, Alberta, pp. 399–407. (Occas. Publ. No. 35).

Wells, D.L., 2007. Domestic dogs and human health: an overview. Br. J. Health Psychol 12, 145–156.

Wells, D.L., 2009. The effects of animals on human health and well-being. J. Soc. Issues 65, 523–543.

Wells, D.L., Hepper, P.G., 2000. Prevalence of behaviour problems reported by owners of dogs purchased from an animal rescue shelter. J. Appl. Anim. Behav. Sci. 69, 55–65.

Willis, M.B., 1992. Practical Genetics for Dog Breeders. Witherby, London.

Willis, M.B., 1995. Genetic aspects of dog behaviour with particular reference to working ability. In: Serpell, J. (Ed.), The Domestic Dog: Its Evolution, Behaviour, and Interactions with People. Cambridge University Press, Cambridge, UK, pp. 51–64.

Wolfensohn, S., 1981. The things we do to dogs. New Scientist, (May 14) 404–407.

World Health Organization., 2004. WHO expert consultation on rabies: first report. Geneva, Switzerland.

Zabek, T., Slota, E., 2007. Bovine achondroplasia. Med. Weter. 63, 512–514.

Zimen, E., 1987. Ontogeny of approach and flight behavior toward humans in wolves, poodles and wolf–poodle hybrids. In: Frank, H. (Ed.), Man and Wolf. Dr. W. Junk, Publishers, Dordrecht, pp. 275–292.

Chapter 7

Behavior Genetics of the Horse (*Equus caballus*)

Mark J. Deesing* and Temple Grandin†
**Grandin Livestock Handling Systems, Inc., Fort Collins, Colorado, USA; †Department of Animal Science, Colorado State University, Fort Collins, Colorado, USA*

INTRODUCTION

Why do horses behave differently from each other? People have debated this question for centuries. Some believe that horses behave as they do according to inborn tendencies; others think it's because they're taught. A recently fast-growing body of evidence makes it clear that both sides are partly right. Nature (genetics) endows horses with inborn abilities and traits, and nurture (experience and upbringing) takes these genetic tendencies and molds them as a horse learns and matures. Now the debate is centered on how much of horse behavior is shaped by genes and how much by the environment. Although genes play a role in shaping behavior, they never determine it. This is because genes only directly code for proteins, not behavior. However, a change in a single protein code can cause a host of downstream effects and may even bring about a distinct phenotype (Keane *et al.*, 2011).

Horses are not very convenient laboratory animals. However, many genetic principles discovered by researchers studying mice or humans in the laboratory are also relevant in horses. In order for behavioral scientists to tease apart the genetic and environmental influences on complex traits such as temperament, it is important to use large numbers of horses controlled for genetics (age, sex, sire) and environment (stable or pasture housing, type of feed, type of work, caretakers, etc.). Three recent studies brilliantly meet these requirements and show the advancements made in the study of behavior genetics.

MOLECULAR BEHAVIOR GENETICS

Momozawa *et al.* (2005) conducted a molecular genetics study on temperament in 136 two-year-old thoroughbred horses (73 males and 63 females) kept at a single training farm in Japan. The horses were at the same stage of

Genetics and the Behavior of Domestic Animals. DOI: http://dx.doi.org/10.1016/B978-0-12-394586-0.00007-X
© 2014 Elsevier Inc. All rights reserved.

training, shared the same training area, and had the same caretakers. Blood samples were collected for DNA analysis, and a questionnaire survey was used to rate the temperament traits of each horse. Interestingly, an association was found between polymorphisms (genes that come in two or more forms) in the dopamine D4 (DRD4) receptor gene and two temperament traits in the horses: curiosity (novelty seeking) and vigilance. The survey had the following descriptions of curiosity and vigilance: "The horse tends to be interested in novel objects and approach them," and "The horse tends to be vigilant about surroundings." Horses with one form of the gene had higher curiosity (novelty seeking). Horses with the other form displayed lower vigilance. The DRD4 gene is also associated with novelty seeking in humans (Benjamin *et al.*, 1996), in primates (Livik *et al.*, 1995), and in dogs (Ito *et al.*, 2004). Momozawa *et al.* (2005) speculates that diversity in the sequence of the DRD4 gene might influence differences in novelty seeking in various species.

In more recent molecular genetics research, Anderson *et al.* (2012) discovered mutations in the DMRT3 genes associated with locomotion in gaited horses. A high frequency of DMRT3 mutations were found in horse breeds such as Tennessee walking horse, Missouri fox trotters, Standardbred trotter, and French trotters. The mutations were not present in non-gaited horses such as Thoroughbreds, Arabians, and Przewalski's horses. In the laboratory, Anderson *et al.* (2012) blocked the normal function of the DMRT3 genes in mice (null mutants) and observed changes in locomotion and gaits similar to those in gaited horses. This research shows that the expression of certain specialized genes in vertebrates as different as mice and horses has remained remarkably conserved over evolution.

GENETIC DIFFERENCES INTERACT WITH ENVIRONMENT

Hausberger *et al.* (2004) highlights how individual horse personalities and temperaments interact with genetic, social, and nonsocial environmental factors. A battery of standardized behavioral tests were performed on 702 horses and ponies stabled at 103 different riding centers, training centers, and national stud and breeding farms in France. Only registered horses were used to enable accurate breed identification. Sex and age were well distributed in the horses (185 mares, 305 geldings, and 212 stallions). Behavioral tests were used to rate different facets of temperament, emotionality, and learning ability. An arena test was used to estimate emotionality. Each horse was released into a familiar arena and its behavior was recorded for 10 minutes. Emotional reactions (whinnying, locomotion, vigilance) were used to evaluate the effects of *social separation* in the arena. This test differed from a standard open-field test used to measure fear because the arena was familiar to the horse. A novel object test was used to evaluate *nervousness*. An object resembling a cage with a ribbon attached to it was placed in the familiar arena. Each horse was released into the arena and behavior was recorded (locomotion, gazes, approaches) for 5 minutes. A bridge test was used to estimate *fearfulness*. A foam mattress covered with checkered cloth (called a bridge) was placed in the familiar arena. Each horse was led to the bridge with a halter and the experimenter tried to encourage the horse to cross the bridge. For the *learning and memory* task, each horse learned how to open a wooden box with food inside, and subsequently it had to remember how to open the box 12 hours later. Given the fact that each test was designed to measure different facets of personality, the results of this study are extremely interesting. The results showed that genetic factors, such as sire or breed, influenced more neophobic reactions (fear of new things or experiences) such as the bridge test. On the other hand, environmental factors, such as type of work, played a more dominant role in learning abilities and reactions to social separation. Show and dressage horses exhibited higher emotional reactions to the social isolation test and the fear test (bridge) compared to horses used for other types of work. This is a notable finding, since a prior study also found a high level of stereotypies in dressage horses (McGreevey *et al.*, 1995). Furthermore, the additive gene–environment effects uncovered are especially significant. In simple terms, additive effects mean that the expression of genes is influenced by differences in the environment. For example, the 98 breeding stallions tested in this study were stabled in two government facilities and kept under similar conditions (food, stall, little or no training for riding). The breeds were similar in both facilities, but environmental differences between the two facilities added to differences in the reactions to the bridge test (*fearfulness*), and the stallion's ability on the memorization test. The researchers suggest that varying handling practices and differences in caretaker behavior probably caused the differences in the behavior of the stallions in the two facilities.

Together, these two studies emphasize how far equine behavior science has come since the first edition of this book was published. The remaining sections cover additional studies on: the effects of early experience on behavior; the difficulties associated with finding easy ways to predict temperament; the relationship between hair whorls and behavior; and lateralization in the nervous system. The emphasis in each section will focus on interactions between the genetic and environmental effects on behavior.

THE EFFECTS OF EARLY EXPERIENCE ON BEHAVIOR

Variability in emotional reactivity between animals is a phenotypic trait defined as temperament (Boissy, 1995). Horse breeders know that some foals are fearful and sensitive, while others are bold and confident. A variety of terms are used to describe temperamental differences in animals, but for simplicity the authors use the term "reactive" to describe animals that are flighty, nervous, or highly emotional. The term "low-reactive" is used to refer to individuals that exhibit low fear and low levels of emotional reactivity. Genetic variability in temperamental traits such as "reactive" or "low-reactive" are examples of the extremes within a normal distribution of temperament. In young foals, genetic influence and pre- and postnatal effects of the environment all contribute to individual differences.

Prenatal Effects on Behavior

A substantial body of evidence in animals and humans suggests that the gestational environment can impact fetal brain structure and function and increase long-term susceptibility to stress (Claessens *et al.*, 2011; O'Connor *et al.*, 2012). This can occur on its own, or in connection with genetic or postnatal factors. Environmental influence during fetal development is especially strong in the brain. Differentiation of major brain structures occur during early gestation, creating more susceptibility to environmental conditions. Brain development involves a number of interactions with the environment, so even small changes from the normal developmental path during fetal life can become progressively exaggerated over time, producing long-lasting or permanent changes (reviewed by Stolp *et al.*, 2012). Recent work in humans shows that high levels of stress in the mother during pregnancy can affect brain function and behavior in her offspring, and that stress hormones are most likely transferred through the placenta (Glover 2011). Although we are not aware of any studies examining the effects of prenatal stress in foals, we speculate that the effects of stress in the mare during gestation are similar to those shown in other animals (Weinstock, 2005).

Early Postnatal Effects on Behavior Induced by Handling

A well-known book titled *Imprint Training of the Newborn Foal* (Miller, 1991) describes a handling procedure that consists of handling foals within 10 minutes of birth. "Imprint training" proposes to offer a singular opportunity to permanently mold a horse's personality. In this procedure, handling of the foals begins before the foals stand up, immediately following routine postnatal care of the umbilical cord. Foals are forcibly held in a lying position on the ground while the experimenter strokes them and exposes them to novel stimuli such as a white towel, a plastic bag, and a spray of water. Each stimulus is repeated until the foals remain immobile. These procedures last approximately 1 hour (Miller, 1991).

"Imprint training" has divided the breeding world into two distinct groups: passionate followers of the technique and those who are strongly opposed to this approach. Some equine behaviorists and veterinarians (including us) were never convinced that this procedure was a good idea—in theory or in practice. Since the publication of Miller's book, several studies have been conducted on the effectiveness of early intensive handling on foal behavior. There are conflicting results. A few of these studies closely followed the procedures of "imprint training" outlined by Miller, while other studies varied considerably. In 14 Arabian, 7 Quarter horse, and 2 Thoroughbred foals, Mal *et al.* (1994) and Mal and McCall (1996) looked at reactions to novelty, as well as individual manageability in minimally handled foals (routine and emergency veterinary care only), intermediately handled (10 minutes twice daily for one week), and extensively handled foals (same as the intermediate group and additionally once weekly for 10 minutes until weaning). During handling sessions, both sides of the foal's body from ears to tail head, including the belly and the legs to the knees or hocks, were rubbed. They found no difference between groups in manageability and response to novelty. Diehl *et al.* (2002) looked at mare–foal interactions in minimally handled foals (routine veterinary care) versus intensive (Miller method) handling of Standardbred foals. The mares were not restrained during the 4-hour post-foaling period, and no differences were observed in the behavior of the mares towards the handler in either group However, mares of intensively handled foals spent less time eating and drinking, and more time sniffing their foals. Using 15 foals of mixed breeds, Simpson (2002) performed an intensive handling protocol at between 2 and 8 hours following birth, and again once daily for 5 days. Reactions to stimuli similar to that used during the intensive handling procedure were measured at four months. Overall, the handled foals had more favorable scores for calmness and friendliness toward handlers, but there were no significant differences in compliance between handled and control foals. Williams *et al.* (2002) followed Miller's imprinting procedure closely, except they did not begin with the handled foals until the foals had stood and nursed. Forty-six Quarter horse foals, and one Thoroughbred foal were handled at various times within the first 72 hours. Subsequent reactions to handling were measured at one, two, and three months of age. They found no differences in handled compared to control foals. Lansade *et al.* (2004) followed a handling procedure close to the procedure used by Miller (1991) on 13 Welsh foals. The first handling session was performed between 1 and 6 hours following birth, the second at 24 hours, and the third at 36 hours following birth. The results showed that neonatal handling had short-term effects on manageability 2 days after the handling period (time to fit a halter, time to pick up feet, and leading on a halter). However, at 12 months of age, there was no significant difference in manageability between the handled and control animals. In a review of these studies, Diehl *et al.* (2002) states that "there have been no consistent positive findings connected to early intensive handling with regard to compliance and long-term benefits in training or in reaction to novelty or potentially fearful situations."

To get a sense of the horse owners' opinions, we read online blog postings from horse owners with "imprinted" foals. Some of the words used to describe the behavior of the foals emphasized the bewildering and variable effects produced by intensive neonatal handling. These words included "trusting," "spoiled brat," "cooperative," "juvenile delinquent," "respectful," "pushy little monster," "drama queen," "lover," "friendly bugger," "tendency to go catatonic," and "potentially dangerous," to list a few (Horse Forum, 2012). It seems that the different responses of foals to early intensive handling are affected by several factors. Genetic factors influencing temperament may be one; another may be differences in the time the handling is performed; and last, differences in the way the handling procedures are performed may also influence responses.

Long-Term Effects of Early Intensive Handling

Henry *et al.* (2009) conducted a groundbreaking study to monitor the behavior of "imprint-trained" foals in normal social groups with adults and peers from the early stages of development to adolescence. The results are both startling and disturbing. The subjects were 19 French Saddlebred mares and their foals (11 females, 8 males), all maintained under the same conditions until birth. Delivery of the foals was not assisted and all foals received minimal care apart from the application of an antiseptic on their umbilical stumps. After birth, the foals were divided into one of two treatment groups: the control group, which included foals and mares left undisturbed (except for antiseptic application), and the experimental group, which included foals that had been handled following the "imprint training" guidelines (Miller, 1991). Apart from the early experimental handling procedure conducted on the handled foals, both handled and non-handled (control) groups received only limited human contact necessary for routine procedures (mainly feeding and changes of pasture). They were kept under identical management, and the groups were mixed on the same pasture. The handled foals were later observed at four stages of life: during the early postnatal period, at six months of age with their mothers, at weaning, and at one year of age.

Early Postnatal Period

Handled foals stood up and suckled significantly later than control foals, and also displayed abnormal suckling activity. This suggested a delay caused by the handling procedure rather than a lowered capacity. However, handled foals displayed short-term disturbances not seen in the non-handled foals, such as fast breathing and excessive trembling, indicating that the procedure significantly delayed these first two developmental stages. The abnormal suckling behaviors included excessive chewing and teat seeking directed "at the air" or at the handler and not toward their mother All of the handled foals struggled during handling (attempting to get up) before lying motionless with high muscle tone (Henry *et al.*, 2009).

Six Months Old

Handled foals were observed on pasture with their mothers and the other mares and foals. Handled foals appeared more dependent on their mothers. They interacted more with their mothers than with the other foals, and played less (especially social play). Handled foals also explored "new" objects in the paddocks less (such as an unfamiliar, motionless human) compared to control foals. Only three out of the nine handled foals approached the unfamiliar human, while all of the control foals approached and investigated the human. All of the observed differences between the handled foals and control foals involved social–emotional behaviors. Other behaviors (exploring, moving around, resting, self-grooming, etc.) did not differ from the control foals.

Seven Months Old (Weaning)

Following separation from their mothers, both control and handled foals experienced similar levels of stress. Increased vocalizations and aggressiveness toward each other was observed in both groups, which indicated that the social stress of weaning was similar in both control and handled foals. However, after the second day, emotional reactions of the control foals decreased, while high levels of vocalizations in the handled foals continued. Handled foals also engaged in less solitary or social play.

One Year Old

Both handled and control foals were separated and housed in same-sex groups. Most normal activities of the foals were not different. However, the experimental foals tended to withdraw socially, spend more time at greater distances from their peers, and exhibit more aggression. Control foals also displayed almost three times more friendly, positive gestures toward each other, while handled foals displayed more fighting, fleeing, or submissive behaviors. In this study, differences between the neonatal experiences of the foals revealed short-, medium-, and long-term effects of early experience on attachment and social competence later in life. As Henry *et al.* (2009) states, "These results underline the importance of this early stage for appropriate social development, despite later experience with peers."

All the studies reviewed here show that forcefully handling foals can have various effects on behavior, mostly negative. The negative effects of intensive handling may be more harmful in highly reactive breeds such as Arabians and Thoroughbreds. In any case, foals that are not handled may also present a problem. At some point they need to be trained to accept handling for hoof care or health care. When these procedures become necessary, forcing foals may cause undue stress and influence the way they respond to handling in the future. A practical and easy way to teach foals to accept handling by people without making them fearful is needed.

A Non-Intrusive Neonatal Handling Method

Over the last 20 years, the first author, Mark J. Deesing, has practiced a self-developed, non-intrusive neonatal handling procedure on his foals with some positive results. Immediately following birth and before the foal stands up, he applies an antiseptic to its umbilical stump. This is the only handling that the foal receives in the first few postnatal days. No more attention is given to the foal—instead, he concentrates his attention on the mare and shows the foal indifference. All of these procedures are conducted either on pasture or in a large paddock where the mare and foal are housed, and not confined in a stall. Each day following the birth of the foal, he spends a short period of time grooming the mare and feeding her carrots, but only if the mare approaches him willingly. As he cares for the mare, the foal becomes interested in the attention given to its mother, and walks around and watches from one side of the mare to the other. At first, the foal is hesitant to get too close, but usually within a day or two it moves closer and closer. Before long, the foal begins to sniff, then nuzzles him while he attends to the mare. When this happens, he continues to behave as if the foal is not there, showing complete indifference until he feels the foal is completely comfortable in his presence. After the foal approaches him several times and stands close by while he cares for the mare, he reaches out and touches the foal for the first time. The first touch is very brief. He gauges the reaction of the foal, and then walks away. If the foal reacts to his touch, he knows not to touch it again until the foal has spent more time standing close by and investigating him. If the foal shows no reaction to his first touch, he knows that he can touch it more the next time. He learned the importance of not hurrying the first touch, and if the procedure is practiced this way, each day the foal allows more frequent and direct touching. Any forced contact results in a foal that avoids subsequent contact. Evidence for this is supported in a study by Henry *et al.* (2006). They found that forced stroking and handling of foals early in life did not improve the foal–human relationship. In a previous study, Henry, *et al.* (2005) also found that softly brushing and feeding mares by hand had positive effects on foal behavior.

The first author practices his procedure in the same way on all his foals. Even so, differences are apparent in the time that it takes for the foals to touch him, or accept being touched by him. Some foals are fearful and take several days to get close and touch him, while others touch, smell, and nuzzle him on the very first day. In the beginning, he avoids touching them on the head or face. They seem to be more protective of their heads than other parts of their bodies. This is especially evident in reactive foals. In order to prevent any negative reactions, he is careful not to try to touch the head or face too quickly. Habituation to being touched on all parts of the body, including the head, takes more time in the reactive/high-fear foals, compared to the low-reactive/low-fear foals.

After the foal allows the second author to touch different parts of its body, he scratches and rubs it on the belly and the area around the top of the tail (Figure 7.1) In his experience, foals seem to find this rewarding, and particularly like to be rubbed and scratched on these areas of the body. However, it may cause some foals to become pushy and begin to seek more and more of this attention. Attention seeking occurs sooner in less reactive foals, and eventually occurs in more reactive foals. In any event, the moment a foal begins to push on him seeking attention, he gently pushes them away, or walks away. This is similar to the way some mares behave toward their foal if she is busy grazing, or doesn't want the foal to nurse. If foals get too rambunctious or demanding, she might swish her tail, threaten, or walk away. The first author has found that handling foals this way teaches them to respect his space, and they also learn that seeking too much attention is unacceptable. The bond that forms is based on trust and respect.

Factors Affecting Attachment to Humans

A recent study conducted by Sankey *et al.* (2011) suggests that food is important in the human–foal attachment process, and that tactile contact is insufficient for bonding to occur. We disagree with this conclusion. The foals handled using the procedure developed by the first author actively seek tactile contact, which is not associated with another primary reinforcement such as food. No food reinforcement is used during the initial bonding process with the foal. In the Sankey *et al.* (2011) study, two groups of horses were reared either "under conventional domestic conditions," or in a forest reserve. No specific information was provided as to the horses' early contact with humans, or the early handling procedures used to habituate the horses to wearing halters. The forest-reared horses were "caught" and put with the

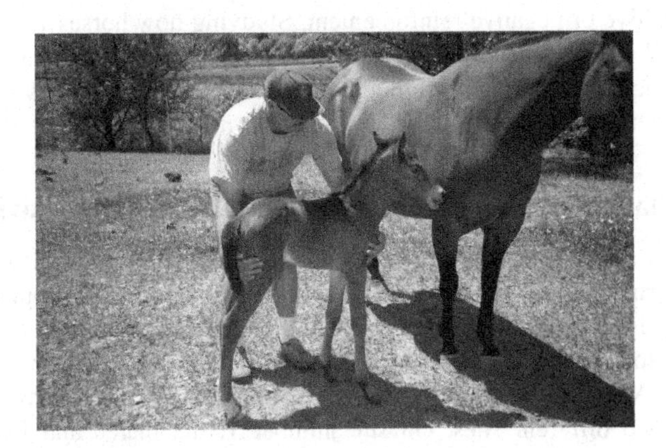

FIGURE 7.1 Initial handling is performed while the mare and foal are not restrained or confined in a stall.

stable-reared horses at about 10 months of age and were 1–2 years old at the time of the experiment. In the methods sections of the paper, Sankey *et al.* (2011) state, "No additional contact with humans took place except for daily tethering for feeding."

We argue that the way tactile contact was perceived may have been influenced by early forced contact for haltering, even if force was kept to a minimum. Very few young horses (or none at all) freely accept being caught, handled, and fitted with halters without prior habituation to this procedure. In the Sankey, *et al.* (2011) experiment, the horses were tethered facing the wall and trained to remain immobile for 5–60 seconds in response to a vocal command "reste!" (French for "stay!"). Training was performed 5 minutes per day for 6 days. The two groups received either tactile contact (scratching the withers), or a food reward (a small piece of carrot) when they responded to the command. After 6 days, the experimenters performed a "motionless human approach" test to assess the human–animal relationship. The results showed that the food reward group approached a stationary human sooner, and spent more time with the human. We do not argue the fact that food is a strong positive reinforcement. However, food motivation may override any reluctance to approach in the food reward group, compared to the group that only received scratching on the withers.

Horses raised by the first author actively seek tactile contact by humans. The second author, Temple Grandin, has noticed that they walk up to people on the pasture and present their belly or rear end for scratching, even towards unfamiliar people. What animals don't like a good belly rub? It is our hypothesis that *any* forced contact at an early age may possibly influence a horse's perception of humans as *potentially threatening*. In reactive/fearful animals, this perception may last a lifetime and be resistant to change. Without knowing the entire history a horse from the day it was born makes it impossible to know how a horse views people and will respond to any type of positive or negative reinforcement. Studying how horses view people, Fureix *et al.* (2009) found that perception of humans by horses may be based on experience. Handling and training can affect horses' memories of human actions either positively or negatively (Henry *et al.*, 2006).

Mare Behavior and Human Behavior Influence Foal Behavior

Both mares and humans can influence foal behavior and their perception of humans during this early period. Henry *et al.* (2005) provided empirical evidence of this in a study of human–mare relationships and the behavior of their foals toward humans. Forty-one French Saddlebreds, Trotters, Arabian, and Anglo-Arab mares and foals were used in this study. The study included horses at two different sites. On site number 1, the mares and foals were kept in 3-m × 4-m box stalls. On site number 2, the mares and foals were also kept in 3-m × 4-m box stalls, but spent several hours each day on pasture with other mares and foals. Half of the mares at each site were brushed and fed by hand for the first 5 days of their foals' lives. The rest of the mares were not handled, and served as the control group. Gently brushing and hand-feeding mares for approximately 1 hour for 5 days following the birth of her foal positively influenced the foal's behavior towards humans. Foals spent longer periods close to the experimenter and initiated more sniffing, and licking contacts, whereas control foals appeared more fearful.

In this study, Henry *et al.* (2005) also observed the influence of mare behavior—protective or not protective—on the foal's behavior towards humans. Protective mares were described as nervous and exhibited behaviors such as always looking at their foals or positioning themselves between their foals and the human. Non-protective mares were less nervous and did not try to block access to their foals. Foal behavior was recorded during tests conducted at 15, 30, and 35 days old. These observations were followed by an approach-stroking test performed by familiar and unfamiliar persons at 11 and 13 months old. At all ages, the approach-stroking tests revealed that foals of protective mares remained further from the experimenter than foals from calm mares. Foals from calm mares also remained closer to the experimenter compared to controls and engaged in more sniffing and licking of the experimenter. In addition, foals of brushed and hand-fed mares exhibited reduced avoidance, reduced flight responses, and faster acceptance of saddle pads on their backs. Finally, handling of mares had effects on foals that were still measurable one year later. Both familiar people (experimenters) and unfamiliar people could rapidly approach and stroke the foals from the handled mares. The results strongly suggest that a mare's behavior can influence their foal's behavior toward humans. In this study, Henry *et al.* (2005) also found an interaction of breed and experience on human–mare relationships and the behavior of their foals toward humans. French Saddlebreds and Trotters were kept on site one, and Arab and Anglo-Arabs on site number two. On site one, French Saddlebreds and Trotter mares and foals were exposed to early handling. The foals were assisted to their feet following birth and moved to the mare to assist first suckling. (This is a common practice on breeding farms around the world.) In addition, their stable boxes were near the riding horses, and the mares and foals were involved in many visual contacts with humans. On site two, the Arabian and Anglo-Arab mares and foals received only minimal early contact by caretakers and were housed far from the riding horses. Overall contact with humans was very limited. Interestingly, breed differences in foal behavior were observed between the two sites. Arab and Anglo-Arab foals at site two were generally calmer and allowed human contact more readily than the foals at site one. This is another genetic–environment interaction. The more reactive Arab and Anglo-Arab foals benefited more from a *lack* of early and later human contact. Brushing and feeding mares by hand has a significant effect on foals' behavior towards humans. Henry *et al.* (2005) concluded that "this procedure is simple, takes little time, and can easily be applied to any dam–foal pair, as it is not intrusive and presents no risks of disrupting mare–foal bonds." In a subsequent study, Henry *et al.* (2007) re-tested the same foals at six months old and found that they remain sensitive to their mares' influences. The study also revealed high individual variability between the foals, suggesting a stronger effect of the foals' own behaviors at this age.

Serotonin Genes and Maternal Behavior

The "protective" behavior of mares in the Henry *et al.* (2005) study represents a complex phenomenon. Naturally occurring variations in maternal care and protection are widespread in mammals, although researchers continue to differ in how much emphasis they place on genetic or environmental factors (McGue, 2010). In recent years, however, the pattern of care given an offspring by its mother has been linked to variation in the serotonin transporter (5-HTT) gene (reviewed by Heiming and Sasher, 2010). Serotonin is an inhibitory neurotransmitter in the central nervous system that is linked to the regulation of mood and emotions. The 5-HTT polymorphism has two common forms: short and long. In human studies, the short 5-HTT gene form is related to increased stress sensitivity in young human females (Gotlieb *et al.*, 2008), and in rats, maternal anxiety associated with the short form has also been found to influence infant temperament in the early postnatal period (Champagne and Curley, 2005). When mothers carrying a short form of the gene live in an environment that they perceive as dangerous or threatening, the effects of the short form are expressed, and their behavior can influence the behavior of their offspring. In a safe environment, the gene is not expressed. Consequently, Sachser *et al.* (2010) argues that the behavioral effects of maternal stress during the postnatal period in mice are not necessarily "pathological," but that the mothers may be adjusting the offspring to their environment in an adaptive way. Mothers may be preparing their offspring to live in a dangerous environment. Whether a mother carries a short form of the 5-HTT gene or not, pre- and postal effects of stress influence sensitivity to stress in offspring.

The effects of maternal stress or other postnatal environmental stressors are also influenced by genetically based individual differences in nervous system reactivity and temperament in foals. The research of Henry *et al.* (2009) on social behavioral development in "imprint trained" foals emphasizes this fact. Individual differences in behavior were observed in the intensively handled foals. Three out of nine foals approached and investigated an unfamiliar human, whereas almost all (nine out of 10) of the control foals left their mother and approached the human. One-third of the intensively handled foals approached the unfamiliar human, suggesting an effect of genetics on this response.

Training Foals to Accept Handling

The neonatal handling procedure that the first author practices may help reduce stress in foals at weaning time by developing a strong human–foal relationship. The procedure begins very early and is continued throughout the time that the foal is with its dam. He handles the foal once or twice a week in the presence of the mare, and only for a few moments each time. All areas of the body are touched, including the legs and feet. By never using force and allowing the foal to initiate contact, the handling is perceived by the foal in a positive and rewarding manner. Teaching the foal to accept a halter begins during this time, but long before he actually halters the foal for the first time. As part of his weekly handling, each time he puts his arm over the foal's neck for a moment and gently pulls it against his body in a "hug-like" position. If the foal resists, he lets it go. Each time he does this, the foal is more likely to accept it if no force is used the first time. During this entire process, the first author uses a calm, soothing voice and firm, but gentle, pressure to hold the foal. During the hug, he puts his free hand on the bridge of the nose, moving it up and down the face. He also places his free hand around and behind the ears. These movements are meant to simulate the movements of fitting a halter. After several short hugs over a few days, he begins to hold the foal for longer and longer periods. Soon, the foal completely accepts gentle restraint (hugs) without showing any signs of fear. Fear is evident if the foal suddenly raises its head, tightens its muscles or widens its eyes. However, at some point the foal may try to pull away, not because of fear, but because it wants to do something else. When this happens, the first author begins to hold more firmly, for only a few seconds at first, and then for a few minutes at a time. If he sees any signs of fear, he lets it go. This allows the foal to predict what is happening and gives it a feeling that it has some control over the situation. Habituation to gentle restraint can occur quickly. The first author's method of positive reinforcement is similar to methods used by Phillips *et al.* (1998) to train flighty Bongo antelope (*Tragelaphus eurycerus*) to accept veterinary and husbandry procedures without chemical or manual restraint. In both these methods, the animal is habituated to gradual increase in the intensity of the stimulus. Allowing an animal to have some control over stimulus presented in a predictable manner has profound effects on behavior in captive marmosets during routine handling and weighing (Bassett, 2003). In an excellent review, Bassett and Buchannan-Smith (2007) examined the link between predictability and control and made recommendations for simple and inexpensive modifications to husbandry routines that are easy to incorporate and can have a profound impact on the welfare of captive animals.

Wearing a Halter and Hoof Handling

After the foal accepts being touched all over and held in the position for haltering, it is time to introduce the halter for the first time. He approaches the foal in the same manner as before, but he carries the halter, tucked into his belt so the foal doesn't see him carrying a novel object, which may cause fear. With the halter tucked into his belt, he gets the foal into the hug position, then slowly pulls the halter from his belt and gently slides it up the foal's nose. If the foal resists and pulls away, he lets go, remembering never to use force when introducing anything for the first time. If the foal does not resist, he slowly slides the halter up and down the nose a few times until the foal begins to show indifference. He then brings the halter fully up the nose and simulates the movements used to buckle the halter before actually buckling it for the first time. After the foal accepts this he buckles the halter and has the foal wear it for a few minutes, and then he takes it off. Every time he handles the foal, he follows the same consistent pattern: the first approach, then hugging with an arm over the neck, then sliding the halter on. In most cases, it takes only a few minutes each day for the foal to accept wearing a halter.

The next step is to teach the foal to lead with a halter and lead rope. Once the foal is habituated to handling and will accept a halter, he clips a lead rope onto the halter. The foal needs to see the lead rope and associate it with the halter, but learning to lead comes later. Holding the lead rope loosely, he stands next to the foal and starts to teach it to accept having its feet handled. Starting with a front leg, he slides a hand down the leg and tries to get the foal to flex the knee. If the foal refuses and pulls away, he lets it go. When the foal is settled down, he again approaches the front leg, slides a hand down the leg and tries to flex the knee. If no force is used the first time, the foal does not associate having the leg handled as something to be afraid of and to subsequently avoid. Once the leg can be flexed and held for a moment, he moves on to the next. Each leg is handled the same way, never using force. Throughout this process, the foal is wearing a halter and the lead rope is held in one hand. Slowly, he begins to pull the foal's head slightly toward the side of the body on which he is standing, using the halter and lead rope to get the foal to watch as he rubs his hand down the leg. This is the first pressure that the foal feels from the halter and lead rope.

Subtle Individual Differences

A note here on individual differences: if a foal is of the reactive type, caution is required to avoid sudden movements during each of these steps in order to prevent causing a fear reaction. He keeps the lead rope loose to avoid any sudden pressure if the foal becomes frightened and tries to pull away. Habituation occurs quickly in less reactive foals, and less caution is required. All foals eventually habituate, but differences in reactivity and fearfulness are important to remember during halter fitting and picking up the feet. A basic principle is that each step of the training procedure must be introduced gradually to more reactive animals (Phillips *et al.*, 1998). It is also important to remember that a time comes when foals are no longer fearful and may refuse to accept the hug, the halter, or having their legs and feet handled. This is a key point to remember, as motivation to resist handling can shift from fear to learned resistance that is not fear-motivated (Phillips *et al.*, 1998). Once the foals overcome their fear, the motivation for self-gratification takes over, and resistance can be shaped by that attitude. Failing to recognize the motivational shift and allowing the foals to pull away is a mistake that, once made, can be difficult to correct. Motivational shifts can occur at any stage of handling, from the initial hug and hold stage, to the haltering stage, to the feet handling stage, and eventually to the learning to lead stage. A firm "NO" at the beginning of a motivational shift is an important reinforcement tool. This teaches the foals that they have to tolerate some things, even if they don't want to. In the early training stages, failure to recognize motivational shifts and deal with them accordingly can cause subsequent resistance in the following stages of training. Learned resistance can stay with foals throughout their lives. However, if too much force is used to correct no-fear resistance and a foal starts to struggle, fear motivation can take over. In less fearful foals, resistance caused by no-fear motivation is stronger and more common. In fearful foals, resistance is primarily motivated by fear.

The next step is teaching the foal to lead and stand tied. After the foal used to wearing a halter and having a lead rope attached, the first author starts by leading the mare and encouraging the foal to follow. He keeps the lead rope loose on the foal so it can see it out front, but does not pull the foal with the rope. Once the foal follows, he begins to apply light pressure on the lead rope. In a short time, the foal can be led along with the mare: short distances at first, then farther and farther. Secondary reinforcement with a food reward begins at this stage. Using the "carrot and stick principle," he rewards the foal with a treat each time it moves forward willingly. He also uses a calm, reassuring voice to help to establish rapid habituation. Learning to lead and stand tied are both done in the presence of the mare. The first author teaches the foal to stand tied by leading both the mare and the foal to the hitching post, tying the mare and wrapping the foal's lead rope over the post. In the beginning, he drapes the lead rope over the hitching post or wraps it around the post once and leaves it untied. This way, the foal learns to see the rope stretched out in front of it, but if it becomes fearful and pulls back, a fear of being tied can develop if is the foal is unable to get away. Staying close to the foal in the beginning is important. First experiences are critical to shaping future responses. A few minutes per week of standing (untied) alongside the mare are usually enough. Each time, the foal can be expected to stand for longer and longer periods. The first author never leaves a foal tied alone until after weaning when more time is spent on teaching the foal to accept being tied. To help the foal accept being tied alone, he starts moving away from the foal throughout this process—a short distance at first, then farther and farther away as time passes. The foal should not be expected to stand alone for any extended periods. Having a mare that calmly accepts handling and being tied also helps the foal accept handling.

More Factors that Affecting Bonding with Humans

As we discussed previously, Henry *et al.* (2005) found that mares can influence the behavior of their foals toward humans. A second hypothesis offered by Henry *et al.* (2005) is that foals can also influence the behavior of their mares toward humans. A person forcibly catching a foal and holding it while it struggles can upset the mare. The non-intrusive approach to foal handling practiced by the first author prevents the mare from becoming upset by seeing her foal in danger. As Henry *et al.* (2005) suggests, social facilitation is involved in the human–mare–foal relationship. To demonstrate this principle, Henry *et al.* (2006) used three procedures to study early handling and human–foal relationships. In the first, foals were restrained and stroked all over their body. In the second, foals were restrained and taken to the dam's teat (first suckling assists). In the third, foals were exposed to a motionless person. The results found that the first two procedures had no effect on the later human–foal relationship, and the third slightly enhanced the foals' reactions to humans. Foals that were assisted during their first suckling avoided human approach and physical contact at 2 weeks of age, and at 1 month of age. This research supports the procedure practiced by the first author. His procedure begins with a human–mare bond, followed by a human–foal bond (only after the foal initiates first contact). The relationship is subsequently enhanced through social facilitation. To quote Henry *et al.* (2007), "It may be important that the young individual is an actor in the establishment of the relation, rather than a passive receiver of stimulations." In any event, foals' perceptions of humans develop from a very early stage. Whether a non-intrusive approach such as that practiced by the first author is used or not, several authors agree that gentle handling, stroking, halter fitting, and picking up the feet of young foals have more positive effects, at least in the short term, than forcibly rubbing the foals' bodies (Heird *et al.*, 1986; Jezierski *et al.*, 1998; Mal and McCall, 1996; and discussed in Henry *et al.*, 2006). But the question of *when* to begin handling foals remains unanswered. Several authors argue the existence of a "sensitive" or "critical period" for learning to occur, and suggest that the weaning period may be similar to those found in other animals at early ages. Lansade *et al.* (2004) addressed this by handling foals at two different life stages: immediately following weaning, and 21 days later. The results of this experiment found that haltering, petting, picking up feet, and leading foals immediately following weaning lowered their reactivity and made them easier to handle compared to non-handled controls. The responses were also stronger in the early handled foals compared to the foals handled 21 days after weaning. Bateson (1979) suggested that stress induced by weaning causes greater sensitivity to external stimuli. Boivin and Braastad (1996) found more ease of handling in goats immediately following weaning and suggested that goats are susceptible to forming new social bonds during this period, and "that the human might serve as a surrogate mother." A complex interaction between genetic factors influencing temperament, pre- and postnatal influences, and other environmental factors affect the behavior of foals at weaning time. This is a stressful time. There are several psychological, physical, and nutritional stressors associated with weaning, along with other stressors, such as maternal deprivation and social isolation (reviewed by Waran *et al.*, 2007). Forced weaning does not occur in nature. Since we are responsible for creating this adverse situation, we need to take the responsibility and make it as easy as possible for the foals to accept.

In summary, the effect of early experience on temperament, learning ability, and cognition is a complex phenomenon requiring further investigation, especially regarding the importance of the presence of stressors. In our opinion, people must make sure first experiences are positive no matter when handling begins. Forcibly catching young foals at any age may be detrimental to forming positive foal–human relationships. In addition, understanding foals' individual differences and motivational states, as well as appropriately handling resistance motivated by fear, or no-fear resistance, sets the stage for a relationship based on mutual trust and respect. This is similar to the social relationships that young foals develop with other horses. Inconsistencies in handling, especially at a young age, may cause conflicts between emotional and motivational states. Many of the assumed idiosyncrasies observed in foals (and mature horses) may in fact be pathological. For example: a common behavior in foals is to approach people and then quickly retreat, only to approach and retreat again. Are the foals curious but afraid (explained in Chapter 1), or are they exhibiting a conflict between a conditioned response of fear toward people and a similarly conditioned response motivated by the positive stimuli that people provide? Imagine for a moment the emotions of a foal that has been handled roughly, captured, haltered, and sometimes forced to stand to have its feet handled. These experiences may remain embedded in a young horse's memory, and may be in conflict with other, more positive memories of people providing affection and food. Therefore, on the one hand, the foal fears people for the capture and forced handling. On the other hand, it likes people for the positive stimulus associated with them. Conflicts between emotional and motivational states are stressful and can compromise health, welfare, and responses later in life. The non-intrusive neonatal handling procedure practiced by the first author is easy and practical. To test its validity, studies similar to those conducted on intensively handling foals could be designed using this non-intrusive handling procedure and comparing the results with non-handled controls. In the following sections we discuss other aspects of foal and horse behavior that possibly influence this vulnerable period of life.

GENETICS AND TEMPERAMENT: ORIGINS AND OUTCOMES

Temperament is defined as differences in behavior between individuals that are consistently displayed when tested under similar situations (Hausberger *et al.*, 2004; Kagan, 2005; Plomin and Daniels, 1997; Zentner and Shiner, 2012). Most scientists agree that temperament consists of early-appearing, biologically based behavior patterns that originate in the horse's biology and appear early in development. The biological foundation is usually genetic, but may be a result of prenatal events. Young horses show considerable variation in temperament (e.g., fearfulness, sociability, reactivity, etc.). Fearfulness is probably the most important factor influencing the suitability of horses for any type of work. Fearfulness is defined as the general propensity of an individual to perceive and react in the same manner to a wide variety of potential threats (Boissy, 1998; Christensen, 2006). Abundant evidence demonstrates that past environmental factors and previous experience have a significant effect on fearfulness. Repeated exposure to stimulus early in life can modulate the development of subsequent fear reactions. Exposing a fearful young horse to an enriched environment and gentle handling procedures may reduce subsequent fear-related reactions. On the other hand, exposing a less fearful horse to a stressful environment and rough handling increase subsequent fear related reactions. Early stimulation and later experience interact throughout the horse's life with the genetic background (Christensen, 2006). The ability to select a horse for a particular type of work, especially from a very young age, is an attractive idea for breeders, trainers, and horse enthusiasts. Behavior tests have been developed to identify temperament traits, but a factor that is usually overlooked is the biological basis of the traits being investigated. For example, by the time a horse is old enough to begin training for riding, it is difficult to detect the biological basis of temperament because the behaviors measured at this time could be a product of experience alone. Not all fearful horses inherit the biological basis of that trait. The origin of temperament has a genetic basis, but the outcome is influenced by environment and upbringing.

A central question in studies of temperament is whether an individual's temperament traits remain consistent over time (Visser *et al.*, 2001). In addition, the age at which temperament can be considered stable is still debatable (Seaman *et al.*, 2002). The reliability of temperament tests of older horses is hindered by the fact that the background is usually unknown and the behavior is more influenced by previous experience. Thus, previous experience may have more influence on responses measured in standardized behavioral tests than any underlying genetic influence. Previous experience affects older horses more than younger horses. In addition, previous experience may influence responses in a familiar setting, whereas response to novelty in an unfamiliar setting may better reflect horses' underlying reactivity (temperament). Standardized behavioral observations that depend on the experience and subjective judgment of handlers or caretakers are not always reliable. A handler's prior expectations of how horses should react may affect a horse's behavior, even if exact instructions are given on how to perform a test. Finally, standardized questionnaire surveys completed by knowledgeable handlers may also have limitations. However, comparing questionnaires with measures from standardized behavioral test assessments produce more valid and reliable results.

Finding easy and reliable ways to identify temperamental differences has two main advantages: the ability to predict behavior, and the means to select horses with suitable work-related characteristics. The following review is not exhaustive, but is intended to show methods used by researchers to assess temperament traits. Some of these tests are practical and can be conducted by a lay person, but others are less practical. Suspending objects from a ceiling would be difficult to apply in typical riding centers or schools, where the ceiling is usually high. However, methods using novel objects such as an umbrella to test reaction to sudden novelty, or an unfamiliar arena to test emotional reactions to social isolation are more practical. The aim of temperament tests for horses is that interested people can perform the tests themselves and make use of the results.

TEMPERAMENT TESTS

Behavioral Measures

Visser *et al.* (2001) measured variation and consistency of horses of the same breed and age, reared under controlled housing and management conditions. Forty-one Dutch Warmblood horses were tested in two behavioral tests (novel object and bridge test) at 9, 10, 21, and 22 months of age. The study identified two temperament traits (flightiness and sensitiveness), but did not show long-term consistency of traits. Seaman *et al.* (2002) recorded the behavior of thirty-three young horses in three behavioral tests (novel arena, human approach, novel object). Tests were repeated three times with an average of 9 days between trials. The responses of horses in a human-approach test and novel object test were similar, but inconsistent: only the open-field test was consistent over time. The authors suggest responses in the open field may reflect a core factor of temperament. A questionnaire completed by the farm manager confirmed the results. Hausberger and Muller (2002) used posture in 224 adult horses to evaluate factors involved in reactions to humans. The results showed breed differences but, more importantly, caretaker involvement revealed differences between the groups. One person cared for every horse in one school, suggesting the important influence of the caretaker on the well-being of the horses and their reactions to an unknown person. Sondergaard and Halekoh (2003) housed 40 young Danish Warmblood horses either alone or in groups of three after weaning. One half of each group received handling three times per week for 10 minutes. At 6, 9, 12, 18, 21, and 24 months of age, the study recorded behavioral observations in a human-approach test performed in the home environment. At 12 and 24 months of age, the study recorded behavioral observations in a novel arena test followed by a human-approach test in a novel arena. Single-housed horses approached sooner and were more easily approached by a human in the home environment, but not in the novel environment. Single-housed horses also expressed less restless behavior, more explorative behavior, and less vocalizations in the novel environment, but not in the home environment. Handled horses also reacted less to the novel environment. Results indicate that handling during the rearing period is a means of avoiding potential dangers associated with handling young horses in novel environments. Horses reared alone were reluctant to approach a person in the novel environment and moved around more in the novel arena. Visser *et al.* (2003) tested 41 young horses in four behavioral tests: a novel object test (lowering an umbrella from the ceiling), a handling test (crossing a bridge), an avoidance-learning test (avoiding a puff of compressed air), and a reward-learning test (choosing a manger to receive a food reward) during the first 2 years of life. Behaviors during each test were recorded, i.e. locomotion, position relative to the novel object or bridge, posture of head and tail, latency times during the novel object or bridge test, and vocalizations. At the age of three, personality traits identified earlier (tail held high, latency time to touch a novel object, standing still in front of the bridge) correlate with early training in jumping performance. The study shows that a large part of jumping performance can be predicted by personality traits evaluated earlier in life. Momozwa *et al.* (2003) used a questionnaire survey to assess caretakers' impressions of temperament in 86 riding horses. The five-point questionnaire revealed three independent factors that they called "anxiety," "novelty seeking," and "understanding." A balloon reactivity test verified the reliability of the survey. Introducing each horse to balloons revolving in the center of an unfamiliar arena and recording heart rate for 5 minutes revealed that horses evaluated as "highly anxious" by caretakers had higher heart rates and defecated more during the balloon test. Also, horses evaluated as having problems with ordinary care or training had higher heart rates, and those evaluated by the survey as having a "long adaptation time to adapt to unfamiliar objects" were unwilling to touch the balloons. Results suggest that questionnaire surveys are a good way to assess temperament traits—especially those related to anxiety. Hausberger *et al.* (2004) used three behavior tests and a learning/memory test on a sample of 702 horses tested at 103 different sites. To establish precise genetic origin, the study used only registered horses. Study results show that determination of behavior/temperament traits varies differentially with genetic influence (see our discussion of this study in the introduction to this chapter). Christensen *et al.* (2005) recorded heart rate and behavioral observations to stationary novel visual, olfactory, and auditory stimuli on 24 untrained stallions. The stallions were habituated to eat from a container placed in an arena. In the visual test, a traffic cone was placed in the arena near the food container; in the auditory test, a white noise was played while the horse ate from the container; and in the olfactory test, eucalyptus oil was applied to the inside of the food container. Results showed no difference between tests in locomotive activity, but heart rate increased during the visual and auditory test. No increase in heart rate occurred during the olfactory test, but the stallions ate more and became more vigilant to their surroundings. During the visual and auditory tests, more time was spent in being alert towards the stimulus. The heart rate responses correlated between tests and reflect a non-differentiated activation of the sympathetic nervous system, while the behavioral responses were linked to the type of stimulus. McCall *et al.* (2006) recorded heart rate and behavioral observations in three tests: isolation, static novel stimulus, and traversing a novel stimulus in a runway. Forty horses performed all three tests on three different days. Both isolation and novel-stimulus tests produced valid measurements, but the novel-stimulus test was considered the more accurate evaluation of reactivity. In the isolation and novel-stimulus tests, walking and defecation frequency were the most precise behavioral measures. The authors suggest that using physiological data alone, combining physiological and behavioral measurements, or using more than one behavioral measurement in reactivity tests may better reflect the reactivity of the horses than a single behavioral measurement. Lloyd *et al.* (2008) assessed personality in 1223 horses of different breeds using a 25-item rating method previously shown to be reliable for personality assessment. The study identified breed differences, and the breed differences shifted between personality traits. Anxiousness and excitability showed the most variation between breeds. Lansade *et al.* (2008) studied fearfulness in 66 Welsh ponies and 44 Anglo-Arabs. The tests included a novel object test, a novel arena test, and a surprise test. Results showed a consistency in fearfulness between all three tests, as well as over time. Von Borstel *et al.* (2010) studied fear reactions in: untrained horses; well-trained dressage and show jumping horses; and horses with mixed genetic lines. An unfamiliar handler led horses into an arena containing a bucket of grain. When the horses began to eat from the bucket, a handler quickly pulled a weighted black plastic bag attached to a string across the arena. Results showed less intensive flight reactions in jumping horses compared with dressage horses or horses of mixed genetic lines. These findings suggest a genetic basis for less strong, though not less long-lasting, fear reactions in jumping horses compared to the dressage horses or the horses with mixed genetic lines.

Physiological Measures

Schmidt *et al.* (2010) measured heart rate and stress hormone levels on seven Warmblood mares and six Warmblood stallions during a 12-week preliminary training program. Results showed that riding and training increased stress hormone levels. Mares had higher stress hormone levels than stallions, and mounting by a rider caused the most pronounced stress reaction. Leiner and Fendt (2011) measured heart rate and behavioral reactions to novel objects on 18 Warmblood stallions. The results show correlations between behavioral and physiological fear responses. Furthermore, after habituation to an object occurs, the fear response to the object is reduced, whereas the fear response to another totally different novel object remains. Habituation to a stationary blue and white umbrella does not transfer to an orange tarp (Leiner and Fendt, 2011). Heart rate increases are a reliable measure of fear intensity in horses, although responses to the novel stimuli differed between individuals.

The purpose of this review was to show some of the methods used to study individual differences in temperament. Many of these tests are complicated and take a considerable amount of time to perform. In addition, a few are not easy to perform and difficult to standardize. For lay persons, it may be difficult to make decisions about a horse based on these results. However, the consistency of responses to sudden novelty, novel objects, and behavior in novel environments, all indicate a core dimension of temperament. As we discussed earlier, fearfulness is probably the most important factor influencing the suitability of horses for any type of work. In the next section we discuss evidence collected by the first author that may provide an easy and practical method for assessing temperament that anyone can use.

THE RELATIONSHIP BETWEEN HAIR WHORLS AND BEHAVIOR

In early 1993, the first author approached me (Temple Grandin) in my office at Colorado State University to discuss some observations he had made while working as a farrier and horse trainer. The anecdotal evidence he collected suggested a possible relationship between hair whorls on horses' foreheads and temperament. I was of course skeptical; however, the story he told was compelling. All my years of working in the industry had taught me that people who work around cattle or horses daily often make valid observations that scientists working in the laboratory often miss. Before we met, Mark approached others in academia. Most said his ideas were too far-fetched to be worth consideration. We began by searching the literature for any evidence to support his observations. At the time, the only reference to horse hair whorls was a questionnaire survey of 1500 horse collected in 1965 by Tellington-Jones and Taylor (1995); however, the conclusions of the authors offered no statistical evidence.

What follows is the anecdotal evidence Mark discussed with me about hair whorls collected over a 15-year period and involving over 10,000 horses. He noticed hair whorls located high on the forehead predicted a flighty temperament. Whorls located below the center of the eyes predicted a calm temperament, and left or right side locations predicted handedness. He also noticed that environment and upbringing had an effect on whether the hair whorl could predict temperament and side preferences. He also noticed some unique and unusual behavior of horses with two side-by-side whorls. Following the anecdotal evidence presented by the first author, in the following section we discuss scientific evidence gathered by us over the last 20 years, as well as a review of more evidence compiled by other researchers around the world.

Personal Observations

I (Mark J. Deesing) first heard about hair whorls in 1982, about one year after graduating from horseshoeing school. I was working full-time, shoeing horses and supplementing my income by breaking young horses to ride. That summer I was training a young Arabian filly, and I commented to her owner about how intelligent I thought her horse was. She said to me, "I know she's smart. My grandpa told me the high swirl her head meant she was smart." I had never heard of this, and I asked her to explain. Her grandfather said that Bedouin tribesmen 2000 years ago placed great significance on hair whorls, and believed the position of the whorls can predict how friendly or smart a horse was. (Hair whorls are located in an area between the eyes, single whorls can be located very high above the eyes, high above the eyes, between the eyes, low, and very low. Some whorls are on the right, others on the left (Figure 7.2). Some horses have two whorls—one on both the right and left sides of the face.)

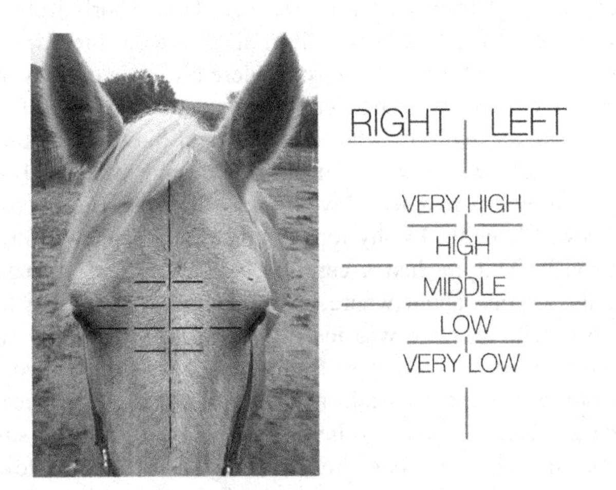

FIGURE 7.2 Hair whorls are located in an area between the eyes, single whorls can be located very high above the eyes, high above the eyes, between the eyes, low, and very low. Some whorls are on the right, others on the left (see Figure 7.1.) Some horses have two whorls—one on both the right and left sides of the face.

I thought it was a silly idea, but I was always curious to learn other people's ideas and opinions about horses. I wanted to understand horses better. My horseshoeing instructor taught me how to shoe horses, but not how to get horses to cooperate. Learning to understand and control horse behavior was a big part of my job. I didn't understand why it is so hard to teach horses something as simple as standing still for shoeing. In any event, what the woman told me was interesting, and I wondered if there was anything to it. I was shoeing 20 or more horses a week. It only took seconds to look at the whorl, and I dealt with horse behavior every day. The position a farrier works in while holding a foot off the ground is close-up and personal. Cooperation is necessary. I thought that if a relationship exists between hair whorls and behavior, I'd know before too long.

I noticed right away that horses with high hair whorl(s) were more likely to resist shoeing . Horses with low whorls resisted less. At first, the relationship was not very strong, but strong enough to keep me interested. I began to value what I was seeing. Looking at the hair whorls on a horse before starting to work made my job a little safer. When I approached an excitable horse with a high whorl, I learned to avoid making quick movements or fighting with them if they resisted. Instead, I learned that a quiet and gentle approach worked better with excitable horses. Using this approach allowed me to finish the job without a fight, and without getting hurt. I was in a unique position to observe hair whorls in many horses. I was shoeing about 20 horses per week, 80 per month. In the first year alone I looked at more than 1000 horses. Even though I handled many horses more than once a year, the number of different horses I worked with far exceeded that of most professional horse trainers. I was learning about whorls from a variety of unfamiliar horses. I handled horses from all the major breeds, ranging in age from six months to 25 years. I still considered it just a curious phenomenon, but I found the relationship between whorls and behavior useful.

It was the behavior and hair whorls of my own three horses that started to convince me to take whorls seriously. Murphy was a 10-year-old Appaloosa gelding with a hair whorl well below the eyes. He was as calm as any horse I knew. I bought Murphy for my nieces and nephews to ride when they came to visit. Murphy had a carefree expression and an independent nature—he only worried about where his next meal was coming from. My other Appaloosa gelding, Dell, was just the opposite of Murphy. His whorl was well above the eyes. Dell was flighty and reactive. He would jump 10 feet if a mouse crossed the trail, and he was insecure and became very nervous when separated from the other horses. He was big and strong, but timid, reactive, and alert. He held his head high and oriented toward any unusual sight, or sound. On the trail, he spotted deer on hillsides long before other horses. He would stop and point with his ears and eyes. My third horse was a race-bred Quarter Horse mare. Jenny had two high side-by-side whorls. Her behavior was similar to Dell's, but less nervous: Dell's vigilance seemed motivated by fear, whereas Jenny's vigilance was more like a curious attention to her surroundings. She didn't always need the company of other horses and was independent. Unlike Murphy's, carefree independence, Jenny had confident independence. As the dominant horse in the herd, she rarely exercised her authority by force: a slight pinning of her ears was all it took to make other horses back off. When challenged, she let out a squeal that caused other horses to scatter. On the trail, she refused to follow another horse. Always the leader, she'd throw a fit if I tried to make her follow.

Behavior differences in these horses was not subtle, neither were the hair whorls. When I saw similar hair whorls in other horses, I would ask the owners if their horses had similar behaviors. Clear patterns of behavior and hair whorls began to emerge. The patterns were still just trends, but it was around this time that I began to notice another, more consistent behavior related to hair whorls.

Hair Whorls and Side Preferences

The awareness of side preferences in horses is common throughout the industry. The preferred side is easier to handle for shoeing, and horses have preferred directions when working in circles, or when ridden. The preference for one side during shoeing was so common that most horse owners would warn me about their horse's side preference before I started shoeing. I faced this problem every day. Before long, I noticed the hair whorl side (left or right) was often the difficult-to-shoe side. A left-side hair whorl predicted left-side resistance to handling for shoeing. A right-side whorl predicted right-side resistance, and double whorls predicted resistance on both sides. Hair whorl and side preference was more obvious than hair whorl height and reactivity. Side preference for shoeing was even obvious in horses with low whorls and a calm temperament. By questioning owners, I found the pattern common in all side preferences, but most common in horses behaving nervously. For example, I found that horses with a left-side hair whorl preferred the right side of a two-horse trailer, and horses with left-side hair whorls approached novel objects with a left-side bias, keeping the left eye on the object. All these behaviors were reversed in horses with right-side whorls, and always more obvious in nervous horses. Interestingly, horses with two side-by-side whorls showed less side preference in general. However, in horses with double whorls, I noticed that one whorl is usually higher than the other. The higher whorl typically shows the preferred side bias. These observations made my job easier, and much safer. For example, when I approached a fearful horse with a left-side whorl for shoeing, I learned to start the job on the horse's right side. This gave horses an opportunity to get comfortable with me before I handled the more difficult hair-whorl side. I also found my observations useful when introducing young horses to the novelty of a saddle, mounting for the first time, and teaching them to rein to the right or left. Strong side preference seemed to be always motivated by fear. Horses showing strong side preferences often stood in defensive postures, holding their heads high, and showing the whites of their eyes. These are all signs of fear.

Hair Whorls, Temperament, Side Preference, and Environmental Interactions

After several years and several-thousand horses, I was finally a believer. Most horses with high whorls were reactive and fearful, but not all. I began to suspect the reason for this must be due to experience and upbringing. For example, I made most of my observations in the home environment, which could have masked fearfulness and reduced reactivity. And for most horses, horseshoeing is not a novel experience. I began to focus my interest on the effects of experience. Horses with high whorls and excitable temperaments often had excitable owners, or owners who handled their horses roughly. Was it possible that reactivity and fearfulness of horses with high hair whorls was being masked by experience, or changed in young horses handled gently? Searching for answers led me to develop the non-intrusive neonatal handling practice described in the first section. I began to handle my foals all in the same way (regardless of hair whorl) and studied their reactions. It soon became clear that experience has a significant effect on behavior. If I handled foals with high hair whorls roughly, they resisted, and resistance continued in the future. On the other hand, gently handling foals with high whorls made them easier to handle in the future. The considerable variation in reactivity and fearfulness I observed in horses with high hair whorls was starting to make sense.

The Paradox of Double Hair Whorls

The puzzle pieces were falling into place—except for one. Horses with double hair whorls showed little variation. This presented a paradox. I never saw double whorls below the eyes. Double whorls above the eyes meant these horses should be similar to horses with single high whorls; however, this was not the case. Doubles were either very reactive and difficult to handle, or non-reactive and easy to handle. Very few individuals showed behavior between the two extremes. I didn't know what to make of this. I also noticed differences in the way some horses resisted during shoeing or training. Horses with single high hair whorls reacted mainly with flight reactions, whereas horses with double whorls were more likely to fight, bite, or kick. On the other hand, when horses with double whorls showed good behavior, it was very good. (Figure 7.3) Many of the good ones also showed leadership and competitive traits, similar to my double-whorled mare, Jenny. I began to wonder about what caused them to be so different. To find an answer, inheritance of hair whorls became the focus of my interest.

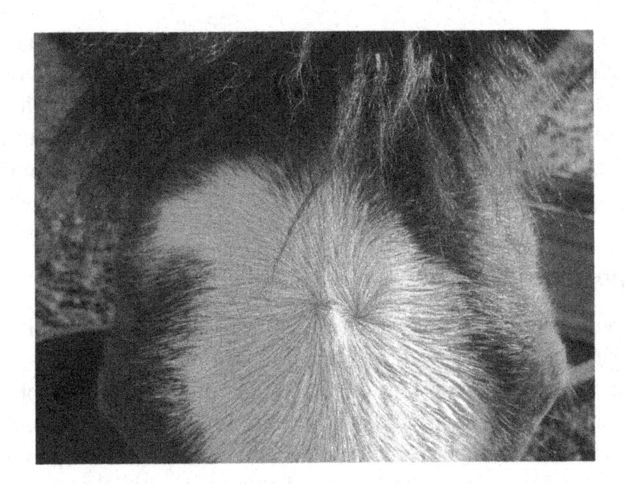

FIGURE 7.3 Double whorls. One whorl is usually higher.

The Heritability of Hair Whorls

Over the years, I noticed hair whorl position was heritable. For example: offspring of horses with high whorls almost always had a high whorl as well. Breeding a low whorl to a high whorl, offspring usually had a hair whorl in between the two but tending toward the high side. Inheritance of double whorls was harder to pin down. I got lucky one day while shoeing a mare. Her two offspring were standing at the fence, a yearling and a three-year-old, both had double whorls. The mare had a high left side whorl. The owner said the sire belonged to the neighbor down the road. When I finished with the mare, I went down the road and asked to see the sire. His whorl was high, like the mare's, but his was on the right side. Surprised by the simplicity, I wondered if it was random. Determined to find out, it took nearly a year to confirm my first observation. After that, I went out of my way to look for other examples. Searching for examples was difficult, but over two or three years I found several following the same pattern. I noticed in all cases that whorls in the sire and dam were always high.

Many Traits are Conserved

After several years of observations and many discussions with owners about their horses, I discovered something I never expected. One afternoon I was shoeing a gelding with double whorls while talking to the owner about my observations. I explained that breeding a left-sided whorl (a "lefty") to a right-sided whorl (a "righty") is where the two whorls (ambidextrous) often come from. He looked at me with an odd expression and said, "My mother is right-handed, my father is left-handed, and I have two whorls on my head. My son also has two whorls and we're both ambidextrous". Frankly, I was shocked. I never considered something similar occurring in humans. I also noticed high heritability of double whorls in horses. I once saw double whorls in fifteen out of eighteen offspring from one stallion that had double whorls. I considered this more than just a coincidence, but was no closer to understanding why double-whorled horses have such unique and unusual behavioral traits.

A Predisposition to Pathology?

I finally found a clue in one of the most unlikely places, and it wasn't a horse that provided the clue. I was in a Laundromat one night washing my clothes and noticed a young boy there with his mother. He was obviously a "problem child", running around wildly, climbing on washers and making noise. Each time his mother tried to settle him down, he was up and running again within minutes. When he happened by me, I noticed two pronounced whorls on the back of his scalp. Unable to contain my curiosity, I struck up a conversation with his mother. I asked the usual questions about handedness in her and her husband. She said she was right handed but her husband was ambidextrous. When she asked what it meant, I explained what I noticed in horses. I also explained the unusual behavior in horses with double whorls. With a disconcerted look on her face, she told me that six months earlier her son was well-behaved, and only recently started to change. When I asked what had happened to make him change, she mentioned being recently separated from her husband and going through a difficult divorce. They were sharing custody and she explained some inconsistencies in the way each of them handled their son. She tried to reason with her son if he was misbehaving, and her husband thought spankings were more appropriate. I thought to myself, "Many children go through the divorce of their parents and accept it well. What was different about him?" Maybe horses with two whorls are over-reactive and more likely to exhibit behavior problems when faced with inconsistent handling or mistreatment? I decided it was time to find out. Soon after my experience with the boy and his mother, I traveled to Colorado to meet with Temple Grandin at Colorado State University. Her reputation as an objective, open-minded, scientist preceded her. If anyone could help me understand, she could. What follows in the next section is the scientific evidence that we collected over the next 20 years, and research conducted by others that supports my observations.

The Science of Hair Whorls in Humans

A review of the human pediatrics literature reveals that hair and brain form from the same fetal cell layer during gestation. At about the 10th week, rapid development of the brain and the hemispheres occurs and the cranial vault rapidly enlarges to a domelike shape (Hall *et al.*, 1989; Kiil, 1948; Smith and Gong, 1973). The expansion of skin over the dome-like outgrowth corresponds to the position of the posterior parietal whorl—this much we know for sure. Interest by geneticists and neuroscientists in recent years focused on associations between hair whorl direction, language dominance, and handedness. The results of these studies are not mutually conclusive. Some investigators found a significant association between scalp-hair whorl direction and handedness (Beaton and Mellor, 2007; Klar, 2003; Friedman *et al.,* 1952), others have questioned these findings (Jansen *et al.*, 2007; Weber *et al.*, 2006). Handedness has long been considered to result from both genetic and environmental factors (Corballis, 1997; Levy and Nagylaki, 1972; McManus, 1985). Klar (2003) was the first to report an association between handedness and scalp-hair whorl rotation, and proposed the role of a single gene controlling both handedness and hair whorl orientation. Other investigators strongly argued against a common genetic basis of handedness or language lateralization with scalp-hair whorl direction (Jansen *et al.*, 2007), and in a more recent report Klar (2009) found that handedness and scalp-hair whorl rotation is random and not genetically determined.

Finding a relationship between hair whorls and behavior in humans is complicated by the difficulty of determining hair rotation in people with long or curly hair. In addition, most studies only focus on whorl rotation (e.g., clockwise, counter-clockwise) and behavior—not whorl *placement* and behavior. Only Schmidt *et al.* (2008) looked at side *placement* and whorl *rotation*. The study revealed a significant association between whorls located on the center of the scalp and counter-clockwise rotation in non-right handed children. None of these studies examined a possible association between handedness and language dominance in people with two or more hair whorls.

Hair Whorls in Cattle

In 1993, we noticed that hair whorl patterns in cattle were similar to those in horses. The second author (Temple Grandin) suggested that we aim our research toward cattle. We could study large numbers of cattle with similar genetics and experiences, a task more difficult with horses. Our first study revealed a significant association between the height of the whorl and behavioral agitation in cattle during restraint (Grandin *et al.*, 1995; see Chapter 4). We studied behavior of 1500 cattle while restrained in a squeeze chute during routine handling for husbandry procedures. We rated the behavior on a scale from one to four (a cow rated "one" was calm, with no movement; a cow rated "four" was violently and continuously shaking the squeeze chute). Many of the cattle given a rating of four reared up in the squeeze chute, vocalized, and violently threw their heads during restraint. Cattle with hair whorls above the eyes were significantly more agitated during restraint when compared to cattle with middle or low hair whorls. The next study revealed an association between positions of the hair whorls (e.g., left, right) and milking parlor side preference (handedness) in dairy cows (Tanner *et al.*, 1994). For this study we chose Holstein dairy cows because large numbers of cows could be monitored for milking parlor side preference with a computerized identification system. All cows were Holsteins and handling experiences were the same. Hair whorl positions and personal identification tag numbers were recorded on 1379 cows. Side preference data was collected for 90 milkings each. At the entrance to the parlor, cows had free choice to enter stalls on the left or right side. (Handlers only force cows to one side if all the stalls on the other side are full.) Electronic ID readers recorded the stall that each cow entered. Results showed that 45 cows with two hair whorls on the forehead were significantly less sided than the rest of the population. These two studies confirmed the anecdotal observations from the first author. Animals with high hair whorls are more reactive, and animals with two hair whorls show less side preference. Randle (1998) assessed the relationship between hair whorl position in 57 *Bos tauras* cattle and responses to humans. The results show the general response of cattle to a familiar human was not associated with hair whorl position, but the general response of cattle to an *unfamiliar* human was. Lanier *et al.* (1999) studied facial hair whorls and temperament in 1636 commercial cattle in the auction ring. Cattle with high hair whorls had higher temperament scores. Temperament scores also showed that Holsteins were calmer than beef cattle. Matson (2004) studied the relationship between hair whorl position and behavior in 146 crossbred heifers. One-half of the heifers were used as controls, and the other half were handled for 2 hours per week for 20 weeks. The handling consisted of walking through a chute, moving in a corral, sorting, and moving through a chute. Behavior of both control and handled heifers were scored in a squeeze chute, during isolation in a pen, and exit time from the isolation pen. Behavior measures were recorded in the beginning of the experiment, at week 10, and at 20 weeks. Results showed that handling treatment decreased in-chute behavior scores of heifers with facial hair whorl positions classified as medium or low, but not in heifers that exhibited a hair whorl high on their face. Broucek *et al.* (2007) investigated associations between hair whorl position and behavior in open-field and maze tests using Holstein heifers. The results did not find any significant differences in hair whorl positions in either test, suggesting an interaction between animals selected for a calm temperament and regular human contact. Holsteins in this study were a more homogenous group compared to the extensively raised (semi-wild) cattle used by Grandin *et al.* (1995). Broucek *et al.* (2007) found the differences among groups with high, middle, and low whorls were not significant, however, heifers with a whorl above the eyes showed more intensive movement in the open-field test and rapid learning and memory in the maze. More recently, Martins *et al.* (2009) found a high percentage of agitated cattle with hair whorls above the eyes compared to cattle with whorls between the eyes, or cattle without whorls. Behavioral agitation was scored in a squeeze chute and the flight speed was measured after exiting the squeeze chute. Silveria *et al.* (2006) evaluated auction ring behavior of 1572 weaned calves with an average age of 210 days. Thirteen-hundred and seventeen calves were crosses between *Bos Indicus* (Brahman) and *Bos Taurus* (Angus, Red Angus, Hereford, and Charlolais), and 255 were pure *Bos Taurus*. The auction ring is similar to tests of social isolation in a novel arena, but accompanied by sounds and movements of both the auctioneer and the public. Results showed that crossbred *Bos* Indicus × *Bos Taurus* cattle had higher hair whorls than pure *Bos Taurus* cattle, and behavioral agitation in the auction ring (escape attempts, restlessness, fast movements) was significantly higher in crossbred cattle compared to the purebred cattle. Florcke *et al.* (2012) studied variability in hair whorl patterns and differences in maternal protective behavior of beef cow–calf pairs 24 hours after calving. Results show that cows with a single high hair whorl or multiple hair whorls oriented toward an approaching vehicle at a further distance, compared to cows in other hair whorl groups. Also, cows with a high hair whorl vocalized at a further distance compared to cows in other groups. These results show that older cows and cows with high hair whorls are more vigilant of their surroundings. Shirley *et al.* (2006) studied heritability of hair whorls in Holstein cattle and reported three main conclusions: hair whorl height, asymmetry, and rotation are moderately to highly heritable polygenic traits; whorl height and asymmetry are the same traits regardless of whorl direction; and different genes are involved in determination of each of these three characteristics.

Hair Whorls in Horses

In 1994, we conducted two preliminary observations on associations between hair whorl patterns and performance in horses. Hair whorl data was collected from 290 Thoroughbred horses during the racing season at Santa Anita Racetrack in Arcadia, California. Performance data was provided by The Daily Racing Form, a statistical service for bettors on horse racing in the United States. Statistical information on starts, wins, and money earnings were compared to hair whorl patterns. The horses were all registered Thoroughbreds, stabled at the track in 11 separate barns and managed by 26 different trainers, each with their own training, handling, and nutritional practices. No statistical correlations were found. In a second study, hair whorls were recorded on 133 Grand Prix jumping horses during the Masters at Spruce Meadows, Calgary, Alberta, Canada. Horses were various breeds from 15 of the world's top show jumping nations. Each horse was managed, handled, and trained with various training preferences. Performance data and rankings were provided by the Fédération Equestre Internationale (FEI), the international governing body for all Olympic equestrian disciplines. No significant correlations between hair whorl position and performance were found in this study either. However, some interesting trends emerged, and of particular interest, two of the biggest money earners in each sample had double hair whorls. One of these was a horse named Big Ben. He was regarded as one of the greatest show jumping horses of all time. Inducted into the Canadian Sports Hall of Fame in 1999, he held two World Cup titles and two Masters Grand Prix wins. Big Ben was ridden by Ian Millar, one of the most successful competitors in the history of Canadian show jumping. Miller once described the gelding as "a skittish elephant," referring to his size and nervous temperament (Equestrian College Advisor, 2012). In the race horse sample, the horse with the most wins and money earnings was a 6-year-old gelding named Sir Beaufort. At the time, his earnings were $1.15 million, and in 1993, he was rated the number one older handicap horse in the nation. His trainer, Charles Whittingham, is one of the most acclaimed trainers in U.S. racing history. Similar in temperament to Big Ben, Sir Beaufort had a slow start in his racing career because of his difficult nature. According to his owner, Victoria Calantoni, "He was a quirky horse. For a long time, he had trouble loading into the [starting] gate. They had to load him twice in the Santa Anita Handicap because his jockey broke a bridle. Besides that, he was always just a little headstrong. He doesn't like the whip a whole lot; if the jockey fought him, he just got worse, not better" (Meixel, 1993). These two horses illustrate the sometimes difficult nature of horses with two whorls observed by the first author. Horses with double or elongated whorls were also characterized by Tellington-Jones and Taylor (1995) as "unpredictable" or "over-reactive.

Increase of Double Whorls in Racing and Jumping Horses

In the sample of racing and jumping horses, we found 16% (68 horses) with two whorls (Swinker *et al.*, 1994). This percentage is considerably higher than the 4.4% (16 horses) in a sample of 362 Konik horses reported by Gorecka-Bruzda *et al.* (2006), and higher than the 10.9% double whorls (six horses) in a second sample of 55 Konik horses (Gorecka-Bruzda *et al.*, 2007). Grandin *et al.* (1995) found 5% double whorls in 1500 cattle; Randle (1998) reported one double whorl out of 57 cows; Tanner *et al.*(1994) found 3.2% double whorls in 1379 dairy cows; Smith and Gong (1973) reported 7% double whorls in 200 normal children; and Lauterbach and Knight (1927) found 1.4% double whorls in 1003 normal children, and 6.5% double whorls in 751 children with developmental disabilities. It is intriguing that the percentage of double whorls in racing and jumping horses is nearly double the percentages found in all other samples cited in the literature.

Hair Whorl Height and Reactivity in Horses

Gorecka-Bruzda *et al.* (2007) published the first scientific evidence on the relationship between hair whorl position and reactivity in horses. The study involved two groups of young Konik horses reared under conventional stable conditions (24 horses), or in a forest reserve under semi-natural conditions (31 horses). The horses were grouped into four classifications based on their hair whorl position and/or shape: a single whorl high above the top of the eye line; a single whorl between the top and the bottom of the eye line; a single whorl below the bottom of the eye line; and four horses had an elongated or double whorl (Figure 7.4). All of the horses were tested only once in three behavioral tests: being led away by a familiar handler (first moving away from the stable, then toward the stable); leg lifting; and a novel object test. In the novel object test, a familiar handler restrained each horse while another familiar handler approached and suddenly opened an umbrella from a distance of 5–6 m. Heart rate was monitored telemetrically during testing. The results revealed that horses with a high hair whorl were significantly more difficult to handle than horses with a medium or low hair whorl, but no significant differences were observed in cardiac reactivity. Horses with an elongated or double hair whorl were the most cautious of the groups when approaching a strange object. They took much longer than the single-whorled horses to approach and touch something new. The lack of a correlation between heart-rate measurements, startle responses, and hair whorl height in this study may be explained by the fact that Konik horses are known for a calm and gentle temperament (PetMd, 2012). This is similar to the lack of correlation between hair whorl height and temperament in

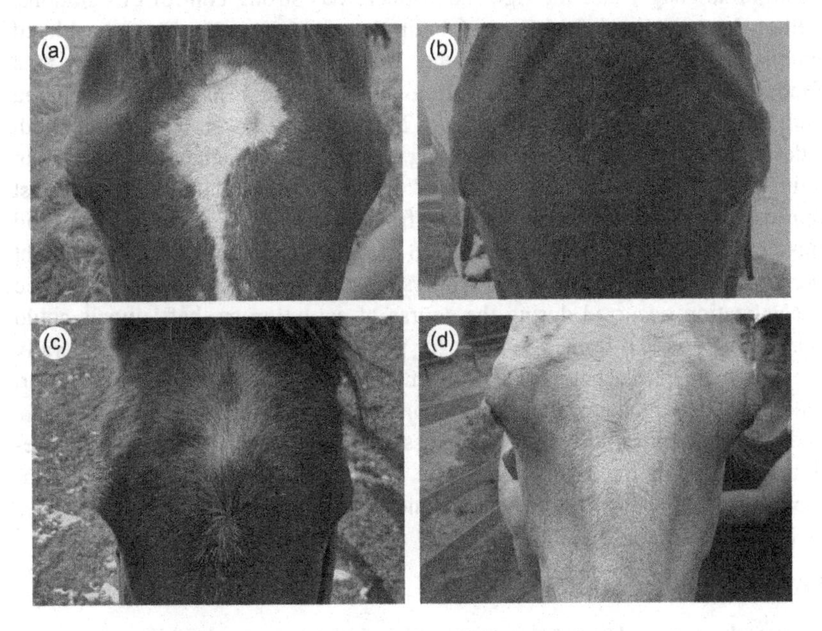

FIGURE 7.4 Classification of hairwhorl positions on the forehead of the horse. (a) High right, (b) High left, (c) Middle, (d) Low.

Holstein cattle reported by Broucek *et al.* (2007). (Holstein cattle are also known for their calm and gentle nature.) Gorecka-Bruzda *et al.* (2007) also found that hair whorl position in Konik horses is highly heritable. The estimates are similar to the heritability estimates of hair whorls in Holstein cattle (Shirley *et al.*, 2006). It would be interesting to repeat the study of Gorecka-Bruzda *et al.* (2006) using breeds known to have highly reactive temperaments, such as Thoroughbreds or Arabians. It is clear that cattle producers, horse breeders, horse trainers, and horse owners could benefit from further studies of hair whorl position and rotation and its relationship to temperament. The evidence presented supports much of the anecdotal evidence collected by the first author. Further evidence of a relationship between hair whorls and side preference is presented in the next section.

LATERALIZATION IN THE NERVOUS SYSTEM

Humans and animals have two types of laterality. One is motor laterality, the other is sensory laterality. The most common examples of motor laterality are right- and left-handedness in humans and side or limb preference in animals. Humans throw a ball or write better with either the right or left hand. Animals prefer to start walking using either the right or left foot, or the right or left paw to hold or manipulate objects. Sensory laterality refers to specialized tasks of both brain hemispheres. In most humans, the left hemisphere controls language, and the right hemisphere has strong control of emotions. Although specialized functions of brain hemispheres were long considered unique to humans, increasing evidence points to behavioral asymmetry as a fundamental principle of organization in vertebrate behavior (Vallortigara and Rogers, 2005; McGreevy *et al.*, 2007). At the most basic level, both sides of vertebrate brains show strong and consistent behavior patterns. In animals as diverse as chickens, fish, lizards, rodents, and humans, most *approach* behaviors (finding food and friends) are left hemisphere tasks, and most *avoidance* behaviors (responses to predators and novel stimuli) are right hemisphere tasks. According to Rogers (2010), complimentary hemispheric specialization occurred during the course of evolution so that animals could search for food and remain vigilant for predators at the same time. Generally the left hemisphere categorizes information and controls routine behavior, whereas the right hemisphere is responsible for responding to novel events, and coordinating rapid responses. Stressed or frightened animals rely on predominant use of the right hemisphere (Rogers, 2010). Genes and environment influence lateralization, but the interaction between them is not always clear (Schaafsma *et al.*, 2009).

LATERALITY IN HORSES

Horses show a preference to stand with one foot forward while grazing, to use either the right or left foreleg to initiate a walk or a trot, to roll on their right or left side, and to gallop on the right or left lead (McGreevy and Thompson, 2006; McGreevy and Rogers, 2005; Murphy *et al.*, 2005; Wells and Blache, 2008; Williams and Norris, 2007). Training horses to work equally well in either direction is important in all equitation disciplines (Manroe, 1992; Maringer, 1983; Sivewright, 1986). Under saddle, horses tend to bend easily to one side and with more difficulty to the other side (Clayton, 1990; Murphy *et al.*, 2005). Horses have long been considered asymmetrical (Steinbrecht, 1886), and many training exercises are aimed at straightening the horse.

Motor Asymmetry

Whether side preference is a result of genetic predisposition or if it results from training and use was unknown until recently. Grizmek (1968) looked into motor asymmetry and found that most horses begin to gallop with the left forelimb first, and that they also turn left more easily. McGreevy and Thompson (2006) found differences in motor laterality between breeds of performance horses. The more temperamental Thoroughbreds showed a higher left foreleg preference during grazing when compared to Standardbred and Quarter Horses. Little or no side preference was shown in Quarter Horses. McGreevy and Rogers (2005) noted the relative position of the horses' forelegs every 60 s for 2 hours in Thoroughbreds grazing on pasture. Results show a population bias of standing with the left foreleg advanced. Forty-three out of 106 horses demonstrated a significant left leg preference, 10 horses demonstrated a significant right leg preference, and 53 horses were ambidextrous in their forelimb preference. In addition, motor laterality bias increased with age, suggesting maturation or influence of training. In contrast to the left foreleg bias observed by McGreevy and Rogers (2005), Murphy *et al.* (2005), found that 52% of horses preferred to start walking or trotting with the right foreleg, while 40% preferred the left foreleg, and 7.5% showed no preference. The horses is this study were of mixed breeds (predominantly Thoroughbred × Irish Draught). The findings show strong sex-related *directions* of laterality, but not *degrees* of laterality. Male horses exhibited significantly more left-lateralized responses and female horses exhibited significantly more right-lateralized responses. Williams and Norris (2007) studied stride lead preferences in racing Arabian, Thoroughbred, and Quarter horse and found that 90% of ridden racehorses show a right lead stride bias. Wells and Blache (2008) studied both laterality in grazing stance and laterality in lunging circle tests in 15 ridden horses (5–20 years of age), and 15 horses (1.5–3 years of age) that had not been ridden. The breeds consisted of 60% Thoroughbred, 2% Arabian, 28% Warmblood, and 10% Riding Ponies.

The results showed that un-ridden horses did not have a grazing stance bias, but the ridden horses showed a right forelimb bias. The lunging test measured three indicators of bias; time spent in canter, correct lead in canter, and disunited canter. A disunited canter was when a horse began a canter on the correct lead, but then started to canter with an incorrect sequence of leads. In both the ridden, and un-ridden horses there was no correlation between grazing stance (relaxed) and any of the lunging indicators (challenged).

One possible reason for the conflicting results of motor bias is that different tests measure different characteristics of motor bias. Grazing stance, initiating a walk or trot, and rolling on the ground are biases exhibited while horses are in a relaxed state. Lunging and galloping biases are exhibited during a stressed (physical or psychological) state. Different conclusions about motor bias may be explained by test conditions, breed and temperamental characteristics, and selective pressures for a particular type of work. McGreevy and Thompson (2006) suggest that some breeds may have inadvertently been selected for laterality, or ambilaterality, due to the nature of the work they perform. It is not known if this is possible in horses, but selective breeding for laterality is routinely done in laboratory mice (Collins, 1975).

Sensory Laterality

Sensory laterality refers to the preferred use of one eye, ear, or nostril over the other under particular circumstances. Science attributes sensory laterality to the processing of different types of information in one or other of the two brain hemispheres (Andrew, 2002; Rogers, 2009, 2010; Villortigara and Rogers, 2005). Rogers (2009) and Proops and McComb (2012) suggest that sensory laterality evolved before motor laterality. Evidence to support this is found in fish without limbs that show lateralized eye preferences to view certain scenes (Biazza *et al.*, 1997), and prey catching on their right side (Takeuchi and Hori, 2008). Scientists that once believed that hemispheric asymmetry is unique in humans now embrace the evidence for lateralization in a wide variety of species (summarized by Rogers, 2009, 2010; Vallortigara and Rogers, 2005).

Until recently, there was no evidence of sensory laterality in horses. Heird and Deesing (1998) were the first to discuss sensory laterality and noticed when horses are suddenly startled or frightened by novel stimulus they focus one eye on the object and make lateral body movements around it. Over the last decade, sensory laterality in horses has attracted considerable attention (Austin and Rogers, 2007; Laroseet *al.*, 2006; McGreevy and Rogers, 2005; Williams and Norris, 2007). Austin and Rogers (2007) and Murphy *et al.* (2005) recorded flight and escape-turning responses in 30 horses of various breeds exposed to a novel object and recorded greater fear responses to fear-inducing stimuli presented to horses' left eyes, and therefore the brains' right hemispheres. In this study, an experimenter opened an umbrella 5 m away from the horse, first to the horse's left side and then to the right side, or *vice versa*. Horses tested first on the left side showed greater reactivity and flight responses for left side approach, whereas horses tested first on the right side displayed less reactivity and shorter flight distance when subsequently approached on the left side. The evidence suggests that stimuli first presented to the right side allowed the horses to learn that the stimulus posed no threat. This information then transferred to the right hemisphere of the horses' brains, leading to a lessened reaction when approached with the umbrella from the left side. Larose *et al.* (2006) studied visual asymmetry during exposure to novel objects in Trotters and French Saddlebreds. The temperaments of the two breeds were similar, but expressions of visual laterality differed between individual horses. Horses with higher emotional scores assessed during behavior tests glanced at a novel object with the left eye (right hemisphere) more than the right eye (left hemisphere). This evidence shows that laterality is stronger in more reactive (fearful) animals. In addition, stimuli that animals fear the most, such as sudden rapid moving objects, or objects that animals have been conditioned to fear influence laterality. De Boyer des Roches *et al.* (2008) found that emotionally laden objects influence lateral responses. Horses showed a slight tendency to use the left eye (right hemisphere) to explore objects with negative emotional value (a veterinarian's white shirt) more than objects with neutral emotional value (a plastic cone) or positive value (a feed bucket). The age of a horse can also influence laterality. Sankey *et al.* (2011) found that yearling horses approached on the left side showed more escape and threat (avoidance) behaviors, as opposed to more positive (approach) behaviors when approached from the right side. Two-year-old horses, on the other hand, reacted positively to approach and contact from both sides. These results show the influence of work and training on laterality in horses. It has been argued by some that asymmetries in horses are caused by humans traditionally handling horses on the left side. In Trotter horses that are, to an extent, handled on both sides, Larouse *et al.* (2006) found no association between eye preference and reactivity. However, field studies of feral horses by Austin and Rogers (2007) shows that lateralization is a characteristic of horses as a species that is not caused by human handling. Observations revealed a left-side bias of reactivity (heads held high and looks to the left or right caused by the presence of humans), a left-side bias in vigilance (raising the head from a grazing position and turning the head to attend to stimuli), and a left-eye bias during aggressive exchanges (eye used by stallion to look at its opponent). The left-eye bias was stronger during attacks than during threats. There was no observed population bias of foreleg preference in grazing stance in feral horses, compared to population biases in domestic horses. The absence of population-level forelimb preference in feral horses suggests that training and work influence forelimb preference in domestic horses (Austin and Rogers, 2007), and that limb preference does not indicate brain lateralization, but may better reflect a horse's attention to its environment.

Auditory and Olfactory Laterality

Basile *et al.* (2009) reported the first evidence of socially dependent auditory laterality in horses. Head and ear orientation responses to whinnies recorded from both familiar and strange horses showed that horses turn their heads to the right (right ear orientation) toward recorded whinnies from familiar horses, whereas no bias was found during calls from unfamiliar horses.

McGreevy and Rogers (2005) investigated olfactory laterality in Thoroughbred horses (32 males, 125 females) less than 4 years old. Horses were presented with olfactory stimuli (stallion feces) and a right nostril preference was found in younger, but not in the older horses.

These studies clearly show that the right and left sides of horses' brains are specialized to handle information in different ways, and to control different classes of behavior. In addition, varying strengths of lateralization have been found between breeds, and in individuals within a breed (De Boyer des Roches *et al.*, 2008; McGreevy and Rogers, 2005; McGreevy and Thompson, 2006; Sankey *et al.*, 2011). Specialization of hemispheric control of behavior during stressful and non-stressful conditions is obvious, and has important implications for horse welfare (Rogers, 2010). In a relaxed state, most horses use the right eye (left hemisphere) to process information. In a stressed state, dominant use of the left eye (right hemisphere) may compromise welfare. This is especially important considering that the right hemisphere also controls endocrine function, heart rate, and blood pressure (Conrad *et al.*, 2011; Wittling and Pfluger, 1990; Wittling *et al.*, 1998). Recent data also suggest that the right and left cerebral hemispheres differ in their ability to regulate autonomic processes, and direct unilateral stimulation of the brain provokes side-dependent endocrine, immune and other visceral reactions. The study of laterality and fearfulness can be used as a new approach to investigate the processes that underlie the inter-individual variability of the vulnerability to stress and stress related diseases.

Hemispheric Dominance

Population asymmetries, such as right-handedness in humans, show consistent directions. Reversed or absent asymmetries occur in some individuals (Corballis, 2009). Ghirlanda *et al.* (2009) note population-level handedness reversals ranging from about 10% in humans to about 35% in chimpanzees (Corballis, 2009). In 326 normal humans, Knecht *et al.* (2000) measured lateralization directly by functional transcranial Doppler sonography and found the incidence of right-hemisphere language ranged from 4% in strong right-handers to 15% in ambidextrous individuals and 27% in strong left-handers. Knecht *et al.* (2000) suggest that these results demonstrate that the relationship between handedness and language dominance is not an artifact of cerebral pathology but a natural phenomenon. Discussing reversals in horses interacting with humans, Farmer *et al.* (2010) writes, "Intriguingly, the right-eyed horses in both groups, although only a very small sample, showed the same trend with regard to the right eye, that is to say they showed stronger lateralization under the stranger (response to an unfamiliar human) condition."

In the studies reviewed in the previous section, the few horses in each sample showing a reverse of left-eye (right hemisphere) processing of emotional information have received little attention. It is interesting to speculate that reversed hemispheric dominance might reflect a distinct cognitive disposition, or bias. An association between cognitive bias, handedness in humans, and limb preference in animals supports this view (Rogers, 2010). Cognitive bias refers to the particular way the brain perceives, forms memories, and makes judgments. Denny (2009) found that left-handed people report more depression symptoms than right-handed people. In chimpanzees, Hopkins and Bennett (1994) found that right-handed chimps interact more with novel objects than left-handed individuals. More evidence suggests that preferred use of the left paw (and right hemisphere) is characteristic of more fearful marmosets (Cameron and Rogers 1999). Rogers (2009) also found that left-handed marmosets are more likely to have higher levels of the stress hormone cortisol when compared with right-handed individuals. Rogers (2010) also suggests that right-handed marmosets are active, and left-handers are reactive. In marmosets, handedness predicts a tendency to adopt a positive or negative cognitive bias.

In horses, no studies exist that specifically address differences in left-eye system *versus* right-eye system dominance and cognitive bias. It may be possible that right-hemisphere-dominant and left-hemisphere-dominant horses respond differently in similar circumstances. Anecdotal evidence collected by the first author suggest that horses with *high* left- or *high* right-side hair whorls respond differently to either stationary or sudden novelty. Horses with *high* right-side whorls display higher levels of arousal such as prancing, vigilance, whinnying, snorting, and tail raising, compared to horses with *high* left-side whorls. To illustrate this point, the first author has two Warmblood mares; Bobbi has a *high* left-side hair whorl, and Sam has a *high* right-side hair whorl. The mares are both 6 years old and share the same sire, but different mothers. The first author bred, raised, and trained both mares himself. Sam and Bobbi are both reactive to sudden novelty and social separation; however, Sam (right-side whorl) vocalizes when separated and makes more attempts to return to the herd. Sam also nickers and whinnies more than Bobbi at feeding time. Emotional responses to positive and negative stimuli between the two are obvious. Recently, we tested both mares on flight and escape-turning responses using the method described by Austin and Rogers (2007). Placing an oat bucket in the center of a large paddock, the first author released the mares one at a time and allowed them to eat from the bucket for 1–2 minutes. He positioned himself in front of each horse to make sure they were looking at him using both eyes, then quickly walked toward each horse while suddenly opening an umbrella. Sam quickly turned and escaped to the left, Bobbi quickly turned and escaped to the right. There was no difference in the immediate response, and both horses ran about 10 m before stopping and turning to look back. After opening the umbrella in front of each horse, the first author moved out of the paddock and noted the time it took for each horse to return to the feed bucket. Bobbi returned to the bucket and resumed feeding within 3 minutes. After 10 minutes Sam had not returned to feed, and for roughly 5 minutes after the umbrella, she snorted, displayed a raised tail, and pranced. Horses snort when startled, in pain, or when frightened (McGreevy, 2004). Prancing, vigilance, and a raised tail are attributed to high levels of emotional arousal (Waring, 1983; Wolff *et al.*, 1997). None of these signs of emotional arousal were shown by Bobbi, and in the following days behavior towards the first author was unchanged. However, Sam was reluctant to approach the first author for several days following this experience. This suggests a strong negative memory of the experience associated with the first author. The escape turning preference of each horse is also the same as their turning preference when lunged in a circle, but only when the horses are nervous. For example, when they have not been lunged for weeks, the first time lunging after time off they get slightly nervous and directional preference is expressed. However, after a few minutes on the lunge line they relax and directional preference disappears.

This preliminary experiment is merely anecdotal, but a controlled experiment conducted using right- and left-side hair whorl horses may provide evidence of cognitive bias. The tests most likely to cause emotional reactions are exposure to sudden novelty, or social isolation in a novel arena. It would be interesting to study heart-rate variables and stress hormone levels following response to novelty in horses with right- or left-side hair whorls.

Hair Whorl Patterns and Lateralization

Murphy and Arkin (2008) found a significant association between hair whorl rotation (direction) and motor asymmetry in horses. The study recorded rotation of single hair whorls in 219 horses (males = 125, females = 94), between 4 and 6 years old. The distribution of whorl patterns were 114 counter-clockwise, 82 clockwise, and 23 radial (no rotation). Experienced trainers rode and jumped each horse to determine motor laterality. Results showed that right-lateralized horses had significantly more clockwise hair whorls, and left-lateralized horses had more counter-clockwise hair whorls. More recently, Ovando (2010) found no association between hair whorl rotation and motor asymmetry. The study recorded rotation of single whorls in 50 Thoroughbred foals (males = 29, females = 21) all under 1 year of age. The distribution of whorl patterns was 16 counter-clockwise, 10 clockwise, and

24 radial (no rotation). The foals were tested 50 times each using the first step described by Murphy *et al.* (2005). Results showed no association between hair whorl rotation and asymmetry, but did find that males were more left-lateralized, and foals with radial whorls were more balanced (ambidextrous). These results contradict those obtained by Murphy *et al.* (2005), and the difference in results may be caused by the difference in the age groups. The yearling foals studied by Ovando (2010) were all the same age and received halter training but otherwise minimal contact with humans. The horses studied by Murphy *et al.* (2005) were all trained to ride. This finding is in agreement with McGreevy and Roger (2005) who found that horses under 2 years of age were less lateralized, compared with a higher percentage of left lateralized horses over 2 years old. It may be that the "motor" laterality observed in ridden horses is actually motor asymmetry influenced by sensory laterality influenced by stress induced by training and riding. It is recognized that the amount of psychological stress (fear) that an animal encounters determines the degree of response of the hypothalamic–pituitary–adrenal (HPA) axis (Cayado *et al.,* 2006). Blood samples collected in Dressage and Jumping horses at their familiar home environment were compared to samples taken before exercise, at the entrance to a schooling area, and post performance over jumps or a dressage course. Plasma cortisol and plasma ACTH were highest in the least experienced horses, high in horses with intermediate experience, and slightly lower in the most experienced horses (Cayado *et al.*, 2006). The fear responses in the most experienced horses were attenuated, but still evident.

The inconclusive results of hair whorl direction in horses and asymmetry are similar to the inconclusive results of studies of hair whorl rotation and handedness in humans (Beaton and Mellor, 2007: Jansen *et al.*, 2007; Klar 2003; Weber *et al.*, 2006). In addition, all these studies only focus on whorl rotation (e.g., clockwise, counter-clockwise) and laterality—not whorl *placement* and laterality. Anecdotal evidence presented by the first author suggests that hair whorl position (e.g., left or right of midline) is a more reliable predictor of laterality in horses. Support for this anecdotal evidence was recently reported by Tomkins *et al.* (2012) in dogs. Specifically, the location of a whorl on the left side of a dog's head and neck was associated with a right visual bias, and a right visual bias was also probable if the hair whorl on the ventral mandibular (back of the bottom jaw) was present in a counter-clockwise direction. The study assessed all three measures of laterality (hair whorl, motor, and sensory) in Labrador retrievers, golden retrievers, and Labrador–golden retriever crosses. The dogs were prospective guide dogs for the blind provided by Guide Dogs NSW/ACT. Dogs' performances on the Kong™ test, the First-Stepping Test, and the Sensory Jump Test determined motor and sensory laterality. Several associations emerged between hair whorls and motor and sensory laterality. The authors suggest that hair whorls may provide a quick and efficient means of determining suitability in prospective guide dogs because a need to work on the left side of trainers often disqualifies dogs that are less flexible when turning right. It also takes a significant amount of time to train guide dogs. The dogs must learn several complex tasks, and it is important to find a quick and efficient means of discovering traits in the early stages of training that decide whether a dog will be a successful guide dog.

Clearly, further research is needed to determine associations between hair whorl direction and laterality in horses. Lateralization is important because it is related to locomotion and subsequent performance during training. Determining lateral trends at a glance could be very useful for designing individual training programs.

Chronic Asymmetry

In his experiences as a horse trainer, the first author has faced several one-sided horses. In some horses, the turning preference was so strong that any attempt to change it caused the horse to prance, throw its head, or rear. Any use of force to change a preference was likely to cause a panic-like response. To our knowledge, no scientific research into the cause of extreme laterality in horses exists; however, various explanations of extreme one-sidedness exist in the popular horse-training literature. The most often-cited reasons are strong motor asymmetry or pain. Lateralized behavior is undesirable in various equitation disciplines, and persistent lateralized behavior has not been thoroughly investigated. Scientists and industry professionals rarely discuss fear as a possible cause of extreme one-sidedness. All the same, evidence of brain asymmetries previously discussed point to many lateral behaviors associated with fear responses. Fearfulness may be more common in horses than is generally assumed. Horses do not require many fearful experiences for fear behavior to be habitually expressed, and many chronic fear behaviors in animals and humans can be traced back to a single event (Lindsay, 2000). One-trial-learning of fearful responses is documented in horses (Kiley-Worthington, 1987). Fear responses are learned at the subcortical level involving the amygdala and hippocampus (LeDoux, 1998), with direct projections to the motor cortex. Lateral behavior and hair whorl side-position has been demonstrated in cattle (Tanner *et al.*, 1994), and fear responses are associated with hair whorl height-position in cattle (Grandin *et al.*, 1995; Silveira Dias Barbosa *et al.*, 2006), and horses (Gorecka-Bruzda *et al.*, 2007). Although hair whorl side-position and lateral fear responses have yet to be demonstrated in horses, anecdotal evidence collected by the first author suggests an association between lateralized fear responses and the side position of hair whorls in such things as handling feet, lunging in circles, riding, reining, trailer loading, and jumping. Left-side hair whorls predict left-side resistance to horseshoeing, and predict a preferred rightward direction of travel when working in circles. In many cases, these lateral behaviors are associated with obvious fear responses, in others the fear responses may become attenuated over time (Cayado *et al.*, 2006), and can easily be mistaken for motor asymmetry or disobedience. For example, young horses trained in a round pen on a lunge line often travel better in either a clockwise, or counter-clockwise direction. If horses have a left-side hair whorl, the preferred direction is clockwise. In the counter-clockwise direction, the horse makes smaller circles and tries to turn its head to the outside. The head is often held high, the speed of travel is increased, and the tail is held high. These are all signs of emotional arousal (McGreevy, 2004; Visser *et al.*, 2001; Waring, 1983; Wolff *et al.*, 1997). Older horses show the same signs of arousal but to a lesser degree. As we discussed earlier, horses with left-side whorls resist horseshoeing on the left side. The resistance may involve attempts to flee, or refusal to allow a leg to be touched. In other instances, the resistance may be subtle and horses stand and tolerate the leg being lifted, but lean on the farrier or try repeatedly to pull the foot and leg away. These are not clear and obvious signs of fear, but may in fact be attenuated fear responses. Leaning on a farrier may be a defense reaction motivated by fear, and more evident on one side. The "assumed" motor asymmetries shown by horses in ridden work (Murphy and Arkin, 2008) may in fact be attenuated lateralized fear responses (Cayado *et al.*, 2006). In early stages of training for riding, fear may motivate horses to prefer to work in a rightward or leftward direction. Horses with left-eye (right hemisphere) control of fearful stimulus may be less fearful when working in a rightward direction. This direction allows the horse to keep the left eye to the outside of a circle where it feels more confident to defend itself against potential threats. Strong one-sidedness motivated by fear in young horses during the early stages of training may be attenuated with experience and mistaken for motor asymmetry.

Research in humans and animals show that trauma experienced early in life, or at any time in life can cause shifts in cognitive bias toward the right hemisphere leading to prolonged anxiety-like behavior (Chamberlain and Sahakian, 2007; Mathews and McLeod, 2005). Adamec *et al.*, (2005) exposed rats to a cat and found a shift to the right hemisphere, and subsequent changes in neural activity and neurotransmitter levels lasting up to 12 days after this stressful event.

Fear stress is highly aversive and subjecting animals to extreme fear stress is very detrimental to welfare (Grandin, 1997).

Experience-Dependent Lateralized Learning

In a previous section, the first author discussed his approach to horses that are difficult to shoe on one side or the other. He found that switching sides and shoeing the "less difficult" side first was helpful. When he finished the less difficult side, horses were often more cooperative when he returned to the difficult side. This may be an example of experience-dependent lateralized learning. Information processed and learned in one hemisphere of the brain transfers to the opposite hemisphere in an experience-dependent manner. Studies report interhemispheric transfer of memory in species as diverse as fish, amphibians, reptiles, birds, and mammals (Robbins and Rogers, 2006; Vallortigara and Rogers, 2005). In cattle, Robbins and Phillips (2010) report the first example of experience-dependent lateralized learning in a mammal. In herd-splitting experiments conducted on dairy and beef cattle, a single experimenter walked repeatedly through herds, either empty-handed or while carrying various novel objects. At first, the cattle viewed the experimenter using their left eyes (right hemisphere). After habituating to the experimenter carrying various objects, the cattle then switched to viewing the experimenter with the right eye (left hemisphere). This directional shift in viewing preference was experience-dependent. Robbins and Phillips (2010) suggest that the reversal of preference from the left eye (when the stimuli is novel) to the right eye (when the stimuli becomes familiar) "frees up" the left eye to simultaneously watch for new threats.

Switching Sides

There is no known experimental evidence of experience-dependent lateralized learning in horses. However, it may be safe to assume that this well-conserved pattern of lateralized learning exists in all mammals (Robbins and Phillips, 2010). During horse training, the first author switches sides when a horse shows a lateralized response. He also uses hair whorl positions (i.e., left or right side of the forehead) to predict a lateralized response before it occurs. For example: if a left whorl horse shows resistance to having a saddle placed on its back from the left side, switching to a right-side presentation speeds habituation. Switching sides can also help when training a horse to lunge in a circle. For example, the first author starts a horse in a circle in the preferred direction, then switches to the non-preferred direction. When the horse shows signs of fear (head high, head turned to the outside, etc.), he turns the back to the preferred direction. Never forcing the horse to continuously travel for long in the non-preferred direction over time has a calming effect. Our hypothesis is that a majority of lateralized behaviors observed during riding, or during ground handling, are motivated by fear.

In summary, preventing lateral behavior before it occurs is easier than trying to correct it once it's been conditioned. Experience-dependent lateralized learning may be the way horses learn about and come to accept handling and training by humans. They first look at novelty using the eye system geared to detect possible danger. When no danger is perceived, the opposite eye system takes over and learning progress without fear inhibiting learning. The non-intrusive neonatal foal handling method used by the first author allows foals to learn in the way they are meant to learn. Foals look at him several times from both sides before approaching and touching him for the first time. By allowing the foal to learn this way, the first author observes reduced or almost non-existent lateralized responses to handling both at this early stage of training, and in subsequent training conducted as the horse matures.

CONCLUSION

The field of behavioral genetics has the potential to uncover both genetic and environmental influences on normal and abnormal behavior (Hausberger *et al.*, 2011; Zentner and Shiner, 2012). Behavioral–genetic methods are based on a foundation of methods that are beginning to explain mechanisms of complex traits. As specific genes are identified, researchers can begin to explore how these interact with environmental factors in development. Only a few decades ago, equine scientists believed that most characteristics of horse behavior were almost entirely the result of environmental influences. Many of these characteristics are now known to be genetically influenced, in many cases to a substantial degree. Learning ability and memory, novelty seeking, activity level, fearfulness and sociability all show some degree of genetic influence. Behavioral geneticists are interested in understanding differences among individuals. Some genetic factors may cause small differences, and others may cause large differences. Such differences may be caused by environmental factors and/or by one or many genes. Studying behavior in this way is especially important when selecting horses for breeding. Understanding how and why behaviors are developed and learned can play a large part in the breeding program. Inheritance of temperament is high, but in comparison to other domestic animals, selection for temperament in horses is less common. Breeding stallions with good performance traits but bad temperament traits is not always a good idea. A difficult temperament can hinder performance. Balanced selection for performance traits, conformation traits, and temperament traits is important. A large body of evidence shows that selection for any single trait has unwanted effects on other traits (see Chapter 1).

Environmental factors can affect fetal brain development; therefore, reducing stress in pregnant mares is important. Evidence showing psychological stress (fear stress) in mothers during gestation has effects on fetal brain development occurring mainly through changes in hormone profiles. Stress hormones enter the fetal bloodstream and change brain structure and function leading to increased sensitivity to stress. When fetal brain circuits are highly plastic, external influences are more pervasive.

Forced contact following birth affects emotional development, causing insecure attachment to the mare, and disrupts normal social behavioral development with peers. Forced contact can also shape how foals perceive humans. Perception of humans as "potentially threatening" may last a lifetime in reactive and fearful foals. Research shows that foals respond differently to intensive handling. Genetic differences in temperament may determine how "threatening" humans are perceived. Even short periods of post-birth handling have long-term effects. Research shows that even brief forced contact in newborn foals (assisting foals to stand and nurse), may cause foals to avoid future contact with humans. The pre- and post-weaning period is a critical period of behavioral development. Conditioned fear of humans during this period by forced handling may, in some circumstances, have long-term negative effects on behavior, health, and welfare.

The ability to select a horse for a particular role, especially from a young age, is an attractive idea for breeders and trainers. Temperament tests have been created by a range of organizations and researchers in order to predict behavioral tendencies. The ultimate aim of temperament tests is that interested groups can perform the tests themselves and make use of the results. Unfortunately, this is not always the case. Some tests are impractical, complicated, and time consuming. Furthermore, in order for tests to be useful, they must be able to predict a consistency of temperament over time. Since consistency of the horse's response is what characterizes temperament, it is important that measures are chosen that clearly demonstrate this consistency. However, tests conducted over long periods of time are not practical for most trainers or people interested in purchasing a horse. In a review of the literature, the most reliable and consistent results are often found by exposing horses to sudden, unexpected novelty, or exposing them to novel objects or novel arenas that provoke sometimes severe reactions. In test situations such as these, consideration should be given to the welfare of the animals being evaluated, particularly when tests provoke fear, anxiety, or aggressive behavior. Creating a situation which causes fear in horses to learn if they have a fearful temperament, may have an effect on their future perception. Horses don't know it's a test, all they know and learn is that humans make them afraid. The umbrella tests used by the first author to illustrate the different reactions of his horses, had negative effects on one of the horses that persisted for several days following the test. For some time after the test, his calm and gentle mare, which was always easy to approach, became distrustful and reluctant to approach. Another problem with temperament tests conducted on adult horses is the influence of previous experience on temperament. Severe stress during adolescence is known to be related to temperament change, but mild stress is also common during adolescence and should not be ignored when studying temperament change. Even mild stressors can influence changes in temperament, especially when multiple evens are experienced. However, evidence for stressful events during adolescence being related to changes in temperament is scarce. Adding to the scarcity of evidence of early environmental influences on temperament, stressful early experiences may be more detrimental to individuals biologically predisposed to fearfulness. All these factors are critical to understanding individual differences in temperament. Clearly, an easy and general means to assess biological factors related to temperament are needed.

A physical characteristic present at birth that remains unchanged throughout life, is easy to see, and predicts the biological foundation of temperament could have enormous implications. Hair whorls could be that characteristic. The fact that hair whorls are linked to early fetal brain development and are similar in type and distribution across at least four mammal species is remarkable. Hair whorls may provide clues to understanding pattern formation mechanisms in nervous system development. All mammal brains are basically similar in both structure and function. Even a superficial examination of the external anatomy of the brains of vertebrates reveals that most vertebrates possess the same number of brain divisions. All vertebrate brains have a brain stem, limbic system, cerebellum, and cerebral cortex. The main difference between the brains of humans and other animals is the size and complexity of the cortex. In addition, basic behaviors are similar in humans and animals, and in more complex human behaviors such as fine motor skill and language, there may be precursors in animals. Animals also respond to the environment much as humans do, reacting emotionally to others and even becoming stressed and anxious in times of danger. Therefore, it requires no stretch of the imagination to assume that hair patterns and brain development follow a similar trajectory. In the past, people read too much into hair whorls in terms of horses' personalities. The way we like to explain it, hair whorls may simply predict how easily an animal gets scared. Personality, like temperament is shaped by a complex interaction between genetic factors and experience. Hair whorls don't predict that a horse will have certain temperament or personality traits, only that the horse is predisposed to those traits and upbringing and experience determine if those traits are expressed. In day-old foals, the first author is fond of saying, "Hair whorls are like a built-in instruction manual."

The recent discovery that the horse brain is lateralized and the two hemispheres are specialized to handle information in different ways, and control different classes of behavior, has the potential to solve many common problems. In our observations in cattle and horses, lateralized behaviors are signs of stress and fear. Repeatedly exposing horses to stimuli they perceive as threatening or dangerous can have profound effects on health, welfare, and productivity. Recognizing fear as the primary motivation of lateral behavior provides an alternative to correcting behaviors once established, but more importantly, a way to prevent the behaviors before they begin. Hair whorl side position and laterality has been shown in cattle and dogs, but further research is needed to find if hair whorl side position is related to laterality in horses.

Good horse trainers are talented observers of horse behavior and respond consistently to the horse's subtle cues during training. Learning the difference between individual horses and how they react differently to different situations is the key.

REFERENCES

Adamec, R.E., Blundell, J., Burton, P., 2005. Neural circuit changes mediating lasting brain and behavioral response to predator stress. Neurosci. Biobehav. Rev. 29, 1225–1241.

Anderson, L.S., Larhammar, M., Memic, F., Wootz, H., Schwochow, D., Rubin, C.-J., et al., 2012. Mutations in the DMRT3 affects locomotion in horses and spinal circuit function in mice. Nature 488, 642–646.

Andrew, R.J., 2002. Behavioral development and lateralization. In: Rogers, L.J., Andrew, R.J. (Eds.), Comparative Vertebrate Lateralization. Cambridge University Press, Cambridge, UK, pp. 157–205.

Austin, N.P., Rogers, L.J., 2007. Asymmetry of flight and escape turning responses in horses. Laterality 12 (5), 464–474.

Basile, M., Boivin, S., Boutin, A., Blois-Heulin, C., Hausberger, M., Lemasson, A., 2009. Socially dependent auditory laterality in domestic horses (Equus caballus). Anim. Cogn. 12, 611–619.

Bassett, L., 2003. Effects of predictability of feeding routines on the behaviour and welfare of captive primates. eTheses from School of Natural Sciences, University of Stirling.

Bassett, L., Buchannan-Smith, H., 2007. The effects of predictability on the welfare of captive animals. Appl. Anim. Behav. Sci. 102, 223–245.

Bateson, P., 1979. How do sensitive periods arise and what are they for? Anim. Behav. 27, 470–486.

Beaton, A.A., Mellor, G., 2007. Direction of hair whorl and handedness. Laterality 12, 295–301.

Benjamin, J., Li, L., Patterson, C., Greenberg, B.D., Murphy, D.L., et al., 1996. Population and familial associations between the D4 dopamine receptor gene and measures of Novelty Seeking. Nat. Genet. 12, 81–84.

Bisazza, A., Pignatti, R., Vallortigara, G., 1997. Detour tests reveal task-and stimulus-specific behavioural lateralization in mosquitofish (Gambusia holbrooki). Behav. Brain Res. 89, 237–242.

Boissy, A., 1995. Fear and fearfulness in animals. Q. Rev. Biol. 70, 165–191.

Boissy, A., 1998. Fear and fearfulness in determining behavior. In: Grandin, T. (Ed.), Genetics and the Behavior of Domestic Animals. Academic Press.

Boivin, X., Braastad, B.O., 1996. Effects of handling during isolation after early weaning on goat kids' later response to humans. Appl. Anim. Behav. Sci. 48, 61–71.

Broucek, J., Kisac, P., Mihina, S., Hanus, A., Uhrincat, M., Tancin, V., 2007. Hair whorls of Holstein Friesian heifers and affects on growth and behavior. Arch. Tierz, Dummerstorf 4, 374–380.

Cameron, R., Rogers, L.J., 1999. Hand preference of the common Marmoset, problem solving and responses in a novel environment. J. Exp. Psychol. 113, 149–157.

Cayado, P., Munoz-Escassi, B., Dominguez, C., Manley, W., Olabarri, B., Sanchez de la Muela, M., et al., 2006. Hormone response to training and competition in athletic horses. Equine exercise physiology 7. Equine. Vet. J. 38 (Suppl. 36), 274–278.

Chamberlain, S.R., Sahakian, B.J., 2007. The neuropsychology of mood disorders. Curr. Psychiatry. Rep. 8, 458–463.

Champagne, F.A., Curley, J.P., 2005. How social experiences influence the brain. Curr. Opin. Neurobiol. 15, 704–709.

Christensen, J.W., 2006. Fear in Horses. Uppsala: Sveriges lantbruksuniv. [Licenciate thesis].

Christensen, J.W., Keeling, L.J., Neilson, B.L., 2005. Responses of horses to novel visual, olfactory, and auditory stimuli. Appl. Anim. Behav. Sci. 93 (1–2), 53–65.

Claessens, S.E., Daskalakis, N.P., van der Veen, R., Oitzl, M.S., de Kloet, E.R., Champagne, D. L., 2011. Development of individual differences in stress reactivity: an overview of factors mediating the outcome of early life experiences. Psychopharmacology (Berl.) 214 (1), 141–154.

Clayton, H.M., 1990. Kinematics of equine jumping. Equine Athlete 3, 17–20.

Collins, R.L., 1975. When left-handed mice live in right-handed worlds. Science 187, 181–184.

Conrad, C.D., Sullivan, R.M., Laplante, F., 2011. Stress, prefrontal cortex asymmetry, and depression. In: Conrad, C.D. (Ed.), The Handbook of Stress: Neuropsychological Effects on the Brain. Wiley-Blackwell, Okford, UK.

Corballis, M.C., 1997. The genetics and evolution of handedness. Psychol. Rev. 104 (4), 714–727.

Corballis, M.C., 2009. The evolution and genetics of cerebral asymmetry. Phil. Trans. R. Soc. B 364, 867–879.

De Boyer des Roches, A., Richard-Yris, M.-A., Henry, S., Ezzaouia, M., Hausberger, M., 2008. Laterality and emotions: visual laterality in the domestic horse (Equus caballus) differs with objects‘ emotional value. Physiol Behav 94 (3), 487–490.

Denny, K., 2009. Handedness and depression: evidence from a large population survey. Laterality 14, 246–255.

Diehl, K.D., Egan, B., Tozer, P., 2002. Intensive, early handling of neonatal foals: Mare-foal interactions. Proceedings of the Havermeyer Foundation Horse Behavior and Welfare Workshop Holar, Iceland, pp. 23–26.

<Equestriancollegeadvisor.wordpresss.com/p.21>, (2012).

Farmer, K., Krueger, K., Byrne, R.W., 2010. Visual laterality in the domestic horse (Equus caballus) interacting with humans. Anim. Cogn. 13, 229–238.

Florcke, C., Engle, T.E., Grandin, T., Deesing, M.J., 2012. Individual differences in calf defense patterns in red angus beef cows. Appl. Anim. Behav. Sci. 139 (3–4), 203–208.

Friedman, J.H., Golomb, J., Mora, M.N., 1952. The hair whorl sign for handedness. Dis. Nerv. Syst. 13 (7), 208–216.

Fureix, C., Jego, P., Henry, S., Hausberger, M., 2009. How horses (Equus caballus) see the world: humans as significant "objects". Anim. Cogn. 12 (4), 643–654.

Ghirlanda, S., Frasnelli, E., Vallortigara, G., 2009. Intraspecific competition and coordination in the evolution of lateralization. Phil. Trans. R. Soc. B 364, 861–866.

Glover, V., 2011. Annual research review: prenatal stress and the origins of psychopathology: an evolutionary perspective. J. Child. Psychol. Psychiatry. 52 (4), 356–367.

Gorecka-Bruzda, A., Sloniewski, K., Golonka, M., Jaworski, Z., Jezierski, T., 2006. Heritablity of hair whorl position on the forehead in Konik horses. J. Anim. Breed. Genet. 123 (6), 396–398.

Gorecka-Bruzda, A., Golonka, M., Chruszczewski, M.H., Jezierski, T., 2007. A note on behaviour and heart rate in horses differing in facial hair whorl. Appl. Anim. Behav. Sci. 105, 244–248.

Gotlieb, I.H., Joormann, J., Minor, K.L., Reale, D., 2008. HPA axis reactivity: a mechanism underlying the associations among 5-HTTLPR, stress, and depression. Biol. Psychiatry. 63, 847–851.

Grandin, T., 1997. Assessment of stress during handling and transport. J. Anim. Sci. 75, 249–257.

Grandin, T., Deesing, M.J., Struthers, J.J., Swinker, A.M., 1995. Cattle with hair whorl patterns above the eyes are more behaviorally agitated during restraint. Appl. Anim. Behav. Sci. 46, 117–123.

Grzimek, B., 1968. On the psychology of the horse. In: Friedrich, H. (Ed.), Man and Animal: Studies in Behavior. St Martin's Press, New York.

Hall, J.G., Froster-Iskenius, U.G., Allanson, J.E., 1989. Handbook of Normal Physical Measurements. Oxford University Press, p. 371.

Hausberger, M., Bruderer, C., LeScolan, N., Pierre, J.S., 2004. Interplay between environmental and genetic factors in temperament/personality traits in horses (Equus caballus). J. Comp. Psychol. 118, 434–446.

Hausberger, M., Muller, C., Lunel, C., 2011. Does work affect personality? A study in horses. PLoS 6 (2).

Heiming, R.S., Sacher, N., 2010. Consequences of serotonin transporter genotype and early adversity on behavioral profile-pathology or adaptation?. Front. Neurosci. 4, 187.

Heird, J.C., Deesing, M.J., 1998. Genetic effects on horse behavior. In: Grandin, T. (Ed.), Genetics and the Behavior of Domestic Animals. Academic Press.

Heird, J.C., Whitaker, D.D., Bell, R.W., Ramsey, C.B., Lokey, C.E., 1986. The effects of handling at different ages on the subsequent learning ability of 2-year-old horses. Appl. Anim. Behav. Sci. 15, 15–25.

Henry, S., Hemery, D., Richard-Yris, M.A., Hausberger, M., 2005. Human–mare relationships and behavior of foals towards humans. Appl. Anim. Behav. Sci. 93 (3), 341–362.

Henry, S., Richard-Yris, M.A., Hausberger, M., 2006. Influence of various early-human-foal interferences on subsequent human-foal relationship. Dev. Psychobiol. 712–718, doi: 10.1002.

Henry, S., Briefer, S., Richard-Yris, M.A., Hausberger, M., 2007. Are 6-month old foals sensitive to dam's influence?. Dev. Psychobiol. 49 (5), 514–521.

Henry, S., Richard-Yris, M.A., Tordjman, S., Hausberger, M., 2009. Neonatal handling affects durably bonding and social development. PLoS One 4 (4), 1–9.

Hopkins, W.D., Bennett, A., 1994. Handdness and approach-avoidance behavior in Chimpanzees. J. Exp. Psychol. 20, 413–418.

<Horseforum.com>, (2012).

Ito, H., Nara, H., Inoue-Murayama, M., Shimada, M.K., Koshimura, A., et al., 2004. Allele frequency distribution of the canine dopamine receptor D4 gene Exon III and I in 23 breeds. J. Vet. Med. Sci. 66, 815–820.

Jansen, A., Lohmann, H., Scharfe, S., Sehlmeyer, C., Deppe, M., Knecht, S., 2007. The association between scalp hair-whorl direction, handedness and hemispheric language dominance: is there a common genetic basis of lateralization?. Neuroimage 35 (2), 853–861.

Jeziersky, T., Jaworshy, Z., Goreka, A., 1998. Effects of handling on behaviour and heart rate in Konic horses: comparisons of stable and forest reared youngstock. Appl. Anim. Behav. Sci. 62, 1–11.

Kagan, J., 2005. Temperament. Encyclopedia on Early Child Development. Center of Excellence for Early Childhood Development.

Keane, T.M., Goodstadt, L., Danecek, P., White, M.A., Wong, K., Yalcin, B., et al., 2011. Mouse genomic variation and its effect on phenotypes and gene regulation. Nature 477, 289–294.

Kiil, V., 1948. Frontal hair direction in mentally deficient individuals, with special reference to mongolism. J. Hered. 39, 281–285.

Kiley-Worthington, M., 1987. The Behavior of Horses. J.A. Allen, London.

Klar, A.J.S., 2003. Human handedness and scalp hair-whorl direction develop from a common genetic mechanism. Genetics 165, 269–276.

Klar, A.J.S., 2009. Scalp hair-whorl orientation of Japanese individuals is random; hence, the trait's distribution is not genetically determined. Semin. Cell. Dev. Biol. 20, 510–513.

Knecht, S., Drager, B., Bobe, L., Lohmann, H., Floel, A., Ringelstein, E.-B., et al., 2000. Handedness and hemispheric language dominance. Brain 123 (12), 2512–2518.

Lanier, J.L., Grandin, T., Green, R.D., Avery, D., Mcgee, K., 1999. Cattle hair whorl position and.temperament in auction houses. J. Anim. Sci. 77 (Suppl. 1), 147 (Abstract).

Lansade, L., Bertrand, M., Boivin, X., Bouissou, M.-F., 2004. Effects of handling at weaning on manageability and reactivity of foals. Appl. Anim. Behav. Sci. 87 (1), 131–149.

Lansade, L., Bouissou, M.F., Erhard, H.W., 2008. Fearfulness in horses: a temperament trait stable across time and situations. Appl. Anim. Behav. Sci. 115 (3), 182–200.

Larose, C., Richard-Yris, M.A., Hausberger, M., Rogers, L.J., 2006. Laterality of horses associated with emotionality. Laterality 11 (4), 355–367.

Lauterbach, C.E., Knight, J.B., 1927. Variation in the whorl of the head hair. J. Hered. 18, 107–115.

LeDoux, J.E., 1998. The Emotional Brain: The Mysterious Underpinnings of Emotional Life. Touchstone, New York.

Leiner, L., Fendt, M., 2011. Behavioral fear and heart rate responses of horses after exposure to novel objects, effects of habituation. Appl. Anim. Behav. Sci. 131, 104–109.

Levy, J., Nagylaki, T., 1972. A model for the genetics of handedness. Genetics 72 (1), 117–128.

Lindsay, S.R., 2000. Handbook of Applied Dog Behavior and Training. Vol. I, Adaptation and Learning. Blackwell, Oxford.

Livik, K.J., Rogers, J., Lichter, J.B., 1995. Variability of dopamine D4 (DRD4) receptor gene sequence within and among nonhuman primate species. Proc. Natl. Acad. Sci. U.S.A. 92, 427–431.

Lloyd, A.S., Martin, J.E., Bornett-gauchi, H.L.I., Wilkenson, R.G., 2008. Horse personality: variation between breeds. Appl. Anim. Behav. Sci. 112, 369–383.

Mal, M.E., McCall, C.A., 1996. The influence of handling during different ages on a halter training test in foals. Appl. Anim. Behav. Sci. 50, 115–120.

Mal, M.E., McCall, C.A., Cummins, K.A., Newland, M.C., 1994. Influence of pre-weaning handling methods on post/weaning learning ability and manageability of foals. Appl. Animal Behav. Sci. 40, 187–195.

Manroe, C.C., 1992. The Horse. Michael Friedman Publishing Group Inc, New York.

Maringer, F., 1983. Horses are Made to be Horses. Simon and Schuster Macmillan Co, New York.

Martins, C.E.N., Quadros, S.A.F., Trindade, J.P.P., Quadros, F.L.F., Costa, J.H.C., Raduenz, G., 2009. Shape and function in braford cows: the body shape as an indicative of performance and temperament. Arch. Zootec. 58 (223), 425–433.

Mathews, A., MacLeod, C., 2005. Cognitive vulnerability to emotional disorders. Annu. Rev. Clin. Psychol. 1, 167–195.

Matson, K.M., 2004. The effect of weekly handling on the temperament of peri-puberal crossbred beef heifers. Masters Theses, Animal and Poultry Sciences, Virginia Tech.

McCall, C.A., Hall, S., McElhenny, W.H., Cummins, K.A., 2006. Evaluation and comparison of four methods of ranking horses based on reactivity. Appl. Anim. Behav. Sci. 96 (1), 115–127.

McGreevy, P.D., 2004. Equine Behavior. A guide for Veterinarians and Equine Scientists. Saunders.

McGreevy, P.D., French, N.P., Nicol, C.J., 1995. The prevalence of abnormal behaviours in dressage, eventing and endurance horses in relation to stabling. Vet. Rec. 137, 36–37.

McGreevy, P.D., Landrieu, J.-P., Malou, P.F.J., 2007. A note on motor laterality in plains zebras (Equus burchelli) and impalas (Aepyceros melampus). Laterality 12, 449–457.

McGreevy, P.D., Rogers, L.J., 2005. Motor and sensory laterality in thoroughbred horses. Appl. Anim. Behav. Sci. 92, 337–352.

McGreevy, P.D., Thompson, P.C., 2006. Differences in motor laterality between breeds of performance horse. Appl. Anim. Behav. Sci. 99 (1), 183–190.

McGue, M., 2010. The end of behavioral genetics? Behav. Genet. 40 (3), 284–296.

McManus, I.C., 1985. Handedness, language dominance and aphasia: a genetic model. Psychol. Med. 8, 3–40.

Meixell 1993 <http://articles.mccall.com/1993-03-25/news/>.

Miller, R.C., 1991. Imprint Training of the Newborn Foal. Colorado Springs, Co: Western Horseman Inc.

Momozawa, Y., Ono, T., Sato, F., Kikusui, T., Mori, Y., Kusunose, R., 2003. Assessment of equine temperament by a questionnaire survey to caretakers and its evaluation its reliability by simultaneous behavior test. Appl. Anim. Behav. Sci. 84 (2), 127–138.

Momozawa, Y., Takeuchi, Y., Kusunose, R., Kikusui, T., Mori, Y., 2005. Association between equine temperament and polymorphisms in dopamine D4 receptor gene. Mamm. Genome. 16, 538–544.

Murphy, J., Arkin, S., 2008. Facial hair whorls (trichoglyphs) and the incidence of motor laterality in the horse. Behav. Processes 79, 7–12.

Murphy, J., Southerland, A., Arkin, S., 2005. Idiosyncratic motor laterality in the horse. Appl. Anim. Behav. Sci. 91, 297–310.

O'Connor, T.G., Bergman, K., Sarkar, P., Glover, V., 2012. Prenatal cortisol exposure predicts infant cortisol response to acute stress. Dev. Psychobiol. doi:10.1002/dev.21007.

Ovando, S.D.N.G.A., 2010. Correlation Between the Swirl Pattern Facial Laterality Thoroughbred Motor in Foals Less than One Year of Age. University of Conception, Faculty of Veterinary, Clinical Science.

<PetMD.com> (2012).

Phillips, M., Grandin, T., Graftam, G., Iribeck, N.A., Cambre, R.C., 1998. Crate conditioning of Bongo (Tragelaphus eurycerus) for veterinary and husbandry procedures at the Denver zoological gardens. Zoo. Biol. 17, 25–32.

Plomin, R., Daniels, D., 1997. Why are children in the same family so different from one another? Behav. Brain. Sci. 10, 1–60.

Randle, H.D., 1998. Facial hair whorl position and temperament in cattle. Appl. Anim. Behav. Sci. 56, 139–147.

Robbins, A., Phillips, C., 2010. Lateralised visual processing in domestic cattle herds responding to novel stimuli. Laterality 15 (5), 514–534.

Robbins, A., Rogers, L.J., 2006. Complimentary and lateralized forms of processing in Bufo marinus for novel and familiar prey. Neurobiol. Learn. Mem. 86, 214–227.

Rogers, L.J., 2009. Hand and paw preference in relation to the lateralized brain. Philos. Trans. R. Soc. Lond., B, Biol. Sci. 364 (1514), 943–954.

Rogers, L.J., 2010. Relevance of brain and behavioural lateralization to animal welfare. Appl. Anim. Behav. Sci. 127 (1–2), 1–11.

Proops, L., McComb, K., 2012. Cross-modal individual recognition in domestic horses (Equus caballus) extends to familiar humans. Proc. R. Soc.B 279 (1741), 3131–3138.

Sachser, N., Hennessey, M.B., Kaiser, S., 2010. Adaptive modulation of behavioral profiles by social stress during early phases of life and adolescence. Neurosci. Biobehav. Rev. in press.

Sankey, C., Henry, S., Clouard, C., Richard-Yris, M.-A., Hausberger, M., 2011. Asymmetry of behavioral responses to a human approach in young naïve vs. trained horses. Physiol. Behav. 104, 464–468.

Schmidt, A., Aurich, J., Mostl, E., Muller, J., Aurich, C., 2010. Changes in cortisol release and heartrate and heartrate variability during the initial training of 3-year-old sport horses. Horm. Behav. 58 (4), 628–636.

Schmidt, H., Depner, M., Kabesch, M., 2008. Medial position and counterclockwise rotation of the parietal scalp hair-whorl as a possible indicator for non-right-handedness. Sci. World. J. 8, 848–854.

Schaafsma, S.M., Riedstra, B.J., Pfannkuche, K.A., Bouma, A., Groothuis, T.G.G., 2009. Epigenis of behavioral lateralization in humans and other animals. Phil. Trans. R. Soc. B. 364, 915–927.

Seaman, S.C., Davidson, H.P.B., Waren, N.K., 2002. How reliable is temperament assessment in the domestic horse (Equus caballus)? Appl. Anim. Behav. Sci. 78 (2–4), 175–191.

Shirley, K.L., Garrick, D.J., Grandin, T., Deesing, M., 2006. Inheritance of facial hair whorls attributes in Holstein cattle. Proc. West. Sect. Am. Soc. Anim. Sci. 57, 62–65.

Silveria, I.D.B., Fischer, V., Mendonca, G.D., 2006. Behavior of beef cattle at an auction ring. Cienc. Rural. Santa. Maria. 36 (5), Sept. –Oct.

Simpson, B.S., 2002. Neonatal foal handling. Appl. Anim. Behav. Sci. 78, 303–317.

Sivewright, M., 1986. Thinking Riding Book 2: In Good Form. J.A. Allen and Co. Ltd, London.

Smith, D.W., Gong, B.T., 1973. Scalp hair patterning as a clue to early fetal development. J. Pediatr. 83, 374–380.

Sondergaard, E., Halekoh, U., 2003. Young horses' reactions to humans in relation to handling and social environment. Appl. Anim. Behav. Sci. 84 (4), 265–280.

Steinbrecht, G., 1886. The Gymnasium of the Horse. Potsdam, Doring.

Stolp, H., Neuhaus, A., Sundramoorthi, R., Molnár, Z., 2012. The long and the short of it: gene and environment interactions during early cortical development and consequences for long-term neurological disease. Front. Psychiatry. 3, 50.

Swinker, A.M., Deesing, M.J., Tanner, M., Grandin, T., 1994. Observation of normal and abnormal hair whorl patterning on the equine forehead. J. Anim. Sci. 72 (Suppl. 1), 207.

Takeuchi, Y., Hori, M., 2008. Behavioural laterality in the shrimp-eating cichlid fish (Neolamprologus faciatus) in Lake Tanganyika. Anim. Behav. 75, 1359–1366.

Tanner, M., Grandin, T., Cattell, M., Deesing, M., 1994. The relationship between facial hair whorls and milking parlor side preference. J. Anim. Sci. 72 (Suppl. 1), 207 (abstr.).

Tellington-Jones, L., Taylor, S., 1995. Die Personalichkeit Ihres Pferdes. Franckh Kosmos Verlag-GmbH & Co., Stuttgart.

Tomkins, L.M., Williams, K.A., Thompson, P.C., McGreevy, P.D., 2012. Lateralization in the domestic dog (Canis familiaris): relationships between structural, motor, and sensory laterality. J. Vet. Behav. 7, 70–79.

Villortigara, G., Rogers, L.J., 2005. Survival with an asymmetrical brain: advantages and disadvantages of cerebral lateralization. Behav. Brain. Sci. 28, 575–589.

Visser, E.K., van Reenen, C.G., Hopster, H., Schilder, M.B.H., Knapp, J.H., Barneveld, A., et al., 2001. Quantifying aspects of young horses' temperament: consistency of behavioural variables. Appl. Anim. Behav. Sci. 74, 241–258.

Visser, E.K., van Reenen, C.G., Engle, B., Schilder, J.H., Barneveld, A., Blokhuis, H.J., 2003. The association between performance in show-jumping and personality traits earlier in life. Appl. Anim. Behav. Sci. 82 (4), 279–295.

von Borstel, U.K., Ian, J.H., Duncan, I.J.H., Claesson Lundin, M., Keeling, L.J., 2010. Fear reactions in trained and untrained horses from dressage and show-jumping breeding lines. Appl. Anim. Behav. Sci. 125 (3), 124–131.

Waran, N.K., Clarke, N., Farnworth, M., 2007. The effects of weaning on the domestic horse (Equus caballus). Appl. Anim. Behav. Sci. doi: 10.1016/.

Waring, G.H., 1983. Horse Behavior: The Behavioral Traits and Adaptations of Domestic and Wild Horses Including Ponies. Noyes, Park Ridge, NJ.

Weber, B., Hoppe, C., Faber, J., Axmacher, N., Fliebach, K., Mormann, F., et al., 2006. Association between scalphair-whorl direction and hemispheric language dominance. Neuroimage 30 (2), 539–543.

Weinstock, M., 2005. The potential influence of maternal stress hormones on development and mental health of the offspring. Brain. Behav. Immun. 19 (4), 296–308.

Wells, A.E.D., Blache, D., 2008. Horses do not exhibit motor bias when their balance is challenged. Animal 2 (11), 1645–1650.

Williams, J.L., Friend, T.H., Toscano, M.J., Collins, M.N., Sisto-Burt, A., Nevill, C.H., 2002. The effects of early training sessions on the reactions of foals at 1, 2, and 3 months of age. Appl. Anim. Behav. Sci. 77, 105–114z.

Williams, D.E., Norris, B.J., 2007. Laterality in stride pattern preferences in racehorses. Anim. Behav. 74, 941–950.

Wittling, W., Pfluger, M., 1990. Neuroendocrine hemisphere asymmetries: salivary cortisol secretion during lateralized viewing of emotion-related and neutral films. Brain. Cogn. 14, 243–265.

Wolff, A., Hausberger, M., LeScholan, N., 1997. Experiemental tests to assess emotivity in horses. Behav. Processes 40, 209–221.

Zentner, M., Shiner, R., 2012. The Handbook of Temperament. Guilford Press.

FURTHER READING

Gorecka-Bruzda, A., Jastrezbska, E., Sosnowska, Z., Jaworski, Z., Jezierski, T., Chruszczewski, M.H., 2011. Reactivity to humans and fearfulness: field validation in polish cold blood horses. Appl. Anim. Behav. Sci. 133, 207–215.

Hsiao, E.Y., Patterson, P.H., 2012. Placental regulation of maternal-fetal interactions and brain development. Dev. Neurobiol. doi:10.1002/dneu.22045.

Reinhardt, V., 2003. Working with rather than against Macaques during blood collection. J. Appl. Anim. Welf. Sci. 6 (3), 189–197.

Smith, D.W., Gong, B.T., 1974. Scalp hair patterning: its origin and significance relative to early fetal brain development. Teratology 9, 17–34.

Chapter 8

Improving the Adaptability of Animals by Selection

Jean Michel Faure and Andrew D. Mills
Station de Recherches Avicoles, INRA de Tours, Novzilly, France

The work of Faure and Mills was groundbreaking research, which was published in the first edition of our book. This is a classic paper in behavioral genetics that everyone in the field should read so we decided to reproduce the original chapter in the new edition. By choosing two easy-to-use behavioral tests, they created four separate genetic lines of quail. Since this chapter was published in 1998, more recent research with these genetic lines has shown that the tonic immobility test does not measure all types of fear and that the social reinstatement trait is linked to other traits. At the end of this chapter, we have a list of new papers that have conducted further research on these genetic lines which show that motivational systems are complex.

INTRODUCTION

Animal welfare is a major issue in Europe. One method frequently used for solving animal welfare problems is to adapt the environment to fit the behavioral needs of the animals. However, environmental modifications are costly because economic factors are presently the predominant determinants of the design of housing systems. Another approach for improving animal welfare is to modify the animals so they are better adapted to intensive husbandry. This can be accomplished by pharmacological, surgical, or genetic methods In this chapter, the use of genetic selection to improve adaptability to intensive rearing systems is outlined and discussed.

In Japanese quail, two types of behavior were subjected to genetic selection: social reinstatement (the tendency to rejoin flockmates) and fearfulness. These two traits interact and determine how a bird may behave in different situations. Information on the inheritance of behavioral traits in quail may provide insights into the inheritance of these traits in other species. Before we discuss genetic selection, we will briefly summarize nongenetic approaches to improving welfare.

Genetics and the Behavior of Domestic Animals. DOI: http://dx.doi.org/10.1016/B978-0-12-394586-0.00008-1
© 2014 Elsevier Inc. All rights reserved.

PHARMACOLOGY

This approach was first tested in the 1950s and 1960s after neuroleptic drugs were discovered (Gilbreath *et al.*, 1959). Pharmacology was used in an attempt to calm animals during harvesting and transport. The early use of neuroleptic drugs as tranquilizers was unsuccessful. Furthermore, consumer awareness of drug residues makes such methods unacceptable for commercial use. However, pharmacological research can provide useful insights into animal behavior, adaptability, and brain function.

SURGERY

Surgical methods such as dehorning, beak trimming, or tail docking in domestic animals can prevent welfare problems such as pecking injuries and tail biting. However, these methods are criticized by welfarists because pain is experienced at the time of the operation. Gentle *et al.* (1991) found that beak trimming in poultry may cause pain for a prolonged period of time after the operation.

ONTOGENY

Some problems which make it difficult for production animals to adapt to their rearing environment can be improved to some degree by ontogenetic processes such as handling or environmental enrichment (Jones, 1982; Jones and Faure, 1981; Jones *et al.*, 1991). Ontogenetic manipulation involves manipulation of the animal's early experiences. The disadvantage of these procedures is that they are environmental rather than genetic and they must be repeated in subsequent generations.

GENETIC SELECTION

Adaptation of animals to their environment can also be improved by genetic selection. This method has the advantage of being applied to relatively small numbers of carefully selected breeding animals with subsequent genetic gain being transmitted to their descendants. This chapter will focus on the use of genetic selection and its applicability to practical problems that occur in intensive husbandry systems. Genetic selection can be used to improve welfare. Although it is unlikely that genetic selection alone can be used to solve all welfare problems, a combination of improvement of the environment and ontogenetic processes, together with the use, at least in emergency cases, of pharmacology or surgery is probably the only realistic way of improving domestic animal welfare (see Chapter 11).

Genetic selection has influenced the processes of domestication in at least two ways (see also Chapter 2). These are natural selection and artificial selection for specific traits. When animals were first domesticated, natural selection was probably the main cause of genetic change. This means that animals showing low levels of adaptation to the new, man-made conditions had poorer fitness and their genes were eliminated from the gene pool of the next generation (Faure *et al.*, 1992). This type of selection is likely to be efficient only when a high proportion of animals are carrying genes of low value in the new artificial environment. For example, animals with high fear and a flighty, nervous temperament may be so stressed that they fail to breed in captivity. Subsequently, the failure of these animals to breed in captivity probably resulted in natural selection for lower levels of fearfulness. During the early stages of domestication, farmers were able to recognize individual traits in their animals and prevented most animals exhibiting undesirable characteristics from reproducing. For example, a very aggressive bull which endangered the farmer would probably be culled. When flocks or herds become too large to allow recognition of individual animals, this method was no longer applicable. For both practical and economic reasons, animal breeding was subsequently concentrated on specific traits, irrespective of their relationships to other traits or concerns for animal welfare.

A more sensible solution would be to select for general adaptability and characteristics that are expressed in a wide variety of environments. High appetite is an example of a trait that can be expressed in a wide range of environments; environmental change cannot mask the effects of genetic selection. On the other hand, perching is a behavior that can only be expressed in certain environments. Obviously, perching behavior cannot be expressed if perches are not available. Faure and Jones (1982a, b) found that perching behavior of chickens differs between strains. This suggests that it should be possible to select lines with high and low levels of perching behavior. Although selection for low levels of perching behavior might improve the adaptation of domestic hens to battery cages, if floor pens or aviaries are used it might be advantageous to promote high levels of perching behavior (Appleby *et al.*, 1988). Either type of selection would lead to very specialized lines of chickens which would be undesirable given the diversity of modern husbandry systems.

We chose to select for two characteristics which are indicators of general adaptability and are behaviorally expressed in most environments: fear and social reinstatement behavior (see Chapter 2 for further information on fear). Social reinstatement is the tendency of animals to flock and stay close to conspecifics (flockmates). Fear and social motivation are basic motivational systems in a wide variety of species. We chose to use Japanese quail as a model of other species.

SELECTION PROCEDURE

The measurement of behavioral traits is often very time consuming, and selection requires behavioral measures of large numbers of animals (Faure, 1981). Therefore, we chose to use relatively short tests (no longer than

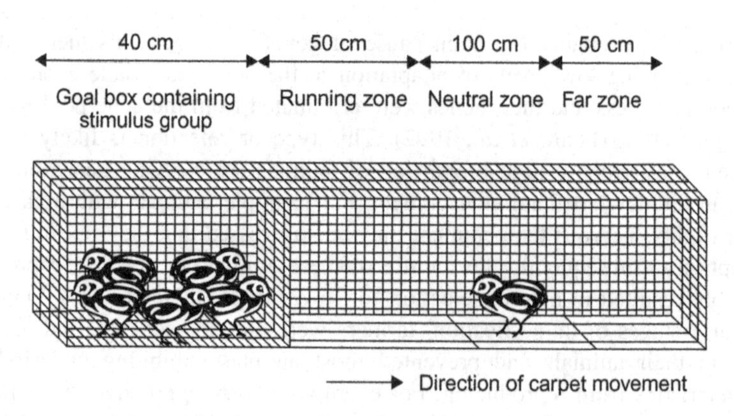

FIGURE 8.1 Schematic plan of the treadmill apparatus used to measure social reinstatement behavior in Japanese quail chicks.

5 minutes). Social reinstatement behavior was measured using the treadmill test (Mills and Faure, 1990). The apparatus has two compartments, one containing flockmates and the other a treadmill on which the test animal is placed (Figure 8.1). If the test animal moves toward the other birds, the treadmill moves in the opposite direction. Movement of the treadmill serves as a measurement of the effort made by the test animal to reinstate social contact. If the test animal moves away from the other birds, the time spent in the "far end" of the apparatus is also measured. Tests were performed when chicks were between 6 and 8 days of age and lasted 5 minutes.

High and low levels of fear were determined using the tonic immobility reaction (Jones, 1987). In this test, the chick is placed on its back in a U-shaped cradle covered with several layers of white cloth and is restrained by the experimenter for 10 seconds. If the chick remains immobile for 10 seconds after the removal of restraint, tonic immobility is considered to have been induced. Animals that remain immobile for "long periods" are considered to be fearful (Jones and Mills, 1982; Mills and Faure, 1986). If the chick rights itself within 10 seconds, it is considered that TI (tonic immobility) has not been induced and a further attempt at induction is made. If five attempts at induction are unsuccessful, the duration of TI is considered to be 0 and the number of attempts at induction is recorded as five. The TI test is performed when the chicks are 9–10 days of age. For the purpose of selection, the maximum duration of TI was set at 5 minutes.

Selection over 20 generations resulted in lines showing either high or low levels of social reinstatement (SR) behavior and long or short duration of TI. Appropriate control lines (CSR and CTI) were also maintained (Mills and Faure, 1991). At each generation, 20 males and 40 females were kept in cages containing one male and two females. These animals were the breeding population of each line. Approximately 200 quail chicks were hatched at each generation. Selected animals from each hatch served as the breeding population for the next generation. The sire of every chick was identified (one male per cage) and within dam families were identified on the basis of egg color, shape, and size. Morphological characteristics of eggs are very constant within females but very different between females. Furthermore, each chick was individually banded at hatching. Wing bands were used to identify the animals throughout their lives. In the SR lines, the measure of SR behavior (measured in arbitrary units) is the difference between the distance run on the treadmill minus the "far" time. It was shown in a preliminary experiment that the distance run and the far time were equal when the animals were tested in the treadmill apparatus without stimulus animals, resulting in a score of 0.

In order to ensure that selection for TI did not influence SR behavior or *vice versa*, all animals were tested for both TI and SR behavior and the scores for each trait were subjected to regression analysis. In the lines selected for high levels of social reinstatement behavior the SR score was calculated on the regression graph with the duration of TI. The selection criterion was the deviation of each individual data point from the regression line. In the tonic immobility lines (high fear), the measure of TI was the time taken by the birds to right themselves. This value was regressed on the chick's score for social reinstatement behavior. Again, the selection criterion was the deviation of each data point from the regression line. This transformation reduced the risk of co-selection for the two traits due to correlation between social reinstatement and tonic immobility. Breeding values were calculated taking into account not only the individual values but also those of full siblings (Mills and Faure, 1991). The two control lines were bred from one son chosen at random from each father and one daughter from each mother. Lines were selected as above until the 16th generation. However, from generation 17 onward, "skip a generation" selection was employed. In the nonselected generation, breeding animals were chosen at random and used for the control lines.

SELECTION RESULTS

Selective breeding for high and low social reinstatement and high and low tonic immobility produced significant changes in behavior. In the first generation, the mean (±SD) value for distance run on the treadmill was 14.4 meters (±8.9). After 20 generations of divergent selection for high or low levels of social reinstatement behavior, values were 59.0 (±27.3) meters run in the high SR line, 7.4 (±11.4) meters run in the control SR line and 4.0 (±8.9) meters run in the low SR line (Figure 8.2). The distribution of values for the distance run were completely different between the first and 20th generation of the high SR and low SR lines (Figure 8.3). From the 10th generation of selection onward, the standard deviations of the high SR and low SR lines ceased to overlap. This would seem to indicate that the high social

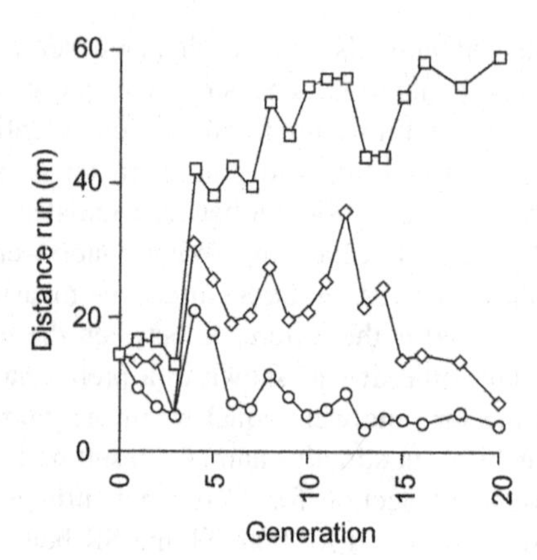

FIGURE 8.2 Distance run on the treadmill in lines of Japanese quail selected for high or low levels of social reinstatement behavior over 20 generations (□) HSR, (◇) CSR, (○) LSR.

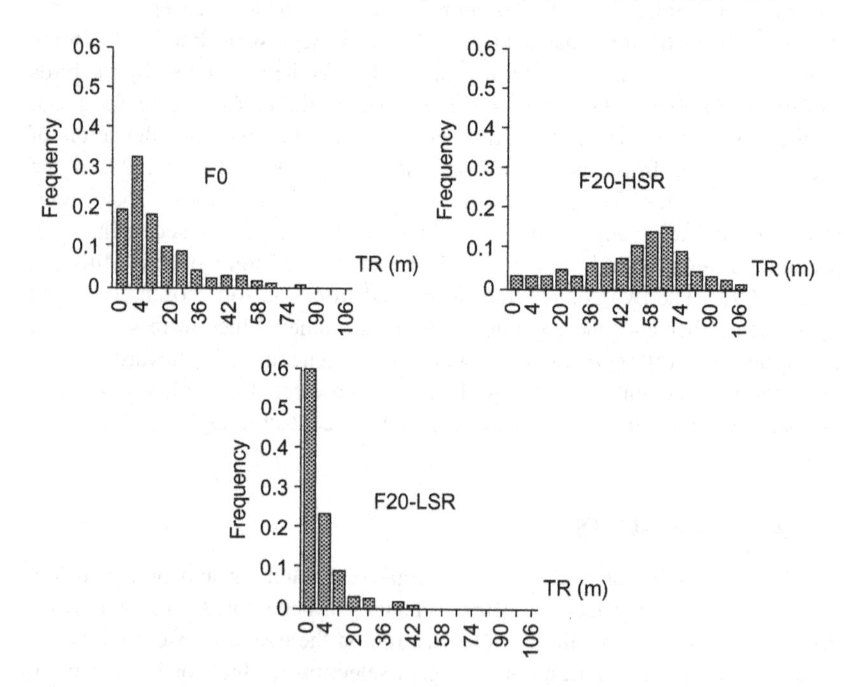

FIGURE 8.3 Distribution of distance run on the treadmill in lines of Japanese quail selected for high or low level of social reinstatement behavior. The graph shows the distributions for the F0 and F20 generations.

reinstatement (HSR) and the low social reinstatement (LSR) birds had become distinct genetic lines.

In the SR lines, the time spent in the far zone of the runway was identical up to the eighth generation of selection. Thereafter, the lines started to diverge with low SR birds showing the longest times and high SR birds the shortest times.

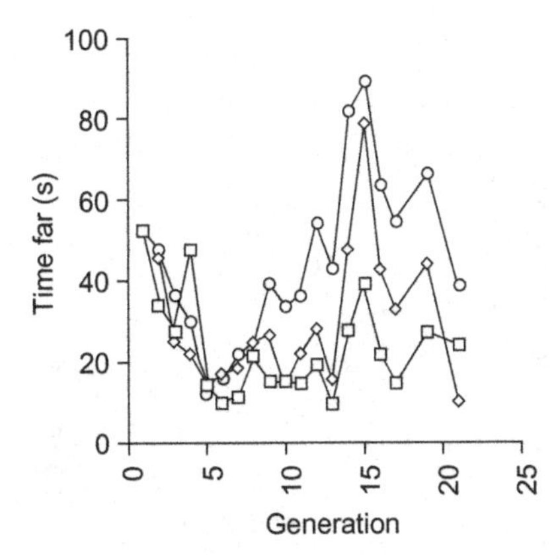

FIGURE 8.4 Time far in the treadmill lines of Japanese quail selected for high or low level of social reinstatement behavior over 20 generations. (□) HSR, (◇) CSR, (○) LSR.

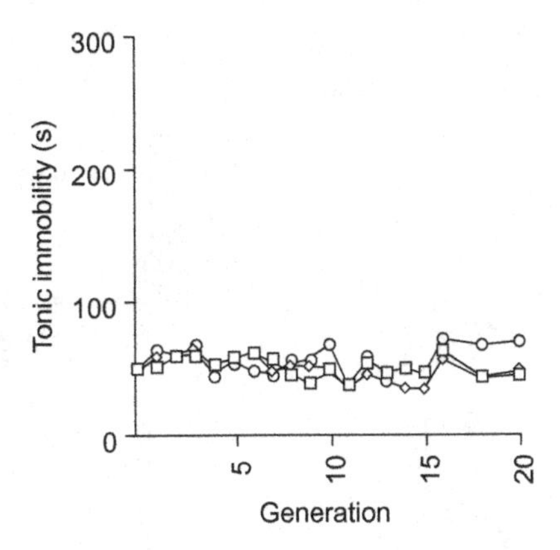

FIGURE 8.5 Duration of tonic immobility in lines of Japanese quail selected for high or low level of social reinstatement behavior over 20 generations. (□) HSR, (◇) CSR, (○) LSR.

Control SR birds had intermediate times (Figure 8.4). Over 20 generations of selection, the duration of the tonic immobility response was remarkably constant and similar in all three of the lines selected for social reinstatement lines (Figure 8.5). This shows that tonic immobility and social reinstatement are separate traits that are genetically independent of each other.

In the first generation (F0) of the lines to be selected for differing duration of TI responses, the mean (±SD) value for the duration of TI was 50.5 (±52.9) seconds. After 20 generations of divergent selection, TI scores were 215.6 (±91.9) seconds in the long TI line, 50.8 (±38.4) seconds in the control TI line, and 9.3 (±11.6) seconds in the short TI line (Fig. 8.6).

The tonic immobility response in the long TI (high fear) line was approximately 3.5 minutes compared to only 9 seconds in the short TI (low fear)

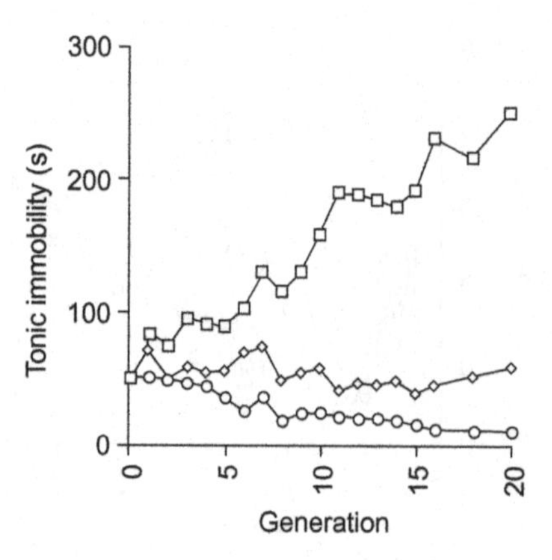

FIGURE 8.6 Duration of tonic immobility in lines of Japanese quail selected for long or short duration of tonic immobility over 20 generations. (□) LTI, (◇) CTI, (○) STI.

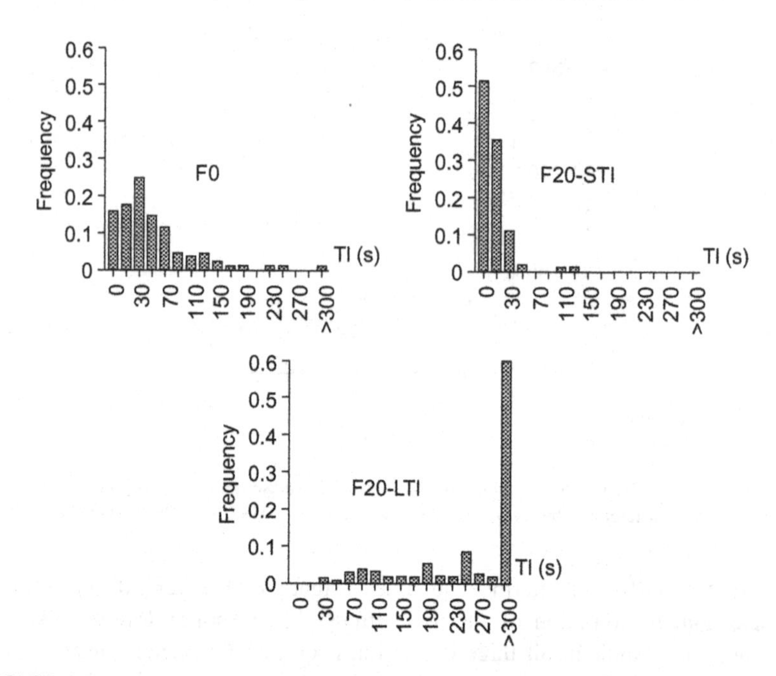

FIGURE 8.7 Distributions of the duration of tonic immobility in lines of Japanese quail selected for long or short duration of tonic immobility. The graph shows the distributions for the F0 and F20 generations.

birds. The distribution of scores for the duration of TI in the first generation was completely different from those in the 20th generation of the long TI and short TI lines (Figure 8.7). Starting from the eighth generation onwards, the standard deviations of the long TI and short TI lines ceased to

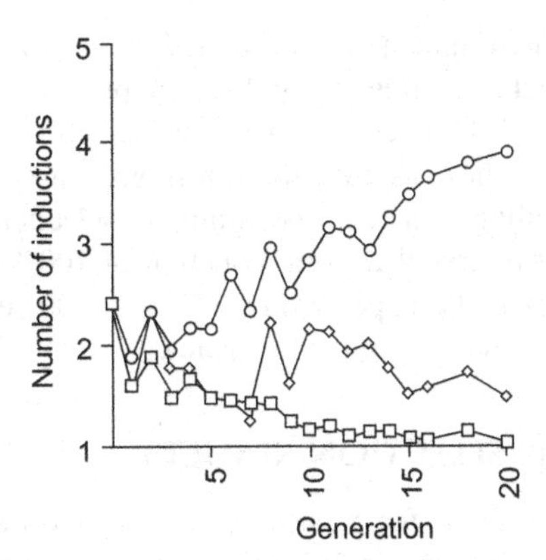

FIGURE 8.8 Number of inductions required to induce tonic immobility in lines of Japanese quail selected for long or short duration of tonic immobility over 20 generations. (□) LTI, (◇) CTI, (o) STI.

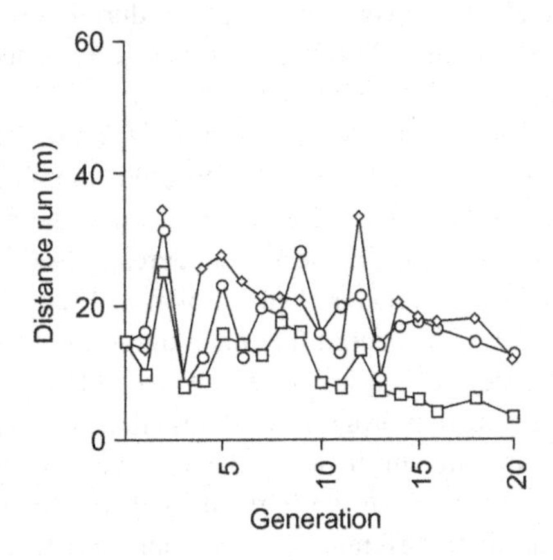

FIGURE 8.9 Distance run on the treadmill in lines of Japanese quail selected for long or short duration of tonic immobility over 20 generations. (□) LTI, (◇) CTI, (o) STI.

overlap. This implies that the long tonic immobility (LTI) and the short tonic immobility (STI) lines had also become distinct genetic lines.

After 20 generations of selection, the number of inductions necessary to obtain tonic immobility increased in the short TI line (3.88 ± 1.47), remained relatively constant in the control TI line (1.47 ± 0.75), and decreased in the long TI line (1.04 ± 0.20) (Figure 8.8).

The distance run to reinstate social contact by the different TI lines remained relatively constant and similar in all three lines up to the eighth generation. Thereafter, distance run started to decrease in the long TI line (Figure 8.9). This finding is not surprising as Jones *et al*. (1991) showed that long tonic immobility chicks (high fear) have reduced activity levels when placed in a novel environment. Reduced activity is due to increased fear levels and leads to increased freezing and the suppression of general activity.

In both the SR and TI lines, responses to selection were asymmetrical (Figures 8.2 and 8.6). These asymmetrical responses are likely to have been due to an artificial ceiling or actual floor limits to selection. For example, the duration of TI cannot be less than 0 (a reduction of 100% relative to the first generation (F0), whereas the upper limit of TI was 300 seconds (an increase of approximately 600% relative to the F0 score).

STABILITY OF THE SELECTION RESULTS

One problem with selection for a specific trait at a certain age is that the response to selection may be specific to the age at which the measure was made. We therefore verified that behavioral differences observed at 6–10 days of age were also present in older animals. Quail were tested for tonic immobility and distance run on the treadmill using the same methods employed for selection of breeding animals. However, the maximum duration of TI measurements were lengthened to 30 minutes. Different animals were tested at 1, 2, 4, 6, or 10 weeks of age Launay *et al.*, 1993; Launay, 1993).

The duration of tonic mobility varied with age (Figure 8.10). However, the scores of the long TI and short TI lines differed significantly at all ages. Comparison of Figure 8.10a and Figure 8.6 shows that the maximum duration of tonic immobility in the long TI line was underestimated because TI testing was terminated after only 5 minutes. Data shown in Figure 8.10 are for quail from the 13th generation of selection; the mean duration of tonic immobility in the long TI line at 1 week of age was over 400 seconds whereas the truncated values obtained in the selection program were less than 200 seconds.

Distance run on the treadmill was also age dependent (Figure 8.10b). However, the differences between the high SR and low SR lines were significant at all ages tested. Since differences in the characteristics selected persisted into adulthood, it seems likely that genetic selection has modified the behavior of the birds throughout their lives rather than only at a specific age.

FEAR REACTIONS

Ratner (1967) proposed that fear (antipredator) reactions may be dependent on the proximity to the predator. When the predator is distant, the animal's first reaction is freezing. Freezing behavior may serve to reduce the likelihood of detection. When the predator approaches, flight is the prevalent reaction. If the bird is not able to escape from the predator, it shows fight responses. If this fails, tonic immobility follows. The open-field test provides a means for treating this hypothesis.

The open-field test is commonly used as a means of measuring fear in birds and mammals (Hall, 1934). The open-field test involves the introduction

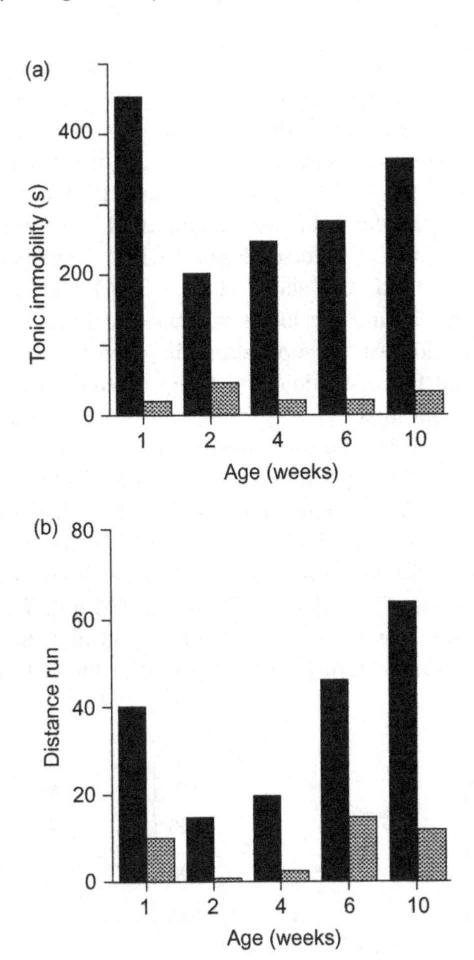

FIGURE 8.10 Duration of tonic immobility in 1-, 2-, 4-, 6-, or 10-week-old quail chicks selected for long (■) or short (▩) duration of tonic immobility over 13 generations (a) and distance run on the treadmill by 1-, 2-, 4-, 6-, or 10-week-old quail chicks selected for high (■) or low (▩) levels of social reinstatement behavior over 13 generations (b).

of an animal into a novel environment. Being placed in an open field induces a low level of fear due primarily to the novelty of the test area. Behavior in the open field may also reflect social reinstatement tendencies because an animal which is isolated in the arena may suffer a motivational conflict between fear and the desire to reinstate social contact with its companions (Faure *et al.*, 1983; Gallup and Suarez, 1980). Therefore, it was important to test the effects of selection for TI and SR behavior on the behavior in the open field. This was performed using a long-duration open-field test (30 minutes split into three periods of 10 minutes; Launay, 1993).

The various parameters used to measure open-field behavior were derived from computer-based trajectometric analysis of video images (see Mills *et al.* 1990), and direction observation of behavior (Launay, 1993). During the first 10-minute period of the open field test, the behavior of short TI (low fear) and high SR lines was of an active type (i.e. short latencies, high activity), whereas the behavior of chicks of the long TI (high fear) and low SR lines was the inverse (Figure 8.11). Immediately after being placed in the open field, the short TI (low fear) and the high SR line birds walked or ran around the arena whereas the long TI (high fear) and the low SR birds did just the opposite. These birds took a long time to start moving around the open field , and they moved very little during the first 10 minutes of the rest.

During the second 10-minute period of the test, discriminant analysis (a powerful statistical test) permitted discrimination between lines. At least for certain behavior patterns, long TI and low SR line chicks began to resemble short TI and high SR line chicks (Figure 8.12). However, during the third 10-minute period of the test, high SR and low SR lines could still be discriminated, but the long TI and short TI lines could not. These results indicate that even though the behavior of short TI and high SR (or of long TI and low SR) were similar during the first period of the test, the motivational

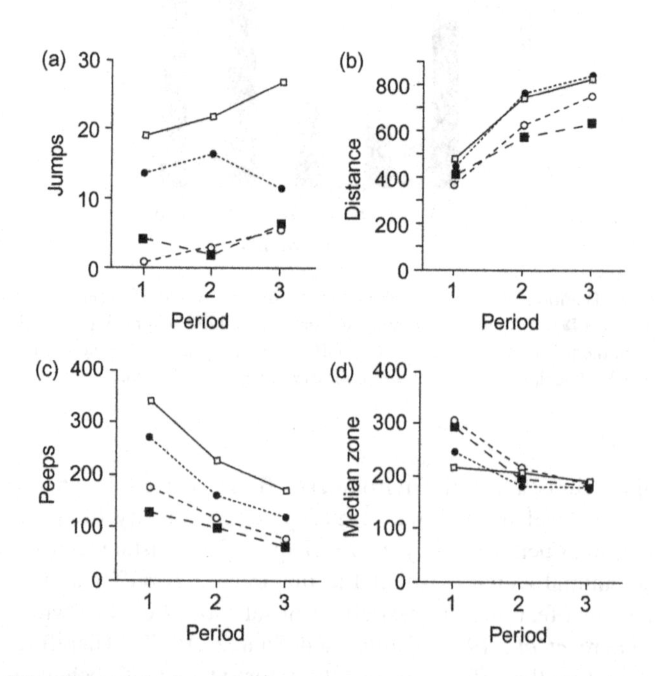

FIGURE 8.11 Changes in four behavior patterns in three successive 10-minute periods in open-field test. (a) Number of jumps, (b) distance run, (c) number of peeps, (d) time in the median zone. (□) HSR, (■) LSR, (○) LTI, (●) STI.

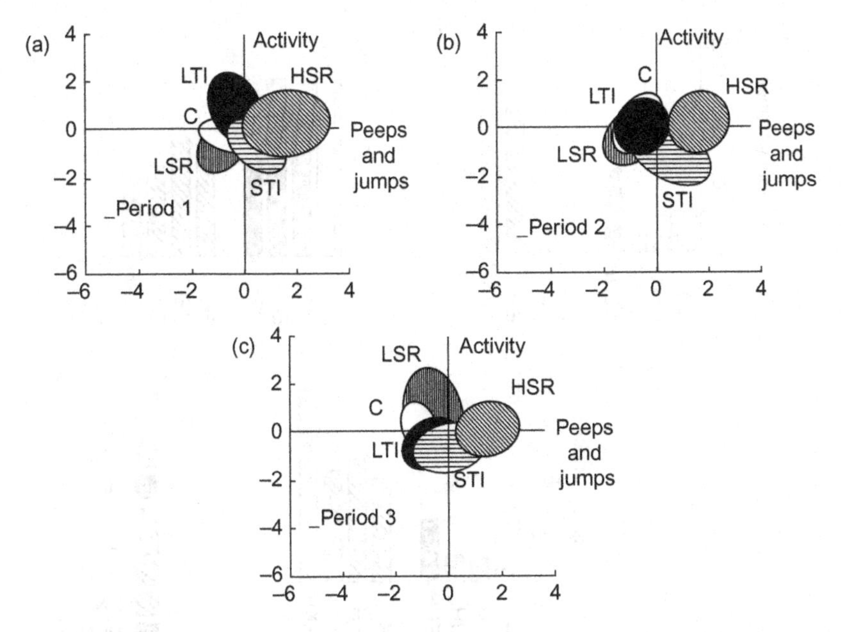

FIGURE 8.12 Discriminant analysis of the open-field behavior of lines of Japanse quail selected for long or short duration of tonic immobility and for high or low levels of social reinstatement behavior over 12 generations.

states of the lines were different. These differences in motivation can be explained by the fact that the behavior of the long TI and short TI lines are primarily motivated by differing levels of fear, but the differences between the two lines wane after habituation to a novel environment. This illustrates why it was important to conduct the open-field test with three different time periods. Fear behavior, which is observed after a short period in the open-field test, disappears after the birds habituate. Conducting the tests in three time periods also provides information on the time requires for habituation in the different lines.

The differences between chicks of the high SR and low SR lines is primarily due to differences in social reinstatement tendencies which do not attenuate or change during the 30-minute test period. As suggested by other authors (Faure *et al.*, 1983; Gallup and Suarez, 1980), these results demonstrate that open-field behavior is the expression of two different motivations: fear which leads to inhibition of activity, and social reinstatement which induces high levels of activity related to the search for social companions (see also Mills *et al.*, 1993). This also demonstrates that short open-field tests may not allow differentiation between fear and social reinstatement tendencies.

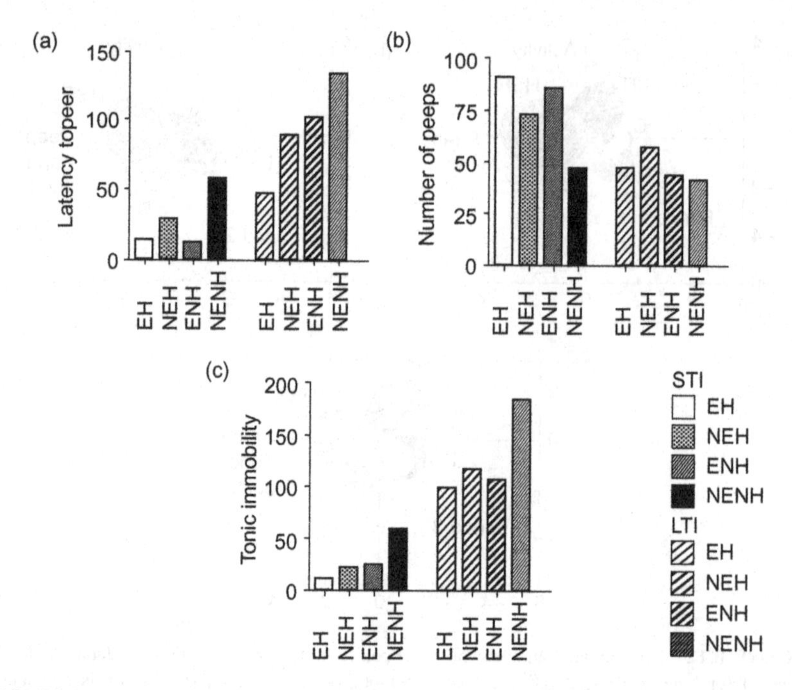

FIGURE 8.13 Variation in measures of fear in response to handling and environmental enrichment in lines of Japanese quail selected for long or short duration of tonic immobility over nine generations (EH: enriched, handled; NEH: non-enriched, handled; ENH: enriched, non-handled; NENH: non-enriched, non-handled). (a) Latency to peep, (b) number of peeps, (c) tonic immobility.

Handling and environmental enrichment are known to reduce fear levels in chickens (Candland *et al.*, 1963; Jones, 1982; Jones and Faure, 1981). When these treatments were applied to chicks of the long TI and short TI lines of quail (Jones *et al.*, 1991) they were effective in reducing fear levels in both the lines (Figure 8.13). However, there was a tendency for handling and environmental enrichment to be more effective in the short TI lines than in the long TI line. This effect might be explained by a reduction in information input in the long TI quail during handling or enrichment, attributable to greater fear-induced behavioral inhibition and fear.

The long TI and the short TI lines also had very different reactions to capture by a person. Two hundred and fifty quail of the long TI and short TI lines were reared together in a floor pen and were caught at three successive ages (2, 4, and 6 weeks of age). Birds were captured at random, and the short TI (low fear) animals were easier to catch, and were captured first at all the ages tested (Figure 8.14).

Ratner (1967) proposed that the fear (antipredator) reaction is dependent on predator proximity. At long distances, the first reaction is freezing which

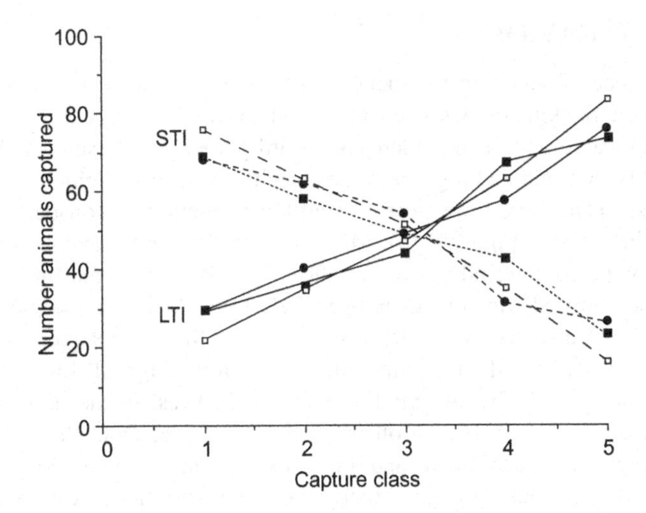

FIGURE 8.14 Proportion of birds of each line in five 100-animal capture classes of quails selected for long or short duration of tonic immobility over 16 generations at three ages: 1, two weeks old; 2, four weeks old; 3, six weeks old. LTI: (■) Age 1, (●) Age 2, (□) Age 3; STI: (●) Age 1, (■) Age 2, (□) Age 3.

may reduce detectability. At closer distances, flight reaction predominates. If flight is unsuccessful, fighting and tonic immobility may follow. The least-studied behavior in this chain of reactions is the fight reaction. To study this reaction we placed quail in a crush (squeeze) cage similar to that described by Satterlee and Johnson (1988). In this apparatus the animal's movement is restricted but bouts of fight reactions occur when the bird struggles to escape from the rush. The average number of struggles was six in the long TI line and 56 in the short TI line. Hence the reactions described by Ratner (1967) as antipredator responses were more difficult to induce in the short TI line. Differences in the threshold for the release of fear-related behavior may explain these line differences.

In the open field, where environmental novelty induces low to moderate levels of fear (see also Turro-Vincent *et al.*, 1995, for a discussion of novelty and fear responses), short TI line chicks initially showed less freezing but the difference between the short TI and long TI lines disappears toward the end of the 30-minute test when the habituation took place. In a capture experiment where man had the quasi-role of predator short TI line birds showed fewer flight reactions and were captured first (Mills and Faure, 1995). In situations which stimulate capture (crush cage and inductions of TI), short TI line birds show strong flight reactions and do not readily show the TI reaction. Furthermore, if the short TI line birds do show tonic immobility, its duration is short.

SOCIAL BEHAVIOR

The influence of selection for social reinstatement behavior on the tendency to approach flockmates has been measured in various ways. Launay *et al.* (1991) compared approach tendencies toward paired goal boxes which either were empty or contained a group of conspecific (flockmate) chicks. High SR line chicks spent more time close to goal boxes containing conspecifics than low SR line chicks (Figure 8.15). Mills *et al.* (1992) and Francois and Mills (1996) compared interindividual distances in mixed or same-line pairs of high SR and low SR line chicks in open-field tests. At 1 and 3 weeks of age, interindividual distances were greatest in the low SR/low SR pairs, intermediate in high SR/low SR line pairs, and lowest in the high SR/high SR pairs, intermediate in high SR/low SR line pairs, and lowest in the high SR/high SR line pairs (Figure 8.16). In other words, low social reinstatement chicks stand away from each other and high social reinstatement chicks group together. At 6 weeks of age, there were no significant differences in

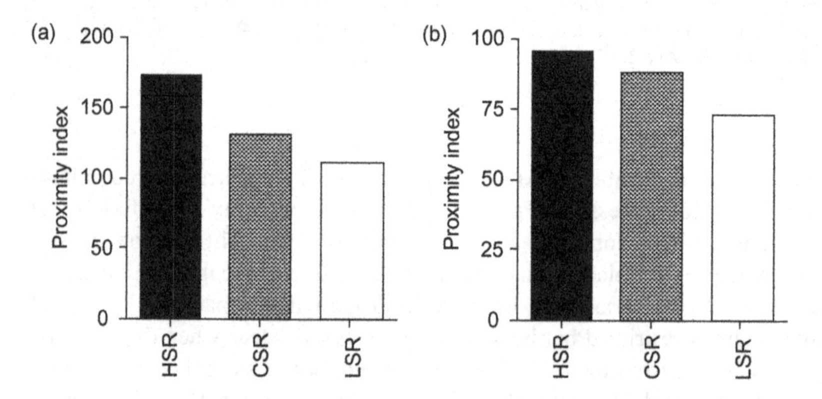

FIGURE 8.15 Proximity indices in lines of Japanese quail selected for high or low level of social reinstatement behavior over five generations. (■) HSR, (▩) CSR, (□) LSR.

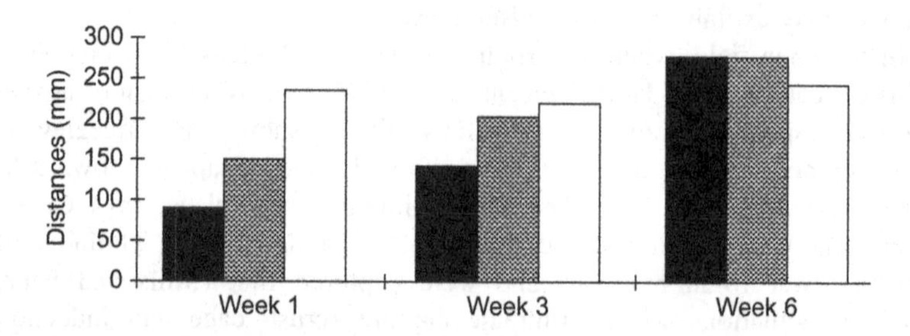

FIGURE 8.16 Interindividual distances in the open field between pairs of 1-, 3-, and 6-week-old quail from lines selected for high or low level of social reinstatement behavior over 18 generations. (■) HSR/HSR, (▩) HSR/LSR, (□) LSR/LSR.

interindividual distances between fixed or same line pairs. Given that genetic line differences in treadmill behavior persist into adulthood, this last finding may appear surprising. However, age-related variations in interindividual distances occurred only in the high SR lines. Low SR line birds appeared to space themselves at random. This implies that social factors influenced spacing in high SR line birds but not in low SR line birds.

In the high and low social reinstatement lines an age-related variation in behavior was found using the time budget test. In this test, quail of the high SR and low SR lines were given a long-term choice between being in social contact with conspecifics (flockmates) without access to food and water, and having access to food and water in the absence of conspecifics. At 1 week of age, chicks of both lines spent most of their time (about 80%) in proximity to conspecifics; the time spent close to flockmates decreased with age in both lines (Figure 8.17). However, this effect was more pronounced in the low SR line than in the high SR line. At 4 and 6 weeks of age high SR line chicks spent more time in close proximity to conspecifics than did low SR line chicks.

Under the three experimental conditions employed, the results show that the social motivation of the high SR animals varied with age and test conditions. In the treadmill test, line differences were stable from hatching until 10 weeks of age. Line differences were also present in the young animals but they disappeared by 6 weeks of age. However, in the time budget rest, line differences were absent in the young birds but appeared at 4 and 6 weeks of age. The results of the time budget experiment can probably be explained by the fact that the birds being tested were familiar with each other and the test environment. This was not the case in the treadmill and open-field tests.

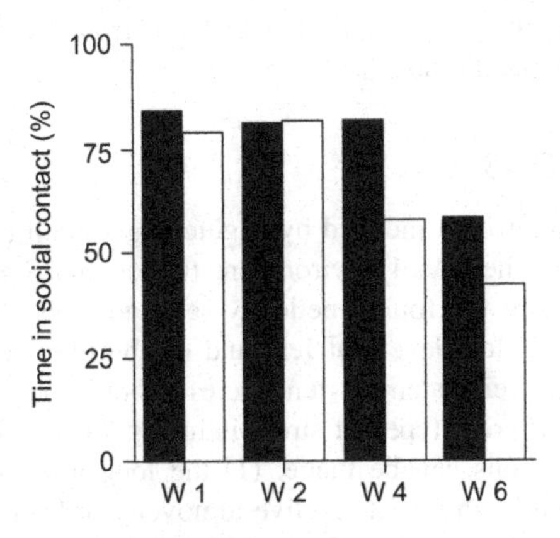

FIGURE 8.17 Time spent in social contact by 1-, 2-, 4- or 6-week-old quail from lines selected for high or low level of social reinstatement behavior over 18 generations. (■) HSR, (□) LSR.

The treadmill and time budget tests did not permit the expression of aggression because the two birds were separated from one another by a wire mesh partition. Under these conditions, these differences between adult high SR and low SR quail. In the open-field test, aggression can be expressed because there is no barrier between birds, and adult interindividual distances are identical in high SR and low SR lines. To test for differences in aggressiveness, paired encounters were staged between low SR and high SR pairs of adult birds. High SR line birds showed more aggression (particularly attack) than the low SR line birds (Magnolon, 1994; Francis, unpublished data).

Another situation in which social motivation can be measured is tests of social facilitation. In many species, social facilitation influences feeding behavior. In quail, social facilitation occurs when one bird sees another bird doing something and then imitates or increases its own expression of that behavior. To test social motivation in each of the four selected lines, groups of three quail were presented with a novel (colored) food, either within sight of birds called teachers, or in their absence. Teacher birds were preadapted to the colored food and readily consumed it. The birds being tested were also familiar with the teacher birds and control tests were run using the normal food of the test birds. There were no differences in the four lines in the control situation (normal food in the presence or absence of teachers), in time elapsed to start eating, nor in the time spent at the feeder. In the novel food without teachers situation, significant differences were found between the long TI and short TI lines (Figure 8.18). Long TI (high fear) birds took longer to eat. The longer latency to begin eating the colored food in the high tonic immobility chicks is possibly explained by higher neophobia (fear of novelty) in this line. In the novel food with teachers present test, the high SR and low SR lines differed. The low social reinstatement line had longer (latency) times to start eating and spent less time at the feeder. In this social situation, high SR line chicks appeared to learn from the teachers whereas the low SR line chicks did not.

STRESS REACTIONS

Stress reactions can also be induced by frightening stimuli (Toates, 1995), or by modifications of the social environment (Gross and Siegel, 1985; Jones and Harvey, 1987). In the four genetically selected lines of quail, the birds selected for high and low levels of fear and the birds selected for high and low levels of social reinstatement tendencies should, theoretically, react in different ways to different types of stress-inducing stimuli. If this model has validity, two predictions can be made: (1) the long tonic immobility (high fear) line is likely to be the most reactive to novelty or frightening manipulation of the test environment, and (2) the high SR lines will be more reactive to manipulations of the social environment.

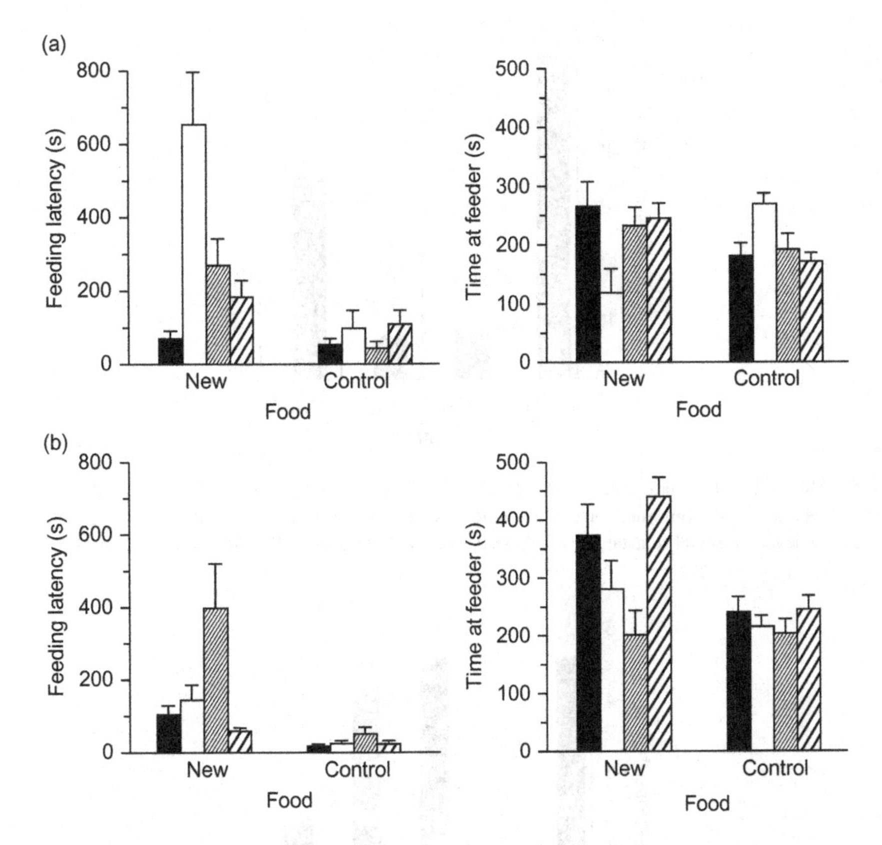

FIGURE 8.18 Feeding latency and time at the feeder during the first contact with familiar or novel food in the presence or absence of animals acquainted with the food tested in lines of quails selected for long or short duration of tonic immobility and for high or low levels of social reinstatement behavior over 14 generations. (■) STI, (□) LTI, (▨) LSR, (▨) HSR. (a) Without teachers, (b) with teachers.

To test these hypotheses, birds from the four lines were subjected to three potentially stressful situations and plasma corticosterone (stress hormone) levels of each animal were measured. The tests were: (1) the introduction of a novel stimulus (an inflated balloon into the home cage), (2) restraint in a crush cage, and (3) isolation in a novel environment (Figure 8.20). The introduction of the balloon into the home cage (Figure 8.19) induced a significant increase in plasma corticosterone levels only in the long TI line (Launay, 1993). Presumably, the more stressful experience of restraint in a crush cage as described by Satterlee and Johnson (1988) induced different behavioral and physiological reactions in the long TI and short TI lines (Jones *et al.*, 1994). During restraint in the crush cage for a period of 15 minutes, the short TI (low fear) birds showed more active struggling reactions (55.7) compared

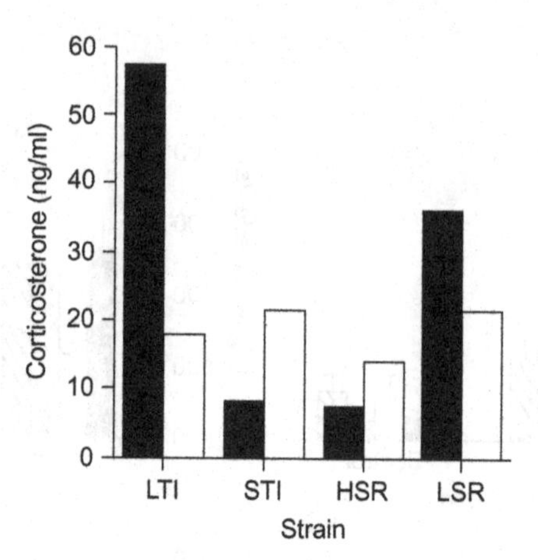

FIGURE 8.19 Plasma corticosterone levels 15 minutes after the introduction of a balloon into the cage of quails from lines selected for long or short duration of tonic immobility and for high or low levels of social reinstatement behavior over 14 generations. (■) Stressed, (□) control.

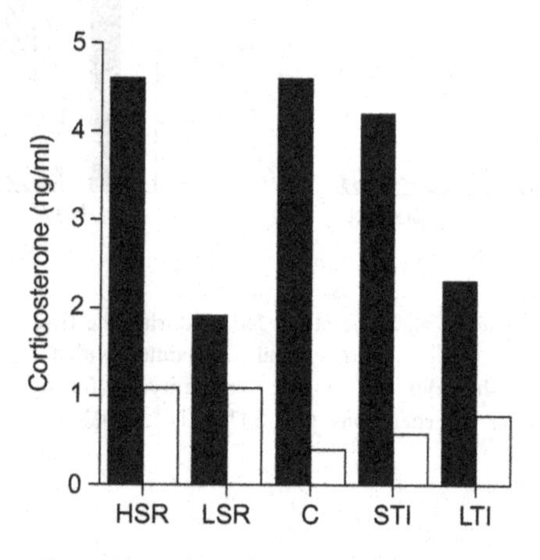

FIGURE 8.20 Plasma corticosterone levels after 15 minutes of social isolation in lines of quails selected for long or short duration of tonic immobility and for high or low levels of social reinstatement behavior over 14 generations. (■) Stressed, (□) control.

to the long TI (high fear) line birds (6.4) which typically show tonic immobility or freezing reactions. In the two selected lines and the control line, restraint induced an increase in plasma corticosterone levels that was greater in the short TI line compared to the control and long TI lines (Figure 8.21) (Jones *et al.*, 1994; Remingnon *et al.*, 1996). The higher level of struggling in the short TI (low fear) line was predictable. However, the relatively low corticosterone response in the long TI (high fear) line was unexpected

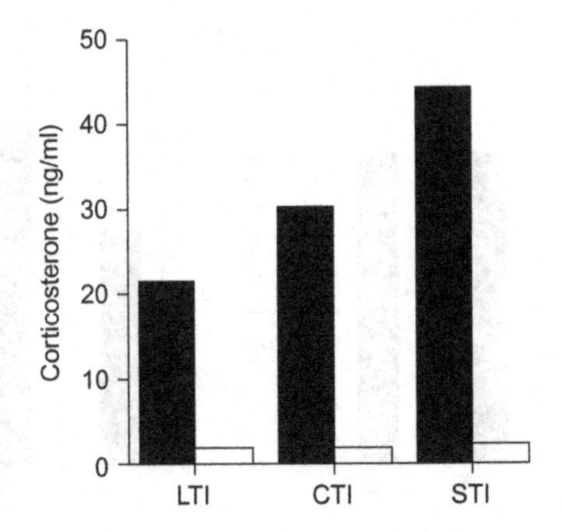

FIGURE 8.21 Plasma corticosterone levels after 15 minutes in a crush cage, in lines of quails selected for long or short duration of tonic immobility and for high or low level of social reinstatement behavior over 18 generations. (■) Stressed, (□) control.

because tonic immobility and freezing were likely to be induced by the highly stressful nature of the crush cage test. The only explanation we can suggest for this is that the TI reaction not only protects and animal from predators, but also allows the animal to cut itself off from noxious or frightening stimuli. It appears that the struggling and fighting displayed by the low TI birds was a more stressful response (as measured by stress more levels in the blood), compared to the freezing response expressed in high TI birds. Each have adaptive significance Freezing behavior protects birds from being detected by the predator, fighting and struggling may also protect the bird once the predator has caught it.

Stress reduces meat quality by increasing water loss and decreasing pH (Remingnon *et al.*, 1996). When water loss and pH decrease 24 hours *post mortem* were measured in the muscle from three lines (short TI, control, and long TI), control values for these two measures were identical. However, following stress muscle from long TI birds showed a higher percentage of water loss and a lower pH decrease than did tissue in their unstressed counterparts, from stressed or unstressed controls, and from the short TI line (Figure 8.22). In quail raised for meat, long TI birds may have reduced meat quality because they may be more likely to be stressed by handling and transport prior to slaughter.

NEUROBIOLOGY

Little is known about the effects of genetic selection on brain function. Selection for behavioral characteristics should theoretically result in neurophysiological traits. We have studied the number and binding activity of

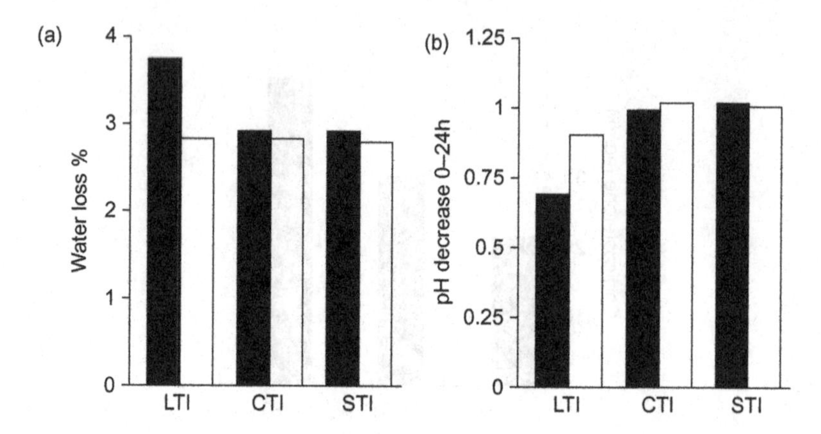

FIGURE 8.22 Meat water loss (a) and 24-hour *post mortem* pH decrease (b) following 15 minutes in a crush cage in lines of quail selected for long or short duration of tonic immobility and for high or low levels of social reinstatement behavior over 18 generations. (■) Stressed, (□) control.

neurotransmitter receptors in the forebrains of long TI (high fear) and short TI (low fear) birds in an attempt to determine neuroanatomical and biochemical differences between lines.

Neurotransmitters are "messenger" substances secreted by nerve endings which cross the synapses (the junctions between nerve endings) and bind to their specific receptor sites on adjoining nerve endings. For example, serotonin transmitters bind to serotonin receptors like a key fitting in a lock. The brain contains many different neurotransmitters and corresponding receptor sites. These transmitters have a wide variety of influences on behavior ranging from calming to excitatory effects. Behavioral effects can be affected by both the number of receptor sites for a certain transmitter and the affinity of the receptor sites. To explain it very simply, a greater number of receptors or an increase in the affinity of binding at each site will influence the "strength" of transmission. Documentation of a relationship between genetic selection for traits, such as high or low levels of fear and the number and affinity of different receptor sites, indicates that genetic selection affects brain mechanisms which influence behavior.

One of the brain areas involved in fear reactions in birds is the archistriatum (Davies *et al.*, 1996); the effects of lesions of this area have been studied (Davies *et al.*, 1997; Grignard, 1996). In young animals, archistriatal lesions have limited effects in the short TI, high SR, and low SR lines. However, such lesions decrease fear reactions in the highly fearful long TI chicks as measured by decreased latency to peep, increased number of peeps in the hole in the wall box test, and a decreased duration of TI (Davies *et al.*, 1997). Lesioned sub-adult long TI line quail were easier to capture and had reduced avoidance of novel objects, people, and unfamiliar food

(Grignard, 1996). The archistriatum has undergone some form of change during selection, and further research will attempt to determine the nature of these changes.

Hoggs *et al.* (1994, 1996) showed that the forebrains of short TI line (low fear) birds contain more benzodiazepine receptors than those of the high TI (high fear) quail and that these receptors were more sensitive to GABA stimulation in the long TI than in the short TI line. GABA is a neurotransmitter. The benzodiazepine receptors are the same receptor sites that are bound by the tranquilizing drug valium (diazepam). These findings indicate that selecting birds for either short or long toxic immobility affects both the number and the affinity of neurotransmitter sites in the brain. The sites affected by selection for either high or low fear are bound by both tranquilizing drugs and brain neurotransmitters that have a calming effect. These studies are evidence that selecting for fearful behavior affects neurotransmitters which modulates fearfulness.

Jacquet and Launay (unpublished data) also demonstrated that there is a tendency for certain plasma cathecolomines to be more abundant in the brains of LSR line birds than in those of HSR line birds. However, none of the differences reported were statistically significant. Isolation induced similar increases in the plasma adrenaline level in both lines.

There has, therefore, been some effect of selection on brain metabolism but again more work needs to be done before it will be possible to elucidate the various mechanisms involved.

CONCLUSIONS

Selection for low fear and high social motivation seems likely to produce animals that show better adaptation and have better welfare under intensive husbandry conditions. Our experiments used Japanese quail; this species appears to be a good model of the domestic chicken, and it is likely that similar selection could be conducted in that species. Furthermore, fear and social motivation are such basic motivational systems that with appropriate modification of the methods used, similar selection techniques could be used in a wide variety of domestic species.

Fear and social reinstatement behavior have general adaptive value. It is unlikely that fear reactions can be completely eliminated from any environment; all environments involve social reactions through either social behavior or social deprivation. The experiments described here show that selection for or against fear and sociality is possible using simple short-term measures and that the traits selected are not specific to a given age or environment.

The short tonic immobility line showed less fear and fewer fear-induced stress responses than did long tonic immobility line chicks in most of the situations investigated. They are therefore probably better adapted to intensive rearing conditions and would be likely to maintain greater productivity if subjected to environmental stress. High social reinstatement line chicks showed greater social affiliation ion most of the situations tested. They would therefore probably be better able to adapt to high stocking densities.

REFERENCES

Appleby, M.C., Duncan, I.J.H., McRae, H.E., 1988. Perching and floor laying by domestic hens, Experimental results and their commercial application. Br. Poult. Sci. 29, 351–357.

Candland, D.L., Nagy, Z.M., Conklyn, D.H., 1963. Emotional behavior in the domestic chicken (White Leghorn) as a function of age and developmental environment. J. Compr. Physiol Psychol. 46, 1069–11073.

Davies, D.C., Mills, A.D., Hamilton, M. Faure, J.M., Gonzales, S.I., 1997. The effect of archistriatal lesions on the behavior of Japanese quail (*Cotunix japonica*) chicks of lines divergently selected for fearfulness and sociality: in preparation.

Faure, J.M., 1981. Behavioural measures for selection. In: Sorenson, L.Y. (Ed.), First European Symposium on Poultry Welfare. Slagelsetryk Slagelse, Denmark, pp. 37–41.

Faure, J.M., Jones, R.B., 1982a. Effect of sex, strain and type of perch on perching behavior in the domestic fowl. Appl. Anim. Ethol. 8, 281–293.

Faure, J.M., Jones, R.B., 1982b. Effect of age, access and time of day on perching behavior in the domestic fowl. Appl. Anim. Ethol. 8, 357–364.

Faure, J.M., Jones, R.B., Bessei, W., 1983. Fear and social motivation in open-field behavior of the domestic chick. A theoretical consideration. Biol. Behav. 8, 103–116.

Faure, J.M., Melin, J.M. Mills, S.D., 1992. Selection for behavioural traits in relation to poultry welfare. Worlds Poultry Science Congress 19th. Amsterdam, vol. 2. pp. 405–408.

Francois, N., Mills, A.D., 1996. Inter individual instances (IID) in Japanese quail (*Coturnix japonica*) selected for high or low levels of sociality, World's Poultry Science Congress 20th New Delhi vol. 4. p. 37.

Gallup Jr., G.G., Suarez, S.D., 1980. An ethological analysis of open-field behavior in chickens. Anim. Behav. 28, 368–378.

Gentle, M.J., Hunter, L.M., Waddington, D., 1991. The onset of pain-related behaviours following partial beak amputation in the chicken. Neurosci. Lett. 128, 113–116.

Gilbreath, J.C., Garwin, L.F., Welch, Q.B., 1959. Effect of orally administered reserpine on egg production and quality. Pooult. Sci. 38, 535–538.

Grignard, L., 1996. Influence de l'archistriatum sur les retains de eur chez la calle japonaise (Cotunix japonica) D.E.A. Universite Tours.

Gross, W.B., Siegel, P.B., 1985. Selective breeding of chickens for corticosterone response to social stress. Poult. Sci. 64, 2230–2233.

Hall, C.S., 1934. Emotional behavior in the rat: defecation and urination as measures of individual differences in emotionality. J. Comp. Physiol. Psychol. 18, 385–403.

Hoggs, S., Paterson, S., Mills, A.D., File, S.E., 1994. Receptor binding in Japanese quail selected for long or short tonic immobility. Pharmacol. Biochem. Behav. 49, 625–628.

Hoggs, S., Mills, A.D., File, S.E., 1996. Differences in benzodiazepine binding in quail selectively bred for differences in tonic immobility. Pharmacol. Biochem. Behav. 54, 117–121.

Jones, R.B., 1982. Effect of early environmental enrichment upon open-field behavior and timidity in the domestic chick. Dev. Psychobiol. 15, 105–111.

Jones, R.B., 1987. The assessment of fear in the domestic fowl. In: Zayan, R., Duncan, I.J.H. (Eds.), Cognitive Aspects of Social Behaviour in the Domestic Fowl. Elsevier, Amsterdam, pp. 40–81.

Jones, R.B., Faure, J.M., 1981. The effect of regular handling on fear responses in the domestic chick. Behav. Processes 6, 135–143.

Jones, R.B., Harvey, S., 1987. Behavioural and adrenocortical responses of domestic chickens to systematic reductions in group size and to sequential disturbance of companions by the experimenter. Behav. Processes 14, 291–303.

Jones, R.B., Mills, A.D., 1982. Estimation of fear in two lines of the domestic chicken, Correlations between various methods. Behav. Process 8, 243–253.

Jones, R.B., Mills, A.D., Faure, J.M., 1991. Genetic and experimental manipulation of fear-related behavior in Japanese quail chicks (Catuni japonica). J. Comp. Psychol. 105, 15–24.

Jones, R.B., Mills, A.D., Faure, J.M., Williams, J.B., 1994. Restraint, fear, and distress n Japanese quail genetically selected for long or short tonic immobility reactions. Physio. Behav. 56, 529–534.

Launay, F., 1993. Consequences comportementales et physiologiques de selections pour l'emotivite et l'attraction sociale chez la calle japonaise (*Coturnix japonica*). These, Universite Rennes.

Launay, F., Mills, A.D., Faure, J.M., 1991. Social motivation in Japanese quail (Coturnix cotunix japonica) chickens selected for high or low levels of treadmill behavior. Behav. Processes 24, 95–110.

Launay, F., Mills, A.D., Faure, J.M., 1993. Effect of test age, line and sex on tonic immobility responses and social reinstatement behavior in Japanese quails, Coturnix japonica. Behav. Processes 19, 1–16.

Magnolon, S., 1994. Etude de l'attraction sociale et de l;agressivite chez deux souches de cailes japonaises selectionnees pour leur fotre/faible motivation sociale, D.E.A. Universite Tours.

Mills, A.D., Faure, J.M., 1986. The estimation of fear in domestic quail, Correlations between various methods and measures. Biol. Behav. 11, 235–243.

Mills, A.D., Faure, J.M., 1990. The treadmill test for the measurement of social motivation in *Phasianidae* chickens. Med. Sci. Res. 18, 179–180.

Mills, A.D., Faure, J.M., 1991. Divergent selection for duration of tonic immobility and social reinstatement behavior in Japanese quail (*Coturnix coturnix japonica*) chickens. J. Comp. Psychol. 105, 25–38.

Mills, A.D., Faure, J.M., Jones, R.B., Clement, P., 1990. Trajectometric analysis of open-field behavior in Japanese quail chicks (*Coturnix japonica*). Biol. Behav. 15, 183–195.

Mills, A.D., Jones, R.B., Faure, J.M., Williams, J.B., 1993. Responses to isolation in Japanese quail genetically selected for high and local sociality. Physio. Behav. 53, 183–189.

Ratner, S.C., 1967. Comparative aspects of hypnosis. In: Fratner, S.C., Thompson, R.W. (Eds.), Handbook in Clinical and Experimental Hypnosis. Macmillan, New York, pp. 550–587.

Remingnon, H., Desrosiers, V., Bruneau, A., Guemene, D., Mills, A.D., 1996. Influence of acute stress on muscle parameters in quails divergently selected for emotivity. World's Poultry Congress 20th, New Delhi, 1996.

Satterlee, D.G., Johnson, W.A., 1988. Selection of Japanese quail for contrasting blood corticosterone response to immobilization. Poult. Sci. 67, 25–32.

Toates, F., 1995. Stress: Conceptual and Biological Aspects. Wiley, Chichester.

Turro-Vincent, I., Launay, F., Mills, A.D., Picard, M., Faure, J.M., 1995. Experimental and genetic influences on learnt food aversion in Japanese quails selected for high or los fearfulness. Behav. Processes 34, 23–41.

ADDITIONAL READING STUDIES PUBLISHED SINCE THE FIRST EDITION

Couty, H.D., Guemene, R.S., 2008. Intensity and duration of corticosterone response to stressful situations in Japanese quail divergent selected for tonic immobility. Gen. Comp. Endocrinol. 155, 288–297.

Formanek, L., Richard-Yris, M.A., Arnould, C., Hourdelier, C., Lumineau, S., 2009. Individual behavioral rhythmicity is linked to social motivation in Japanese quail. J. Appl. Behav. Sci. 121, 126–133.

Formanek, L., Houdelia, C., Lumineau, S., Bertine, A., Cabenes, G., Richard-Yris, M.S., 2008. Selection of social traits in juvenile Japanese quail affects adult's behavior. Appl. Anim. Behav. Sci. 112, 174–186.

Ghareeh, K., Niebuhr, K., Award, W.A., Warblinger, S., Troxler, J., 2008. Stability of fear and sociality in two strains of laying hens. Br. Poult. Sci. 49, 502–508.

Guzman, D.A., Satterlee, D.G., Kembro, J.M., Schmidt, J.B., Martin, R.H., 2009. Effect of the density of con-specification on running social reinstatement behavior of male Japanese quail genetically selected for contrasting adrenocortical responsiveness to stress. Poult. Sci. 88, 2482–2490.

Hazard, D., Couty, M., Richard, S., Guemene, D., 2008. Intensity and duration of corticosterone response to stressful situations in Japanese quail divergently selected for tonic immobility. Gen. Comp. Endocrinol. 155, 288–297.

Richard, S., Wacrenler, C.N., Hazard, D., Saint-Dizier, H., Arnould, C., Faure, J.M., 2008. Behavioral and endocrine fear responses in Japanese quail upon presentation of a novel object in their home cage. Behav. Processes 77, 313–319.

Richard, S., Land, N., Saint-Dizier, H., Leterrier, C., Faure, M., 2010. Human handling and presentation of a novel object evoke independent dimension of fear in Japanese quail. Behav. Processes 85, 18–23.

Saint-Dizier, H., Leterrier, C., Levy, F., Richard, S., 2008. Selection for tonic immobility duration does not affect response to novelty in quail. Appl. Anim. Behav. Sci. 112, 297–306.

Schweitzer, C., Arnould, C., 2010. Emotional reactivity of Japanese quail chicks with high and low social motivation reared under unstable social conditions. Appl. Anim. Behav. Sci. 125, 143–150.

Schweitzer, C., Houdelier, C., Lumineaus, S., Levy, F., Arnould, C., 2010. Social maturation does not go hand-in-hand with social bonding between two familiar Japanese quail chicks. Anim. Behav. 79, 571–578.

Chapter 9

Genetic Influences on the Behavior of Chickens Associated with Welfare and Productivity

William M. Muir* **and Heng Wei Cheng**[†]
**Purdue Agriculture, West Lafayette, IN, USA; †Purdue Agriculture, Livestock Behavior Research Unit, USDA-ARS, West Lafayette, IN, USA*

INTRODUCTION

Behavior is the observable actions, mannerisms, and means by which animals cope with their environment, including other animals in that environment. The environment may be considered in terms of both the internal or physiological environment and the external environment. The latter may be subdivided, with some overlapping, into social and physical aspects. When failure to cope with changes is detected in chickens, the question arises as to whether there should be an attempt to adapt the environment to the bird or the bird to the environment (Muir, 1996, 2003; Muir and Cheng, 2004). Similarly, Cheng (2007) questions "Should we change housing to better accommodate the animal or change the animal to accommodate the housing?" Attention in this chapter is focused on genetic modification to improve the chickens to adaptability to their environments and consequently to enhance their welfare.

In the last half of the 20th Century the greatest change in the layer industry was that from floor pens to batteries of multiple-hen cages. The popularity of multiple-hen cages resulted in part because commercial producers found that multiple-hen cages offered several advantages in both economic and animal welfare aspects, including higher feed efficiencies (Cunningham, 1992), improvement of biosecurity and food safety (De Vylder *et al.*, 2011; Hannah *et al.*, 2011; Holt *et al.*, 2011; Jansson *et al.*, 2010; Jones *et al.*, 2011), and reduced housing and labor costs (Craig, 1982; de Boer and Cornelissen, 2002; Sumner *et al.*, 2011; Van Horne, 1996). Ironically, the biggest change in the 21st century may be away from cages and a return to floor pens or alternative rearing systems (California proposition 2, 2008; United Egg Producers, 2011). Craig and Swanson (1994), Goodwin (2001), Mellor and Stafford (2001), Hemsworth (2007), Swanson (2008), and Mench *et al.* (2011) examined both historical and current ethical perspectives and attitudes towards husbandry

Genetics and the Behavior of Domestic Animals. DOI: http://dx.doi.org/10.1016/B978-0-12-394586-0.00009-3
© 2014 Elsevier Inc. All rights reserved.

conditions of livestock. Although there is considerable disparity among different social and professional groups, it appears that during the last three decades there has been a definite shift in the direction of greater concern in matters of pain and suffering of farm animals and a tendency to favor protectionist legislation (Mench, 2008; Mench *et al.*, 2011; Millman *et al.*, 2010). Public pressure for increasing the well-being of chickens in Western Europe led to codes of recommendations for "humane" husbandry and, in the extreme, legislation was enacted phasing out the use of conventional cages for layers starting in 2012 in the European Union (CEC, 1999) and some states of the United States (California proposition 2, 2008). It is of special interest that the prohibition of caging hens occurred even though cages have been shown to have some advantages over alternative housing systems (Appleby *et al.*, 1992, p. 207; de Wit, 1992; Lay *et al.* 2011; Mench *et al.*, 2011), for example, cannibalism is more common in non-cage systems than in dimly lit cage houses.

Behavioral problems exist not only among chickens kept in cages but also in alternative systems (Appleby and Hughes, 1991; Craig and Adams, 1984; Lay *et al.*, 2011; Mench, 1992; Siegel, 1989). Recent reviews examined the future of the poultry industry and included issues related to politics, housing, ethical, and well-being concerns (Hester, 2005; Lay *et al.*, 2011; Mench, 2008; Mench *et al.*, 2011; Millman *et al.*, 2010; Newberry and Tarazona, 2011) but perhaps the most striking conclusion given this move away from cages was that of Lay and associates (Lay *et al.*, 2011) who concluded that there is no one clear solution to the housing issue and that no housing system is ideal from a hen-welfare perspective. In floor pens and outdoor systems, hens are exposed to litter and soil, providing a greater opportunity for disease and parasites with similar consequences from adding environmental complexity. Further, the more complex the system, the more difficult it is to manage and with larger group sizes, disease is more likely to spread. Additionally, increasing the space per hen increases incidence of cannibalism and smothering and introduces difficulties in terms of disease and pest control while adding environmental complexity increases undesirable behaviors that are detrimental to welfare (Lay *et al.*, 2011).

A promising resolution was suggested by Muir and associates (Muir, 1996, 2003; Muir and Cheng, 2004) and others (Cheng, 2007, 2010; Laible, 2009; Lay *et al.*, 2011; Newman, 1994; Cheng and Muir, 2005) who concluded that selective breeding for desired traits may help to improve welfare in domesticated settings and decrease or eliminate instances of undesirable behaviors in cages or alternative housing systems. The focus of this chapter is toward changing the nature rather than the nurture of chickens to improve welfare in domestic environments.

ASSESSMENT OF WELFARE

The words welfare and well-being are widely used but connotations vary. Dictionary definitions reflect an anthropomorphic viewpoint, e.g.,

Merriam-Webster, Inc. (1983) defines welfare as "the state of doing well esp. in respect to good fortune, happiness, well-being, or prosperity" and well-being is defined as "the state of being happy, healthy, or prosperous." To animal researchers, prosperity and happiness are unlikely to be meant, but rather absence of chronic pain and suffering, good health, ability to respond to stressors, and components of fitness (including productivity in some contexts) are likely to be included. There is a tendency, especially among Western Europeans, to place emphasis on mental conditions such as fear, frustration, anxiety, anger, and similar emotional states (Barnett and Hemsworth, 2003; Duncan and Dawkins, 1983; Duncan and Petherick, 1991; Green and Mellor, 2011; Mellor, 2012; Yeates and Main, 2008). Although most agree that emotional states should be included, it needs to be recognized that certain temporary stressors, likely to be associated with unpleasant mental states, prepare chickens to cope better with more severe stressors later in life (Gross, 1983; Zulkifli and Siegel, 1995). Life in general is a series of stressors, for humans and animals, in the quest for food, shelter, security, and reproduction. In each of these basic quests, even in natural conditions, most animals experience periods of food deprivation, disease, predation, competition, and stress. Our quest should not be to provide as natural conditions as possible, lest we expose animals to the undesirable aspects of nature as well, but rather to protect animals from such natural stressors, while ensuring a safe healthy food supply, free of disease, for human consumption.

Criteria of Welfare

Having defined, in very general terms, what is meant by well-being, it is then necessary to examine how it may be evaluated. In what follows we adopt the general categories of evidence suggested in the review of Craig and Adams (1984). More detailed descriptions may be found in other reviews (Craig and Swanson, 1994; Duncan, 2005; Mench, 1992; Webster, 2001; Yeates and Main, 2008).

Overt Signs

Bodily injury, morbidity, extreme weight loss, heavy feather loss (not associated with natural molting), and death of individuals are widely accepted as indicators that welfare is impaired. When the well-being of groups in production settings is considered, longevity, i.e. days' survival over the production period, can be used as an overt indicator. This measure takes into account all aspects of stress (Hurnik, 1990; Muir and Craig, 1998; Quinteiro-Filho *et al.*, 2012). Lifetime stress as the result of adverse environments will increase susceptibility to diseases and will result in other physiological manifestations which shorten the life-span. Vices such as feather and cannibalistic pecking also increase exposure to disease or cause death. Thus, longevity, if measured over a long period, is a direct measure of welfare.

Physiological and Immune Responses

The general adaptation syndrome (GAS) can be delineated into short-term fight or flight responses and longer duration accommodation responses involving the hypothalamic–pituitary–adrenal (HPA) and the sympathetic–medullary–adrenal (SMA) axes and is associated with immunocompetence and health. Siegel (1981) stated that, based on general metabolic effects ascribed to stress, there is a direct link between stress and susceptibility to disease and that social stress is an effective initiator of non-specific stress response in fowl. Increases in social stress reduce antibody levels and thereby lower the resistance of fowl to several important viral infections and to bacterial infections of the respiratory system. Non-specific environmental stressors can have significant influences on growth, reproduction, and the ability of the bird to resist diseases (Siegel, 1981). Dantzer (1994) indicated that animal response to stressor by changing neurohormonal homeostasis does not depend on the physical nature of the stressor. Corticosteroids, as a final compound, are produced in response to stressors. These produce many of the symptoms associated with long-term stress, including cardiovascular disease, hypercholesteremia, and modifications of the immune function (Siegel, 1995).

Although physiological (including immune) responses frequently reflect the presence of known stressors (Freeman, 1985; Gross, 1983; Hill, 1983; Hoehn-Saric and McLeod, 2000; Piazza *et al.*, 2010; Siegel, 1994; Theorell, 2012; Zulkifli and Siegel, 1995), it has been shown also that such measures can occasionally be misleading or difficult to interpret. For example, plasma corticosteroid concentrations were found to be elevated similarly for hens in high-density cages and spacious floor pens even though other measures indicated that those in high-density cages were under greater stress (Craig and Craig, 1985, Craig *et al.*, 1986).

Functional Genomics

At a recent Poultry Science Keynote Symposium—Tomorrow's Poultry: Genomics, Physiology, and Well-Being (Muir *et al.*, 2010), a novel approach to objectively quantify stress and well-being in animals and to identify management practices that necessarily compromise well-being was discussed. Toward that goal we proposed using the RNA transcriptome as a holistic tool to formulate a metric. The transcriptome represents the expression of all genes in an organism at a given time in response to environmental stimuli. The output from each gene can be quantified on a microarray, or more recently using whole-genome sequencing techniques, commonly termed RNAseq. The methodology for developing this technique was described in a landmark paper by Golub *et al.* (1999) and recently further reviewed by Myers (2011) and Kolpfleisch and Gruber (2012). They demonstrated the power of supervised clustering to diagnose cancer type in humans. In the first step cancer types were classified by a panel of experts. In the second step the transcriptome was used in a discriminate analysis to predict each cluster. In the third step the transcriptome was sampled on a new set of individuals, and the samples assigned to clusters based on the discriminate function in the previous step. The results were verified by the same team of experts and were found to be extremely accurate. Knowledge of which genes were up- and down-regulated by each cancer type was not needed, only the pattern was important. Transcritomic profiling has also been used widely in improving human well-being (Chaussabel *et al.*, 2010; Elstner and Turnbull, 2012; Hindmarch and Murphy, 2010; Mestan *et al.*, 2011).

Similarly, application of this novel approach for assessing well-being has the potential to provide common ground and address housing and other issues allowing the animals to speak for themselves through their transcriptome. The approach would include known stressors and animal-friendly environments. Exposure of the animals to those environments for short- and long-term durations, and then collection of their transcriptome from tissues in which emotions are known to be controlled, such as the hippocampus and the HPA axis. The transcriptome would then be isolated and quantified, using the latest technology, such as RNAseq. A supervised cluster analysis would then be performed into known treatments to create a discriminate function. The discriminate function classifies observations by the metric used to create the function. Next, an independent set of animals would be exposed to contentious alternative housing and management practices for short- and long-term duration, the same tissues isolated, the transcriptome quantified, and finally clustered using the previously trained discriminate function. The function will classify questionable housing and management practices into those that are most similar to those recognized as stressors and those that are animal friendly.

Alternatively, if training environments cannot be agreed on by a panel of experts, an unsupervised clustering analysis can be performed on the transcriptome of animals housed in alternative and questionable environments. An unsupervised cluster analysis will at least show if the alternative environments are really different **as perceived by the animals** in the environmental conditions imposed for whatever reason. If two environments do not induce a different transcriptome profile, it would be difficult to argue that the environments are inducing different levels of stress.

Productivity

The use of productivity or performance data as indexes of well-being in chickens has been criticized (Hill, 1983). However, as Craig and Adams (1984) made clear, early comparisons confounded profitability and productivity. When productivity of the individual laying hen is considered in terms of number of eggs, feed efficiency, and mortality, it proves to be a fairly sensitive indicator of well-being (Adams and Craig, 1985; Appleby and Hughes, 1991, pp. 116–117; Cunningham *et al.*, 1988). Productivity indicators may be useful measures of the continuing effects of stressors when hens have made an adjustment to the "stage of resistance" as defined by physiological criteria. For example, Beuving and Vonder (1978) reported and reviewed by Ruszler (1998): little or no behavior evidence of the stressful conditions of high temperature or water and feed deprivation after the first few days, though their hens stopped laying.

The relationship between stress and production also supports the notion that production is a good indicator of stress. Siegel (1995) and Scheele (1997) pointed out that stress has important consequences on the birds' well-being, especially those that affect energy and mineral metabolism and interactions with the immune system. Social or behavioral environments can also activate stress responses in birds just as physical stressors do (Cheng *et al.*, 2002, 2003). Results may manifest themselves in reduced growth in juveniles, reproductive capacity in adults and increased food consumption.

When management of broiler breeder stock is considered, a problem with interpretation of their welfare arises. Because of excessive appetites, birds to be kept as broiler breeders are severely restricted in feed intake. This practice prevents obesity of adults which if uncontrolled reduces health, longevity, and reproductive performance, and increases the incidence of leg problems (de Jong et al., 2005; Hocking *et al.*, 2004; Siegel, 1984; Tolkamp *et al.*, 2005). By the criteria cited above, productivity indicates improved well-being with broiler breeders on feed restriction. Nevertheless, feed deprivation has been shown to cause increased aggressive pecking of chickens (Duncan and Wood-Gush, 1971; King, 1965; Mench *et al.*, 1986; Shea *et al.*, 1990). Nevertheless, from the standpoint of physiological indicators, it appears that broiler breeders on feed restriction adjust rapidly (Benyi *et al.*, 2009; Katanbaf *et al.*, 1989).

Behavior

European researchers often use behavioral criteria as the primary, or sometimes the only, indicator of hens' welfare (Baxter, 1994; Faure and Mills, 1989; Savory and Hughes, 2010; Sorensen, 1981; Wegner, 1985). Dawkins (1983) was critical of such an approach and wrote:

At first sight, the growing awareness that an animal's behavior may be important evidence to have when assessing its welfare is entirely to be welcomed. But there is, amid the encouraging trend of a greater use of ethological data, also a disturbing one of possible misuse of ethology.

For example, Dawkins (1988) considered the absence of behavior patterns seen in the wild as an invalid criterion of animal suffering although such evidence is frequently used by critics of high-density cage environments.

Assessment of chronic pain by behavioral criteria appears to be a valuable technique. For example, behavioral evidence suggested that pain and heightened beak sensitivity persist for several weeks or even months following beak trimming in hens trimmed at 5 weeks of age or older (Breward and Gentle, 1985; Craig and Lee, 1990; Duncan *et al.*, 1989; Gentle, 1986b, 1989, 1992; Gentle *et al.*, 1990, 1991; Lee and Craig, 1990, 1991; Mench and Duncan, 1998). Nevertheless, interpretations of behavioral evidence can be controversial. Thus, Eskeland (1981) observed that beak-trimmed hens spent more time "resting" and concluded that fearful running by hens of low social rank was almost eliminated by beak trimming. However, Gentle (1986a) suggested that the significant increase in resting behavior among trimmed-beak birds was similar to the inactivity following injury seen in both human beings and animals (Cheng, 2005; Wall, 1979). Because of interpretation difficulties, it is valuable to supplement behavioral observations with other types of evidence as was done by Breward and Gentle (1985) and Gentle (1986a, 1989). They presented evidence that following partial beak amputation, growth of neuromas occurred in the part of the beak remaining and that spontaneous firing of afferent neurons associated with pain occurred up to at least 12 weeks after beaks were trimmed.

Several researchers, including Dawkins (1983) and Faure (1986), have used the willingness of hens to express preferences or "work" for environmental conditions allowing the expression of certain behaviors as evidence of the relative desirability of alternative environments from the hen's viewpoint. Nevertheless, possible confounding factors must be avoided or compensated for in such tests because previous experience with an environment may have a powerful effect on motivation to choose or work for it so that birds with different experiences choose differently (Byholm and Kekkonen, 2008; Campos *et al.*, 2008; Cornil and Ball, 2010; Mumme *et al.*, 2006; Petherick, *et al.*, 1990).

Behaviors seen as indicating problems with the environment, such as vacuum activities and stereotypies, are frequently cited as indicators that the environment impairs welfare in animals (Mason *et al.*, 2007; Martins *et al.*, 2012). However, Hughes (1980), Dantzer (1993), and Mason and Latham (2004) argue that the occurrence of such activities may allow animals to cope.

General

Difficulties of interpretation of the various criteria of welfare have been discussed (Clark *et al.*, 1997; Consortium, 1988; Craig and Adams, 1984; Craig and Swanson, 1994; Dawkins, 1998; Hetts, 1991; Hill, 1983; Hurnik, 1990; Izzo *et al.*, 2011; Kumar *et al.*, 2012; Mench, 1992; Muir and Craig, 1996), and there is a tendency to agree that, except for overt indicators, no single measure has previously been shown to be adequate, at least when long-term stressors are involved. Therefore, so far as possible, multiple indicators are preferable. One such multiple indicator is the transcriptome, which includes all expressed genes in the genome and thus includes all physiological responses to the environment.

DOMESTICATION AND BEHAVIOR

Behaviors associated with fitness in a natural environment may not differ greatly from those found in domesticated stock under relatively primitive husbandry conditions. Thus "native fowl" in villages of Southeastern Asia obviously cope with their environment although they may receive only minimal care such as being fed unusable by-products of subsistence farming or leftover food scraps and some protection from predators.

Those behaviors favoring domestication of a species have been described (Cheng, 2010; Hale, 1969), and chickens fit the pattern well. Behaviors that were essential in the wild may be unnecessary or even counterproductive in terms of well-being under modern systems of husbandry. Thus, extreme alertness and flighty behavior may lead to waste of energy and injuries in attempting to escape from otherwise harmless stimuli when predators are not present. Aggressiveness in obtaining desired resources may result in low-status birds' failure to receive adequate necessities even when enough is supplied for all. Broody behavior, essential to fitness in natural environments, is undesirable in domestic conditions and has been all but eliminated by breeders in egg-production stocks.

GENETIC *VERSUS* ENVIRONMENTAL INFLUENCES

As species-specific activity, behavior is under genetic control through regulating physiological homeostasis. In artificial environments, it may be difficult to know, in the absence of experimental data, whether differences between stocks are of genetic or environmental origin or if both kinds of influence are at work. For example, age at onset of egg laying, egg weight, intensity of egg production, and duration of laying have been under effective genetic selection for many generations but the stimulus of day length and artificial control of photoperiods can also have large effects. Attempts have been made to rule out environmental effects by standardizing environmental conditions, as when stocks are compared for egg production in random sample tests. Even so, stocks being compared may have different requirements for maximum performance or special characteristics which make them superior in some environments and inferior in others.

The existence of genotype by environment interactions for behavioral traits (as for production traits also) complicates the assessment of behavioral problems and possible approaches for improving the welfare of chickens by genetic means. Thus, if a particular egg-laying stock should possess a limited appetite and therefore require a more concentrated ration for optimal well-being and productivity, it would be penalized if compared with other stocks having greater appetites if all are fed a ration more suitable for the latter stocks.

BEHAVIORAL DIFFERENCES AMONG POPULATIONS

Behavioral differences, usually of a quantitative nature, are found among domesticated stocks of chickens. Most of these can be demonstrated to be of genetic origin. Often it is difficult or impossible to know how these differences arose. Likely contributors consist of genetic differences in foundation populations, random genetic drift in relatively small and isolated populations, adaptation to local environments by natural selection, and different goals under artificial selection. The impact of selection by poultry breeders is clear when comparing breeds developed specifically for beauty, fighting ability, meat production, or table-egg production. As an example of how selection for an economic trait can affect behaviors, Siegel (1989) described how long-term selection for heavy juvenile body weights in broilers resulted in their having more docile behavior, greater appetite, and reduced motor activity when compared with egg-production stock.

Comparisons among stocks are complicated in some cases because of developmental stage effects. Studies by Tindell and Craig (1959) and Bellah (1957) are especially instructive in this regard. Tindell and Craig found differences in peck-order status among hens of four breeds (White Leghorn, Australorp, Rhode Island Red, and White Plymouth Rock) when placed together in intermingled flocks and observed to determine relative peck-order status at about 5 months of age. Bellah used hens of the four breeds from the same experiment that had been kept in separated-strain flocks and determined their relative social dominance ability beginning at about 8 months of age. Strikingly different results were obtained from those of Tindell and Craig when relative social dominance ability of the breeds was compared at the later age. In the most extreme case, Rhode Island Red hens, which ranked lowest among the breeds when compared in intermingled flocks at 5 months, were the most dominant when tested in pair contests at 8 months of age.

Inconsistencies have also been found for social dominance ability among experimental stocks within the White Leghorn breed when compared at different ages. Using two populations subjected to long-term selection for number of eggs laid before 40 weeks of age, mostly in competitive floor-pen environments, and the unselected control from which they were derived, Craig *et al.* (1975) found that the selected stocks had decreased age at sexual maturity and increased aggressiveness and social dominance ability during the adolescent period. However, for one of the selected strains, rank for the agonistic traits was reversed relative to the control stock between adolescence and full maturity.

Social inertia can confound accurate determinations of relative social dominance ability after peck orders have become well established (Guhl, 1968). Lee and Craig (1981a) demonstrated this using the same stocks compared in the Craig *et al.* (1975) study. They found that the selected strain had social dominance over the control at 5 months (before egg laying), but the selected stock dominance ability had diminished or disappeared when strangers of the strains were placed together at 17 months of age. Nevertheless, selected strain pullets maintained undiminished dominance over control pullets for the entire laying year when they were kept together from time of housing at 5 months.

Comparisons among breeds, in which a single stock within each breed was used, have revealed behavioral differences of genetic origin. Such differences were found for chicks' open-field activity (Faure, 1979), presence or absence of hysteria among hens in large-group-size cages (Elmslie *et al.*, 1966), duration of tonic immobility (Campo and Alverez, 1991; Campo *et al.*, 2007; Gallup *et al.*, 1976; Jones and Faure, 1981), and pre-laying behavior of individually caged hens (Mills and Wood-Gush, 1985), and eating (Howie *et al.*, 2011). Such studies are of value in demonstrating whether genetic stock differences exist at all but could be misleading in terms of identifying breeds as having particular characteristics. For example, large strain differences have been shown within the White Leghorn breed for behavioral traits. Thus, Hansen (1976) found hysteria-susceptible and hysteria-resistant strains. Al-Rawi *et al.* (1976) and Buitenhuis *et al.* (2009) detected differences in frequency of agonistic acts; and hens of different stocks were found to differ in feather-loss and mortality from cannibalism when their beaks were left intact (Craig and Lee, 1990).

INBREEDING DEPRESSION AND RANDOM GENETIC DRIFT

Expected general consequences of inbreeding within a population are: there will be subdivision into distinctive lines or subpopulations (because of random genetic drift), uniformity will increase within subpopulations, and there will be a general loss of vigor. Although there are widespread observations over species confirming the consequences of inbreeding in general, the evidence for behavioral traits in chickens is sparse.

A study by Craig and Baruth (1965) examined the effects of inbreeding on vigor as indicated by social dominance ability of cockerels and pullets. Because of small numbers available within each of five partially inbred lines, differences among lines and uniformity within lines were not examined. However, mean performance over the lines when compared with that of a sample of the non-inbred and unselected population from which the lines were derived provided the evidence needed. Partially inbred cockerels were inferior in ability to win pair contests when matched with non-inbreds, and pullets' peck-order status was negatively associated with their inbreeding coefficients.

McGibbon (1976) observed that the incidence of floor laying in pens increased as inbreeding increased in lines originating from random-bred populations of White Leghorns (Cornell Control) and Rhode Island Reds (Regional Reds). This evidence suggests that homozygous recessive genes were involved. However, environmental conditions during the rearing period were also identified as being influential; floor laying was more prevalent among confinement-reared pullets than in range-reared pullets.

Differentiation associated with keeping relatively small subpopulations isolated over several generations was illustrated for fear-related escape and avoidance behavior by Craig *et al.* (1983). One of two subpopulations selected alike required about 10-times as long to recover from fear as did the other (Craig *et al.*, 1983). Likewise, differences in "righting time" following the induction of tonic immobility were present within some pairs of similarly selected lines (Craig *et al.*, 1984).

MENDELIAN TRAITS

Genes having major effects on morphology and the nervous system are typically deleterious in terms of the well-being of affected individuals and often affect their behavior (Axelsson and Ellegren, 2009; Eyre-Walker and Keightley, 2009; Siegel and Dunnington, 1990). Such genes are usually either sex-linked recessives, which may be expressed in non-inbred stocks, or autosomal recessives, most likely to be seen following inbreeding. Because of the generally deleterious effects of these major genes they are selected against and are not seen in commercial stocks except at very low frequency for homozygous recessive phenotypes.

Some major genes, seemingly having adverse effects, may not be as disadvantageous as they seem at first consideration. For example, results of a study carried out by Ali and Cheng (1985) are relevant. They compared genetically blind hens (rc/rc) and sighted hens (Rc/rc) segregating from matings of Rc/rc × rc/rc chickens. During a 2-month experimental period, blind hens were less active, had better feather cover, produced more eggs, required less feed, and did not differ in body weight from sighted hens.

SELECTION INVOLVING BEHAVIORAL TRAITS

Behaviors associated with welfare and productivity in commercial stocks are typically under polygenic control and are measured in quantitative terms. However, some, such as "broody" behavior, are more of an all-or-none nature. Both autosomal and sex-linked genes may be involved. Because selection studies are costly and time consuming, breeders often depend on heritability estimates in guiding choices as to whether selection is likely to change behaviors of interest. Heritability estimates often have large standard errors and are most relevant to the population in which the data for their estimation are gathered.

Effects of both direct and indirect (genetically correlated) selection responses on several relevant behaviors have been carried out and will be reviewed below. Although selection to improve behaviors associated with well-being has been effective in most experimental studies, commercial breeders are under economic constraints that may prevent their adoption of programs in which behavioral problems are addressed (Faure, 1980, 1981b; Faure *et al.*, 1992; Hocking, 1994). Nevertheless, results of some recent studies (Cheng, 2010; Cheng and Muir, 2005, 2007; Kuo *et al.*, 1991; Muir, 1996) show that major advances may be made within a reasonable number of generations in reducing behavioral problems having an economic impact.

Some behaviors are easily identified and therefore amenable to selection procedures, as in the case of broodiness, or because they leave "traces" (Faure, 1981a), as in the case of feeding behavior which can be measured from feed consumed. However, as Faure (1981a) and Hocking (1994) indicated, selection involving traits such as pre-laying pacing (see below) must be measured by laborious and time-consuming observations that are likely to be impractical for breeders who must evaluate hundreds or thousands of birds. Also, breeders know that each additional trait added to a breeding program reduces the effectiveness of selection for other traits of importance.

Broodiness

Behavior related to incubation of eggs is no longer evident or occurs with very low frequency in major commercial stocks in which broodiness has been selected against. Incubation behavior of hens was a common characteristic in most chicken stocks until a few decades ago. As the need for natural incubation and maternal care of chicks became obsolete, because of artificial incubators and chick brooding equipment, broodiness came under closer scrutiny.

Broodiness is a trait involving sex-linked genes as suggested by Warren (1930) and confirmed by later studies (Dunn *et al.*, 1998; Jiang *et al.*, 2005; Ohkubo *et al.*, 1998; Saeki, 1957). Early on it became evident that within the White Leghorn breed incidence was relatively low whereas other breeds usually had higher levels. Although Saeki (1957) obtained estimates of heritability of less than 20% in the Japanese Nagoya fowl, an early selection study by Goodale *et al.* (1920) indicated that realized selection progress was substantial in the Rhode Island Red stock used. The fact that broodiness was successfully selected against by commercial breeders is not surprising; it appears to be at least moderately heritable (Romanov *et al.*, 1999); most experiments indicated that it is negatively associated with total egg production; and it is easily identified during the first year of egg production (Saeki, 1957).

Should there be concern about the loss of a previously important behavior from chickens? We believe the absence of a behavior previously expressed is weak evidence that suffering results. Craig and Swanson (1994) discussed this as follows:

With the example of broodiness in mind, one may ask, why should behaviors that are no longer required be expressed? Presumably, most or all such behaviors could be eliminated by genetic selection between or within stocks, so that hens would not be motivated to show them or be frustrated by the absence of conditions allowing their expression.

Behavioral exhibition is controlled by both internal and external factors. Internal factors are mainly neuroendocrine factors. Neurons of the central nervous system have the capacity to change structure and functions (neuronal plasticity) to adapt to a given environment (Joseph, 1999; Kolb and Whishaw, 1998; Smith, 1993), which, in turn, affects animals' behavioral exhibition (Rosenzweig and Bennett, 1996; Sausa *et al.*, 2000). Through domestication, farm animals are quite different from their wild counterparts in many ways, including behavior. In addition, the degree of welfare could be dramatically different between strains of a species and individuals of the same strain even housed at the same environment, which is dependent on each individual's behavioral and physiological characteristics (genotypic and phenotypic variations). Adaptation of an individual, but not another, to its surrounding environment can be achieved through environmental stimulation-associated behavioral and physiological plasticity in the animal (a genetic–environmental interaction) and the experiences received during the animal's lifetime (allostasis, i.e. allostatic load or overload) (Clark *et al.*, 2007; McEwen, 2001).

Stereotyped Pacing During the Pre-Laying Period

Breeds, strains within breeds, and individuals within strains vary in stereotyped pre-laying pacing seen in single-hen cages. Such behavior may be interpreted alternatively as indicating frustration (Duncan and Wood-Gush, 1972) or the alleviation of stress (Hughes, 1980; Dantzer, 1993). The absence of nests in single-hen cages is apparently responsible for stereotyped pacing before an egg is laid in some stocks but not in others (Mills and Wood-Gush, 1985). Instead of pacing, some stocks show quiet sitting behavior during the pre-laying period. Thus, it may be conjectured that those stocks showing pacing are frustrated whereas other stocks are not. If convincing evidence becomes available that well-being is compromised in stocks that pace in the absence of nests, then choices need to be made. For example, either cages should be modified by providing nests, stocks should be used that are not frustrated by the absence of nests, or within-strain selection should be directed towards selecting for quiet sitting behavior rather than stereotyped pre-laying behavior. Mills *et al.* (1985) demonstrated, in a brief selection study, that such selection could be effective.

Increased pacing before egg laying was not seen in high-density, multiple-bird cages, as shown by Ramos and Craig (1988). Crowding in multiple-hen cages reduces the opportunity for increased pacing behavior. Even so, the question arises as to whether the environment in crowded cages differs enough from that in single cages so that results obtained in single-bird cages are relevant. The greater "enclosure" of space which exists among hens in a group may result in less anxiety than would be seen in an individual-hen cage. Therefore, the question arises as to whether selection against stereotyped pre-laying pacing, even if found to reduce stress in that environment, would also reduce stress in multiple-hen environments.

Feather and Cannibalistic Pecking

A behavioral problem occurring especially among hens kept for table-egg production is feather loss and cannibalism. In most production settings these vices are greatly reduced by beak-trimming. However, behavioral and physiological evidence, reviewed earlier, indicated that beak-trimming causes pain which persists for weeks and perhaps even months. The Brambell committee (Bramble, 1965) concluded with reluctance that de-beaking of birds should be permitted for a limited time to control outbreaks of "vice" (feather pecking and cannibalism). However, within a short period of time they hoped that suitable strains would be available and in adequate supply so that de-beaking was no longer necessary.

Craig and Muir (1993) showed that direct selection to reduce beak-inflicted injuries could be rapid when group selection was used. They kept intact-beak hens separately by sire families in multiple-hen cages and selected sire families to produce succeeding generations on the basis of hen-days without beak-inflicted injuries for 168 days from 16 to 40 weeks of age. Selected lines achieved 75% of the possible selection differential by the second generation.

Agonistic Behavior

Gamecock matches frequently terminate in the death of the loser. Selection of winners in game breeds to sire progeny of future generations has without doubt included selection for strength, agility, stamina, and other elements of behavior necessary for that outcome. Game breeds also have accelerated blood prothrombin and coagulation times (Mohapatra and Siegel, 1969) and arteries and veins that are stronger, smaller in diameter, and thicker walled as compared with White Rocks (Steeves and Siegel, 1968). Social dominance of males was strongly associated with fitness in small flocks of non-game breeds (Guhl and Warren, 1946). However, in somewhat larger flocks with higher densities, Kratzer and Craig (1980) found only a moderate correlation of male social status and completed matings. With even larger flocks and sex ratios similar to those used in commercial breeding flocks, Craig *et al.* (1977) found that social status of White Leghorn cockerels appeared to have little effect on frequency of mating.

Bidirectional selection for high and low social dominance ability, based primarily on the outcomes of initial pair contests in both sexes, was carried out over four generations in White Leghorns by Guhl *et al.* (1960). In a similar study Craig *et al.* (1965) selected on the basis of pair contests between males only over five generations within White Leghorns and Rhode Island Reds. Although realized heritabilities were only moderate, very large differences between high and low strains were produced by the end of the studies in strains derived from each of the three foundation stocks. In the more comprehensive study of Craig *et al.* (1965), there was no obvious decrease in additive genetic variation over generations for the selected trait. The availability of unselected random breeds from the foundation stocks and testing of both males and females by Craig *et al.* (1965) revealed that the responses were essentially symmetrical and were not sex specific. Peck-order status of selected strain hens in intermingled-strain flocks within breeds differed significantly, but the differences appeared to be of reduced magnitude as compared with relative status based on initial pair contests.

McBride (1962), noting the results of Tindell and Craig (1959), postulated that high aggressiveness in a genetic stock would have the same effect on productivity as he had hypothesized for poor husbandry (McBride, 1960). Specifically, he predicted that more aggressive strains would be under greater stress and would have lower mean performance levels and greater variances. Further, he postulated that strains with higher levels of aggressiveness would have a greater negative skew to their frequency distribution curves for productivity traits than found in strains having lower aggressiveness (McBride, 1968).

McBride's hypothesis was tested by Craig and Toth (1969) using hens of the high and low social dominance strains from the White Leghorn and Rhode Island Red strains developed by Craig *et al.* (1965). Strains were kept separately in floor pens and social instability was assured in half of the pens of each strain by randomly redistributing birds among flocks weekly. Although the unstable flocks had more fights, peck-avoidances, and threat-avoidances than the stable flocks (Craig *et al.*, 1969), there was no evidence of loss of productivity in the unstable flocks. A possible explanation advanced was that the greater frequency of agonistic interactions in unstable flocks were not necessarily associated with greater stress for all hens because the frequent changes in group membership would benefit those individuals that otherwise would have been low in social status over the entire test period if they had been kept in stable organizations.

Mating Behavior of Males

Two bidirectional selection studies for frequency of matings by males (Tindell and Arze, 1965; Wood-Gush, 1960) indicated that large differences could be produced readily. However, a long-term bidirectional study, in which the foundation population was maintained as a control, by Siegel and his colleagues (Dunnington and Siegel, 1983) indicated complex results. It became evident early in the study that results were not symmetrical. Although the initial response to selection in the high line was near zero in the first few generations, responses then began and accumulated until at least 20 generations had elapsed. Realized heritability was 0.18 when calculated over 23 generations. In the low line very different results were obtained; a realized heritability of 0.32 was obtained through generation 11 (Siegel, 1972), but thereafter a very large additional response appeared to be associated with little change in effective selection differentials (Dunnington and Siegel, 1983). Some males failed to mate at all during the test situation. The percentage of non-maters decreased in the high line, was unchanged in the control, and increased dramatically in the low line. During generations 21, 22, and 23 the incidence was 84, 89, and 60%, respectively in the low line.

A surprising lack of correlated responses occurred in Siegel's long-term selection study and an important genotype by rearing interaction was found. Although very large differences were present among the high, control, and low lines for the trait of selection (Dunnington and Siegel, 1983), fertility in natural matings did not differ when males were present for extended periods of time or when artificial insemination was practiced (Bernon and Siegel, 1981). Females of the lines in generations 16 through 23 were tested for mating behavior, and when analyses were carried out within generations no consistent line differences were found (Dunnington and Siegel, 1983).

The question of whether males of the high-, control-, and low-mating frequency lines would respond similarly if reared with females as compared to separate sex rearing after 6 weeks of age was answered by Cook and Siegel (1974). A large genotype by rearing environment interaction was found. Rearing males of the unselected and low lines in heterosexual flocks increased the frequency of maters by about 20% whereas essentially all males of the high line mated regardless of how they were raised.

Appetite

From a controlled five-generation study of single-trait selection on body-weight gain, food consumption, and food conversion ratio (Pym and Nicholls, 1979) it was learned that all three selected lines responded with realized heritability estimates of 0.37, 0.44, and 0.27, respectively. Of particular interest from the behavioral standpoint is that food consumption and weight gain had a realized genetic correlation of 0.71. In a further study, the similar results were reported from their 10 generations of chickens (Pym *et al.*, 1984). Therefore, selection for weight gain in broilers also involves considerable selection for appetite and *vice versa*. Food consumption records, collected in addition to body weight information, involve considerable expense in data collection but would allow selection for the economic trait of food-conversion ratio also. Presumably because of the extra expense involved and the fact that body weight and food-conversion ratio are correlated in the desired direction (−0.40 in the Pym and Nicholls study), commercial breeders have chosen to select for increased gains in body weight only. Obviously such selection has resulted in increased appetite in broiler stocks. From the considerations above, it comes as no surprise to learn (Nir *et al.*, 1978) that in comparisons of a light and heavy breed to 18 days of age, heavy breed chicks consumed only 11% more feed and failed to gain more weight than *ad libitum* fed controls of their own breed when force-fed to gut capacity as compared to light-breed chicks that consumed 43% more feed and gained 30% more body weight when force-fed to capacity as compared to *ad libitum* fed controls of their breed. Because of their huge appetites, meat-strain birds kept as broiler breeders are severely restricted in feed intake to prevent obesity of adults which would otherwise reduce health, longevity, and reproductive performance, and increase the incidence of leg problems (Decuypere *et al.*, 2010; Sandilands *et al.*, 2005; Siegel, 1984; Su *et al.*, 1999; Yu *et al.*, 1992).

Profound effects of selection for food intake and for body weight to a given age in changing appetite and associated physiological phenomena have been demonstrated by several investigators, and especially by Siegel and his colleagues (Siegel and Dunnington, 1990). The latter selected bidirectionally for 8-week body weight over more than a quarter-century. Their high and low lines diverged dramatically not only for body weight (Dunnington and Siegel, 1985) but also for specific aspects of food intake behavior (Dunnington *et al.*, 1987). Long-term selection for increased body weight apparently resulted in "genetic lesions" of brain satiety centers whereas selection for decreased body weight was offset by electrolytic lesioning of the satiety centers (Burkhart *et al.*, 1983), respectively, involving the melanocortin circuit of the brain (Hen *et al.*, 2006).

Because of poor appetite and associated low feed intake, minimal body weights and fatness required for the onset of egg laying may not be met (Bornstein *et al.*, 1984; Zelenka *et al.*, 1988). In the low-weight line described above, some females are anorexic, eating so little feed *ad libitum* that onset of lay is delayed or prevented (Dunnington *et al.*, 1984); thus over 50% of pullets did not mature in generations 25 and 26 (Siegel and

Dunnington, 1987). Zelenka *et al.* (1988) were able to induce egg production in non-layers of this line by force feeding.

Walking Problems and Tibial Dyschondroplasia

Tibial dyschondroplasia (TD), involving abnormal cartilage development in the proximal end of the tibia, occurs primarily in broiler chickens being fed *ad libitum* and the incidence is higher in males than in females. TD seriously impairs walking ability when present in advanced stages and may reduce weight gains due to the inability of birds to move freely in obtaining feed and water (Yalcin *et al.*, 1995; Whitehead, 1997; Farquharson and Jefferies, 2000; Leach and Monsonego-Ornan, 2007). It has been shown that susceptibility is genetically influenced and that selection procedures are effective in separating lines selected bidirectionally for incidence (Leach and Nesheim, 1965, 1972; Ray *et al.*, 2006; Sheridan *et al.*, 1978; Shirley *et al.*, 2003; Wong-Valle *et al.*, 1993a). Wong-Valle *et al.* (1993a) included an unselected control in their selection study and after seven generations of selection incidences at 7 weeks of age were 92.0, 5.4, and 13.2% in the high selected, low selected, and control lines, respectively.

Primary breeders are concerned about the presence of TD in their stocks as indicated by their use of an instrument which quickly and accurately diagnoses the disease in live broilers (Bartels *et al.*, 1989; Wong-Valle *et al.*, 1993a). Deleterious effects of the disease among prospective broiler breeders are minimized by weight-control management. Although results are not always consistent among studies, it appears from the selection study that body weight at 7 weeks did not differ between the high- and low-incidence lines in three of the first four generations (Wong-Valle *et al.*, 1993b). Kuhlers and McDaniel (1996) and Zhang *et al.* (1998), using data from seven generations of selection in the same stocks, indicated that TD expression and body weights at 4 and 7 weeks of age are genetically and phenotypically independent traits.

Fear-Associated Behavior

Fear is an emotional internal state which can cause physiological indicators of stress (Bronson and Eleftheriou, 1965; Lang and McTeague, 2009; Ohmura and Yoshioka, 2009; Selye, 1956). Reliable behavioral indicators of fear are difficult to establish. For example, Duncan and Filshie (1979) showed that genetic stocks that appeared to be less fearful as indicated by escape and avoidance behavior, exhibited as much fearfulness as another stock when heart rate was used as the criterion of fearfulness. Also, different behavioral indicators may yield inconsistent results when genetic stocks are compared. Murphy (1978) outlined problems in classifying genetic stocks as "flighty" and "docile". Difficulties in classification of relative fearfulness were also encountered by Craig *et al.* (1983, 1984, 1986), who found that duration of tonic immobility as a criterion of fearfulness of genetic stocks gave results that were inconsistent with those obtained when the same stocks were compared in terms of escape and avoidance behavior. Therefore, caution is appropriate in interpretation of behaviors associated with fearfulness as indicating welfare problems unless other kinds of evidence are also present.

Nervousness, feather loss, and hysteria are likely to occur in some genetic stocks, but not in others, when group size is "large" and hens have been kept for a period of several months in barren, high-density cages (Buijs *et al.*, 2009; Craig *et al.*, 1983; Lay *et al.*, 2011 Elmslie *et al.*, 1966; Hansen, 1976; Shimmura *et al.*, 2010;). Minimum group size associated with hysteria in susceptible stocks probably varies but the risk appears to rise rapidly when more than 12 are present. Repeated and severe episodes of hysteria are associated with reduced well-being because of resulting scratches, torn skin, feather loss, and reduction in feed consumption and egg production (Campo *et al.*, 2001; Hansen, 1976).

Open-Field Activity

Open-field activity has been used as a criterion of fearfulness in rodents (Bellavite *et al.*, 2009; Overstreet, 2012). The test in livestock has been recently reviewed by Forkman *et al.* (2007). Faure (1981b) selected bidirectionally for differences in open-field activity of 2- and 3-day-old chicks. The technique used was highly automated and Faure (1980) estimated that one person could measure between 500 and 1000 chicks per day. Large and significant differences were established between active and inactive lines over an eight-generation study. The inactive or more fearful line had a higher resting level of plasma corticosterone at two weeks and about twice as high a level at five weeks when stressed, but differences were not found at 6 or 25 weeks of age (Faure, 1981a). Also, hens of the more fearful line consumed less feed for a few days when a different kind of feeder was used.

Tonic Immobility

The phenomenon of induced tonic immobility (TI) has been known for centuries and its duration in chickens is generally interpreted as an indicator of fearfulness (Campo *et al.*, 2012; Dávila et al., 2011; Gallup, 1974a; Jongren *et al.*, 2010). That TI is influenced genetically was evident from large and moderate realized heritability estimates obtained in single-generation selection studies, respectively, by Gallup (1974b) with 21-day-old "Production Reds", by Campo and Carciner (1993), and Anderson and Jones (2012) with White Leghorn hens. Also, Craig and Muir (1989) obtained a moderately large heritability estimate on White Leghorn hens which had been kept in three-bird cages for over 20 weeks. That the housing environment can have a significant effect on the duration of TI has been shown by Jones and Faure (1981), Kujiyat *et al.* (1983), Campo *et al.* (2008), and Lay *et al.* (2011).

Escape and Avoidance ("Flighty") Behavior

Consistent and significant differences in escape and avoidance behavior of hens from two genetic stocks of White Leghorns were found for each of three stimulus situations believed to cause fear (Craig *et al.*, 1983). Human observers were involved in two of the tests and a mechanical device was the primary source of stimulation in the third. In addition to stock differences, significant sire family effects, indicative of genetic variation within strains, were also found. Within genetic stocks, negative phenotypic correlations were detected between criteria used as indicators of fearfulness and feather loss; cages with more fearful hens lost more feathers presumably because of abrasion with cages and trampling and scratching. Cages held from 10 to 14 birds each. In a later study, involving hens of a random-bred White Leghorn stock kept in three-hen cages, Craig and Muir (1989) obtained heritability estimates ranging from a nonsignificant value of 0.08 to a significant 0.34.

Potential Problems with Selection on Behavioral Traits

Wegner (1990) suggests that welfare could be improved through adaptation by selecting against frustration, restlessness and stereotyped pacing before laying, and a greater tendency to sit during the pre-laying period. However, direct selection on either behavior or physiological objectives should be viewed with caution. The intended results may not be as expected. For example, Webster and Hurnik (1991) showed that traits associated with non-aggression, such as sitting and resting, were negatively correlated with productivity. Furthermore, the link between behavior and stress is misinterpreted. For example, Duncan and Filshie (1979) showed that a flighty strain of birds that exhibited avoidance and panic behavior following stimulation returned to a normal heart-beat sooner than a line of more docile birds, implying that docile birds may be too frightened to move. Therefore, is flightiness good or bad for well-being?

An example of problems that can occur if selection is directed at the physiological responses to stress was provided by Gross and Siegel (1985). They were successful in selecting lines of birds for high and low plasma corticosterone in response to social strife. Further testing (Siegel, 1993) showed that the birds did not differ in their corticosterone response to a non-social stressor. Siegel concluded that genetic selection altered the perception of the animal to stress rather than involving the GAS directly. Even worse, Siegel (1993) noted that the low line in a low-strife environment was more susceptible to infections from endemic bacteria and external parasites while the high line in a high-stress environment was more susceptible to viral infections. Thus direct selection for low or high immune response would compromise the birds' welfare as measured by longevity. Also, correlated responses on productivity cannot be ignored. Birds with a high genetic potential for immunorespossiveness must divert energy from growth and reproduction to the immune system.

SELECTION INVOLVING PRODUCTION TRAITS

Antagonisms

There is a natural antagonism between some undesirable behaviors and egg production. Research has shown that selection to improve productivity based on individual bird productivity was associated with increased aggressiveness (Bhagwat and Craig, 1977, 1978; Craig *et al.*, 1975; James and Foenander, 1961; Lowry and Abplanalp, 1972; Lee and Craig, 1981a; Tindel and Craig, 1959). From the reverse perspective, Craig (1994) has shown that selection for social dominance will reduce performance when hens are housed in a large group, but when housed singly, performance is increased. Bhagwat and Craig (1978) found that social dominance increased in response to selection for egg mass. Alternatively, direct selection for social dominance has been shown to reduce performance when hens are housed in groups (Craig, 1994). Craig *et al.*, 1965 and Craig and Toth (1969) showed that hens of White Leghorn lines selected for high social dominance had lower rates of egg production and higher mortality than did hens of the same line selected for low social dominance. In addition, Craig (1970) showed that the high-social-dominance line withstood crowing less well than the low-social-dominance line. However, in single bird cages egg production of the high line was superior to that of the low. Biswas and Craig (1970) also showed that the high-strain hens had much lower production than the low line in floor pens or multiple-bird cages, but were equally productive in single-bird cage. Craig *et al.* (1975) demonstrated that stocks selected for part record egg production, based on individual performance, exhibited more social dominance and aggressiveness.

Other associations that are relevant include the finding that a stock selected for increased productivity had greater feather loss than its unselected control when kept in three-bird cages (Lee and Craig, 1981b). Also, Choudary *et al.* (1972) compared four commercial lines of poultry and found that the line which had the highest day rate of lay had the lowest hen-housed rate of lay due to high mortality. Results from Lowry and Abplanalp (1970, 1972) showed that strains selected under floor-flock conditions became socially dominant to both those selected in single-bird cages and unselected controls. Craig and Lee (1989) detected a strong genotype by beak-treatment interaction for egg mass per hen housed among three commercial lines. The genotype that produced the greatest egg mass with beak treatment produced the least with intact beaks. The re-ranking was shown to be due to mortality from beak-inflicted injuries.

Results from Emsley *et al.* (1977) suggest that flightiness is associated with higher productivity. They estimated genetic correlations between egg production and flightiness score which showed that greater excitability was mildly associated with higher rates of lay. Kashyap *et al.* (1981), attempting to break the correlation between production and flightiness, selected layers using a selection index for aggregate economic gain based on genetic parameters estimated by Emsley *et al.* (1977), which gave positive weight to egg number and negative weight to excitability or flightiness. Nevertheless, their results showed a positive response in excitability. Bennett *et al.* (1981) reanalyzed those with retrospective indices and showed that genetic changes in excitability were actually greater than what would have been predicted by theory.

The Environment of Selection

Biswas and Craig (1970) compared White Leghorn hens of lines selected for high- and low-social-dominance strains in single-hen cages and floor pens and in crowded and un-crowded conditions in both cages and floor pens. Significant interactions between genetic stock and environment were present with reversals in ranking between environments. In the first study, high social dominance hens matured earlier and laid more eggs per hen housed when kept in single-hen cages but were later in maturity and laid fewer eggs when kept in within-strain social groups in floor pens. In the second study, changes in rank of performance were also found for age at first egg and hen-housed egg production when the strains were compared in cage and floor pen environments. High-strain hens were lower in performance than low-strain hens when compared in intra-strain groups in floor pens or in multiple-hen cages but were more or equally productive when kept in individual cages.

A significant genotype by cage-environment (single *versus* multiple) interaction for days' survival was observed in a random-bred population of White Leghorns (Muir, 1985). However, such an interaction can be caused by a change in variance or a change in ranking. Muir *et al.* (1992) showed that the interaction was mainly due to re-ranking of genotypes. Thus, the bird which does the best in one environment is much poorer in the other and *vice versa*.

Although there was no significant interaction for eggs per hen housed, Muir (1985) showed that this result was due to a quadratic relationship of surviving group size with rate of lay, whereby cages with seven surviving birds had a higher rate of lay than those with eight or nine. Estimation of genetic correlations assumes a linear relationship between genotype and phenotype. Therefore, to improve overall eggs per hen housed it is not only necessary to measure production in multiple-bird cages but to select separately on rate of lay and days' survival.

Therefore the problem of adaptation is a joint one: the environment of selection and the traits of selection. The usual solution to eliminate, or guard against, genotype by environment interactions in production animals, is to select the animal in the environment in which it is to perform, i.e. to make the basis of selection performance in multiple-bird cages rather than in single-bird cages. However, that solution will not work in this case and could actually make the situation worse. Selecting the highest performing individual in a group environment will select for the more dominant birds.

From a genetics perspective, the environment is not constant but is constantly changing because the environment is defined not only as that of a group but also of the individuals within that group. Individual genotypes in groups are constantly changing as selection progresses, i.e. the associate effects of other genotypes in the population are involved. Thus, theory predicts that selection based on the individual will be antagonistic to group performance.

Although the studies cited indicated major effects of agonistic behavior on performance in egg-production stocks, most breeders select their elite stocks on the basis of records obtained in single-hen cages (Hunton, 1990). The genotype by housing environment interactions found in these studies emphasize the risk taken by breeders who select for egg production on the basis of hens' performance in single-bird cages for stocks to be housed commercially in social groupings.

MULTI-LEVEL SELECTION

Siegel (1989) considered adaptability to be an individual's ability to adapt to its environment. He concluded that individuals that adapt have a higher probability of contributing genes to subsequent generations than those that do not. This concept emphasizes the individual. What if an individual adapts to its environment by eating its cagemates? Survival of the individual is maximized, but what of that of the group?

There are numerous ways that performance of one individual can influence that of another. Accommodation for such interactions presents an insurmountable dilemma from the point of view of classical (non-interaction) quantitative genetic methodology. Griffing (1967) recognized that with competition, the usual gene model for a given genotype must be extended to include not only the direct effects of its own genes, but also the associate contributions from other genotypes in the group. The problem is to optimize production of a given genotype in a competitive environment. As a consequence of interacting genotypes, the same genotype can have different expressions in populations having different population structures.

Griffing (1967, 1977) extended selection theory to take into consideration interactions of genotypes. The conceptual biological model was extended to define the group and the usual model was extended to include not only direct effects of its own genes, but also associate contributions from other genotypes in the group. It is also of special note that Griffing (1967) shows that selection only on associate components cannot guarantee a positive response to selection, i.e. selection for reduced aggression will not ensure that production will increase.

The modern general classification of methods that use group selection or some combination of group and individual selection is now termed multi-level selection. Multi-level selection is the direct analogy of the classical index selection results based on the non-interaction model. Let Y_{kl} be the phenotype of the lth individual in the kth group (or family), then, the between and within group deviations can be combined in an index, $I_{kl} = B_1\tau_k + B_2\gamma_{(k)l}$, where $\tau_k = (\overline{Y}_{k.} - \overline{Y}_{..})$ and $\gamma_{(k)l} = (Y_{kl} - \overline{y}_{k.})$ are respectively the between- and within-group deviations, and B_1, B_2 are weights. Classical group selection occurs when $B_1 = 1$ and $B_2 = 1$, within group selection results when $B_1 = 0$ and $B_2 = 1$, and multilevel selection results when $B_1 \neq 0$ and $B_2 \neq 0$.

Group Selection Experiments

There is limited, but ample, experimental evidence to support Griffing's (1967) theory. The first experiment was that of Goodnight (1985) who showed that leaf area of *Arabidopsis thaliana* would respond to group but not individual selection. The first experiment to use group selection with chickens was unsuccessful (Craig *et al.*, 1982). Craig (1994) reflected that the reason for lack of response was because he had not provided an environment in which the hidden genetic variability could be expressed, i.e. the beak trimmed hens, density was low, and only a part record was used.

The first successful experiment with group selection in poultry was initiated in 1981 at Purdue University. After four generations of group selection based on half-sister families housed initially in groups of nine, and later in groups of 12, performance was compared between the selected and control lines in six-hen cages with 387 cm^2 floor space per bird from 16 to 40 and 16 to 36 weeks of age, respectively, by Kuo *et al.* (1991) and Craig and Muir (1991). In the second study a stock derived from a competitive commercial stock by two generations of relaxed selection was included also. In both experiments, hens had their beaks left intact or beaks were trimmed to two different lengths.

Results of the first study showed highly significant beak-treatment by genetic stock by age interactions for hen-housed rate of lay and daily egg mass; as mortality from cannibalism increased dramatically with age in the control, but not in the selected stock, differences between the selected and control became greater. Somewhat similar results were obtained in the second study; with intact beaks, the selected line had significantly higher egg production, egg mass, and survival than its control. However, with two-thirds of the beak removed, differences in egg production, egg mass, and survival were no longer as evident. Differences in production were presumably due to stress induced by birds with intact beaks. In comparisons between the selected and the commercially derived stock, Craig and Muir (1991) found that egg production was at about the same level with all three levels of beak trimming. However, the selected stock had significantly better survival when beaks were left intact. Craig and Muir (1991) hypothesized that kin selection favored cooperative or at least tolerant behavior as suggested by Crow and Kimura (1970) and concluded that selection on family means when families are reared as family groups provides a method of improving traits in which behavioral interactions influence overall well-being and productivity.

Muir (1996) reported that after six generations, in comparison to the unselected control, annual per cent mortality of the selected line in multiple-bird cages decreased from 68% in the initial generation to 8.8% in the sixth generation. Per cent mortality in the sixth generation of the selected line in multiple-bird cages was similar to that of the non-selected control in single-bird cages (9.1%). Annual days' survival improved from 169 to 348 days and rate of lay improved from 52 to 68%. Annual egg mass improved from 5.1 to 14.4 kg per bird. The dramatic improvement in livability demonstrates that adaptability and well-being of these birds were improved by group selection. The similar survival of the selected line in multiple-bird cages and the control in single-bird cages suggests that beak-trimming of the selected line would not further reduce mortalities which implies that group selection can eliminate the need to beak-trim. An independent study by Craig and Muir (1993), involving different foundation stock, adds confirmatory evidence; kin-selection for days survival of intact-beak hens was dramatically increased relative to the unselected control over a two-generation study. In the Muir (1996) study, corresponding improvements in rate of lay and egg mass demonstrated that such changes can also be profitable.

Craig and Muir (1996) compared the selected and control lines to a commercial line in generation 7. Birds were not beak-trimmed and lights during the laying period were set to high intensity. Birds that died were replaced with extra birds of the same line. In single-bird cages performance, as measured by eggs per hen housed, the commercial line was superior to that of the selected line but in 12-bird cages performances were reversed. The difference in ranking was due to both an improved rate of lay and viability. In the same study, Craig and Muir (1996) observed that feather scores did not differ in single-bird cages among genetic stocks. However, in 12-bird cages, the selected line had significantly better feather score than the other lines.

The lines, described by Craig and Muir (1996), were subjected to the stress of housing at about 17 weeks, to cold stress at 36 weeks, and to heat stress at 47 weeks of age, with results as reported by Hester *et al.* (1996a, 1996b). Blood physiology and egg production were monitored before, during, and after each of these periods. Packed cell volume immediately after housing indicated that the selected line adapted to the new watering system more quickly than the other lines. During cold stress the commercial and control lines showed an increase in heterophil to lymphocyte ratio in 12-bird cages while the selected line did not. Egg production before, during, and after stress indicated that the selected line withstood social, handling, and environmental stress better than the control and in some case the selected line. Similar observations with heat stress showed that the selected line withstood heat stress better as indicated by a lower mortality than the control or commercial lines. Egg production before, during, and after heat stress indicated that the selected line withstood social, handling, and environmental stress better than the control line and in some cases the commercial. Cheng *et al.* (2001a, b) studied the same line compared with the revised selected line (line with low production and high mortality) housed in single-bird cages. At 21 weeks, the selected line had a higher CD4 to CD8 lymphocyte ratio and lower concentrations of dopamine and epinephrine and lower ratios of epinephrine to norepinephrine and heterophil to lymphocyte ratios. The bidirectional selection has resulted in the change in the immunity and the expression of the neuroendocrine systems. In social stress studies using the same lines plus a commercial line, DeKalb XL, Cheng *et al.* (2002, 2003) and Cheng and Fahey (2009) reported that the selected line has higher resistance to various social stressors than the other two lines. These changes magnify the differences in productivity and survivability observed in the lines under basal and challenged conditions.

Conclusions from Group-Selection Studies

General conclusions were summarized by Muir (2003) and Muir and Cheng (2004). Selection to reduce the major economic and welfare problems of feather pecking and cannibalism among laying hens may be successfully achieved by either direct selection against pecking injuries or by means of correlated responses associated with group (kin) selection for adaptation to multiple-hen cages as indicated by mean hen-day rate of lay and survival.

Multi-Level Selection Experiments

Although (Griffing 1977) had developed equations to determine optimal weights for between- and within-group deviations, those methods were only published in a conference proceedings and were largely unknown. Further, the results were academic at the time because methods to estimate genetic parameters associated with the model did not exist as Henderson's (Henderson 1975, 1984; Henderson & Quaas, 1976) mixed model methods were just being developed and was limited to models with one random effect.

In order to estimate genetic parameters associated with the model, Muir and colleagues (Muir 2005; Muir & Schinckel 2002) recast Griffing's model in terms of a mixed model methodology, with two random effects, one for direct effect of the allele on the phenotype and another for effects on associates. Griffing termed the social or competitive effects, as "associative effects", which are now commonly referred to as indirect genetic effects (IGEs) (Agrawal *et al.*, 2001; Bijma 2010a, b; Bijma and Wade, 2008; McGlothlin and Brodie, 2009; Wade *et al.*, 2010). An important result of the mixed model approach of Muir (2005) and later (Bijma *et al.* 2007a) is that genetic variances and covariances associated with direct (σ_D^2) indirect (σ_A^2), and their covariance (σ_{DA}) could be estimated by use of REML (Patterson and Thompson, 1971). These genetic parameters, in combination with weights for the between- and within-group deviations, unified the concepts of multilevel (kin and non-kin), group, and within group selection (Bijma *et al.*, 2007a, b; Griffing, 1977). Bijma *et al.* (2007a, b) used the term g to define the strength of multi-level selection, ranging from 0 to 1 corresponding respectively to within and between group selections.

As a result of the two-component model of Muir (2005), the direct and IGEs can be directly estimated for each individual using best linear unbiased prediction (BLUP) (Henderson, 1975, 1984; Henderson and Quaas, 1976). As a result, any combination of weights could be placed on those effects resulting in optimal selection (Bijma *et al.* 2007b). Muir and colleagues (Muir, 2005; Muir and Schinckel, 2002) used this approach, in comparison with individual selection, to select for increased 6-week weight in Japanese quail housed in multiple random groups of size 16. We used a base index in which the IGEs were summed over all $(n-1)$ interacting genotypes in a group of size n, plus the direct effect known as the total breeding value [TBV, (Bijma *et al.*, 2007a, b)]. We demonstrated that selection for TBV was effective in increasing 6 week weight while reducing mortality. In contrast, individual selection did not increase 6 wk wt while having the negative effect of increasing the rate of mortality.

An interesting consequence of defining the index as the trait of selection, in terms of the TBV, is that the total heritable variance can exceed the observed phenotypic variance (Bijma *et al.*, 2007a) indicating that the response to selection can be greater than the selection differential, and explains the results we observed with poultry where the realized heritability in the initial generations were >1. Basically the response to selection includes the heritable social environment, not just the direct effects

Muir and Schinckel (2012) examined multi-level selection for quail housed in kin groups for 43 days weight and survival in Japanese quail. We showed that multi-level selection occurs whenever BLUP selection is applied if quail are housed in groups because BLUP weights all information from relatives, in addition to that of the individual. If families are housed in kin groups, then BLUP selection has elements of both kin and group selection. We showed that multi-level selection in kin groups was effective in reducing detrimental social interactions, and reduced mortality, which contributed to improved weight gain. Simple multi-level selection using standard animal model BLUP was easy to implement, only requiring that animals be housed in kin groups.

GENERAL CONCLUSIONS

The wild progenitors of domesticated chickens possessed a number of traits, including absence of highly specific environmental requirements that preadapted them to domestication. They continue to show adaptability and have responded to man's selection as seen by wide diversity among breeds and types. Nevertheless, because of large and rapid changes in housing systems and husbandry practices during the last 50 years, questions arise as to whether their welfare is compromised and whether genetic adaptation has been adequate. Criteria used in assessment of chickens' well-being include overt indicators (bodily injury, morbidity, weight loss, non-molt feather loss, death, and mean days survival of groups), physiological and immune responses, productivity, and behavior. With the exception of overt indicators, difficulties of interpretation of the various criteria lead to the conclusion that, so far as possible, multiple indicators should be used. Nevertheless, because behavior is the means by which animals attempt to cope with their environments, behavioral observations can yield valuable clues as to what is causing impaired welfare, and it is clear that behavior-related problems can themselves have major adverse consequences.

Most behavioral differences found among domesticated stocks of chickens can be shown to be determined genetically and are under polygenic influence. However, it is often difficult to know whether those differences arose because of foundation population differences, random genetic drift, adaptation to local environments, or different goals under artificial selection. The tendency to assign behavioral differences to breeds needs to be moderated when only a single strain within a breed has been used for characterization. Long-term selection studies for behavioral or economic traits that succeed have shown that profound correlated responses are likely to accompany the primary responses.

Although selection to improve behaviors associated with well-being has been effective in most experimental studies, commercial breeders are under economic constraints that may prevent their ready adoption of programs to address those problems. Behaviors that are easily identified (such as broodiness) or that leave traces (such as feed consumed) are more amenable to selection than those that must be measured by time-consuming observations (e.g., pre-lay pacing, fear-related and aggressive behavior). It has also been noted that direct selection could be disappointing if such selection reduced efficiency of production. It has been shown that kin selection was effective in reducing a major social-behavior problem when the consequences were readily evident. Thus, cannibalistic behavior by quail was reduced by selecting individuals on the basis of weight when kin were caged together. This method is easy to implement and does not require behavior to be directly measured. The impacts of behavior are evident in performance traits. An alternative approach, also utilizing kin selection for improvement of well-being of birds in multiple-hen cages, was based on the hypothesis that when productivity is high, behavioral stresses must be reduced or absent. Thus, increased adaptation of layers to multiple-hen cages was attained by selecting families that had high egg production and survival even when kept together under conditions that would stress less-well-adapted stocks (hens with intact beaks in high-density, larger-group-size cages). That an impressive response was obtained within a few generations indicates that this approach deserves further attention commercial conditions and that eventually management practices such as peak trimming will no longer be necessary.

REFERENCES

Adams, A.W., Craig, J.V., 1985. Effect of crowding and cage shape on productivity and profitability of caged layers: a survey. Poult. Sci. 64, 238–242.

Agrawal, A.F., Brodie, E.D., Wade, M.J., 2001. On indirect genetic effects in structured populations. Am. Nat. 158, 308–323.

Ali, A., Cheng, K.M., 1985. Early egg production in genetically blind (rc/rc) chickens in comparison with sighted (Rc ± rc) controls. Poult. Sci. 64, 789–794.

Al-Rawi, B., Craig, J.V., Adams, A.W., 1976. Agonistic behavior and egg production of caged layers: genetic strain and group-size effects. Poult. Sci. 55, 796–807.

Anderson, K.E., Jones, D.R., 2012. Effect of genetic selection on growth parameters and tonic immobility in Leghorn pullets. Poult. Sci 91, 765–770.

Appleby, M.C., Hughes, B.O., 1991. Reviews: welfare of laying hens in cages and alternative systems: environmental, physical and behavioural aspects. World's Poult. Sci. J. 47, 109–128.

Appleby, M.C., Hughes, B.O., Elson, H.A., 1992. Poultry Production Systems: Behaviour, Management and Welfare. CAB International, Wallingford, Oxon, England.

Axelsson, E., Ellegren, H., 2009. Quantification of adaptive evolution of genes expressed in avian brain and the population size effect on the efficacy of selection. *Mol* Biol. Evol. 26, 1073–1079.

Barnett, J.L., Hemsworth, P.H., 2003. Science and its application in assessing the welfare of laying hens in the egg industry. Aust Vet J. 81, 615–624.

Bartels, J.E., McDaniel, G.R., Hoerr, F.J., 1989. Radiographic diagnosis of tibial dyschondroplasia in broilers: a field selection technique. Avian. Dis. 33, 254–257.

Baxter, M.R., 1994. The welfare problems of laying hens in battery cages. Vet Rec. 134, 614–619.

Bellah, R.G., 1957. Breed Differences Among Chickens as Related to Compatibility When Reared Together (M. S. Thesis). Kansas State University, Manhattan, KS.

Bellavite, P., Magnani, P., Marzotto, M., Conforti, A., 2009. Assays of homeopathic remedies in rodent behavioural and psychopathological models. Homeopathy 98, 208–227.

Bennett, G.L., Dickerson, G.E., Kashyap, T.S., 1981. Effectiveness of progeny test index selection for filed performance of strain-cross layers. II. Predicted and realized responses. Poult. Sci. 60, 22–33.

Bernon, D.F., Siegel, P.B., 1981. Fertility of chickens from lines divergently selected for mating frequency. Poult. Sci. 60, 45–48.

Benyi, K., Acheampong-Boateng, O., Norris, D., Mathoho, M., Mikasi, M.S., 2009. The response of Ross 308 and Hybro broiler chickens to early and late skip-a-day feed restriction. Trop Anim Health Prod. 41, 1707–1713.

Beuving, G., Vonder, G.M.A., 1978. Effect of stressing factors on corticosterone levels in the plasma of laying hens. Gen. Comp. Endocrinol. 35, 153–159.

Bhagwat, A.L., Craig, J.V., 1977. Selecting for age at first egg: effects on social dominance. Poult. Sci. 56, 361–363.

Bhagwat, A.L., Craig, J.V., 1978. Selection for egg mass in different social environments. 3. Changes in agonistic activity and social dominance. Poult. Sci. 57, 883–891.

Bijma, P., 2010a. Estimating indirect genetic effects: precision of estimates and optimum designs. Genetics 186, 1013–1028.

Bijma, P., 2010b. Multilevel selection 4: modeling the relationship of indirect genetic effects and group size. Genetics 186, 1029–1031.

Bijma, P., Muir, W.M., Ellen, E.D., Wolf, J.B., Van Arendonk, J.A.M., 2007a. Multilevel selection 2: estimating the genetic parameters determining inheritance and response to selection. Genetics 175, 289–299.

Bijma, P., Muir, W.M., Van Arendonk, J.A.M., 2007b. Multilevel selection 1: quantitative genetics of inheritance and response to selection. Genetics 175, 277–288.

Bijma, P., Wade, M.J., 2008. The joint effects of kin, multilevel selection and indirect genetic effects on response to genetic selection. J. Evol. Biol. 21, 1175–1188.

Biswas, D.K., Craig, J.V., 1970. Genotype–environment interactions in chickens selected for high and low social dominance. Poult. Sci. 44, 681–692.

Bornstein, S., Plavnik, I., Lev, Y., 1984. Body weight and/or fatness as a potential determinants of the onset of egg production in broiler breeder hens. Brit. Poult. Sci. 25, 323–341.

Bramble, F.W.R., 1965. The welfare of animals. In: Report of the Technical Committee to enquire into the welfare of animals kept under intensive livestock husbandry sysems, F. W. R. Bramble (Chairman). Her majesty's Stationary Office, London, England (Chapter 4).

Breward, J., Gentle, M.J., 1985. Neuroma formation and abnormal afferent nerve discharges after partial beak amputation (beak trimming) in poultry. Experientia 41, 1132–1134.

Bronson, F.H., Eleftheriou, B.E., 1965. Adrenal response to fighting in mice: seperation of physical and psychological causes. Science 147, 627–628.

Buijs, S., Keeling, L., Rettenbacher, S., Van Poucke, E., Tuyttens, F.A., 2009. Stocking density effects on broiler welfare: identifying sensitive ranges for different indicators. Poult. Sci. 88, 1536–1543.

Buitenhuis, B., Hedegaard, J., Janss, L., Sorensen, P., 2009. Differentially expressed genes for aggressive pecking behaviour in laying hens. BMC Genomics 10, 544.

Burkhart, C.A., Cherry, J.A., Van Krey, H.P., Siegel, P.B., 1983. Genetic selection for growth rate alters hypothalamic satiety mechanisms in chickens. Behav. Gen. 13, 295–300.

Byholm, P., Kekkonen, M., 2008. Food regulates reproduction differently in different habitats: experimental evidence in the Goshawk. Ecology 89, 1696–1702.

California proposition 2, 2008. Shall certain farm animals be allowed, for the majority of every day, to fully extend their limbs or wings, lie down, stand up and turn around?

Campo, J.L., Alvarez, C., 1991. Tonic immobility of several Spanish breeds of hens. Arch. Geflugelk. 55, 19–22.

Campo, J.L., Carnicer, C., 1993. Realized heritability of tonic immobility in White Leghorn hens: a replicated single generation test. Poult. Sci. 72, 2193–2199.

Campo, J.L., Gil, M.G., Torres, O., Davila, S.G., 2001. Association between plumage condition and fear and stress levels in five breeds of chickens. Poult. Sci. 80, 549–552.

Campo, J.L., Gil, M.G., Davila, S.G., Muñoz, I., 2007. Effect of lighting stress on fluctuating asymmetry, heterophil-to-lymphocyte ratio, and tonic immobility duration in eleven breeds of chickens. Poult. Sci. 86, 37–45.

Campo, J.L., Prieto, M.T., Dávila, S.G., 2008. Effects of housing system and cold stress on heterophil-to-lymphocyte ratio, fluctuating asymmetry, and tonic immobility duration of chickens. Poult. Sci. 87, 621–626.

Campo, J.L., Dávila, S.G., Prieto, M.T., Gil, M.G., 2012. Associations among fluctuating asymmetry, tonic immobility duration, and flight distance or ease of capture in chickens. Poult. Sci. 91, 1575–1581.

Campos, D., Llebot, J.E., Méndez, V., 2008. Limited resources and evolutionary learning may help to understand the mistimed reproduction in birds caused by climate change. Theor. Popul. Biol. 74, 16–21.

CEC, 1999. Council Directive 1999/74/EC laying down minimum standards for the protection of laying hens. Off. J. L. 203, 53–57.

Chaussabel, D., Pascual, V., Banchereau, J., 2010. Assessing the human immune system through blood transcriptomics. BMC Biol. 8, 84.

Cheng, H.W., 2005. Acute and chronic pain in beak trimmed chickens. In: Glatz, P. (Ed.), Poultry Welfare Issues – Beak trimming. Nottingham University Press, UK, pp. 31–49.

Cheng, H.W., 2007. Animal welfare: Should we change housing to better accommodate the animal or change the animal to accommodate the housing. CAB Reviews: Perspectives in Agriculture, Veterinary Science, Nutrition and Natural Resources review <http://www.cababstractsplus.org/cabreviews/reviews.asp>.

Cheng, H.W., 2010. Breeding of tomorrow's chickens to improve well-being. Poult. Sci. 89, 805–813.

Cheng, H.W., Eicher, S.D., Chen, Y., Singleton, P., Muir, W.M., 2001a. Effect of genetic selection for group productivity and longevity on immunological and hematological parameters of chickens. Poult. Sci. 80, 1079–1086.

Cheng, H.W., Dillworth, G., Singleton, P., Chen, Y., Muir, W.M., 2001b. Effects of group selection for productivity and longevity on blood concentrations of serotonin, catecholamine and corticosterone of laying hens. Poult. Sci. 80, 1278–1285.

Cheng, H.W., Fahey, A.G., 2009. Effects of repeated social disruption on the serotonergic and dopaminergic systems in two genetic lines of white leghorn layers. Poult. Sci. 88, 2018–2025.

Cheng, H.W., Muir, W.M., 2005. The effects of genetic selection for survivability and productivity on chicken physiological homeostasis. World's Poult. Sci. J. 61, 383–398.

Cheng, H.W., Muir, W.M., 2007. Mechanisms of aggression and production in chickens: genetic variations in the functions of serotonin, catecholamine, and corticosterone. World's Poult. Sci. J. 63, 233–254.

Cheng, H.W., Singleton, P., Muir, W.M., 2002. Social stress in laying hens: differential dopamine and corticosterone responses following intermingling of different genetic strain chickens. Poult. Sci. 81, 1265–1273.

Cheng, H.W., Singleton, P., Muir, W.M., 2003. Social stress in laying hens: differential effect of genetic-environmental interactions on plasma dopamine concentrations and adrenal function in genetically selected chickens. Poult. Sci. 82, 192–198, 2003.

Choudary, M.R., Adams, A.W., Craig, J.V., 1972. Effects of strain, age at flock assembly, and cage arrangement on behavior and productivity in White Leghorn type chickens. Poult. Sci. 51, 1943–1950.

Clark, J.D., Rager, D.R., Calpin, J.P., 1997. Animal well-being. III. An overview of assessment. Lab. Anim. Sci. 47, 580–585.

Clark, M.S., Bond, M.J., Hecker, J.R., 2007. Envuronmental stress, psychological stress and allostatic load. Psychol., *Health Med.* 12, 18–30.

Consortium, 1988. Guide for the Care and Use of Agricultural Animals in Agricultural Research and Teaching. Association Headquarters, Champaign, IL.

Cook, W.T., Siegel, P.B., 1974. Social variables and divergent selection for mating behavior of male chickens. Anim. Behav. 22, 390–396.

Cornil, C.A., Ball, G.F., 2010. Effects of social experience on subsequent sexual performance in naïve male Japanese quail (Coturnix japonica). Horm. Behav. 57, 515–522.

Craig, J.V., 1970. Interactions of genotype and housing environment in White Leghorn chickens selected for high and low social dominance. Proceedings of the Fourteenth World's Poultry Congress, Madrid, Spain vol 2, pp. 37–42.

Craig, J.V., 1982. Behavioral and genetic adaptation of laying hens to high-density environments. BioScience 32, 33–37.

Craig, J.V., 1994. Genetic influences on behavior associated with well-being and productivity in livestock. Proceedings of the Fifth World Congress on Genetics Applied to Livestock Production, Guelph, Canada. 20, 150–157.

Craig, J.V., Adams, A.W., 1984. Behaviour and well-being of hens (*Gallus domesticus*) in alternative housing environments. World's Poult. Sci. J. 40, 221–240.

Craig, J.V., Baruth, R.A., 1965. Inbreeding and social dominance ability in chickens. Anim. Behav. 13, 109–113.

Craig, J.V., Craig, J.A., 1985. Corticosteroid levels in White Leghorn hens as affected by handling, laying-house environment, and genetic stock. Poult. Sci. 64, 809–816.

Craig, J.V., Lee, H.-Y., 1989. Research note: genetic stocks of White Leghorn type differ in relative productivity when beaks are intact versus trimmed. Poult. Sci. 68, 1720–1723.

Craig, J.V., Lee, H.-Y., 1990. Beak trimming and genetic stock effects on behavior and mortality from cannibalism in White Leghorn-type pullets. *Appl. Anim. Behav.* Sci. 25, 107–123.

Craig, J.V., Muir, W.M., 1989. Fearful and associated responses of caged White Leghorn hens: genetic parameter estimates. Poult. Sci. 68, 1040–1046.

Craig, J.V., Muir, W.M., 1991. Research note: genetic adaptation to multiple-bird cage environment is less evident with effective beak trimming. Poult. Sci. 70, 2214–2217.

Craig, J.V., Muir, W.M., 1993. Selection for reduction of beak-inflicted injuries among caged hens. Poult. Sci. 72, 411–420.

Craig, J.V., Muir, W.M., 1996. Group selection for adaptation to multiple-hen cages: beak-related mortality, feathering, and body weight responses. Poult. Sci. 75, 294–302.

Craig, J.V., Swanson, J.C., 1994. Review: Welfare perspectives on hens kept for egg production. Poult. Sci. 73, 921–938.

Craig, J.V., Toth, A., 1969. Productivity of pullets influenced by genetic selection for social dominance ability and by stability of flock membership. Poult. Sci. 48, 1729–1736.

Craig, J.V., Al-Rawi, B., Kratzer, D.D., 1977. Social status and sex ratio effects on mating frequency of cockerels. Poult. Sci. 56, 767–772.

Craig, J.V., Biswas, D.K., Guhl, A.M., 1969. Agonistic behaviour influenced by strangeness, crowding and heredity in female domestic fowl (*Gallus gallus*). Anim. Behav. 17, 498–506.

Craig, J.V., Craig, J.A., Vargas, J.V., 1986. Corticosteroids and other indicators of hen's well-being in four laying-house environments. Poult. Sci. 65, 856–863.

Craig, J.V., Dayton, A.D., Garwood, V.A., Lowe, P.C., 1982. Selection for egg mass in different social environments. 4. Selection response in Phase I. Poult. Sci. 61, 1786–1798.

Craig, J.V., Jan, M.-L., Polley, C.R., Bhagwat, A.L., Dayton, A.D., 1975. Changes in relative aggressiveness and social dominance associated with selection for early egg production in chickens. Poult. Sci. 54, 1647–1658.

Craig, J.V., Kujiyat, S.K., Dayton, A.D., 1983. Duration of tonic immobility affected by housing environment in White Leghorn hens. Poult. Sci. 62, 2280–2282.

Craig, J.V., Kujiyat, S.K., Dayton, A.D., 1984. Tonic immobility responses of White Leghorn hens affected by induction techniques and genetic stock differences. Poult. Sci. 63, 1–10.

Craig, J.V., Ortman, L.L., Guhl, A.M., 1965. Genetic selection for social dominance ability in chickens. Anim. Behav. 13, 114–131.

Crow, J.F., Kimura, M., 1970. An Introduction to Population Genetics. Harper and Row, New York, NY (Chapter 5: Selection).

Cunningham, D.L., 1992. Beak trimming effects on performance, behavior and welfare of chickens: a review. J. Appl. Poult. Res. 1, 129–134.

Cunningham, D.L., van Tienhoven, A., Gvaryahu, G., 1988. Population size, cage area, and dominance rank effects on productivity and well-being of laying hens. Poult. Sci. 67, 399–406.

Dantzer, R., 1993. Research perspectives on farm animal welfare: the concept of stress. J. Agric. Environ. Ethics 6 (Suppl. 2), 86–92.

Dantzer, R., 1994. Animal welfare methodology and criteria. Rev. Sci. Tech. 13, 277–302.

Dávila, S.G., Campo, J.L., Gil, M.G., Prieto, M.T., Torres, O., 2011. Effects of auditory and physical enrichment on 3 measurements of fear and stress (tonic immobility duration, heterophil to lymphocyte ratio, and fluctuating asymmetry) in several breeds of layer chicks. Poult. Sci. 90, 2459–2466.

Dawkins, M.S., 1983. Battery hens name their price: consumer demand theory and the measurement of ethological 'needs'. Anim. Behav. 31, 1195–1205.

Dawkins, M.S., 1988. Behavioural deprivation: a central problem in animal welfare. Appl. Anim. Behav. Sci. 20, 200–225.

Dawkins, M.S., 1998. Evolution and animal welfare. Q. Rev. Biol. 73, 305–328.

de Boer, I.J., Cornelissen, A.M., 2002. A method using sustainability indicators to compare conventional and animal-friendly egg production systems. Poult. Sci. 81, 173–181.

de Jong, I.C., Enting, H., van Voorst, A., Blokhuis, H.J., 2005. Do low-density diets improve broiler breeder welfare during rearing and laying?. Poult. Sci. 84, 194–203.

De Vylder, J., Dewulf, J., Van Hoorebeke, S., Pasmans, F., Haesebrouck, F., Ducatelle, R., Van Immerseel, F., 2011. Horizontal transmission of Salmonella enteritidis in groups of experimentally infected laying hens housed in different housing systems. Poult. Sci. 90, 1391–1396.

de Wit, W., 1992. The welfare of laying hens kept under various housing systems. Proceedings. Nineteenth World's Poultry Congress, Amsterdam, The Netherlands 2, 320–323.

Decuypere, E., Bruggeman, V., Everaert, N., Li, Y., Boonen, R., De Tavernier, J., Janssens, S., Buys, N., 2010. The broiler breeder paradox: ethical, genetic and physiological perspectives, and suggestions for solutions. Br. Poult. Sci. 51, 569–579.

Duncan, I.J., 2005. Science-based assessment of animal welfare: farm animals. Rev. Sci. Tech. 24, 483–492.

Duncan, I.J.H., Dawkins, M.S., 1983. The problem of assessing "well-being" and "suffering" in farm animals. In: Smidt, E. (Ed.), Indicators Relevant to Farm Animal Welfare. Martinus Nijhoff, The Hague, pp. 13–24.

Duncan, I.J.H., Filshie, 1979. The use of radio telemetry devices to measure temperature and heart rate in domestic fowl. In: Amlaner, C.J., MacDonald (Eds.), A Handbook on Telemetry and Radio Tracking. Pergamon Press, Oxford, England, pp. 579–588.

Duncan, I.J.H., Petherick, J.C., 1991. The implications of cognitive processes for animal welfare. J. Anim. Sci. 69, 5017–5022.

Duncan, I.J.H., Wood-Gush, D.G.M., 1971. Frustration and aggression in the domestic fowl. Anim. Behav. 19, 500–504.

Duncan, I.J.H., Wood-Gush, D.G.M., 1972. Thwarting of feeding behavior in the domestic fowl. Anim. Behav. 20, 444–451.

Duncan, I.J.H., Slee, G.S., Seawright, E., Breward, J., 1989. Behavioural consequences of partial beak amputation (beak trimming) in poultry. Br. Poult. Sci. 30, 479–488.

Dunn, I.C., McEwan, G., Okhubo, T., Sharp, P.J., Paton, I.R., Burt, D.W., 1998. Genetic mapping of the chicken prolactin receptor gene: a candidate gene for the control of broodiness. Br. Poult. Sci. (39 Suppl), S23–24.

Dunnington, E.A., Siegel, P.B., 1983. Mating frequency in male chickens: long-term selection. Theor. Appl. Genet. 64, 317–323.

Dunnington, E.A., Siegel, P.B., 1985. Long-term selection for eight-week body weight in chickens—direct and correlated responses. Theor. Appl. Genet. 71, 305–313.

Dunnington, E.A., Nir, I., Cherry, J.A., Jones, D.E., Siegel, P.B., 1987. Growth-associated traits in parental and F_1 populations of chickens of chickens under different feeding programs. 3. Eating behavior and body temperatures. Poult. Sci. 66, 23–31.

Dunnington, E.A., Siegel, P.B., Cherry, J.A., 1984. Delayed sexual maturity as a a correlated response to selection for 56-day body weight in White Plymouth Rock pullets. Arch. Geflugelk. 48, 111–113.

Emsley, A., Dickerson, G.E., Kashyap, T.S., 1977. Genetic parameters in progeny-test selection for field performance of strain-cross layers. Poult. Sci. 56, 121–146.

Elmslie, L.J., Jones, R.H., Knight, D.W., 1966. A general theory describing the effects of varying flock size and stocking density on the performance of caged layer. Proceedings of the thirteenth World's Poult Cong., Kiev, USSR pp. 490–495.

Elstner, M., Turnbull, D.M., 2012. Transcriptome analysis in mitochondrial disorders. Brain. Res. Bull. 88, 285–293.

Eskeland, B., 1981. Effects of beak trimming. In: Sorensen, Y. (Ed.), Proceedings of the First European Symposium on Poultry Welfare. Copenhagen, Denmark. pp. 193–200.

Eyre-Walker, A., Keightley, P.D., 2009. Estimating the rate of adaptive molecular evolution in the presence of slightly deleterious mutations and population size change. Mol. Biol. Evol. 26, 2097–2108.

Farquharson, C., Jefferies, D., 2000. Chondrocytes and longitudinal bone growth: the development of tibial dyschondroplasia. Poult. Sci. 79, 994–1004.

Faure, J.-M., 1979. Influence de la souche et du sexe sur le comprtement en open-fiel du jeune poussin. Biol. Behav. 1, 19–24.

Faure, J.-M., 1980. To adapt the environment to the bird or the bird to the environment? In: Moss, R. (Ed.), The Laying Hen and its Environment. Martinus Nijhoff, The Hague, The Netherlands, pp. 19–30.

Faure, J.-M., 1981a. Behavioural measures for selection. Proceedings of the First European Symposium on Poultry Welfare, Copenhagen, Denmark. pp. 37–41.
Faure, J.-M., 1981b. Bidirectional selection for open-field activity in young chicks. Behav. Genet. 11, 135–144.
Faure, J.-M., 1986. Operant determination of the cage and feeder size preferences of the laying hen. Appl. Anim. Behav. Sci. 15, 325–336.
Faure, J.-M., Mills, A.D., 1989. Proceddings of the Third European Symposium on Poultry Welfare, Tours, France, p. 284.
Faure, J.-M., Melin, J.M., Mills, A.D., 1992. Selection for behavioural traits in relation to poultry welfare. Proceedings of the Nineteenth World's Poultry Congress, Amsterdam, The Netherlands 405–408.
Forkman, B., Boissy, A., Meunier-Salaün, M.C., Canali, E., Jones, R.B., 2007. A critical review of fear tests used on cattle, pigs, sheep, poultry and horses. Physiol. Behav. 92, 340–374.
Freeman, B.M., 1985. Stress and the domestic fowl: physiological fact or fantasy? World's Poult. Sci. J. 41, 45–50.
Gallup Jr., G.G., 1974a. Animal hypnosis: factual status of a fictional concept. Psychol. Bull. 81, 836–853.
Gallup Jr., G.G., 1974b. Genetic influence on tonic immobility in chickens. Anim. Learn, Behav. 2, 145–147.
Gallup Jr., G.G., Ledbetter, D.H., Maser, J.D., 1976. Strain differences among chickens in tonic immobility: evidence for an emotionality component. J. Comp. Physiol. Pschy. 90, 1075–1081.
Gentle, M.J., 1986a. Beak trimming in poultry. World's Poult. Sci. J. 42, 268–275.
Gentle, M.J., 1986b. Neuroma formation following partial beak amputation (beak trimming) in the chicken. Res. Vet. Sci. 41, 383–385.
Gentle, M.J., 1989. Cutaneous sensory afferents recorded from the nervus intramandibularis of Gallus gallus var domesticus. J. Comp. Physiol. [A] 164, 763–774.
Gentle, M.J., 1992. Pain in birds. Anim. Wel. 1, 235–247.
Gentle, M.J., Waddington, D., Hunter, L.N., Jones, R.B., 1990. Behavioural evidence for persistent pain following partial beak amputation in chickens. Appl. Anim. Behav. Sci. 27, 149–157.
Gentle, M.J., Hunter, L.N., Waddington, D., 1991. The onset of pain related behaviours following partial beak amputation in the chicken. Neurosci. Lett. 128, 113–116.
Golub, T.R., Slonim, D.K., Tamayo, P., Huard, C., Gaasenbeek, M., Mesirov, J.P., et al., 1999. Molecular classification of cancer: class discovery and class prediction by gene expression monitoring. Science 286, 531–537.
Goodale, H.D., Sanborn, R., White, D., 1920. Broodiness in domestic fowl. Data concerning its inheritance in the Rhode Island Red breed. Mass. *Agr. Expt. Sta. Bull.* 199, 94–116.
Goodnight, C.J., 1985. The influence of environmental variation on group and individual selection in cress. Evolution 39, 545–558.
Goodwin, J.L., 2001. The ethics of livestock shows—past, present, and future. J. Am. Vet. Med. Assoc. 219, 1391–1393.
Green, T.C., Mellor, D.J., 2011. Extending ideas about animal welfare assessment to include 'quality of life' and related concepts. N. Z. Vet. J. 59, 263–271.
Griffing, B., 1967. Selection in reference to biological groups. I. Individual and group selection applied to populations of unordered groups. Aust. J. Biol. Sci. 20, 127–139.
Griffing, B., 1977. Selection for populations of interacting phenotypes. In: Pollak, E., Kempthorne, O., Bailey., T.B. (Eds.), Proceedings of the 2nd International Conference on Quantitative Genetics. Iowa State University. Press, Ames, pp. 413–434.

Gross, W.B., 1983. Chicken–environment interactions. In: Miller, H.B., Williams, W.H. (Eds.), Ethics and Animals. The Humana Press, Inc., Clifton, NJ, pp. 329–337.
Gross, W.B., Siegel, P.B., 1985. Selective breeding of chickens for corticosterone response to social stress. Poult. Sci. 64, 2230–2233.
Guhl, A.M., 1968. Social inertia and social stability in chickens. Anim. Behav. 16, 219–232.
Guhl, A.M., Warren, D.C., 1946. Number of offspring sired by cockerels related to social dominance in chickens. Poult. Sci. 25, 460–472.
Guhl, A.M., Craig, J.V., Mueller, C.D., 1960. Selective breeding for aggressiveness in chickens. Poult. Sci. 39, 970–980.
Hale, E.B., 1969. Domestication and the evolution of behaviour. In: Hafez, E.S.E. (Ed.), The Behaviour of Domestic Animals, second ed. The Williams and Wilkins Co., Baltimore, MD.
Hemsworth, P.H., 2007. Ethical stockmanship. Aus. Vet. J. 85, 194–200.
Hannah, J.F., Wilson, J.L., Cox, N.A., Richardson, L.J., Cason, J.A., Bourassa, D.V., et al., 2011. Horizontal transmission of Salmonella and Campylobacter among caged and cage-free laying hens. Avian Dis. 55, 580–587.
Hansen, R.S., 1976. Nervousness and hysteria of mature female chickens. Poult. Sci. 55, 531–543.
Hen, G., Yosefi, S., Simchaev, V., Shinder, D., Hruby, V.J., Friedman-Einat, M.J., 2006. The melanocortin circuit in obese and lean strains of chicks. Endocrinol 190, 527–535.
Henderson, C.R., 1975. Best linear unbiased estimation and prediction under a selection model. Biometrics 31, 423–447.
Henderson, C.R., 1984. Applications of Linear Models in Animal Breeding. University of Guelph, Guelph, Ontario, Canada.
Henderson, C.R., Quaas, R.L., 1976. Multi-Trait selection using relatives records. J. Anim. Sci. 43, 218.
Hester, P.Y., 2005. Impact of science and management on the welfare of egg laying strains of hens. Poult. Sci. 84, 687–696.
Hester, P.Y., Muir, W.M., Craig, J.V., 1996a. Group selection for adaptation to multiple-hen cages: humoral immune response. Poult. Sci. 75, 1315–1320.
Hester, P.Y., Muir, W.M., Craig, J.V., Albright, J.L., 1996b. Group selection for adaptation to multiple–hen cages: Hematology and adrenal function. Poult. Sci. 75, 1295–1307.
Hetts, S., 1991. Psychologic well-being: conceptual issue, behavioral measures, and implications for dogs. Vet. Clin. North Am. Small Anim. Pract. 21, 369–387.
Hill, J.A., 1983. Indicators of stress in poultry. World's Poult. Sci. J. 39, 24–32.
Hindmarch, C.C., Murphy, D., 2010. The transcriptome and the hypothalamo–neurohypophyseal system. Endocr. Dev. 17, 1–10.
Hocking, P.M., 1994. Role of genetics in safeguarding and improving the welfare of poultry. In: Proceedings of the nineth European Poultry Conference, Glasgow, U.K. pp. 233–236.
Hocking, P.M., Zaczek, V., Jones, E.K., Macleod, M.G., 2004. Different concentrations and sources of dietary fibre may improve the welfare of female broiler breeders. Br. Poult. Sci. 45, 9–19.
Hoehn-Saric, R., McLeod, D.R., 2000. Anxiety and arousal: physiological changes and their perception. J. Affect. Disord. 61, 217–224.
Holt, S.P., Davies, R.H., Dewulf, J., Gast, R.K., Huwe, J.K., Jones, D.R., et al., 2011. The impact of different housing systems on egg safety and quality. Poult. Sci. 90, 251–262.
Howie, J.A., Avendano, S., Tolkamp, B.J., Kyriazakis, I., 2011. Genetic parameters of feeding behavior traits and their relationship with live performance traits in modern broiler lines. Poult. Sci. 90, 1197–1205.

Hughes, B.O., 1980. The assessment of behavioural needs. In: Moss, R. (Ed.), The Laying Hen and its Environment. Martinus Nijhoff, The Hague, The Netherlands, pp. 149–166.

Hunton, P., 1990. Industrial breeding and selection. In: Crawford, R.D. (Ed.), Poultry Breeding and Genetics. Elsevier Science Publishers B. V., Amsterdam, The Netherlands, pp. 985–1028.

Hurnik, J.F., 1990. Animal welfare: ethical aspects and practical considerations. Poult. Sci. 69, 1827–1834.

Izzo, G.N., Bashaw, M.J., Campbell, J.B., 2011. Enrichment and individual differences affect welfare indicators in squirrel monkeys (Saimiri sciureus). J. Comp. Psychol. 125, 347–352.

James, J.W., Foenander, F., 1961. Social behavior studies on domestic animals. 1. Hens in laying cages. Aust. J. Agric. Res. 12, 1239–1252.

Jansson, D.S., Nyman, A., Vågsholm, I., Christensson, D., Göransson, M., Fossum, O., et al., 2010. Ascarid infections in laying hens kept in different housing systems. Avian. Pathol. 39, 525–532.

Jiang, R.S., Xu, G.Y., Zhang, X.Q., Yang, N., 2005. Association of polymorphisms for prolactin and prolactin receptor genes with broody traits in chickens. Poult. Sci. 84, 839–845.

Jones, R.B., Faure, J.M., 1981. Sex and strain comparisons of tonic immobility ("righting time") in the domestic fowl and the effects of various methods of induction. Behav. Processes 6, 47–55.

Jones, D.R., Anderson, K.E., Musgrove, M.T., 2011. Comparison of environmental and egg microbiology associated with conventional and free-range laying hen management. Poult. Sci. 90, 2063–2068.

Jongren, M., Westander, J., Natt, D., Jensen, P., 2010. Brain gene expression in relation to fearfulness in female red junglefowl (Gallus gallus). Genes Brain. Behav. 9, 751–775.

Joseph, R., 1999. Environmental influences on neural plasticity, the limbic system, emotional development and attachment a review. Child. Psychiatry Hum. Dev. 29, 189–208.

Kashysap, T.S., Dickerson, G.E., Bennett, G.L., 1981. Effectiveness of progeny test multiple trait index selection for filed performance of strain-cross layers. I. Estimated responses. Poult. Sci. 60, 1–21.

Katanbaf, M.N., Dunnington, E.A., Siegel, P.B., 1989. Restricted feeding in early and late feathering chickens. 1. Growth and physiological responses. Poult. Sci. 68, 344–351.

King, M.G., 1965. Disruptions in the pecking order of cockerels concomitant with degrees of accessibility to feed. Anim. Behav. 13, 504–506.

Klopfleisch, R., Gruber, A.D., 2012. Transcriptome and proteome research in veterinary science: what is possible and what questions can be asked? Sci. World J. 2012, 254962.

Kolb, B., Whishaw, I.Q., 1998. Brain plasticity and behavior. Annu. Rev. Psychol. 49, 43–64.

Kratzer, D.D., Craig, J.V., 1980. Mating behavior of cockerels: effects of social status, group size and group density. Appl. Anim. Ethol. 6, 49–62.

Kuhlers, C.L., McDaniel, G.R., 1996. Estimates of heritabilities and genetic correlations between tibial dyschondroplasia expression and body weight at two ages in broilers. Poult. Sci. 75, 959–961.

Kujiyat, S.K., Craig, J.V., Dayton, A.D., 1983. Duration of tonic immobility affected by housing environment in White Leghorn chickens. Poult. Sci. 62, 2280–2282.

Kumar, B., Manuja, A., Aich, P., 2012. Stress and its impact on farm animals. Front Biosci. (Elite Ed) 4, 1759–1767.

Kuo, F.-L., Craig, J.V., Muir, W.M., 1991. Selection and beak-trimming effects on behavior, cannibalism, and short-term production traits in White Leghorn pullets. Poult. Sci. 70, 1057–1068.

Laible, G., 2009. Enhancing livestock through genetic engineering—recent advances and future prospects. Comp. Immunol. Microbiol. Infect. Dis. 32, 123–137.

Lang, P.J., McTeague, L.M., 2009. The anxiety disorder spectrum: fear imagery, physiological reactivity, and differential diagnosis. Anxiety Stress Coping 22, 5–25.

Lay Jr., D.C., Fulton, R.M., Hester, P.Y., Karcher, M.D., Kjaer, J.B., Mench, J.A., et al., 2011. Hen welfare in different housing systems. Poult. Sci. 90, 278–294.

Leach Jr., R.M., Nesheim, M.C., 1965. Nutritional, genetic and morphological studies of an abnormal cartilage formation in young chicks. J. Nutr. 86, 236–244.

Leach Jr., R.M., Nesheim, M.C., 1972. Further studies on tibial dyschondroplasia (cartilage abnormality) in young chicks. J. Nutr. 102, 1673–1680.

Leach Jr., R.M., Monsonego-Ornan, E., 2007. Tibial dyschondroplasia 40 years later. Poult. Sci. 86, 2053–2058.

Lee, H.-Y., Craig, J.V., 1990. Beak trimming effects on the behavior and weight gain of floor-reared, egg-strain pullets from three genetic stocks during the rearing period. Poult. Sci. 69, 568–575.

Lee, H.-Y., Craig, J.V., 1991. Beak trimming effects on behavior patterns, fearfulness, feathering, and mortality among three stocks of White Leghorn pullets in cages or floor pens. Poult. Sci. 70, 211–221.

Lee, Y.-P., Craig, J.V., 1981a. Agonistic and nonagonistic behaviors of pullets of dissimilar strains of White Leghorns when kept separately and intermingled. Poult. Sci. 60, 1759–1768.

Lee, Y.-P., Craig, J.V., 1981b. Evaluation of egg-laying strains of chickens in different housing environments: role of genotype by environment interactions. Poult. Sci. 60, 1769–1781.

Lowry, D.C., Abplanalp, H., 1970. Genetic adaptation of White Leghorn hens to life in single cages. Br. Poult. Sci. 11, 117–131.

Lowry, D.C., Abplanalp, H., 1972. Social dominance differences, given limited access to common food, between hens selected and unselected for increased egg production. Br. Poult. Sci 13, 365–376.

Martins, C.I., Galhardo, L., Noble, C., Damsgård, B., Spedicato, M.T., Zupa, W., et al., 2012. Behavioural indicators of welfare in farmed fish. Fish Physiol. Biochem. 38, 17–41.

Mason, G.J., Latham, N., 2004. Can't stop, won't stop: is stereotypy a reliable animal welfare indicator? Anim. Welf. 13, S57–S69.

Mason, G., Clubb, R., Latham, N., Vickery, S., 2007. How and why should we use enrichments to tackle stereotypic behavior? Appl. Anim. Behav. Sci. 102, 163–188.

McBride, G., 1960. Poultry husbandry and the peck order. Br. Poult. Sci. 1, 65–68.

McBride, G., 1962. Behaviour and a theory of poultry husbandry. Proceedings of the Twelfth World's Poultry Congress (Symposia), Sydney, Australia, pp. 102–105.

McBride, G., 1968. Behavioral measurement of social stress. In: Hafez, E.S.E. (Ed.), Adaptation of Domestic Animals. Lea and Febiger, Philadelphia, PA, pp. 360–366.

McEwen, B.S., 2001. From molecules to mind. Stress, individual differences, and the social environment. Ann. N. Y. Acad. Sci. 935, 42–49.

McGibbon, W.H., 1976. Floor laying — a heritable and environmentally influenced trait of the domestic fowl. Poult. Sci. 55, 765–771.

McGlothlin, J.W., Brodie, E.D., 2009. How to measure indirect genetic effects: the congruence of trait-based and variance-partitioning approaches. Evolution 63, 1785–1795.

Mellor, D.J., 2012. Animal emotions, behaviour and the promotion of positive welfare states. N. Z. Vet. J. 60, 1–8.

Mellor, D.J., Stafford, K.J., 2001. Integrating practical, regulatory and ethical strategies for enhancing farm animal welfare. Aust. Vet. J. 79, 762–768.

Mench, J.A., 1992. The welfare of poultry in modern production systems. Poult. Sci. Rev. 4, 107–128.

Mench, J.A., 2008. Farm animal welfare in the USA: farming practices, research, education, regulation, and assurance programs. Appl. Anim. Behav. Sci. 113, 298–312.

Mench, J.A., Duncan, I.J., 1998. Poultry welfare in North America: opportunities and challenges. Poult. Sci. 77, 1763–1765.

Mench, J.A., Van Tienhoven, A., Marsh, J.A., McCormick, C.C., Cunningham, D.L., Baker, R.C., 1986. Effects of cage and floor management on behavior production and physiological responses of laying hens. Poult. Sci. 65, 1058–1069.

Mench, J.A., Sumner, D.A., Rosen-Molina, J.T., 2011. Sustainability of egg production in the United States—The policy and market context. Poult. Sci. 90, 229–240.

Merriam-Webster, Inc., 1983. Webster's Ninth New Collegiate Dictionary. Merriam-Webster Inc., Publishers, Springfield, MA.

Mestan, K.K., Ilkhanoff, L., Mouli, S., Lin, S., 2011. Genomic sequencing in clinical trials. J. Transl. Med. 9, 222.

Millman, S.T., Mench, J.A., Malleau, A.E., 2010. The Future of poultry welfare. Wel. Domest. Fowl Other Captive Birds 9, 279–302.

Mills, A.D., Wood-Gush, D.G.M., 1985. Pre-laying behaviour in battery cages. Br. Poult. Sci. 26, 247–252.

Mills, A.D., Wood-Gush, D.G.M., Hughes, B.O., 1985. Genetic analysis of strain differences in pre-laying behaviour in battery cages. Br. Poult. Sci. 26, 187–197.

Mohapatra, S.C., Siegel, P.B., 1969. Selection for Prothrombin time at 18 weeks of age in chickens. J. Hered. 60, 289–299.

Muir, W.M., 1985. Relative efficiency of selection for performance of birds housed in colony cages based on production in single bird cages. Poult. Sci. 64, 2239–2247.

Muir, W.M., 1996. Group selection for adaptation to multiple-hen cages. Selection program and direct responses. Poult. Sci. 75, 447–458.

Muir, W.M., 2003. Indirect selection for improvement of animal well-being. In: Muir, W.M., Aggrey, S. (Eds.), Poultry Genetics, Breeding and Biotechnology. CABI Press, Cambridge MA, pp. 247–256. (Chapter 14).

Muir, W.M., 2005. Incorporation of competitive effects in forest tree or animal breeding programs. Genetics 170, 1247–1259.

Muir, W.M., Cheng, H., 2004. Breeding for productivity and welfare. In: Perry, G. (Ed.), Welfare of the Laying Hen: Poultry Science Symposium Series, No. 27. CABI Press, Cambridge, MA, pp. 123–138.

Muir, W.M., Craig, J.V., 1998. Improving animal well-being through genetic selection. Poultry Science 77, 1781–1788.

Muir, W.M., Schinckel, A., 2012. Simple and effective methods of addressing competitive effects in animal breeding programs. In: Proceedings of the Fourth International Conference on Quantitative Genetics: Understanding Variation in Complex Traits, Edinburgh UK, pp. 54–55.

Muir, W.M., Schinckel, A.S., 2002. Incorporation of competitive effects in breeding programs to improve productivity and animal well being. Proceedings of the Seventh World Congress of Genetics Applied to Livestock Breeding 32, 35–36.

Muir, W.M., Nyquist, W.E., Xu, S., 1992. Alternative partitioning of the genotype-by-environment interaction. Theor. Appl. Genet. 84, 193–200.

Muir, W.M., Cheng, H.W., Nguyen, T., 2010. Keynote Symposium Poultry Science Association: Tomorrow's Poultry: Genomics, Physiology, and Well-Being, Genetics and genomic approaches to address both breeding and management issues of poultry well-being. Raleigh NC. July 20, 2009. In Modification of animals versus modification of the production environment to meet welfare needs, (Ed.) Aggrey, S. E. Poultry Science 89, 852–854.

Mumme, R.L., Galatowitsch, M.L., Jabłoński, P.G., Stawarczyk, T.M., Cygan, J.P., 2006. Evolutionary significance of geographic variation in a plumage-based foraging adaptation: an experimental test in the slate-throated redstart (Myioborus miniatus). Evolution 60, 1086–1097.

Murphy, L., 1978. The practical problems of recognizing and measuring fear and exploration behaviour in the domestic fowl. Anim. Behav. 26, 422–431.

Myers, A.J., 2011. The age of the "ome": genome, transcriptome and proteome data set collection and analysis. Brain Res Bull 88, 294–301.

Newberry, R.C., Tarazona, A.M., 2011. Behavior and welfare of laying hens and broiler chickens. Rev. Colomb. Cienc. Pecuarias 24, 301–302.

Newman, S., 1994. Quantitative- and molecular-genetic effects on animal well-being: adaptive mechanisms. J. Anim. Sci. 72, 1641–1653.

Nir, I., Nitsan, Z., Dror, Y., Shapira, N., 1978. Influence of overfeeding on growth, obsesity and intestinal tract in young chicks of light and heavy breeds. Br. J. Nutr. 39, 27–35.

Ohkubo, T., Tanaka, M., Nakashima, K., Talbot, R.T., Sharp, P.J., 1998. Prolactin receptor gene expression in the brain and peripheral tissues in broody and nonbroody breeds of domestic hen. Gen. Comp. Endocrinol. 109, 60–68.

Ohmura, Y., Yoshioka, M., 2009. The roles of corticotropin releasing factor (CRF) in responses to emotional stress: is CRF release a cause or result of fear/anxiety? CNS Neurol. Disord. Drug. Targets 8, 459–469.

Overstreet, D.H., 2012. Modeling depression in animal models. Methods Mol. Biol. 829, 125–144.

Patterson, H.D., Thompson, R., 1971. Recovery Of Inter-Block Information When Block Sizes Are Unequal. Biometrika 58, 545.

Petherick, J.C., Duncan, I.J.H., Waddington, D., 1990. Previous experience with different floors influences choice of peat in a Y-maze by domestic fowl. Appl. Anim. Behav. Sci. 27, 177–182.

Piazza, J.R., Almeida, D.M., Dmitrieva, N.O., Klein, L.C., 2010. Frontiers in the use of biomarkers of health in research on stress and aging. J. Gerontol. B. Psychol. Sci. Soc. Sci. 65, 513–525.

Pym, R.A.E., Nicholls, P.J., 1979. Selection for food conversion in broilers: direct and correlated responses to selection for body-weight gain, food consumption and food conversion ratio. Br. Poult. Sci. 20, 73–86.

Pym, R.A., Nicholls, P.J., Thomson, E., Choice, A., Farrell, D.J., 1984. Energy and nitrogen metabolism of broilers selected over ten generations for increased growth rate, food consumption and conversion of food to gain. Br Poult. Sci. 25, 529–539.

Quinteiro-Filho, W.M., Rodrigues, M.V., Ribeiro, A., Ferraz-de-Paula, V., Pinheiro, M.L., Sa, L.R., et al., 2012. Acute heat stress impairs performance parameters and induces mild intestinal enteritis in broiler chickens: role of acute hypothalamic–pituitary–adrenal axis activation. J. Anim. Sci. 90, 1986–1994.

Ramos, N.C., Craig, J.V., 1988. Pre-laying behavior of hens kept in single-or multiple-hen cages. Appl. Anim. Behav. Sci. 19, 305–315.

Ray, S.A., Drummond, P.B., Shi, L., McDaniel, G.R., Smith, E.J., 2006. Mutation analysis of the aggrecan gene in chickens with tibial dyschondroplasia. Poult. Sci. 85, 1169–1172.

Romanov, M.N., Talbot, R.T., Wilson, P.W., Sharp, P.J., 1999. Inheritance of broodiness in the domestic fowl. Br. Poult. Sci.(40 Suppl), S20–21.
Rosenzweig, M.R., Bennett, E.L., 1996. Psychobiology of plasticity: effects of training and experience on brain and behavior. Behav. Brain. Res. 78, 57–65.
Ruszler, P.L., 1998. Health and husbandry considerations of induced molting. Poult. Sci. 77, 1789–1793.
Saeki, Y., 1957. Inheritance of broodiness in Japanese Nagoya Fowl, with special reference to sex-linkage and notice in breeding practice. Poult. Sci. 36, 378–383.
Sandilands, V., Tolkamp, B.J., Kyriazakis, I., 2005. Behaviour of food restricted broilers during rearing and lay—effects of an alternative feeding method. Physiol. Behav. 85, 115–123.
Savory, C.J., Hughes, B.O., 2010. Behaviour and welfare. Br. Poult. Sci. 51 (Suppl 1), 13–22.
Sausa, N., Lukoyanov, N.V., Madeira, M.D., Almeida, O.F., Paula-Barbosa, M.M., 2000. Reorganization of the morphology of hippocampal neuritis and synapses after stress-induced damage correlates with behavioral improvement. Neuroscience 97, 253–266.
Scheele, C.W., 1997. Pathological changes in metabolism of poultry related to increasing production levels. Vet. Q. 19, 127–130.
Selye, H., 1956. The Stress of Life. McGraw-Hill Book Company, Inc., New York, NY.
Shea, M.M., Mench, J.A., Thomas, O.P., 1990. The effect of dietary tryptophane on aggressive behavior in developing and mature broiler breeder males. Poult. Sci. 69, 1664–1669.
Sheridan, A.K., Howlett, C.R., Burton, R.W., 1978. The inheritance of tibial dyschondroplasia in broilers. Br. Poult. Sci. 19, 491–499.
Shimmura, T., Hirahara, S., Azuma, T., Suzuki, T., Eguchi, Y., Uetake, K., et al., 2010. Multi-factorial investigation of various housing systems for laying hens. Br. Poult. Sci. 51, 31–42.
Shirley, R.B., Davis, A.J., Compton, M.M., Berry, W.D., 2003. The expression of calbindin in chicks that are divergently selected for low or high incidence of tibial dyschondroplasia. Poult. Sci. 82, 1965–1973.
Siegel, H.S., 1981. Adaptation of poultry to modern production practices. In: Scheele, C.W., Veerkamp, C.H. (Eds.), World Poultry Production Where and How? Spelderholt Institute for Poultry Research, Beekbergen, The Netherlands.
Siegel, H.S., 1995. Stress, strains, and resistance. Br. Poult. Sci. 36, 3–22.
Siegel, P.B., 1972. Genetic analysis of male mating behavior in chickens (*Gallus domesticus*). I. Artificial selection. Anim. Behav. 20, 564–570.
Siegel, P.B., 1984. The role of behavior in poultry production: a review of research. Appl. Anim. Ethol. 11, 299–316.
Siegel, P.B., 1989. The genetic–behaviour interface and well-being of poultry. Br. Poult. Sci. 30, 3–13.
Siegel, P.B., 1993. Behavior–genetic analyses and poultry husbandry. Poult. Sci. 72, 1–6.
Siegel, P.B., Dunnington, E.A., 1987. Selection for growth in chickens. CRC Crit. Rev. Poult. Biol. 1, 1–24.
Siegel, P.B., Dunnington, E.A., 1990. Behavioral genetics. In: Crawford, R.D. (Ed.), Poultry Breeding and Genetics. Elsevier Science Publishers B. V., Amsterdam, The Netherlands, pp. 877–895.
Smith, D.D., 1993. Brain, environment, heredity, and personality. Psychol Rep. 72, 3–13.
Sorensen, L.Y., 1981. Poceedings of the First European Symposium on Poultry Welfare, Copenhagen, Denmark, p. 238
Steeves III, H.R., Siegel, P.B., 1968. Comparative histology of the vascular system in the domestic and game cock. Experientia 14, 937–938.

Su, G., Sorensen, P., Kestin, S.C., 1999. Meal feeding is more effective than early feed restriction at reducing the prevalence of leg weakness in broiler chickens. Poult. Sci. 78, 949–955.

Sumner, D.A., Gow, H., Hayes, D., Matthews, W., Norwood, B., Rosen-Molina, J.T., et al., 2011. Economic and market issues on the sustainability of egg production in the United States: analysis of alternative production systems. Poult. Sci. 90, 241–250.

Swanson, J.C., 2008. The ethical aspects of regulating production. Poult. Sci. 87, 373–379.

Tindell, D., Arze, C.G., 1965. Sexual maturity of male chickens selected for mating ability. Poult. Sci. 44, 70–72.

Tindell, D., Craig, J.V., 1959. Effects of social competition on laying house performance in the chicken. Poult. Sci. 38, 95–105.

Theorell, T., 2012. Evaluating life events and chronic stressors in relation to health: stressors and health in clinical work. Adv. Psychosom. Med. 32, 58–71.

Tolkamp, B.J., Sandilands, V., Kyriazakis, I., 2005. Effects of qualitative feed restriction during rearing on the performance of broiler breeders during rearing and lay. Poult. Sci. 84, 1286–1293.

United Egg Producers (UEP), 2011. Historic Agreement Hatched to Set National Standard for Nation's Egg Industry HSUS and UEP Find Common Ground and a New Way Forward—Ballot Measures Suspended in Oregon and Washington. <http://www.unitedegg.org/homeNews/UEP_Press_Release_7-7-11.pdf>.

Van Horne, P.L., 1996. Production and economic results of commercial flocks with white layers in aviary systems and battery cages. Br. Poult. Sci. 37, 255–261.

Wade, M.J., Bijma, P., Ellen, E.D., Muir, W.M., 2010. Group selection and social evolution in domesticated animals. Evol. Appl. 3, 453–465.

Wall, P.D., 1979. Review article: on the relation of injury to pain. Pain 6, 253–264.

Warren, D., 1930. Crossbred poultry. Kansas Agr. Exp. Sta. Bull., 252.

Webster, A.J., 2001. Farm animal welfare: the five freedoms and the free market. Vet. J. 161, 229–237.

Webster, A.B., Hurnik, J.F., 1991. Behavior, production, and well-being of the laying hen. 2. Individual variation and relationships of behavior to production and physical condition. Poult. Sci. 70, 421–428.

Wegner, R.-M., 1985. Proceedings of the Second European Symposium on Poultry Welfare, Celle, Germany, p. 359

Wegner, R.-M., 1990. Poultry welfare—problems and research to solve them. World's Poult. Sci. J. 46, 19–33.

Whitehead, C.C., 1997. Dyschondroplasia in poultry. Proc. Nutr. Soc. 56, 957–966.

Wong-Valle, J., McDaniel, G.R., Kuhlers, D.L., Bartels, J.E., 1993a. Divergent genetic selection for incidence of tibial dyschondroplasia in broilers at seven weeks of age. Poult. Sci. 72, 421–428.

Wong-Valle, J., McDaniel, G.R., Kuhlers, D.L., Bartels, J.E., 1993b. Correlated responses to selection for high or low incidence of tibial dyschonroplasia in broilers. Poult. Sci. 72, 1621–1629.

Wood-Gush, D.G.M., 1960. A study of sex drive of two strains of cockerels through three generations. Anim. Behav. 8, 43–53.

Yalcin, S., Zhang, X., McDaniel, G.R., Kuhlers, D.L., 1995. Effect of selection for high or low incidence of tibial dyschondroplasia for seven generations on live performance. Poult. Sci. 74, 1411–1417.

Yeates, J.W., Main, D.C., 2008. Assessment of positive welfare: a review. Vet J. 175, 293–300.

Yu, M.W., Robinson, F.E., Robblee, A.R., 1992. Effect of feed allowance during rearing and breeding on female broiler breeders. 1. Growth and carcass characteristics. Poult. Sci. 71, 1739–1749.

Zhang, X., McDaniel, G.R., Roland, D.A., Kuhlers, D.L., 1998. Response to ten generations of divergent selection for tibial dyschondroplasia in broiler chickens: growth, egg production, and hatchability. Poult. Sci. 77, 1065–1072.

Zelenka, D.J., Dunnington, E.A., Cherry, J.A., Siegel, P.B., 1988. Anorexia and sexual maturity in female White Rock chickens. I. Increasing the feed intake. Behav. Genet. 18, 383–387.

Zulkifli, I., Siegel, P.B., 1995. Is there a positive side to stress? World's Poult. Sci. J. 51, 63–76.

FURTHER READING

Craig, J.V., 1981. Domestic Animal Behavior. Prentice-Hall, Inc., Englewood Cliffs, NJ, p. 364.

Chapter 10

Genetics of Domesticated Behavior in Dogs and Foxes

Anna V. Kukekova*, Lyudmila N. Trut†, and Gregory M. Acland‡
**Department of Animal Sciences, University of Illinois at Urbana-Champaign, Urbana, IL, USA; †Institute of Cytology and Genetics of the Russian Academy of Sciences, Novosibirsk, Russia; ‡Baker Institute for Animal Health, Cornell University, Ithaca, NY, USA*

INTRODUCTION

This chapter offers an overview of recent studies in behavior and behavioral genetics of dogs, wolves, and foxes with the aim of providing insight into the complex structure of domesticated behavior in canids.

All domesticated animals have common behavioral features, becoming not only tame to humans but showing lower levels of aggression toward each other. It is not too much of a stretch to wonder if these behavioral features have something in common with the species-specific differences between the behavior of humans, and their two closest extant "cousin species"—the chimpanzee and bonobo, leading to the suggestion that the evolution of modern humans may have been a process of self domestication (Brüne, 2007; Hare *et al.*, 2012; Shipman, 2010).

Behavior is a preeminent phenotype that drives and is driven by the selective pressures of both domestication and evolution. Domestication is a microcosm of evolution. Driven by directed or natural selection, it causes divergence of a founder population into new populations that can exploit new and/or different environmental niches. Once selection takes hold, the genetic aspects of behavioral differences are reinforced, as specialized subpopulations become better adapted to specific niches. Domestication thus provides a window in which the processes involved in the genetics of behavior can be observed at close range providing phenotypes that are accessible to the modern tools of molecular genetics.

The intraspecific and interspecific (i.e. human directed) social behavior of domesticated animals differs dramatically from their wild ancestors (Price, 2002). The paradigmatic example is provided by the domesticated dog (*Canis familiaris*) and its progenitor species, the gray wolf (*Canis lupus*). Although

Genetics and the Behavior of Domestic Animals. DOI: http://dx.doi.org/10.1016/B978-0-12-394586-0.00010-X
© 2014 Elsevier Inc. All rights reserved.

the differences in social behavior of dogs and wolves are undeniably genetic, little is known about the evolutionary processes that led to these behavioral modifications. In the mid-20th century, the Russian geneticist, Dmitry Belyaev, argued for a primary role of selection for behavior in animal domestication. To test this hypothesis, experimental selection of the silver fox (color morph of the red fox, *Vulpes vulpes*), a close relative of the dog, was started in 1959 (Belyaev, 1969, 1979; Trut, 1999, 2001, 2009). The experiment led to the development of a strain of foxes that show friendly, dog-like behavior to humans. Selecting foxes for behavior constituted a completely novel long-range approach to the genetics of behavioral modification that preceded the demonstration that dogs evolved from wolves (Axelsson *et al.*, 2013; Leonard *et al.*, 2002; Savolainen *et al.*, 2002; Vilà *et al.*, 1997; vonHoldt *et al.*, 2010).

THE DOG IS THE FIRST DOMESTICATED SPECIES

The dog was the first species to be domesticated. In fact, this "event" took place so early that it is arguable that the dog self-domesticated itself, and provided the example that humans subsequently followed in deliberately domesticating other species.

Archeological evidence for the coexistence of dogs with humans has been identified from as early as 14,000–17,000 ybp in Russia (Sablin and Khlopachev, 2002); 14,000 ybp in Germany (Nobis, 1979); 12,000 ybp in Israel (Dayan, 1994; Davis and Valla, 1978; Tchernov and Valla, 1997); and 5000 ybp (Olsen, 1985) in China. Recent findings identified remains of dog-like canids living prior to the Last Glacial Maximum (*c.* 26,500–19,000 cal BP) in southern Siberia (Altai Mountains, Russia) dated *c.* 33,000 cal BP (Ovodov *et al.*, 2011), and Western Europe (Goyet, Belguim) dated 36,000 cal BP (Germonpré *et al.*, 2009). Although these incipient dogs may represent domestication events that were terminated by climatic changes associated with the Last Glacial Maximum, and thus may have not given rise to modern dogs, molecular analysis of these remains may still shed light on the evolution of domestication in the dog. Identification of futher dog-like canid remains in different parts of Eurasia supports a hypothesis of recurrent, multiregional dog domestication (Ovodov *et al.*, 2011).

Molecular data suggest that the divergence of dogs from wolves took place as recently as 15,000 years ago (Savolainen *et al.*, 2002; Skoglund *et al.*, 2011). Initial molecular analyses, based on mitochondrial DNA and Y-chromosome markers from dog breeds and wolf populations from different geographical locations, suggested an East-Asian origin for canine domestication (Brown *et al.*, 2011; Ding *et al.*, 2012; Pang *et al.*, 2009; Savolainen *et al.*, 2002; Oskarsson *et al.*, 2012); while genomic DNA analysis supports the Middle East (vonHoldt *et al.*, 2010) as the most likely center. Studies of ancient dog remains from Latin America and Alaska, and dog breeds indigenous to North America, establish the origin of New World dogs as from Old World wolves (Leonard *et al.*, 2002; Vilà *et al.*, 1999; vonHoldt *et al.*, 2010). Village dogs from Africa that did not experience the population bottleneck associated with modern breed formation, and the Australian dingo and New Guinea singing dogs, were also sourced to ancient dogs domesticated in Eurasia (Ardalan *et al.*, 2012; Boyko *et al.*, 2009; Larson *et al.*, 2012; Oskarsson *et al.*, 2011; Savolainen *et al.*, 2004). The multiple recent studies of dog ancestry differ in their estimate of how many domestication events led to the emergence of modern dog lines (Larson *et al.*, 2012; Savolainen *et al.*, 2002; Vilà *et al.*, 1997), and further document recent admixture events between dogs and wolves (Andersone *et al.*, 2002; Anderson *et al.*, 2009; Vilà *et al.*, 1997, 2003; Vilà and Wayne, 1999; vonHoldt *et al.*, 2010; Wayne and vonHoldt, 2012). For example, shared genomic regions between some European breeds and European wolves indicate admixture events between the two species after the divergence of dogs from wolves in the course of dog domestication (vonHoldt *et al.*, 2010; Wayne and vonHoldt, 2012).

Despite the differences in detail cited above, it is clear that dog domestication occurred at a time when modern *Homo sapiens* were hunter-gatherers, and soon after the first entry of modern humans into geographical regions populated by the wolf (Clutton-Brock, 1995). A long co-existence of dogs with humans is supported by recent findings of genomic signatures in the genome of the dog, but not the wolf, associated with the adaptation to a starch-rich diet (Axelsson *et al.*, 2013). Although just how humans and pre-dog wolves benefitted mutually from their new-found coexistence is still an open question, several scenarios have been proposed (Coppinger and Coppinger, 2001; Lorenz, 1954). Significant modification of the dog domestication picture came with recognition of the genetic basis of differences in the social behavior of dogs and wolves (Acland and Ostrander 2003; Coppinger and Coppinger, 2001; vonHoldt *et al.*, 2010). The current hypothesis proposes that dog domestication was a genetic selection for specific behaviors in the course of a long-term co-existence of pre-dog wolves with humans (Acland and Ostrander, 2003; Coopinger and Coopinger, 2001; Hare *et al.*, 2002; Miklosi *et al.*, 2003). Formation of this genetics-centered view of dog domestication was influenced in large part by the farm-fox domestication experiment (Trut 1999, 2001; Trut *et al.*, 2009) which is reviewed in this chapter.

The importance of dogs to human society is reflected in ancient art (Olsen, 1985; Zeuner, 1963) and by ancient co-burial of humans with dogs (Davis and Valla, 1978). Wall paintings of dogs in ancient Egypt show that distinct types of dogs had already existed 2500–4500 ybp (Zeuner, 1963). The morphotypic range of such ancient dogs remain with us today among modern breeds such as mastiffs, hounds, and toy breeds, indicating that distinctly different dog types first appeared thousands of years ago

(Cesarino, 1975; Clutton-Brock, 1981, 1995; Harcourt, 1974; Pionnier-Capitan *et al.*, 2011; Zeda *et al.*, 2006). Significantly, many of these ancient paintings and other works of art portray dogs and humans in contexts of social interaction between the two species, and evidence exists from prehistoric times for task-related interactions between humans and dogs. Dogs have been and remain present in almost every human culture and population as our close companions, assisting us in many tasks. This has led to the concept that dogs and humans are co-evolved species, each influencing the development of the other (Hare *et al.*, 2012; Shipman, 2010).

DOGS ARE MORE COMPETENT THAN WOLVES IN SOCIAL INTERACTION WITH HUMANS

When we talk about dogs we often describe their behavior in anthropomorphic terms. Dogs actively participate in our lives using their unique abilities to understand words from our language, the tone of our voice, and our body language (Grassmann *et al.*, 2012; Kaminski *et al.*, 2004, 2009; Pilley and Reid, 2011; Pongrácz *et al.*, 2001a, 2004, 2005a). Humans and dogs have developed mutual skills for cross-species communication (Ellingsen *et al.*, 2010; Pongrácz *et al.*, 2001a, 2001b, 2005b) that have led to development of the unique social interaction between the dog and its master. No other species is so routinely used as guides for blind people (Naderi *et al.*, 2001; Serpell and Hsu, 2001) and as assistants for people with other disabilities (Eddy *et al.*, 1988; Guest *et al.*, 2006; Lanea *et al.*, 1998; Rintala *et al.*, 2008).

Dogs develop attachment to their owners after a short period of interaction (Gacsi *et al.*, 2001, 2004; Topál *et al.*, 2005), in contrast to home-raised wolves who do not show patterns of attachment to humans similar to those observed in dogs, even after extensive socialization from an early age (Gacsi *et al.*, 2004; Topal *et al.*, 2005). Attachment is one of the basic phenomena of human social relationships (Bowlby, 1958) and one of the obvious features of the human–dog attachment is its resemblance to the human parent–child relationship (Hare and Tomasello, 2005; Scott and Fuller, 1965; Topál *et al.*, 1998, 2005).

In the laboratory, the social skills of dogs and wolves and their ability to communicate with humans are commonly tested using the object-choice paradigm. In this test the canid should choose between two containers using the experimenter's tap, point, or gaze cues. The correct choice is rewarded. The work of Brian Hare and colleagues (2002) has shown that dogs are more skillful than hand-reared wolves at using human social cues. Dogs presented with the object-choice task showed competence in reading proximal point, proximal point and gaze, and all three types of cues (tap, point, and gaze) combined. The wolves in the same test demonstrated abilities to identify a correct container above the random chance level only when provided with combined gaze and proximal point cues (Hare *et al.*, 2002). The number of successful trials in this test was lower for the wolves than for the dogs (Hare *et al.*, 2002).

Subsequent studies showed that wolf performance in the object-choice test is strongly influenced by the level of wolf socialization with humans (Miklósi *et al.*, 2003; Udell *et al.*, 2008; Virányi *et al.*, 2008). Udell *et al.* showed that highly socialized hand-reared wolves from a population bred in captivity for many generations can successfully use distal point cues to make a correct choice. In this particular test, the wolves tested outdoors in a familiar area outperformed dogs tested outdoors in an unfamiliar area (Udell *et al.*, 2008). A group in Hungary studied highly socialized hand-reared wolves from a different population and demonstrated that these wolves had the skills to identify the correct container using tapping cues and some of these wolves were competent in reading proximal point cues (Gácsi *et al.*, 2009; Miklosi *et al.*, 2003; Virányi *et al.*, 2008). Virányi *et al.* have found that wolves can reach a level of untrained dogs in following distal pointing gestures after several months of formal training (Virányi *et al.*, 2008). Importantly, this study has also demonstrated a variation among the hand-reared wolves in their ability to follow human cues, suggesting that the variability in the wolves' communicative behavior might have provided the raw material for selection during domestication (Virányi *et al.*, 2008).

Several studies have shown that dogs are competent in the recognition of human attention status (Bräuer *et al.*, 2004; Call *et al.*, 2003; Gacsi *et al.*, 2004). Dogs use the orientation of the body, the orientation of the head, and the visibility of the eyes to distinguish between "attentive" and "inattentive" humans (Gacsi *et al.*, 2004; Viranyi *et al.*, 2004). Comparison of dog behavior across several tests led to the conclusion that dogs can rely on the same set of human facial cues that are used by humans to understand the attention status of other humans (Gacsi *et al.*, 2004).

Dogs also look to humans for assistance and direction more than wolves. When human-reared dogs and wolves trained to solve a simple manipulative task were presented with an impossible version of the same test (e.g., opening a locked box with food inside), dogs looked at the human almost immediately and then directed their gaze between the human and the box, while wolves did not look to the human but continued to attempt to solve the problem on their own (Miklosi *et al.*, 2003). The ability and desire of dogs to look at a human's face is one of the key differences between dogs and wolves in initializing and maintaining communicative interaction with humans (Miklosi *et al.*, 2003).

Differences in the behavior of dogs and wolves in social interactions with humans are recognized early in the postnatal period. The amount of interaction required for socialization of a wolf is dramatically different from and greater than the amount of interaction required for the socialization of a dog (Coppinger and Coppinger, 2001). Successful socialization of wolf puppies requires the puppy's separation from the family before the eyes open and intense socialization with a human caregiver (Coppinger and Coppinger, 2001). Competence of dog and hand-reared wolf puppies in their interaction with humans was tested in multiple tests at 3, 4, and 5 weeks of age (Gacsi *et al.*, 2005). Significant differences in their behavior have been recorded at each time point. Dog puppies showed more communicative behavior with humans. Wolf puppies were more often aggressive to humans and more often demonstrated avoidance behavior (Gacsi *et al.*, 2005).

These reviewed experiments argue for a domestication hypothesis (Hare *et al.*, 2002) proposing that canine competence in interaction with humans has been acquired in the course of long-term selection for communicative skills. Differences between dogs and wolves in their skills of communication with humans strongly suggest a genetic basis for these behaviors. Early socialization improves the competence of wolves for interaction with humans but a dramatic difference in wolf and dog behavior remains (Coppinger and Coppinger, 2001). Variation in social skills among individuals in each species, domesticated and wild, is caused by a complex interplay of genetic determinants, life experiences, and environmental factors. Selection for specific behaviors can lead to genetically determined modifications of these behaviors, furthermore, it can lead to unexpected modifications of other behavioral aspects as well.

ANALYSIS OF TEMPERAMENT TRAITS IN DOGS

Selection for breed-specific behaviors has clearly played an important role in the course of the formation of the working breeds of dogs. Although reliable historical records on breed origins barely exist for most breeds, there is substantial evidence that breed formation was associated with cultural and economical developments in human societies (Clutton-Brock, 1995; Willis, 1999). Modern breeds were officially registered relatively recently, less than 200 years ago, but a gamut of breeds recognized for their unique behavioral skills had existed for a much longer period of time (Leighton, 1910). Dogs with similar working skills but differences in terms of the specific functions they served as well as differences in appearance were recognized as independent breeds and received official breed standards when registries were created by dog clubs (AKC, 2006; Dalziel, 1897). The migration of Europeans and their dogs to the New World promoted the formation of new working breeds descended from the old breeds or their crosses, incorporating new combinations of desirable skills which were better adapted to new needs and places (Clark, 2003; Cummins, 2002; Huson *et al.*, 2012).

The development of herding breeds is inevitably linked to the growth of livestock farming. Herding breeds differ among the geographical regions, as well as the species they herd and the types of tasks they perform (Combe, 1987). Genetically, working sheepdog breeds that originated in Great Britain are more related to each other than to any other breed category selected for different working tasks in Great Britain or other countries (Parker *et al.*, 2004, vonHoldt *et al.*, 2010).

The use of dogs for hunting is reflected in paintings and historical records from ancient times (Clutton-Brock, 1995). Although, we do not have evidence that modern hunting breeds are descendants of these ancient dogs, it is instructive to note that the sight hound breeds, perhaps the oldest type of hunting dogs, are genetically closely related to each other (vonHoldt *et al.*, 2010), even those from different geographical locations. Likewise, hunting breeds sharing highly specialized behavioral repertoires, such as retrieving breeds, are more closely related genetically to each other than to breeds such as scent hounds selected for different hunting skills (vonHoldt *et al.*, 2010).

It is well established that humans have deliberately preserved specific valuable genetic variants in domestic animals from ancient times (Andersson, 2009; Andersson *et al.*, 2012; Rosengren Pielberg *et al.*, 2008). For example, all breeds characterized by the chondrodysplastic dwarfism phenotype carry the same, identical by descent, *FGF4* gene mutation (Parker *et al.*, 2009) since prehistoric times. These dwarf breeds were developed in different countries, at different times, and do not share recent common ancestry (Parker *et al.*, 2004, 2007). For example, the Pekinese is an ancient Chinese breed while Basset Hounds are traced back to Medieval Western Europe. This *FGF4* allele would clearly be disadvantageous for free-living canids, and has been maintained in the dog population by artificial selection favoring the phenotype.

To assume a similar scenario for behavioral traits is more difficult. In contrast to some morphological and appearance traits that are often determined in dogs by a small number of genes with large effects (Boyko *et al.*, 2010; Cadieu *et al.*, 2009; Hoopes *et al.*, 2012; Sutter *et al.*, 2007), behavior is most likely influenced by a large number of genes, each of which has a relatively small effect on the trait (Anholt and Mackay, 2009; Flint, 2003). However, the close relationships among dog breeds with specific working skills argues for the preservation of genetic variants for behavioral genes among these breeds.

Selection for breed-specific behaviors in the course of breed development was inevitably linked to selection for morphology, appearance, and other characteristics that were important for the dog's ability to perform specific tasks (Coppinger and Schneider, 1995). Guard dogs had to have tough bodies to fight enemies or predators. Sight hounds had to have gracile bodies and high endurance for fast running and long hours of hunting. A white coat color was favored by shepherds because it helped to distinguish herding dogs from wolves (Coppinger and Coppinger, 2001). A thick insulating coat was important for dogs in many geographical regions in order to survive the cold weather outside.

These days, each dog breed has a strictly defined standard that is much more focused on morphologic appearance than on dog behavior. In fact, many behaviors for which working breeds were originally selected became inconvenient in modern life and are not supported by current breeding practices. Behavioral characteristics for which the breeds were known historically thus often do not correspond to the behavioral characteristics of these breeds today. For example, Kubinyi *et al.*, analyzed temperament traits in different breeds in Europe and found significant differences for some breeds between behavioral characteristics provided by the breed standard and behavioral profiles of same breeds based on the owners reports (Kubinyi *et al.*, 2009).

In the Kubinyi *et al.* (2009) study, 93.3% of the respondents representing 14,004 dog owners indicated the "family member" category as the function of their dogs. These results are consistent with other studies in Europe, Australia, and the US, indicating that the majority of dog owners acquire dogs not for specific working tasks but for companionship (Bennett and Rohlf, 2007; Ellingsen *et al.*, 2010; Jagoe and Serpell, 1996; Serpell, 2004). Although the strict demand for working skills is reduced, some breed clubs are making an effort to preserve these skills through behavioral tests and incorporating the results of these tests into breeding programs. This sometimes leads to the development of two independent lines of breeding, one for show and pet dogs, but another one for working dogs, as in, for example, the Border Collie breed.

Programs implemented by organizations to evaluate specific aspects of dog behavior (e.g., herding, protection, detection of drugs or explosives) or for general behavioral assessment (e.g., to screen potential breeding populations for dogs with undesirable behavioral characteristics), have yielded valuable data for scientific evaluation of dog temperament. Svartberg and Forkman (2002) identified five factors (dimensions) underlying canine temperamental traits using data from 1175 dogs of 47 different breeds tested in the dog mentality assessment (DMA) test designed by the Swedish Working Dog Association (Svartberg and Forkman, 2002). The factors were labeled as "Playfulness," "Curiosity/Fearlessness," "Chase-proneness," "Sociability," and "Aggressiveness," and accounted for 37% of common variation in the test. The first four factors were related to one other higher-order factor, which was interpreted as shyness–boldness. The same broad factors with slight variations were identified in an analysis of 15,329 dogs from 164 breeds performed for each of eight FCI (Fédération Cynologique Internationale) breed groups (Svartberg and Forkman, 2002). These data argue that personality factors underlying dog behavior in the DMA test are universal among breeds and not related to specific characteristics of FCI breed groups. Consistency of these behaviors over repeated tests for individual dogs was observed (Svartberg *et al.*, 2005). A heritability of 0.25 for the shyness–boldness dimension was reported (Saetre *et al.*, 2006).

Another common approach for assessment of dog behavior is the collection of information from dog owners by questionnaires (Arhanta *et al.*, 2010; Bennett and Rolf, 2007; Ellingsen *et al.*, 2010; Hsu and Serpell, 2003; Kubinyi *et al.*, 2009; Serpell and Hsu, 2001; Svartberg *et al.*, 2005; Vas *et al.*, 2007). This approach assumes that the dog owner knows its dog's typical behavior and that by asking appropriate questions this information can be extracted in a reasonably accurate, quantitative, and reliable manner (Hsu and Serpell, 2003; Serpell and Hsu, 2001). Using 132 items from a questionnaire filled by owners of 2054 dogs, Hsu and Serpell identified 11 factors that accounted for 57% of common variance in the scored items (Hsu and Serpell, 2003). The factors have been labeled based on the most significant traits contributing to each factor. The factors identified by Hsu and Serpell (2003) arguably describe narrower behavioral categories than the factors identified in the study by Svartberg and Forkman (2002). For example, the study by Hsu and Serpell identified three factors related to aggression: stranger-directed aggression, owner-directed aggression, and dog-directed fear or aggression.

Kubinyi used a 48-item questionnaire adapted for the dog from a questionnaire for human personality assessment (Kubinyi *et al.*, 2009) and, in parallel, a second questionnaire collecting demographic information for the owners and their dogs. In total, the study included owner reports for 14,004 dogs. Factor analysis identified four dimensions of dog personality: calmness, trainability, sociability, and boldness. Each dimension was shown to be influenced by other variables: calmness was influenced by age, neutered status, and the number of professional training courses the dog had experienced; trainability was influenced by training experience, age, and the reason for keeping the dog; sociability to other dogs was mainly determined by age, sex, training experience, and time spent together; boldness was affected by the sex, age of the dog, and the age of the dog at acquisition (Kubinyi *et al.*, 2009). Differences in trainability and boldness among the dog groups were identified: herding dogs were more trainable than Hounds, Working dogs, Toy dogs, and Non-sporting dogs, while Terriers were bolder than Hounds and Herding dogs (Turcsán *et al.*, 2011).

Assessment of dog behavioral phenotypes using data from either tests conducted either by dog organizations or by owner-filled questionnaires has some pitfalls (for a review, see Hall and Wynne, 2012; Jones and Gosling, 2005). The former, biased towards dog populations whose owners are interested in dog training, are often focused on specific behavioral characteristics and target-limited set of breeds. The latter do not provide results of direct observation of dog behavior in a defined period of time; and are sensitive to owner experience in judging dog behavior, and to the validity of the questionnaire. There are practical difficulties in observing dogs in their owners' homes. It is rare for such studies to be undertaken on dogs maintained in research facilities where their behavior is observed in a controlled familiar environment in a systematic manner. There are almost no dog studies that take into account developmental aspects of behavior and the dogs' life experiences (Coppinger and Coppinger, 2001). One exception stands out: the long-term study of dog behavior conducted at Jackson Laboratory in Bar Harbor in the mid-20th century (Scott and Fuller, 1965).

THE BAR HARBOR EXPERIMENT

In this 13-year-long study in Bar Harbor, the behavior of five different dog breeds (Basenji, Beagle, Cocker Spaniel, Shetland Sheepdog, and Wire-haired Fox Terrier) was studied in great detail from birth to approximately one year of age. The experiment was designed to gain insight into the genetics of social behavior; the dog was selected as a model because dog social behavior was regarded as resembling human behavior, and there was a large amount of variation in behavior among the dogs (Hahn and Wright, 1998; Scott and Fuller, 1965). The results of this experiment had a profound effect not only on our understanding of dog behavior but on other disciplines including human psychology, genetics, and evolutionary biology. Here we describe a few of the Bar Harbor experiments and findings as examples that clearly illustrate the complexity of dog behavior and the necessity of applying a systematic approach for understanding basic biological principles underlying it.

A detailed analysis of canine behavioral development from the early postnatal period identified a critical period in dog socialization (Freedman *et al.*, 1961). The importance of this discovery to the field of dog behavior is paramount. The role of the environment during the socialization period was subsequently shown to be important for the development of other aspects of dog behavior, such as herding behavior (Coppinger and Coppinger, 2001).

Analysis of social relationships among littermates revealed different behavioral mechanisms of aggression. By 11 weeks of age, 95% of litters developed a food-related dominance hierarchy. Fox terriers formed more dominance relationships than cocker spaniels or beagles. Significantly more male fox terriers and basenjis formed dominance relationships than did females, whereas beagles and cocker spaniels did not show gender-specific differences in dominance relationships. Shelties (Shetland sheep dogs) developed strong territorial rather than food-related dominance relationships, whereas basenjis showed the opposite pattern: absence of territorial aggression but escalated aggression over food (Hahn and Wright, 1998; Scott and Fuller, 1965).

Measuring emotional reactivity showed significant differences among the breeds: terriers, beagles, and basenjis were consistently more emotional than shelties and cockers. Although significant differences in breed reactivity were observed, a variation in reactivity was also very apparent among individuals from the same breed. This reactivity showed very little change with age and no differences between males and females.

Problem-solving behavior was studied in different tests including a maze test. Fox terriers and shelties made more errors and required a longer time in moving through the maze. Basenjis had the best time in the first day of learning but beagles appeared superior on overall scores through multiple trials. The authors suggested that the beagles' success was attributable to their avoidance of stereotyped habits and their demonstrated random exploration of their surroundings.

Differences in trainability were observed among both breeds and individuals from an early age. Some individuals showed response to training at 5 weeks of age (forced training to be quiet on the weight platform); by the age of 16 weeks nearly 70% of the cocker spaniels were remaining quiet in the test while only 10% of the fox terriers were. All dogs learned to walk appropriately on a leash within a 10-day training period, but different breeds differed in the course of their learning: basenjis were outstanding in their vigorous resistance to restraint, shelties interfered excessively with the handler by leaping on him and winding between his legs; and beagles demonstrated excessive vocalization during the initial stages of training. These experiments show that the basic trainability characteristics of the different breeds tend to be specific to particular test situations and that they are based on a large variety of capabilities. The authors concluded,

In short, the effect of heredity upon trainability is highly complex, both because of the number of specific basic abilities involved and because of the complicated interaction between them made possible by behavioral adaptation.

Scott and Fuller (1965)

Genetic inheritance of several behavioral characteristics was demonstrated by experimental crossbreeding among breeds with distinctive behavioral repertoires. Fearful behavior was analyzed in crosses between basenjis (all tested individuals exhibited some fearful behavior) and cocker spaniels (62% of tested individuals exhibited fearful behavior). F1 dogs exhibited behavior broadly similar to that of the basenji parents, although the inheritance of different specific behavioral characteristics in this cross varied significantly. In several training and spatial-orientation tests the F1 offspring were superior relative to both parents suggesting a complex interplay in inheritance of different behavioral characteristics.

Furthermore, this study clearly demonstrated the influence of genotype-by-environment interaction on dog behavior. As summarized in a recent review:

The most intriguing result was the interaction of genotype and experience (Freedman, 1958). Puppies were raised in isolation and exposed to either tolerant humans or disciplinary humans who made the puppies sit and wouldn't play with them. Later the puppies were presented one by one with a bowl of food. If they approached, the handler would clap his hands to frighten them away. The handler then left the room so the puppies were free to eat the food. The basenjis did, no matter how they had been handled. The Shetland sheepdog never did, no matter how they had been handled. The disciplined beagles ate the food, but not the tolerantly treated ones. The conclusion is that identical handling will have different effects on different genotypes of dogs.

Houpt (2007)

The Bar Harbor experiment clearly shows the complexity of dog social behavior in terms of its development and inheritance. A strong maternal effect was shown for several behavioral characteristics. In particular, Scott and Fuller demonstrate how easily behavior can be misjudged in terms of its genetic origin or gene–environment interaction nature. Underestimation of critical parameters underlying the formation of dog behavioral traits can lead to serious pitfalls in an attempt to identify genes implicated in these behaviors.

GENETIC ANALYSIS OF BEHAVIORAL TRAITS IN DOGS

In contrast to the genetic analysis of morphological traits (Boyko *et al.*, 2010; Cadieu *et al.*, 2009; Chase *et al.*, 2002; Hoopes *et al.*, 2012; Sutter *et al.*, 2007) and inherited diseases (Miyadera *et al.*, 2012; Ostrander and Beale, 2012; Parker *et al.*, 2010; Rimbault and Ostrander, 2012) success in the identification of loci and genes influencing dog behavior has been noticeably more modest.

Breed stereotypes, characteristics for which dog breeds were selected for (i.e. size, skull shape, and ear type), were successfully used as phenotypes for mapping morphological traits (Boyko *et al.*, 2010; Schoenebeck *et al.*, 2012; Sutter *et al.*, 2007). One study used a similar approach for genetic mapping of breed behavioral characteristics (i.e. herding, pointing, boldness, and trainability) defined by a dog trainer as phenotypes (Jones *et al.*, 2008). Several genomic regions that showed statistically significant association with breed behavioral characteristics were identified. Several candidate genes were suggested but no follow up analyses of involvement of these genes in dog behavior were performed.

Recent studies of performance in Alaskan sled dogs identified genetic loci associated with two main racing styles segregating in this breed: sprinting *versus* long-distance racing (Huson *et al.*, 2012). The Alaskan sled dog is a relatively recent breed originating from Alaskan Malamute, Siberian Husky, German Shorthaired Pointer, and Borzoi-type dogs bred for performance, rather than appearance. The racing style in this breed was shown to correlate with the contribution of parental breeds: increase in Alaskan Malamute and Siberian Husky ancestry correlated with increased endurance while contribution from Pointer and Borzoi-type dogs was associated with enhanced speed (Huson *et al.*, 2010). Genetic comparison between sprint-type versus long-distance racing Alaskan sled dogs identified several genomic regions associated with differences in racing style and pinpointed a variant of *MYH9* gene significantly associated with increased heat tolerance in sprint-type dogs (Huson *et al.*, 2012).

Multiple efforts have adopted a candidate gene approach to identify genes associated with behavioral traits in dogs, with mostly negative results (recently reviewed in Hall and Wynne, 2012 and Houpt, 2007). Positive associations were identified for: Activity–Impulsivity with *DRD4* and *TH* in German Shepherds; with *DAT*, *DBH*, *DRD4* in Belgian Tervuren (Hejjas *et al.*, 2007, 2009; Kubinyi *et al.*, 2012); activity level with *SLC1A2* and *COMT* in Labrador Retrievers (Takeuchi *et al.*, 2009a); human-directed aggression with *DRD1*, *HTR1D*, *HTR2C*, and *SLC1A1* in English Cocker Spaniels (Våge *et al.*, 2010); aggression with *SLC1A2* in Shiba Inu (Takeuchi *et al.*, 2009b), but with *AR* in Akita Inu (Konno *et al.*, 2011). All these studies were performed in relatively small dog samples and none has been validated in independent data sets. Taking into account the complexity of dog behavior, the validity of this single gene approach may be questionable.

Comparison of dog and wolf genomes identified several regions where the two species differ most significantly (Axelsson *et al.*, 2013; vonHoldt *et al.*, 2010). From purely genomic data it is almost impossible to determine whether any of these regions are causally associated with behavioral differences between dogs and wolves. One such region identified in the vonHoldt *et al.* (2010) study is homologous to the region on human chromosome 7, where a deletion causes Williams–Beuren syndrome in humans, a complex disorder associated with exceptional gregariousness (Doyle *et al.*, 2004; Jarvinen-Pasley *et al.*, 2008). Identification of a selection signal in the corresponding canine genomic region does suggest that this region may harbor genes influencing behavior in both species. However, as dogs diverged from wolves so long ago and have been selected for so many different traits, one would certainly expect their genomes to harbor multiple species-specific selection signals not necessarily associated with behavioral differences. What is clearly needed to complement such studies as these, is evidence of co-segregation of genomic changes with behavior, and expression differences of implicated genes that correlate with differences of behavior. The opportunity to obtain such information is provided in studies that recapitulate canine domestication in the silver fox.

THE FARM-FOX EXPERIMENT

The silver fox is a coat color morph of the red fox (*Vulpes vulpes*). In contrast to the dog that was domesticated prehistorically, the fox was domesticated in controlled farm conditions. Experimental domestication of farm-bred foxes was started by Dmitry Belyaev and Lyudmila Trut at the Institute of Cytology and Genetics (ICG) of the Russian Academy of Sciences, in Novosibirsk, Russia, in the late 1950s and is still ongoing, as has been reviewed in several publications (Trut, 1999; 2001; Trut *et al.*, 2004, 2009). Belyaev proposed that selection for behavior was the primary force in the course of animal domestication. He further hypothesized that selection for an unusual behavior in wild species, that is, a tame response to humans, may affect many physiological processes in animals under selection (i.e. activity of hypothalamic-pituitary-adrenal axis) and lead to the emergence of variation in morphology, a common feature of domesticated species (Belyaev, 1979; Belyaev and Trut, 1989; Trut 1988). The fox was selected as a model to test this hypothesis.

When the fox domestication experiment began, foxes had been bred in captivity for over 50 years. Fox-fur farming was pioneered on Prince Edward Island in Southeastern Canada, beginning in the 1890s (Westwood 1989). Farmers on Prince Edward Island primarily raised the silver–black coat color variant of red foxes, which had the greatest economic value and were subsequently used to stock fur farms in many areas of North America and Eurasia (Bespyatih, 2009; Nes *et al.*, 1988; Westwood 1989; Petersen 1914). Mitochondrial DNA analysis has identified Eastern Canada as the primary, if not sole, source of ancestry for farm-bred fox populations maintained at the ICG in Novosibirsk (Statham *et al.*, 2011).

SELECTION OF FOXES FOR TAME BEHAVIOR

In the beginning of the experiment, 130 foxes that showed less fearful and aggressive responses to humans were identified at several commercial fox farms across the former Soviet Union and brought to the experimental farm at the ICG to become the founders for the experimental population. The response to human presence was the sole selection criteria in this population. At first the selection was focused against aggressive responses to humans, as summarized in Trut *et al.*, 2004:

The main task at this stage of selection was eliminating defensive reactions to humans. In order to reveal variability in the expression of these reactions more completely, the animals in the selected population were subjected to more intensive contacts with humans than in usual practice. During these contacts, the pups were subjected to a number of tests: the experimenter attempted to hand feed, stroke or handle them. This type of human–animal communication continued for the first three to four months of life of the animals. As a result, the emotionally negative defensive reactions to humans in these foxes weakened, disappeared or, in some of the animals, emotionally positive reactions were formed. The foxes that retained aggressive–fearful reactions to humans in spite of the 3-month period of human contacts with them, were eliminated by selection from the population as soon as in 2 to 3 generations (Trut, 1980a, 1980b; Trut, 1999). In generation 4 of selection, the first pups appeared that did not form aggressive–fearful reactions to humans as a result of positive contacts with them. On the contrary, these pups demonstrated emotionally positive response to humans: when the experimenter approached them, they whined and wagged their tails anticipating a positive contact.

Results of the selection against defensive aggression became apparent after a few generations of selective breeding. Subsequently, selection for an emotionally positive ("friendly") response to humans was applied. Behavior of all foxes was evaluated in a test with five consecutive stages: (1) observer approaches the cage; (2) observer stands near the closed cage; (3) observer opens the cage door and stands near by; (4) observer attempts to touch the fox; (5) observer closes the cage door, then stays near the closed cage (Trut, 1980a; Trut *et al.*, 2004). Fox behavior in this test was scored based on their response to humans using a categorical system (Table 10.1A). Behavior of young animals was tested several times during development and the final test was giving to all pups at 7–8 month of age. Only the tamest individuals in each generation, less than 10%, were bred. At the same time a deliberate effort was made to avoid inbreeding (Kukekova *et al.*, 2004; Trut, 1999, 2001; Trut *et al.*, 2004). The response to selection was extremely rapid (Table 10.2). Improvement of behavioral scores recorded in fox pedigrees clearly showed genetic inheritance of tame behavior. It is important to note that except for the first few generations of selection, foxes from the population selected for tame behavior did not receive special handling. All human contacts were, and continue to be limited to maintenance and testing procedures. Foxes from the tame population show a friendly response to humans as early as one month postanatal and remain friendly throughout their entire lives. Tame foxes are eager to establish human contact, whimper to attract attention, and sniff and lick similar to dogs (Figure 10.1A). The behavior of tame foxes can be observed on the webpage: http://cbsu.tc.cornell.edu/ccgr/behaviour/Index.htm.

TABLE 10.1A System for Scoring Behavior in Tame Fox Population and Selecting the Most Tame Foxes for the Breeding Program

Animal Reaction	Scores
Passive-protection response; fox avoids experimenter or bites if stroked or handled, comes if offered food.	0.5–1.0
Foxes let themselves be petted and handled, but show no emotionally friendly response to experimenter.	1.5–2.0
Foxes show emotionally positive, friendly, response to experimenter, wagging tails and whining.	2.5–3.0
Foxes are eager to establish human contact, whimpering to attract attention and sniffing and licking experimenters like dogs. They start displaying this kind of behavior before one month postnatal age.	3.5–4.0

Tame behavior is scored from zero (representing "neutral" behavior; an absence of both active aggressive and tame responses directed towards the observer) to 4 (representing the most tame behavior).
Source: Trut (1980a).

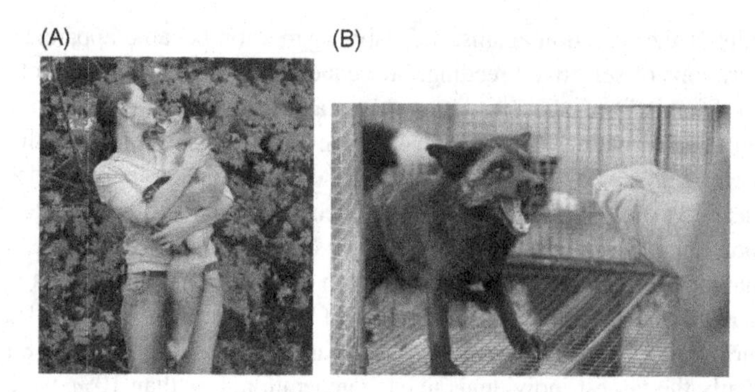

FIGURE 10.1 (A) Behavior of a fox from the tame population. (B) Behavior of a fox from the aggressive population. *(Photographs: Darya Shepeleva)*

SELECTION OF FOXES FOR AGGRESSIVE BEHAVIOR

In parallel with selection for tameness, selective breeding of farm foxes for aggressive behavior began in 1970s. Because there is deliberate selection on commercial farms against animals that show excessively aggressive responses to humans, selection of the aggressive strain at ICG was undertaken to preserve this behavior for research. Fifty farm-bred silver foxes with the most aggressive responses to humans were selected from several fox farms and used as founders of the aggressive population. The critical distance between experimenter and caged animal and the intensity of fox aggressive response were the major criteria for measuring aggression in the fox population, and to select animals for producing the next generation (Table 10.1B). Progress in selection for aggressive behavior did not follow the same pattern as selection for tame behavior, being slower in comparison (Trut 1980a, b). Foxes from the aggressive population are, however, consistently and distinctly aggressive towards humans and difficult to handle (Figure 10.1B).

BEHAVIORAL AND PHYSIOLOGICAL CHANGES ASSOCIATED WITH SELECTION FOR TAME BEHAVIOR

The genetic basis of fox tame and aggressive phenotypes has been clearly demonstrated in multiple experiments which included experimental cross-breeding of tame and aggressive animals, cross-fostering of newborn pups, and embryo transplantation (Kukekova *et al.*, 2008, 2011a; Trut, 1980a, b, 2001). Although selection of foxes for tame behavior was strictly limited to defined behavioral criteria, further developmental and behavioral differences emerged in tame foxes. Opening of the eyelids and the external auditory canal was accelerated; the sensitive period for socialization persisted past 60 days of age (compared to less than 45 days in unselected foxes); and play activity (normally only seen in infantile wild type foxes) extended into

TABLE 10.1B System for Scoring Behavior in Aggressive Fox Population and Selecting the Most Aggressive Foxes for the Breeding Program

Animal Reaction	Scores
Fox shows teeth, snarls, growls **at first sight of human**. When experimenter is near **closed cage** fox attacks experimenter and other objects in field of view. Bared teeth and fixed dilated pupils.	−4.0
When experimenter is near **closed cage**, fox shows teeth, snarls, growls, tries to attack both the experimenter and other objects in field of view. Bared teeth and fixed dilated pupils.	−3.5
When experimenter is near **open cage**, fox shows teeth, snarls growls, attacks experimenter and other objects in the field of view. Bared teeth and fixed dilated pupils.	−3.0
When experimenter is near the **open cage**, fox growls but does not attack.	−2.5
When experimenter, near the **open cage**, moves protected arm towards fox, it growls and tries to bite.	−2.0
As experimenter opens cage, fox is calm, but attempts to touch the fox provoke it to show teeth and snarl.	−1.5

The score is based on the critical distance between the experimenter and the caged fox when the animal first demonstrates an aggressive reaction to the experimenter's presence. Originally the scoring system was based on a range from zero (least aggressive) to − 4 (most aggressive). However, after multiple generations of selection, no current individuals exhibit behavior scoring between − 1.5 and 0.
Source: Trut (1980a).

TABLE 10.2 The Number and Proportion of Elite-Behavior Progeny at Different Stages of Selection

Year of Study (Generation of selection)	Number of Progeny Scored	Out of them, Elite Animals	
		Number	*Proportion (%)*
1965 (F_6)	213	4	1.8
1970 (F_{10})	370	66	17.8
1980 (F_{20})	1438	503	35.0
1990 (F_{30})	1641	804	49.0
2002 (F_{42})	902	642	71.2

Source: Trut *et al.* (2004).

adulthood (Belyaev *et al.*, 1985; Plyusnina *et al.*, 1991; Trut, 2001). Tame foxes developed a novel repertoire of vocalizations toward humans (Gogoleva *et al.*, 2009, 2011). Significant differences in corticosteroid and neurotransmitter levels were found between tame and control foxes

(Oskina and Tinnikov, 1992; Popova *et al.*, 1997). Significant differences in hypothalamic–pituitary–adrenocortical (HPA) system reactivity in ontogenesis were observed between foxes from tame and unselected populations (Oskina, 1996; Trut and Oskina, 1985). Significantly lower density of serotonin 5-HT_{1A} receptors was observed in the hypothalamus; and significantly higher levels of serotonin and tryptophan hydroxylase were detected in the midbrain and hypothalamus of domesticated strain of foxes (Popova *et al.*, 1997, 2007; Trut, 2001). The transformation of the seasonal reproductive pattern was also observed. Some foxes from the tame population showed sexual activity outside of the regular breeding season and a few females have been mated twice in a year, a pattern which was not recorded for foxes in nature or in commercial bred populations (Belyaev and Trut, 1983; Trut 1980b).

Several *de novo* traits were detected in the fox population selected for tameness. In particular, coat color changes such as the appearance of a white spot on the head (Star phenotype) and loss of pigment in other areas began to appear in the eighth selected generation (Belyaev *et al.*, 1981). It is intriguing that white spotting, which appeared without direct selection or inbreeding in the tame fox population, is frequently observed as a distinctive difference between domesticated animals of several species (dogs, cats, cattle, horses, etc.) and their wild progenitors. Other morphological characteristics also arose in the same manner: some foxes had floppy ears until significantly older age than foxes from commercial populations, some had rolled tails, some had changes in the skull shape that made them look more like dogs (Trut *et al.*, 1991, 2009).

The basis of the morphological and physiological changes that occurred in the course of selection for tame behavior in the farm fox population is unclear. Genetic drift is plausible, as is genetic hitchhiking, especially if genes involved in development or physiology are located on the same chromosomes and in close proximity to genes involved in behavior. Selection acting on behavioral genes would inevitably act as well on tightly linked neighboring genes. In-phase alleles for both behavioral genes and neighboring genes on the same haplotype would be inherited together until sufficient generations passed for the haplotype to be broken by recombination events. Selection for rare alleles of genes involved in development and physiology could promote the development of novel phenotypes or increase the frequency of rare ones. Indeed, the higher rate of atypical phenotypes was observed during the early stages of selection for behavior. Alternatively, genes involved in behavior could have a pleiotroptc effect and be involved in regulatory processes not only restricted to behavior, but that play an important role in animal development and physiology (Belyaev, 1879; Trut, 1999, 2001; Trut *et al.*, 2009). Support for this hypothesis comes from the observation that multiple behavioral and morphological changes observed in tame foxes are associated with delayed or truncated development (Trut, 1999, 2001; Trut *et al.*, 2009). This concept of neoteny, the retention of puppy-like traits into adulthood, is widely used to explain behavioral, morphological, and physiological changes associated with dog domestication as well (Coppinger and Coppinger, 2001; Wayne, 1986).

The multifaceted changes observed in the course of selection of foxes for tame behavior affirm the destabilizing hypothesis of Dmitry Belyaev (1979). The molecular mechanisms underlying these phenomena are best addressed by analyzing the genetic architecture of fox behavior.

IDENTIFICATION OF LOCI AND GENES IMPLICATED IN FOX BEHAVIOR

The strains of tame and aggressive foxes provide a robust model for identification of genes and loci involved in behavioral differences between the two populations. Unlike modern dogs, the strain of domesticated foxes was created rapidly by selection focused exclusively on specific behavioral traits. Although these fox strains have been carefully studied for several decades, only recently has it become possible to consider a systematic approach to identify the loci and molecular mechanisms controlling these behaviors. Classical genetic approaches to genome mapping had derived a rudimentary map of the fox genome by 1998, with a well-defined fox karyotype (Graphodatsky *et al.*, 1981, 1995; Yang *et al.*, 1999) and sparsely populated linkage groups (for a review, see Rubtsov, 1998). To begin an attack on the molecular genetics of behavior in these foxes, however, further resources were essential, including a set of suitable pedigrees for mapping, an adequate set of suitable molecular markers, and a robust method for measuring behavior in resegregating pedigrees.

RESOURCES FOR MAPPING FOX BEHAVIOR

Accordingly, a program was instituted at ICG to crossbreed foxes from the tame and aggressive strains, producing an F1 population, and subsequently generating informative backcross and intercross pedigrees. Simultaneously, the canine genome sequence and linkage map were exploited to identify a set of microsatellite markers shared between the dog and the fox (Kukekova *et al.*, 2004). Approximately 60% of canine markers proved useful for interrogating the fox genome.

To construct a meiotic linkage map of the fox genome, 286 individual foxes (180 animals in the third generation) from 37 pedigrees were genotyped, for a total of 320 markers (Kukekova *et al.*, 2007). A second generation of the fox meiotic linkage map was then developed using an extended set of fox three-generation pedigrees, including a total of 916 progeny in informative generations, and adding a further 93 microsatellite markers adapted from dog genome sequence and the recently published genetic maps of the dog genome (Sargan *et al.*, 2007; Wong *et al.*, 2010). This increased the total number of markers on the map to 408 (Kukekova *et al.*, 2011a, 2012), 405 of which could be uniquely identified in the 7.6x genome sequence of the dog. The resulting fox linkage map is thus directly anchored to the dog genome sequence, enabling detailed comparisons to be made between corresponding chromosomal fragments of the two species and indirect comparisons between fox and human chromosomes. The resulting sex-averaged map comprises 1548.5 cM (Kukekova *et al.*, 2011a) that covers 16 fox autosomes and the X chromosome. Alignment of the fox meiotic map against the 7.6x canine genome sequence revealed high conservation of marker order between homologous regions of the two species and provides a robust method for predicting the chromosomal location of the fox orthologs of genes identified in the canine or human genome sequences (Kukekova *et al.*, 2007, 2011a, 2012).

MEASUREMENT OF FOX BEHAVIORAL PHENOTYPES

In the course of the selective breeding program, the behavioral phenotypes in the selected tame and aggressive populations quickly diverged, and separate scoring systems for assignment of fox behavioral phenotypes were developed for the tame (Table 10.1A) and the aggressive population (Table 10.1B) (reviewed in Kukekova *et al.* 2005; Trut 1980a, b, 1999, 2001). However, a single, unified scoring system was required for measuring behavior in experimental pedigrees descended from crosses of the tame and aggressive strains. Individuals in such pedigrees have a wide range of behaviors and often exhibit fragmented or reshuffled elements of the behavioral patterns characteristic of the founder populations.

A new method was devised that could measure the variation in quantitative behavioral phenotypes between different fox populations as well as the resegregation of behaviors in experimental pedigrees. This new system was rooted in the traditional behavioral test used for selective breeding (Vasilieva and Trut, 1990; Trut 1980b, Trut *et al.*, 2004, see Table 10.1). Fox responses to humans were evaluated in a standard series of four sequential steps that were videotaped (Kukekova *et al.*, 2008, 2011a).

- Step A: observer stands calmly near the closed cage but does not deliberately try to attract the animal's attention;
- Step B: observer opens the cage door, remains nearby but does not initiate any contact with the fox;
- Step C: observer attempts to touch the fox;
- Step D: observer closes the cage door, then stays calmly near the closed cage.

Because all foxes live under consistent conditions including similar interactions with humans and because their behavior is tested at precise time points using standard tests under constant conditions, the environmental factors that might influence behavior were held to a strict minimum.

An ethological survey of these videotaped tests identified over 300 discrete behavioral observations (traits) which could be reproducibly scored from video records in a binary fashion, e.g., presence or absence. Examples, each simply scored Yes or No, include: "Wagging Tail?"; "Stays at the front wall of the cage?"; and "Ears pinned?" (Kukekova *et al.*, 2008). Evaluation of these traits for informativeness, redundancy, reproducibility, and consistency identified two overlapping sets of traits: (i) a minimal set of 50 traits that reliably distinguished fox populations along a tame–aggressive axis (Kukekova *et al.*, 2008); and (ii) a larger, 98-trait set (Kukekova *et al.*, 2011a), that includes all traits from the minimal trait set plus additional traits that capture other dimensions of fox behavior (Kukekova *et al.*, 2011a).

Behahior of foxes from parental (tame and aggressive) and crossbred pedigrees was evaluated using this standardized testing protocol and videotaped. The resulting videotape dataset comprised a total of 1003 foxes (83 tame foxes, 80 aggressive, 93 F1, 293 backcross-to-tame, 202 backcross-to-aggressive, and 252 F2 foxes). Video records were analyzed to deconstruct fox behavior by scoring for the presence or absence of each of 98 specific binary traits (Kukekova *et al.*, 2011a). Principal-components (PC) analysis was then used (Kukekova *et al.*, 2011a), to identify independent (i.e. uncorrelated) underlying factors (i.e. the principal components) that accounted for decreasing amounts of the total variance in observed behavior (Kukekova *et al.*, 2011a). Specific methodological aspects of the PC analysis are described in Kukekova *et al.*, 2008 and 2011a.

In this large data set, the first two principal components, PC1 and PC2, accounted for 33% and 9% of the total behavioral variation, respectively (Kukekova *et al.*, 2011a). PC1 clearly distinguished tame foxes from aggressive foxes; F1 foxes yield intermediate values that extend into the ranges of both the tame and aggressive foxes. The scores of the backcross generations resegregate (Figure 10.2A) such that mean values of PC1 in the different populations defined a linear gradient of heritable behavior, ranging from aggressive to tame, that clearly corresponds to the relative proportions of aggressive to tame ancestry in each population (Figure 10.2A).

PC2 did not follow a similar gradient (Figure 10.2B). Review of the discrete behavioral traits that contribute to these two principal components demonstrated that PC1 and PC2 are comprised of very different aspects of behavior (Figure 10.3). PC1 is comprised of traits that distinguish overall aggressiveness from tameness, whereas most of the traits important to PC2 can be interpreted as distinguishing bold from shy behavior (independent of the degree to which a fox was tame or aggressive).

QTL MAPPING OF FOX BEHAVIOR

These two principal components of behavior were used as phenotypes to identify associated quantitative trait loci (QTL) in informative fox pedigrees

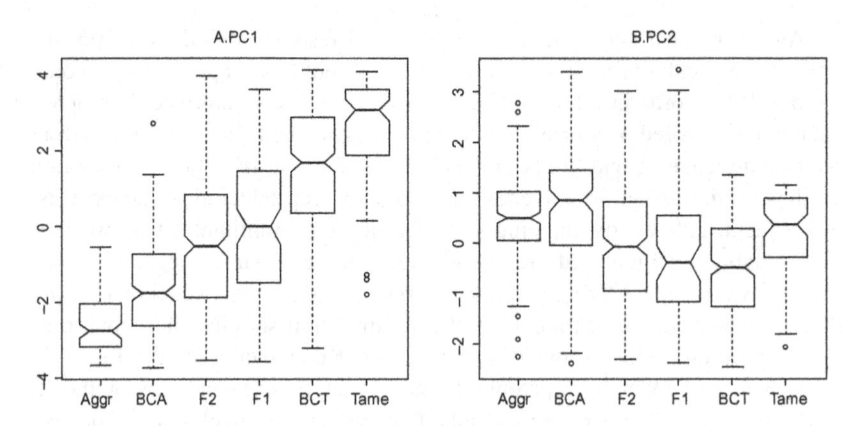

FIGURE 10.2 Population distributions for the first two principal components of silver fox behavior. (A) Distributions for principal component 1 (PC1). (B) Distributions for PC2. Aggr = ''aggressive'' founder population; BCA = backcross-to-aggressive; F2 = F2 population (F1 × F1); F1 = F1 population (''tame'' × ''aggressive''); BCT = backcross-to-tame population; Tame = ''tame'' founder population. Horizontal bars within each box indicate the population median. Confidence intervals for the medians are shown as notches such that two distributions with non-overlapping notches are significantly different (at $P < 0.05$). The bottom and top edges of the boxes indicate the 25th and 75th percentiles. The whiskers indicate the range of data up to 1.5-times the interquartile range. Outliers are shown as individual circles. In each of the populations, the PC1 distribution pattern conforms to that expected for a heritable trait reflecting the proportional contributions from ''Tame'' and ''Aggressive'' ancestry. This is clearly not the case for PC2.

by interval mapping. Fox informative pedigrees included three populations: backcross-to-tame (293 progeny), backcross-to aggressive (202 progeny), and intercross (F2) (250 progeny) (Kukekova *et al.*, 2011a).

Interval mapping in the combined data set, which included all the experimental pedigrees, identified a locus for PC1 on fox chromosome 12 (VVU12), located in a region between 10 and 60 cM (Figure 10.4A). Intriguingly, part of this QTL interval on VVU12 is homologous to canine chromosome 5 (Figure 10.5) and includes a region corresponding to a genomic location recently implicated in domestication of dogs from wolves (VonHoldt *et al.*, 2010). On the fox meiotic map, this conserved syntenic region lies between markers CM5.41 and CM5.60, or between 27.2 and 28.4 cM, respectively. Among individual behavioral traits that map precisely to this interval are B3 (touches observer's hand) and C17 (allows observer to touch belly), both traits of trustfulness that will be immediately familiar to persons familiar with dogs and other domesticated species.

In contrast to PC1, PC2 did not yield a significant peak on VVU12 using the combined data set. However, PC2 also mapped to VVU12 when only the backcross-to-tame segregants were used for interval mapping (Figure 10.4B). This result provides insight into the complex expression of behavioral phenotypes in different populations. It suggests that, although independent by

A. (Sorted by C.PC1)

	Trait	Trait description	C.PC1	C.PC2
*	C37	Aggressive sounds	–86	8
	C34	Follows the hand (aggr.)	–84	14
	C31	Attack alert	–83	12
	C32	Pinned ears (aggr.)	–56	8
	C36	Triangle ears directed back (aggr.)	–42	4
	C33	Trying to bite	–33	12
	C30	Attack	–33	20
	C55	Leans on back or side walls in zones 5-6	–5	–14
	C3	Fox is in zones 3-4-5-6 at the beginning of step C	–4	–24
	C4	Spends more than 30 seconds in zones 3-4-5-6	–3	–33
**	C7	Observer can first touch fox in zones 5-6	–2	–41
**	C38	Fox remains only in zones 3-5-6	–1	–39
	C50	Tail is up for at least for 3 seconds	–1	1
	C35	Narrow ears directed back	3	–26
***	C39	Moved forward at least one zone during the step	3	29
***	C2	Fox is in zones 1-2-3-4 at the beginning of step C	6	24
	C6	Observer can first touch fox in zones 3-4	18	4
	C204	Tame sounds (combined)	18	10
	C19	Comes into zone 2 at the end of step C	21	19
	C25	Wagging tail	25	13
	C18	Fox holds observer's hand with its mouth	26	14
	C17	Fox rolls onto its side, inviting observer to touch its belly	27	20
	C29	Comes to and sniffs observer's hand at the end of step C	32	14
	C24	Loud breathing	36	20
	C13	Fox allows observer to touch the rear part of its back	49	–18
	C14	Fox allows observer to touch its back	73	–15
	C8	Lies down during a contact for at least 5 seconds	88	0
	C16	Fox allows observer to touch its head	105	–12
	C12	Tame ears	107	7
****	C15	Fox allows observer to touch its nose	115	–10

B. (Sorted by C.PC2)

Trait	Trait description	C.PC1	C.PC2	
C7	Observer can first touch fox in zones 5-6	–2	–41	**
C38	Fox remains only in zones 3-5-6	–1	–39	**
C4	Spends more than 30 seconds in zones 3-4-5-	–3	–33	
C35	Narrow ears directed back	3	–26	
C3	Fox is in zones 3-4-5-6 at the beginning of step C	–4	–24	
C13	Fox allows observer to touch the rear part of its back	49	–18	
C14	Fox allows observer to touch its back	73	–15	
C55	Leans on back or side walls in zones 5-6	–5	–14	
C16	Fox allows observer to touch its head	105	–12	
C15	Fox allows observer to touch its nose	115	–10	****
C8	Lies down during a contact for at least 5 seconds	88	0	
C50	Tail is up for at least for 3 seconds	–1	1	
C6	Observer can first touch fox in zones 3-4	18	4	
C36	Triangle ears directed back (aggr.)	–42	4	
C12	Tame ears	107	7	
C37	Aggressive sounds	–86	8	*
C32	Pinned ears (aggr.)	–56	8	
C204	Tame sounds (combined)	18	10	
C33	Trying to bite	–33	12	
C31	Attack alert	–83	12	
C25	Wagging tail	25	13	
C18	Fox holds observer's hand with its mouth	26	14	
C29	Comes to and sniffs observer's hand at the end of step C	32	14	
C34	Follows the hand (aggr.)	–84	14	
C19	Comes into zone 2 at the end of step C	21	19	
C30	Attack	–33	20	
C17	Fox rolls onto its side, inviting observer to touch its belly	27	20	
C24	Loud breathing	36	20	
C2	Fox is in zones 1-2-3-4 at the beginning of step C	6	24	***
C39	Moved forward at least one zone during the step	3	29	***

FIGURE 10.3 Comparison of behavioral traits contributing to the first two principal components of silver fox behavior at test step C (C.PC1 and C.PC2). All observations from one step of the test (Step C "Observer attempts to touch the fox") are shown. Behavioral observations (traits) contributing to C.PC1 and C.PC2 are ranked according to their loadings (eigenvalues) from principal component analysis. The significances for each trait loading for C.PC1 and C.PC2 are shown as the number of standard errors from zero (negative or positive) as established by bootstrap trials. Note in particular the difference in ranking for the traits marked with asterisks. C.PC1 and C.PC2 are very different behavioral gestalts. PC1 is comprised of traits that contribute to overall aggressiveness (*) or tameness (****), whereas PC2 represents traits that are marked by either passivity (**) or activity (***).

definition, the phenotypes measured by PC1 (tameness *vs* aggressiveness) and PC2 (bold *vs* shy) are not entirely unrelated, as PC2 can enhance the observed expression of PC1. That is, if an animal is aggressive, passive behavior will reduce the perceived expression of that trait whereas active behavior will enhance the expression, and the same effect applies if an animal is tame. In backcross populations the distribution of behavior is skewed toward the extreme of the recurrent parent, reducing the range of tame *vs* aggressive behaviors. Under these circumstances, PC2 acts to increase that range. We would therefore expect that whereas PC1 and PC2 are distinct principal components in a matrix composed of all populations, they could be correlated in particular backcross populations. This is in fact the case for the backcross-to-tame populations ($r = 0.75–0.8$). In contrast, in F2 populations where the behaviors are more normally distributed, this is not the case ($r = -0.06$). It could well be, therefore, that the PC2 QTL on VVU12 in the backcross-to-tame population reflects enhancement of the expression of PC1.

TRANSCRIPTOME ANALYSIS

A good example of an approach that can be applied for identification of genes and regulatory elements involved in fox behavioral phenotypes is the

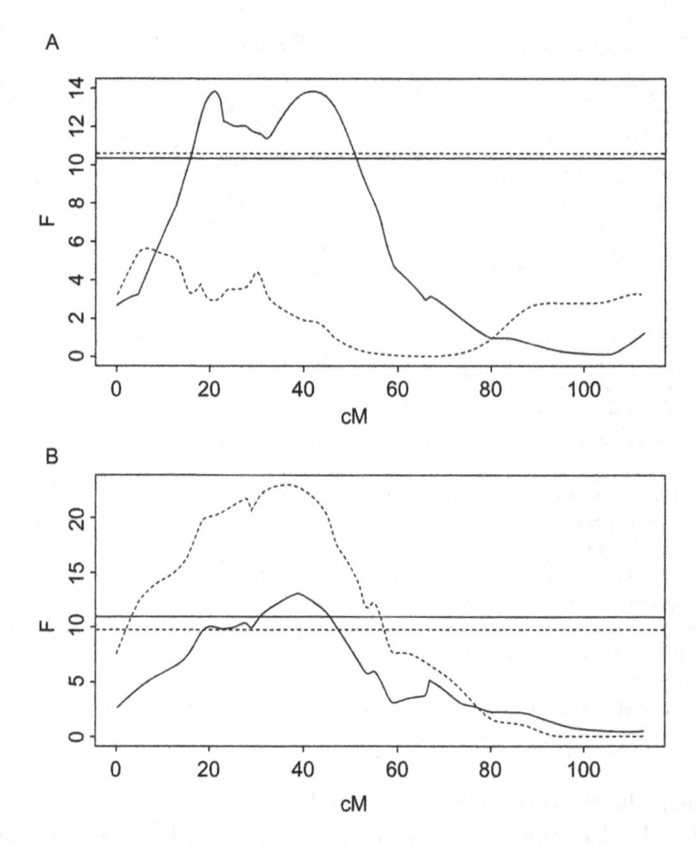

FIGURE 10.4 Interval mapping of the first two principal components of silverfox behavior (PC1, PC2) to Fox chromosome 12 (VVU12). Interval mapping using GridQTL software, was undertaken on (A) a combined data set including all experimental silver fox populations; and (B) Backcross-to-tame population only (i.e. excluding backcross-to-aggressive, and F2). Solid lines = PC1, dashed lines = PC2. The F stat (y-axis) is graphed as a function of cM distance across VVU12 (data and map distances from: Kukekova *et al.* 2011). Horizontal lines (solid = PC1, dashed = PC2) = thresholds for genome wide significance at $P < 0.01$. Interval mapping across all populations (A) yields support for PC1-associated loci on VVU12, located broadly between 10 and 60 cM, that exceeds the threshold for genome-wide significance; but support for PC2 does not achieve significance. Interval mapping restricted to backcross-to-tame populations (B) yields support for PC2-associated loci on VVU12, located broadly between 10 and 60 cM, that exceeds the threshold for significance; but support for PC1 does not achieve significance.

analysis of the transcriptome of tame and aggressive foxes. High-throughput genetic sequencing of all the gene transcripts expressed in a specific tissue sample yields that sample's transcriptome, and provides quantitative information about gene expression in different tissues. Comparative analysis of gene expression in tame and aggressive foxes will highlight the expressed parts of the genome that differentiate the strains. Furthermore, comparison of gene expression profiles and loci (QTL) identified in the course of genetic

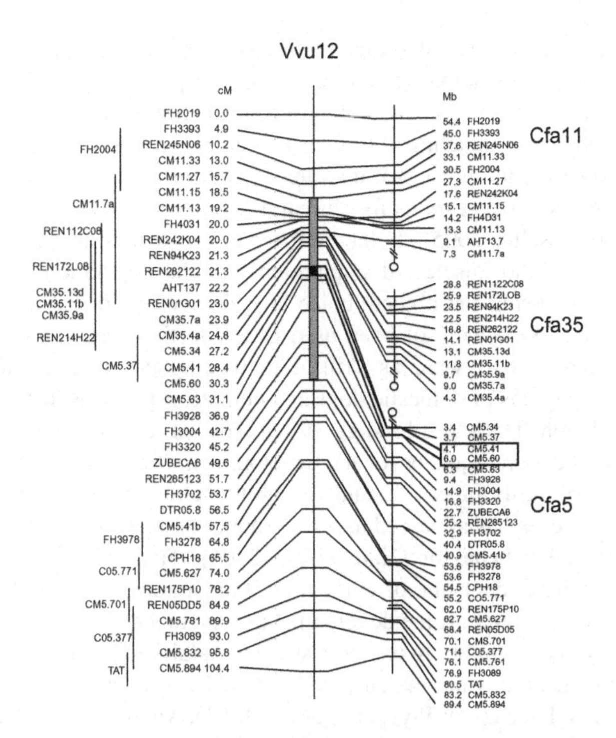

FIGURE 10.5 A locus for the first principal component of behavior (PC1) on Fox chromosome 12 (VVU12). The meiotic linkage map of VVU12 (left side of figure) is aligned to the genome sequence of the homologous canine chromosomes (CFA11, CFA35, CFA5) on the right side of the figure (data and map distances from Kukekova *et al.*, 2011). In the middle of the figure, the gray interval indicates the broad support interval for PC1 in the silver fox, and the black interval a region homologous to a locus on CFA 5 identified with domestication of the wolf (VonHoldt *et al.*, 2010).

mapping will help to identify genes under selection for behavior. Transcriptome sequencing of brain samples from tame and aggressive foxes has recently been initiated (Kukekova *et al.*, 2011b). Although only preliminary results have so far been reported, this has provided significantly expanded genomic resources for the fox, a species without a sequenced genome; as well as significant insights into the gene expression profile of the fox prefrontal cortex, expression differences between fox samples, and a catalog of potentially important gene-specific sequence variants (Kukekova *et al.*, 2011b). This approach can provide valuable insights into the molecular mechanisms implicated in behavioral differences between the two strains.

WHAT THE FARM-FOX EXPERIMENT TELLS US ABOUT BEHAVIOR

One of the most striking results of the farm-fox experiment is the clear demonstration of the influence of genes on behavior (Kukekova *et al.*, 2008, 2011a; Trut, 1980a, b, 2001).

The farm-fox experiment showed that modification of a specific behavior (friendly response to humans) can affect other aspects of behavior as well. Tame foxes are eager to establish human contact from a very early age (Trut, 1999), develop close attachment to their owners when raised in human homes (Trut, personal communication), and are as skillful as dog puppies in using human point-and-gaze gestures for finding the hidden food (Hare *et al.*, 2005; Hare and Tomasello, 2005). Behavioral testing of kits from two farm-bred populations (i.e. the unselected population and the population selected for tame behavior), using the object-choice test, demonstrated that although fox communicative skills had not been used as a selection criterion (Trut *et al.*, 2009), the tame foxes were as skillful in reading proximal point cues as age-matched puppies of domestic dogs and outperformed foxes from the unselected population. These results suggest that reduced fear and/or aggression to humans may be a prerequisite for the development of advanced inter-specific communicative skills in canids and the ability and wiliness of tame foxes to read human cues has appeared as a by-product of, or is associated with, the selection for tame behavior (Hare *et al.*, 2005). Thus, the fox experiment provides an independent line of evidence for the argument that reducing fear and aggression can support the development of interactive behaviors.

Play activity of tame foxes persists into adulthood and they actively seek communication with humans and other foxes (Belyaev *et al.*, 1985; Kukekova and Trut, 2010; Plyusnina *et al.*, 1991). Many other differences of behavior between tame and aggressive foxes that are not directly related to their response to humans remain to be investigated.

Selection of foxes for positive responses to humans led rapidly to the development of a behavioral repertoire with significant parallels to that of domestic dogs. The prehistoric domestication of the dog from the wolf may have taken a longer time and more generations but in all probability was similarly achieved through selection for behavioral modifications. The similarities between the behaviors of domesticated foxes and dogs suggest that the behavioral response to humans in these two species could involve similar sets of genes.

SUMMARY

Archeological and molecular data have well documented the long history of coexistence of humans and dogs. Although the full scenario of historical dog domestication remains to be determined, the behavioral differences between dogs and wolves strongly argue that selection for behavior was the force that created “Man’s best friend”. The fox domestication experiment demonstrated that many behavioral characteristics that differentiate dogs from wolves can be obtained by selection solely for a friendly response to humans. The rapid progress in selection of foxes for tame behavior strongly suggests that selection was acting on genetic variation pre-existing in the founder fox population. These results support dog domestication hypotheses that suggest that pre-dog wolves underwent natural selection for behavioral traits that allowed them to coexist with humans. Genetic variants pre-existing in wolf populations were likely targets of this selection. Support for this theory of selection for pre-existing variants also arises from recent studies of size in dogs. While all small breeds of dogs are fixed for a specific variant of the *IGF1* gene (Sutter *et al.*, 2007) the same allele was found to segregate in a modern population of gray wolves from the Middle East (Gray *et al.*, 2010) and in Portugese Water dogs, a breed in which the size variation takes place (Chase *et al.*, 2002).

Identification of loci that both influence tame behavior in foxes and are homologous to regions in the dog genome supporting genetic signals related to selection of dogs from wolves leads to the intriguing hypothesis that domesticated behavior in dogs and foxes may have similar genetic bases.

Not only do dogs differ in their behavior from wolves, but different breeds of dog are characterized by genetically determined differences in behavior. Although the latter differences are highly characteristic and readily recognized, in fact each such behavioral pattern likely represents a synthesis of many discrete behaviors. This synthesis has been clearly demonstrated in the fox study. For example, the apparently simple response to human presence can be influenced by many behavioral characteristics at once: interplay of discrete mechanisms captured by the tame–aggressive (PC1) and bold–shy (PC2) dimensions are just two of the most obvious ones that influence the test outcome. However, a detailed understanding of the genetic basis of behavior in dogs requires well-defined phenotypes, as well as control of environmental factors and gene–environment interactions that can strongly influence behavioral outcomes. These are not as easily implemented as in the fox study.

Although domestication has, for many years, been of great interest to scientists, most notably including Darwin, until recently basic questions concerning the genetic mechanisms involved have been difficult to address. With modern advances in molecular genetic technologies, however, it is now feasible to pursue research on diverse organisms to find the links between genes, brain function, and a wide range of social behaviors (Robinson *et al.*, 2005, 2008).

We expect that identification of the molecular mechanisms underlying domesticated behavior in foxes and dogs will help us to understand basic biological principles guiding social behavior in mammals and to provide insights into our own behavior as well.

ACKNOWLEDGMENTS

We thank Dr. K. Gordon Lark, Kevin Chase, and Anastasiya V. Kharlamova for insightful discussions and to Dr. Lark for critical reading of the manuscript. We would like to express profound gratitude to members of Dr. Trut, Dr. Acland, and Dr. Kukekova's groups and all personnel of the ICG experimental farm for many years of dedicated work on this project. We gratefully acknowledge NIH grants RO1MH077811, R24GM082910, RO1EY006855, and FCP84-74 for support.

REFERENCES

Acland, G.M., Ostrander, E.A., 2003. The dog that came in from the cold. Heredity 90, 201–202.

American Kennel Club, The Complete Dog Book: Official Publication of the American Kennel Club (Howell Book House, New York, ed. 20, 2006).

Anderson, T., vonHoldt, B.M., Candille, S.I., et al., 2009. Molecular and evolutionary history of melanism in North American gray wolves. Science 323 (5919), 1339–1343.

Andersone, Z., Lucchini, V., Randi, E., et al., 2002. Hybridisation between wolves and dogs in Latvia as documented using mitochondrial and microsatellite DNA markers. Mamm. Biol. 67, 79–90.

Andersson, L., 2009. Studying phenotypic evolution in domestic animals: a walk in the footsteps of Charles Darwin. Cold Spring Harb. Symp. Quant. Biol. 74, 319–325.

Andersson, L.S., Larhammar, M., Memic, F., Wootz, H., Schwochow, D., Rubin, C.J., et al., 2012. Mutations in DMRT3 affect locomotion in horses and spinal circuit function in mice. Nature 488 (7413), 642–646, Aug 30.

Anholt, R.R.H., Mackay, T.F.C., 2009. Principles of Behavioral Genetics. Elsevier Academic Press, Oxford, UK.

Ardalan, A., Oskarsson, M., Natanaelsson, C., Wilton, A.N., Ahmadian, A., Savolainen, P., 2012. Narrow genetic basis for the Australian dingo confirmed through analysis of paternal ancestry. Genetica 140 (1–3), 65–73, Mar.

Arhanta, C., Bubna-Littitzb, H., Bartelsc, A., Futschikd, A., Troxlera, J., 2010. Behaviour of smaller and larger dogs: effects of training methods, inconsistency of owner behaviour and level of engagement in activities with the dog. Appl. Anim. Behav. Sci. 123 (3–4), 131–142.

Axelsson, E., Ratnakumar, A., Arendt, M.-L., Maqbool, K., Webster, M.T., Perloski, M., et al., 2013. The genomic signature of dog domestication reveals adaptation to a starch-rich diet. Nature. Published online January 23.

Belyaev, D.K., 1969. Domestication of animals. Science (Russ.) 5, 47–52.

Belyaev, D.K., 1979. Destabilizing selection as a factor in domestication. J. Hered. 70, 301–308.

Belyaev, D.K., Trut, L.N., 1983. Reorganization of the seasonal rhythm of reproduction in the silver black foxes during their selection for the capacity to domestication. Zh. Obshch. Biol. 42, 739–752, In Russian.

Belyaev, D.K., Trut, L.N., 1989. The convergent nature of incipient forms and the concept of destabilizing selection. In: Ovchinnikov, Y.A., Rapoport, I.A. (Eds.), Vavilov's Heritage in Modern Biology. Nauka, Moscow, pp. 155–169.

Belyaev, D.K., Ruvinsky, A.O., Trut, L.N., 1981. Inherited activation–inactivation of the star gene in foxes: its bearing on the problem of domestication. J. Hered. 72 (4), 267–274, Jul–Aug.

Belyaev, D.K., Plyusnina, I.Z., Trut, L.N., 1985. Domestication in the silver fox (*Vulpes fulvus Desm*): changes in physiological boundaries of the sensitive period of primary socialization. Appl. Anim. Behav. Sci. 13, 359–370.

Bennett, P.C., Rohlf, V.I., 2007. Owner–companion dog interactions: relationships between demographic variables, potentially problematic behaviours, training engagement and shared activities. Appl. Anim. Behav. Sci. 102, 65–84.

Bespyatih, O.Y., 2009. The consequences of amber acid feeding in different genotypes of farm-bred foxes. VOGIS (Russian) 13 (3), 639–646.

Bowlby, J., 1958. The nature of the child's tie to his mother. Int. J. Psychoanal. 39, 350–371.

Boyko, A.R., Boyko, R.H., Boyko, C.M., Parker, H.G., Castelhano, M., Corey, L., et al., 2009. Complex population structure in African village dogs and its implications for inferring dog domestication history. Proc. Natl. Acad. Sci. U.S.A. 106 (33), 13903–13908.

Boyko, A.R., Quignon, P., Li, L., Schoenebeck, J.J., Degenhardt, J.D., Lohmueller, K.E., et al., 2010. A simple genetic architecture underlies morphological variation in dogs. PLoS Biol. 8 (8), e1000451, Aug 10.

Bräuer, J., Call, J., Tomasello, M., 2004. Visual perspective taking in dogs (Canis familiaris) in the presence of barriers. Appl. Anim. Behav. Sci. 88 (3), 299–317.

Brown, S.K., Pedersen, N.C., Jafarishorijeh, S., Bannasch, D.L., Ahrens, K.D., Wu, J.T., et al., 2011. Phylogenetic distinctiveness of Middle Eastern and Southeast Asian village dog Y chromosomes illuminates dog origins. PLoS ONE 6 (12), e28496.

Brüne, M., 2007. On human self-domestication, psychiatry, and eugenics. Philos. Ethics. Humanit. Med. 2, 21.

Cadieu, E., Neff, M.W., Quignon, P., Walsh, K., Chase, K., Parker, H.G., et al., 2009. Coat variation in the domestic dog is governed by variants in three genes. Science 326 (5949), 150–153, Oct 2.

Call, J., Bräuer, J., Kaminski, J., Tomasello, M., 2003. Domestic dogs (Canis familiaris) are sensitive to the attentional state of humans. J. Comp. Psychol. 117 (3), 257–263, Sep.

Cesarino, F., 1975. Il Molosso: Viaggio Intorno Al Mastino Napoletano. Casa editrice Faustino Fiorentino, Napoli, p.203.

Chase, K., Carrier, D.R., Adler, F.R., Jarvik, T., Ostrander, E.A., Lorentzen, T.D., et al., 2002. Genetic basis for systems of skeletal quantitative traits: principal component analysis of the canid skeleton. Proc. Natl. Acad. Sci. U.S.A. 99 (15), 9930–9935, Jul 23.

Clark, N.R.A., 2003. 1840–2000, Dog Called Blue: The Australian Cattle Dog and the Australian Stumpy Tail Cattle Dog. WrightLight Publishers, Wallacia, NSW.

Clutton-Brock, J., 1981. Domesticated Animals from Early Times. British Museum (Natural History)/Heinemann, London.

Clutton-Brock, J., 1995. Origins of the dog: domestication and early history. In: Serpell, J. (Ed.), The Domestic Dog: Its Evolution, Behaviour, and Interactions with People. Cambridge Univ Press, Cambridge, UK, pp. 7–20.

Combe, I., 1987. Herding Dogs, Their Origins and Development in Britain. Faber & Faber Ltd., London.

Coppinger, R., Schneider, R., 1995. Evolution of Working Dogs. In: Serpell, J. (Ed.), The Domestic Dog: Its Evolution, Behaviour and Interactions with People. Cambridge University Press, pp. 21–50.

Coppinger, R., Coppinger, L., 2001. Dogs: A Startling New Understanding of Canine Origin, Behavior & Evolution. Scribner, New York, NY.

Cummins, B.D., 2002. First Nations, First Dogs : Canadian Aboriginal Ethnocynology. Detselig Enterprises, Calgary, p.351.

Dalziel, H., 1897. British Dogs: Their Varieties, History, Characteristics, Breeding, Management, and Exhibition. "The Bazaar" Office, London.

Dayan, T., 1994. Early domesticated dogs of the Near East. J. Archaeol. Sci. 21, 633–640.

Davis, S.J.M., Valla, F.R., 1978. Evidence for the domestication of the dog 12,000 years ago in the Natufian of Israel. Nature 276, 608–610.

Ding, Z.L., Oskarsson, M., Ardalan, A., Angleby, H., Dahlgren, L.G., Tepeli, C., et al., 2012. Origins of domestic dog in southern East Asia is supported by analysis of Y-chromosome DNA. Heredity (Edinb) 108 (5), 507–514, May.

Doyle, T.F., Bellugi, U., Korenberg, J.R., Graham, J., 2004. "Everybody in the world is my friend": hypersociability in young children with Williams syndrome. Am. J. Med. Genet. 124A, 263–273.

Eddy, J., Hart, L.A., Boltz, R.P., 1988. The effects of service dogs on social acknowledgments of people in wheelchairs. J. Psychol. 122 (1), 39–45.

Ellingsen, K., Zanella, A.J., Bjerkås, E., Indrebø, A., 2010. The relationship between empathy, perception of pain and attitudes toward pets among norwegian dog owners. A Multidisciplinary Journal of The Interactions of People & Animals 23 (3), 231–243.

Flint, J., 2003. Analysis of quantitative trait loci that influence animal behavior. J. Neurobiol. 54 (1), 46–77.

Freedman, D.G., King, J.A., Elliot, O., 1961. Critical period in the social development of dogs. Science 133 (3457), 1016–1017, Mar 31.

Gácsi, M., Miklósi, Á., Varga, O., Topál, J., Csányi, V., 2004. Are readers of our face readers of our minds? Dogs (Canis familiaris) show situation-dependent recognition of human's attention. Anim. Cogn. 7, 144–153.

Gácsi, M., Győri, B., Miklósi, Á., Virányi, Z., Kubinyi, E., Topál, J., et al., 2005. Species-specific differences and similarities in the behavior of hand-raised dog and wolf pups in social situations with humans. Dev. Psychobiol. 47, 111–122.

Gácsi, M., Győri, B., Virányi, Z., Kubinyi, E., Range, F., et al., 2009. Explaining dog wolf differences in utilizing human pointing gestures: selection for synergistic shifts in the development of some social skills. PLoS ONE 4 (8), e6584.

Germonpré, M., Sablin, M.V., Stevens, R.E., Hedges, R.E.M., Hofreiter, M., Stiller, M., et al., 2009. Fossil dogs and wolves from Palaeolithic sites in Belgium, the Ukraine and Russia: Osteometry, ancient DNA and stable isotopes. J. Archaeol. Sci. 36 (2), 473–490.

Gogoleva, S.S., Volodin, I.A., Volodina, E.V., Kharlamova, A.V., Trut, L.N., 2009. Kind granddaughters of angry grandmothers: the effect of domestication on vocalization in cross-bred silver foxes. Behav. Processes 81, 369–375.

Gogoleva, S.S., Volodin, I.A., Volodina, E.V., Kharlamova, A.V., Trut, L.N., 2011. Explosive vocal activity for attracting human attention is related to domestication in silver fox. Behav. Processes 86, 216–221.

Graphodatsky, A.S., Beklemisheva, V.R., Dolf, G., 1995. High-resolution GTG-banding patterns of dog and silver fox chromosomes: description and comparative analysis. Cytogenet. Cell. Genet. 69 (3–4), 226–231.

Grassmann, S., Kaminski, J., Tomasello, M., 2012. How two word-trained dogs integrate pointing and naming. Anim. Cogn. 15 (4), 657–665, Jul.

Gray, M.M., Sutter, N.B., Ostrander, E.A., Wayne, R.K., 2010. The IGF1 small dog haplotype is derived from Middle Eastern grey wolves. BMC Biol. 8, 16. 10.1186/1741-7007-8-16, Feb 24.

Guest, C.M., Collis, G.M., McNicholas, J., 2006. Hearing dogs: a longitudinal study of social and psychological effects on deaf and hard-of-hearing recipients. J. Deaf. Stud. Deaf. Educ. 11 (2), 252–261.

Hahn, M.E., Wright, J.C., 1998. The influence of genes on social behavior of dogs. Chapter 10. In: Grandin, T. (Ed.), Genetics and the Behavior of Domestic Animals. Academic Press, San Diego.

Hall, N.J., Wynne, C.D., 2012. The canid genome: behavioral geneticists' best friend? Genes. Brain. Behav. Sep 14.
Harcourt, R.A., 1974. The dog in prehistoric and early historic Britain. J. Archeol. Sci. 1, 151–175.
Hare, B., Brown, M., Williamson, C., Tomasello, M., 2002. The domestication of social cognition in dogs. Science 298 (5598), 1634–1636.
Hare, B., Plyusnina, I., Ignacio, N., Schepina, O., Stepika, A., Wrangham, R., et al., 2005. Social cognitive evolution in captive foxes is a correlated by-product of experimental domestication. Curr. Biol. 15 (3), 226–230.
Hare, B., Tomasello, M., 2005. Human-like social skills in dogs? Trends. Cogn. Sci. 9, 439–444.
Hare, B., Wobber, V., Wrangham, R., 2012. The self-domestication hypothesis: evolution of bonobo psychology is due to selection against aggression. Anim. Behav. 83, 573–585.
Hejjas, K., Vas, J., Topal, J., Szantai, E., Ronai, Z., Szekely, A., et al., 2007. Association of polymorphisms in the dopamine D4 receptor gene and the activity–impulsivity endophenotype in dogs. Anim. Genet. 38, 629–633.
Hejjas, K., Kubinyi, E., Ronai, Z., Szekely, A., Vas, J., Miklósi, Á., et al., 2009. Molecular and behavioral analysis of the intron 2 repeat polymorphism in the canine dopamine D4 receptor gene. Genes. Brain. Behav. 8, 330–336.
Hoopes, B.C., Rimbault, M., Liebers, D., Ostrander, E.A., Sutter, N.B., 2012. The insulin-like growth factor 1 receptor (IGF1R) contributes to reduced size in dogs. Mamm. Genome. 23 (11–12), 780–790, Dec.
Houpt, K., 2007. Genetics of canine behavior. Acta. Vet. Brno. 76, 431–444.
Hsu, Y., Serpell, J.A., 2003. Development and validation of a questionnaire for measuring behavior and temperament traits in pet dogs. J. Am. Vet. Med. Assoc. 223 (9), 1293–1300.
Huson, H.J., Parker, H.G., Runstadler, J., Ostrander, E.A., 2010. A genetic dissection of breed composition and performance enhancement in the Alaskan sled dog. BMC Genet. 11, 71, Jul 22.
Huson, H.J., vonHoldt, B.M., Rimbault, M., Byers, A.M., Runstadler, J.A., Parker, H.G., et al., 2012. Breed-specific ancestry studies and genome-wide association analysis highlight an association between the MYH9 gene and heat tolerance in Alaskan sprint racing sled dogs. Mamm. Genome. 23 (1–2), 178–194, Feb.
Jagoe, A.J., Serpell, J., 1996. Owner characteristics and interactions and the prevalence of canine behaviour problems. Appl. Anim. Behav. Sci. 47, 31–42.
Jarvinen-Pasley, A., Bellugi, U., Reilly, J., Mills, D.L., Galaburda, A., Reiss, A.L., et al., 2008. Defining the social phenotype in Williams syndrome: a model for linking gene, the brain, and behavior. Dev. Psychopathol. 20, 1–35.
Jones, A.C., Gosling, S.D., 2005. Temperament and personality in dogs (*Canis familiaris*): a review and evaluation of past research. Appl. Anim. Behav. Sci. 95, 1–53.
Jones, P., Chase, K., Martin, A., Davern, P., Ostrander, E.A., Lark, K.G., 2008. Single-nucleotide-polymorphism-based association mapping of dog stereotypes. Genetics 179 (2), 1033–1044, Jun.
Kaminski, J., Call, J., Fischer, J., 2004. Word learning in a domestic dog: evidence for fast mapping. Science 304, 1682–1683.
Kaminski, J., Tempelmann, S., Call, J., Tomasello, M., 2009. Domestic dogs comprehend human communication with iconic signs. Dev. Sci. 12 (6), 831–837, Nov.
Konno, A., Inoue-Murayama, M., Hasegawa, T., 2011. Androgen receptor gene polymorphisms are associated with aggression in Japanese Akita Inu. Biol. Lett. 7, 658–660.

Kubinyi, E., Turcsán, B., Miklósi, Á., 2009. Dog and owner demographic characteristics and dog personality trait associations. Behav. Processes 81, 392–401.

Kubinyi, E., Vas, J., Hejjas, K., Ronai, Z., Brúder, I., Turcsán, B., et al., 2012. Polymorphism in the tyrosine hydroxylase (TH) gene is associated with activity–impulsivity in German Shepherd Dogs. PLoS ONE 7 (1), e30271.

Kukekova, A.V., Trut, L.N., Oskina, I.N., Kharlamova, A.V., Shikhevich, S.G., Kirkness, E.F., et al., 2004. A marker set for construction of a genetic map of the silver fox (*Vulpes vulpes*). J. Hered. 95, 185–194.

Kukekova, A.V., Oskina, I.N., Kharlamova, A.V., Chase, K., Erb, H.N., Aguirre, G.D., et al., 2005. In: Ostrander, E.A., Giger, U., Lindblad-Toh, K. (Eds.), The Dog and Its Genome. Cold Spring Harbor Laboratory Press, Woodbury NY, pp. 515–538.

Kukekova, A.V., Trut, L.N., Oskina, I.N., Johnson, J.L., Temnykh, S.V., Kharlamova, A.V., et al., 2007. A meiotic linkage map of the silver fox, aligned and compared to the canine genome. Genome. Res. 17 (3), 387–399.

Kukekova, A.V., Trut, L.N., Chase, K., Shepeleva, D.V., Vladimirova, A.V., Kharlamova, A.V., et al., 2008. Measurement of segregating behaviors in experimental silver fox pedigrees. Behav. Genet. 38 (2), 185–194.

Kukekova A.V., Trut L.N., 2010. Analysis of fox interspecific and intraspecific behavior in selectively bred strains of silver foxes (*Vulpes vulpes*). 15th European Conference in Personality. Brno, Czech Republic, July 20–24, 2010.

Kukekova, A.V., Trut, L.N., Chase, K., Kharlamova, A.V., Johnson, J.L., Temnykh, S.V., et al., 2011a. Mapping Loci for fox domestication: deconstruction/reconstruction of a behavioral phenotype. Behav. Genet. 41, 593–606.

Kukekova, A.V., Johnson, J.L., Teiling, C., Li, L., Oskina, I.N., Kharlamova, A.V., et al., 2011b. Sequence analysis of prefrontal cortical brain transcriptome from tame and aggressive strains of the silver fox (*Vulpes vulpes*). BMC Genomics 12 (1), 482.

Kukekova, A.V., Temnykh, S.V., Johnson, J.L., Trut, L.N., Acland, G.M., 2012. Genetics of behavior in the silver fox. Mamm. Genome 23 (1–2), 164–177.

Lanea, D.R., McNicholasb, J., Collisb, G.M., 1998. Dogs for the disabled: benefits to recipients and welfare of the dog. Appl. Anim. Behav. Sci. 59 (1–3), 49–60.

Larson, G., Karlsson, E.K., Perri, A., Webster, M.T., Ho, S.Y., Peters, J., et al., 2012. Rethinking dog domestication by integrating genetics, archeology, and biogeography. Proc. Natl. Acad. Sci. U.S.A. 109 (23), 8878–8883, Jun 5.

Leighton, R., 1910. Dogs and All About Them. Cassell and Company, Ltd., London.

Leonard, J.A., Wayne, R.K., Wheeler, J., Valadez, R., Guillén, S., Vilà, C., 2002. Ancient DNA evidence for old world origin of New World dogs. Science 298, 1613–1616.

Lorenz K. Man Meets Dog. 1954.

Miklósi, Á., Kubinyi, E., Topál, J., Gácsi, M., Virányi, Z., Csányi, V., 2003. A simple reason for a big difference: wolves do not look back at humans but dogs do. Curr. Biol. 13, 763–766.

Miyadera, K., Acland, G.M., Aguirre, G.D., 2012. Genetic and phenotypic variation of inherited retinal diseases in dogs: the power of within- and across-breed studies. Mamm. Genome 23 (1–2), 40–61, 2012 Feb.

Naderi, S., Miklósi, Á., Dóka, A., Csányi, V., 2001. Co-operative interactions between blind persons and their dogs. Appl. Anim. Behav. Sci. 74, 59–80.

Nes, N., Einarsson J., Lohi O., Jarosz S., Scheelje R., 1988. Beautiful Fur Animals and their Color Genetics, Scientifur, 60 Langagervij, DK-2600 Glostrup, Denmark.

Nobis, G., 1979. Der alteste Haushunde lebte. Umschau 79, 610.

Olsen, S.J., 1985. Origins of the Domestic Dog: The Fossil Record. Univ. of Arizona Press, Tucson, USA.

Oskarsson, M.C., Klütsch, C.F., Boonyaprakob, U., Wilton, A., Tanabe, Y., Savolainen, P., 2012. Mitochondrial DNA data indicate an introduction through Mainland Southeast Asia for Australian dingoes and Polynesian domestic dogs. Proc. Biol. Sci. 279 (1730), 967–974.

Oskina, I.N., 1996. Analysis of the function state of the pituitary–adrenal axis during postnatal development of domesticated silver foxes (Vulpes vulpes). Scientifur 20, 159–161.

Oskina, I.N., Tinnikov, A.A., 1992. Interaction between cortisol and cortisol-binding protein in silver foxes (Vulpes fulvus). Comp. Biochem. Physiol. Comp. Physiol. 101, 665–668.

Ostrander, E.A., Beale, H.C., 2012. Leading the way: finding genes for neurologic disease in dogs using genome-wide mRNA sequencing. BMC Genet. 13, 56. 10.1186/1471-2156-13-56.

Ovodov, N.D., Crockford, S.J., Kuzmin, Y.V., Higham, T.F.G., Hodgins, G.W.L., van der Plicht, J., 2011. A 33,000-year-old incipient dog from the Altai Mountains of Siberia: evidence of the earliest domestication disrupted by the last glacial maximum. PLoS ONE 6 (7), e22821, Open Access.

Pang, J.F., Kluetsch, C., Zou, X.J., Zhang, A.B., Luo, L.Y., Angleby, H., et al., 2009. mtDNA data indicate a single origin for dogs south of Yangtze River, less than 16,300 years ago, from numerous wolves. Mol. Biol. Evol. 26 (12), 2849–2864.

Parker, H.G., Kim, L.V., Sutter, N.B., Carlson, S., Lorentzen, T.D., Malek, T.B., et al., 2004. Genetic structure of the purebred domestic dog. Science 304 (5674), 1160–1164, May 21.

Parker, H.G., Kukekova, A.V., Akey, D.T., Goldstein, O., Kirkness, E.F., Baysac, K.C., et al., 2007. Breed relationships facilitate fine-mapping studies: a 7.8-kb deletion cosegregates with Collie eye anomaly across multiple dog breeds. Genome. Res. 17 (11), 1562–1571, Nov.

Parker, H.G., VonHoldt, B.M., Quignon, P., Margulies, E.H., Shao, S., Mosher, D.S., et al., 2009. An expressed fgf4 retrogene is associated with breed-defining chondrodysplasia in domestic dogs. Science 325 (5943), 995–998.

Parker, H.G., Shearin, A.L., Ostrander, E.A., 2010. Man's best friend becomes biology's best in show: genome analyses in the domestic dog. Annu. Rev. Genet. 44, 309–336.

Petersen, M., 1914. The Fur Traders and Fur Bearing Animals. The Hammond Press, Buffalo, New York, USA, 364.

Pilley, J.W., Reid, A.K., 2011. Border collie comprehends object names as verbal referents. Behav. Process. 86, 184–195.

Pionnier-Capitan, M., Bemilli, C., Bodu, P., Célérier, G., Ferrié, J.-G., Fosse, P., et al., 2011. New evidence for Upper Palaeolithic small domestic dogs in South-Western Europe. J. Archaeol. Sci. 38 (9), 2123–2140.

Plyusnina, I.Z., Oskina, I.N., Trut, L.N., 1991. An analysis of fear and aggression during early development of behavior in silver foxes (Vulpes vulpes). Appl. Anim. Behav. Sci. 32, 253–268.

Pongrácz, P., Miklósi, Á., Csányi, V., 2001a. Owners' beliefs on the ability of their pet dogs to understand human verbal communication. A case of social understanding. Curr. Psychol. Cogn. 20, 87–107.

Pongrácz, P., Miklósi, Á., Kubinyi, E., Gurobi, K., Topál, J., Csányi, V., 2001b. Social learning in dogs: the effect of a human demonstrator on the performance of dogs (Canis familiaris) in a detour task. Anim. Behav. 62, 1109–1117.

Pongrácz, P., Miklósi, Á., Timár-Geng, K., Csányi, V., 2004. Verbal attention getting as a key factors in social learning between dog (Canis familiaris) and human. J. Comp. Psychol. 118, 375–383.

Pongrácz, P., Miklósi, Á., Vida, V., Csányi, V., 2005a. The pet-dogs' ability for learning from a human demonstrator in a detour task is independent from the breed and age. Appl. Anim. Behav. Sci. 90, 309–323.

Pongrácz, P., Miklósi, Á., Molnár, C., Csányi, V., 2005b. Human listeners are able to classify dog barks recorded in different situations. J. Comp. Psychol. 119, 136–144.

Popova, N.K., Kulikov, A.V., Avgustinovich, D.F., Voitenko, N.N., Trut, L.N., 1997. Effect of domestication of the silver fox on the main enzymes of serotonin metabolism and serotonin receptors. Genetika 33, 370–374.

Price, E.O., 2002. Animal Domestication and Behavior. CABI Publishing, Wallingford, United Kingdom, p.297.

Rimbault, M., Ostrander, E.A., 2012. So many doggone traits: mapping genetics of multiple phenotypes in the domestic dog. Hum. Mol. Genet. 21 (R1), R52–R57.

Rintala D.H., Matamoros R., Seitz L.L. 2008. Effects of assistance dogs on persons with mobility or hearing impairments: a pilot study. 45(4):489–504.

Robinson, G.E., Grozinger, C.M., Whitfield, C.W., 2005. Sociogenomics: social life in molecular terms. Nat. Rev. Genet. 6, 257–270.

Robinson, G.E., Fernald, R.D., Clayton, D.F., 2008. Genes and social behavior. Science 322 (5903), 896–900.

Rosengren Pielberg, G., Golovko, A., Sundström, E., Curik, I., Lennartsson, J., Seltenhammer, M.H., et al., 2008. A cis-acting regulatory mutation causes premature hair graying and susceptibility to melanoma in the horse. Nat. Genet. 40 (8), 1004–1009, Aug.

Rubtsov N.B. (1998) The fox gene map. ILAR Journal 39(2/3). <http://dels-old.nas.edu/ilar_n/ilarjournal/39_2_3/39_2_3Fox.shtml>.

Sablin, M.V., Khlopachev, G.A., 2002. The earliest Ice Age dogs: evidence from Eliseevichi I. Curr. Anthropol. 43, 795–799.

Saetre, P., Strandberg, E., Sundgren, P.E., Pettersson, U., Jazin, E., Bergström, T.F., 2006. The genetic contribution to canine personality. Genes. Brain. Behav. 5 (3), 240–248, Apr.

Sargan, D.R., Aguirre-Hernandez, J., Galibert, F., Ostrander, E.A., 2007. An extended microsatellite set for linkage mapping in the domestic dog. J. Hered. 98, 221–231.

Savolainen, P., Zhang, Y.P., Luo, J., Lundeberg, J., Leitner, T., 2002. Genetic evidence for an East Asian origin of domestic dogs. Science 298, 1610–1613.

Savolainen, P., Leitner, T., Wilton, A.N., Matisoo-Smith, E., Lundeberg, J., 2004. A detailed picture of the origin of the Australian dingo, obtained from the study of mitochondrial DNA. Proc. Natl. Acad. Sci. U.S.A. 101 (33), 12387–12390.

Schoenebeck, J.J., Hutchinson, S.A., Byers, A., Beale, H.C., Carrington, B., Faden, D.L., et al., 2012. Variation of BMP3 contributes to dog breed skull diversity. PLoS Genet. 8 (8), e1002849. 10.1371/journal.pgen.1002849.

Scott, J.P., Fuller, J.L., 1965. Genetics and the Social Behavior of the Dog. The University of Chicago Press, Chicago, Illinois.

Serpell, J.A., 2004. Factors influencing human attitudes to animals and their welfare. Anim. Welf. 13, S145–S151.

Serpell, J.A., Hsu, Y., 2001. Development and validation of a novel method for evaluating behavior and temperament in guide dogs. Appl. Anim. Behav. Sci. 72 (4), 347–364, Jun 1.

Shipman, P., 2010. The animal connection and human evolution author. Curr. Anthropol. 51 (4), 519–538.

Skoglund, P., Götherström, A., Jakobsson, M., 2011. Estimation of population divergence times from non-overlapping genomic sequences: examples from dogs and wolves. Mol. Biol. Evol. 28 (4), 1505–1517, Apr.

Statham, M.J., Trut, L.N., Sacks, B.N., Kharlamova, A.V., Oskina, I.N., Gulevich, R.G., et al., 2011. On the origin of a domesticated species: identifying the parent population of Russian silver foxes (Vulpes vulpes). Biol. J. Linn. Soc. Lond. 103, 168–175.

Sutter, N.B., Bustamante, C.D., Chase, K., Gray, M.M., Zhao, K., Zhu, L., et al., 2007. A single IGF1 allele is a major determinant of small size in dogs. Science 316 (5821), 112–115.

Svartberg, K., Forkman, B., 2002. Personality traits in the domestic dog (Canis familiaris). Appl. Anim. Behav. Sci. 79 (2), 133–155.

Svartberg, K., Tapper, I., Temrin, H., Radesäter, T., Thorman, S., 2005. Consistency of personality traits in dogs. Anim. Behav. 69 (2), 283–291.

Takeuchi, Y., Hashizume, C., Arata, S., Inoue-Murayama, M., Maki, T., Hart, B.L., et al., 2009a. An approach to canine behavioural genetics employing guide dogs for the blind. Anim. Genet. 40, 217–224.

Takeuchi, Y., Kaneko, F., Hashizume, C., Masuda, K., Ogata, N., Maki, T., et al., 2009b. Association analysis between canine behavioural traits and genetic polymorphisms in the Shiba Inu breed. Anim. Genet. 40, 616–622.

Tchernov, E., Valla, F.F., 1997. Two new dogs, and other Natufian dogs, from the Southern Levant. J. Archaeol. Sci. 24, 65–95.

Topál, J., Miklósi, Á., Csányi, V., 1998. Attachment behaviour in dogs: a new application of Ainsworth's (1969) Strange Situation Test. J. Comp. Psychol. 112, 219–229.

Topál, J., Gácsi, M., Miklósi, Á., Virányi, Z., Kubinyi, E., Csányi, V., 2005. Attachment to humans: a comparative study on hand-reared wolves and differently socialized dog puppies. Anim. Behav. 70, 1367–1375.

Trut L.N., 1980a. The genetics and phenogenetics of domestic behaviour. In: Problems in General Genetics (Proceeding of the 14th International Congress of Genetics), vol. 2, book 2, pp. 123–136.

Trut L.N., 1980b. The role of behavior in domestication-associated changes in animals as revealed with the example of silver fox. Doctoral (Biol.) Dissertation, Novosibirsk: Inst. Cytol. Genet.

Trut, L.N., 1988. The variable rates of evolutionary transformation and their parallelism in terms of destabilizing selection. J. Anim. Breed. Genet. 105, 81–90.

Trut, L.N., 1999. Early Canid domestication: the Farm Fox Experiment. Am. Sci. 87, 160–169.

Trut, L.N., 2001. Experimental studies of early Canid domestication. In: Ruvinsky, A., Sampson, J. (Eds.), The Genetics of the Dog. CABI, New York, pp. 15–43.

Trut, L.N., Oskina, I.N., 1985. Age-related changes in blood corticosteroids in foxes differing in behavior. Dokl. Akad. Nauk SSSR 281 (4), 1010–1014.

Trut, L.N., Dzerzhinskii, F.Y., Nikol'skii, V.S., 1991. Intracranial Allometry and Craniological Measurements in Silver Fox during Domestication. Genetika (Moscow) 27 (9), 1605–1612.

Trut, L.N., Pliusnina, I.Z., Oskina, I.N., 2004. An experiment on fox domestication and debatable issues of evolution of the dog. Genetika (Russ.) 40, 794–807.

Trut, L., Oskina, I., Kharlamova, A., 2009. Animal evolution during domestication: the domesticated fox as a model. Bioessays 31 (3), 349–360.

Turcsán, B., Kubinyi, E., Mikósi, Á., 2011. Trainability and boldness traits differ between dog breed clusters based on conventional breed categories and genetic relatedness. Appl. Anim. Behav. Sci. 132, 61–70.

Udell, M.A.R., Dorey, N.R., Wynne, C.D.L., 2008. Wolves outperform dogs in following human social cues. Anim. Behav. 76, 1767–1773.

Våge, J., Wade, C., Biagi, T., Fatjó, J., Amat, M., Lindblad-Toh, K., et al., 2010. Association of dopamine-and serotonin-related genes with canine aggression. Genes. Brain. Behav. 9, 372–378.

Vas, J., Topál, J., Péch, É., Miklósi, Á., 2007. Measuring attention deficit and activity in dogs: a new application and validation of a human ADHD questionnaire. Appl. Anim. Behav. Sci. 103, 105–117.

Vasileva, L.L., Trut, L.N., 1990. The use of the method of principal components for phenogenetic analysis of the integral domestication trait. Genetika 26, 516–524.

Vilà, C., Savolainen, P., Maldonado, J.E., et al., 1997. Multiple and ancient origins of the domestic dog. Science 276, 1687–1689.

Vilà, C., Wayne, R.K., 1999. Hybridization between wolves and dogs. Conserv. Biol. 13 (1), 195–198.

Vilà, C., Maldonado, J.E., Wayne, R.K., 1999. Phylogenetic relationships, evolution, and genetic diversity of the domestic dog. J. Hered. 90 (1), 71–77.

Vilà, C., Walker, C., Sundqvist, A., et al., 2003. Combined use of maternal, paternal and bi-parental genetic markers for the identification of wolf-dog hybrids. Heredity 90 (1), 17–24.

Virányi, Z., Topál, J., Gácsi, M., Miklósi, Á., Csányi, V., 2004. Dogs respond appropriately to cues of humans' attentional focus. Behav. Processes 66, 161–172.

Virányi, Z., Gácsi, M., Kubinyi, E., Topál, J., Belényi, B., Ujfalussy, D., et al., 2008. Comprehension of human pointing gestures in young human-reared wolves (Canis lupus) and dogs (Canis familiaris). Anim. Cogn. 11, 373–387.

vonHoldt, B.M., Pollinger, J.P., Lohmueller, K.E., Han, E., Parker, H.G., Quignon, P., et al., 2010. Genome–wide SNP and haplotype analyses reveal a rich history underlying dog domestication. Nature 464 (7290), 898–902, 8.

Wayne, R.K., 1986. Cranial morphology of domestic and wild canids: the influence of development on morphological change. Evolution 40, 243–259.

Wayne, R.K., vonHoldt, B.M., 2012. Evolutionary genomics of dog domestication. Mamm. Genome. 23 (1–2), 3–18.

Westwood, R.E., 1989. Early fur-farming in Utah. Utah. Hist. Q. 57, 320–339.

Willis, M.B., 1999. Genetic aspects of dog behaviour with particular reference to working ability. In: Serpell, J. (Ed.), The Domestic Dog. Cambridge University Press, Cambridge, pp. 52–64.

Wong, A.K., Ruhe, A.L., Dumont, B.L., Robertson, K.R., Guerrero, G., Shull, S.M., et al., 2010. A comprehensive linkage map of the dog genome. Genetics 184, 595–605.

Yang, F., O'Brien, P.C., Milne, B.S., Graphodatsky, A.S., Solanky, N., Trifonov, V., et al., 1999. A complete comparative chromosome map for the dog, red fox, and human and its integration with canine genetic maps. Genomics 62 (2), 189–202, Dec 1.

Zedda, M., Manca, P., Chisu, V., Gadau, S., Lepore, G., Genovese, A., et al., 2006. Ancient pompeian dogs—morphological and morphometric evidence for different canine populations. Anat. Histol. Embryol. 35 (5), 319–324, Oct.

Zeuner, F.E., 1963. A History of Domesticated Animals. Harper and Row, NY.

Chapter 11

Behavioral Genetics in Pigs and Relations to Welfare

Lotta Rydhmer* and Laurianne Canario†
*Department of Animal Breeding and Genetics, Swedish University of Agricultural Sciences, Uppsala, Sweden; †French National Institute for Agricultural Research, Animal Genetics Division, Castanet-Tolosan, France

INTRODUCTION

The human global population is growing and there is an increased demand for pork. At the same time, the competition for agricultural land is increasing and the negative environmental influence of pig production must be decreased. Thus, there is a need for increased efficiency at all levels of the production chain, starting with the pig's genotype. But intense selection for production traits such as feed efficiency may decrease welfare, for example, by decreasing the capacity to adapt to stress. The challenge is to increase efficiency without decreasing pig welfare.

The domestication of the pig started approximately 10,000 years ago. Since then the animal has developed from a small wild boar sow giving birth to a litter of six slow-growing fat pigs once a year to today's large sow, which produces 25 fast-growing lean pigs a year. During the long process of domestication, pigs' behavior has changed genetically. Fear of humans has decreased and the pigs have become much easier to handle. More recently, the genetic progress in production traits has been enormous. However, breeding may threaten animal welfare due to undesirable correlated effects in behavior, metabolism, reproduction, and health traits. Fortunately, the genetic tools that have been used to increase production can also be used to decrease negative side effects of selection and improve traits important for welfare.

PIG BEHAVIOR

Explorer and Generalist

Pigs have extensive social skills and relatively low sensitivity to confinement; traits that favored domestication. After generations of successful

Genetics and the Behavior of Domestic Animals. DOI: http://dx.doi.org/10.1016/B978-0-12-394586-0.00011-1
© 2014 Elsevier Inc. All rights reserved.

genetic selection to increase pigs' performance, modern pigs differ from their wild counterparts not only in production and reproduction, but also in behavior. The relaxation of natural selection has reduced the degree of fear of predators and made pigs easier to handle. It is, however, important to remember that only the frequency and the threshold at which behaviors are triggered have changed. No new behavior has appeared and none has disappeared.

Pigs are sentient animals. They express strong behavioral needs relating back to the behavior of the wild boar in nature. Welfare problems arise if those needs are not met. When the production system does not permit a sow to build a nest for farrowing (because there is no building material) it may redirect the behavior into bar-biting, which may become a behavioral disorder (Lawrence and Terlouw, 1993). The behavior is a good indicator of welfare because it refers to strong decision-making and motivational processes in the pig (Dawkins, 2004; Kittawornrat and Zimmerman, 2011).

Pigs are probably the smartest animals among livestock (Broom *et al.*, 2009). They are able to make contextual associations and to memorize them for a long time. Following the example of wild boars that have their home range in memory (Spitz, 1986), De Jonge *et al.* (2008) showed that piglets were capable of associating music to access to a playroom, and also suggested that music can elicit play behavior. Pigs' hearing range is similar to that of humans and they show aversion to loud noise. From birth, piglets rely on olfactory cues from the dam and its udder. In adulthood, recognition between individuals also depends on olfactory cues (Curtis *et al.*, 2001). With its highly developed senses and its large cognitive abilities, the pig has a very good perception of its surroundings.

Being a generalist, the pig is a forager that can eat almost anything. Domestic pigs fed from an automatic feeder eat around seven times a day and spend almost 1 hour feeding per day during the growing–finishing phase (von Felde *et al.*, 1996). On pasture, feed-related behaviors like rooting, grazing, and exploring substrate account for 75% of daily activity (Stolba and Wood-Gush, 1989). The explorative behavior is promoted by their capacity to detect odors that they memorize for several hours (Signoret *et al.*, 1975). In parallel, pigs show a high level of curiosity. Given the choice, they choose to enter environments that contain new objects to investigate, and this investigation is not only related to feeding motivation (Wood-Gush and Vestergaard, 1991). The investigation of novelties is rewarding in itself.

Adaptation and Learning

The pig is a generalist that can adapt in many different environments (Figure 11.1). When domestic pigs are released into the wild, they show a large capacity for behavioral adaptation. Feral pigs have colonized many different types of habitats, e.g., in Australia (Edwards *et al.*, 2004) and in the

FIGURE 11.1 The pig is a generalist. Feral pigs are common, e.g., in Australia, and in Europe wild boar pigs thrive in the city center. These pigs live in Berlin. *Photograph: Florian Möllers/ wildesBerlin*

U.S.A. (Wyckoff *et al.*, 2009) and Dzieciolowski *et al.* (1992) report that under the pressure of natural selection, characteristics of feral pigs return to those of wild boars.

Two common tests used for genetic studies of individual temperament are the open-field test and the novelty test. The open-field test was developed for rodents. Animals are tested in an open field to induce a conflict between aversion and voluntary exploration of a novel environment. Locomotion and exploration are used as indicators of adaptation to change. A short period to the first move in the open field and a high level of activity may reflect a low level of anxiety (Réale *et al.*, 2007), but the interpretation of pigs' reactions has been questioned since pigs and rodents do not exhibit the same motivational processes. The response to novelty can be measured by introducing a novel object in the home pen, or in a test arena after a period of acclimatization to the novel environment. Usually, the first reaction to the novel object is to freeze. The latency until the animal reacts, i.e. moves or gets in contact with the object, is used as a measure of its boldness (Réale *et al.*, 2007).

Even though pigs show the ability to modify their behavior to adapt to the present environment, many behaviors are repeatable across situations and some of them are inherited. Behavior traits do, however, differ from many other traits in that they are changed by experience and learning. For instance, there is a strong effect of age on the interest for a given object used in the novelty test (Docking *et al.*, 2008). Early contact with humans influences subsequent levels of fear (Hemsworth and Barnett, 1992). Thus, individuals in a group of young pigs may enter the growing–finishing facility with very different experiences. They may also have different genotypes, for example, regarding explorative and aggressive behavior.

The Social Pig

The group size is quite similar for wild boars, feral pigs and farmed pigs, although these animals live in a wide range of habitat types and resource availability. Under natural conditions, the groups are based on matriarchal hierarchies (Stolba and Wood-Gush, 1989) and males are loosely associated with these groups (Mauget, 1981). On the farm, pigs born in different families are often grouped together in small pens. Whereas aggression is avoided in the wild, where different groups seldom meet, mixing with unacquainted pigs is frequent in pig production. When unacquainted pigs meet they fight to establish dominance. During the first 24 hours after mixing, most pigs are involved in many fights which leads to energy expenditure and injuries such as skin lesions. Turner *et al.* (2006a) counted skin lesions on pigs after mixing as a measurement of aggressive behavior. Ten per cent of the pigs had more than 50 skin lesions. The frequency and intensity of aggressive interactions decline over time after mixing, until social relationships stabilize. However, an ongoing lower level of aggression persists to maintain social relationships.

Aggressiveness in pigs is known to be repeatable over time and across different situations and it is partly influenced by the genotype. A pig's decision to engage in an aggressive interaction or not may be made according to the relative costs and benefits of the behavior, which will vary depending on resource scarcity and the other pigs' behavior (Enquist and Leimar, 1983). Arey and Franklin (1995) studied groups of 15 pigs and found that in 60% of the dyads (pairs) there were no fights.

Pigs can identify unfamiliar individuals in large groups of up to 80 pigs (Turner *et al.*, 2001). Gilts and sows are able to remember their group mates and identify them when they meet again after several weeks (Arey, 1999). Thus fighting can be avoided. The stability of the group of pregnant sows is maintained by a dominance hierarchy that depends on subordinates avoiding the dominant sows (Jensen, 1982). The dominance order is, however, resource relative; group members may have different dominance orders for different resources (Lindberg, 2001). Whether it is worth it or not to initiate a fight depends on group size and the predictability of the resource, e.g. feeding. In large groups, a higher number of competitors dilutes the effectiveness of aggression and increases its energetic cost (Fraser *et al.*, 1995).

Individual differences in aggressiveness can be measured in a resident–intruder test (Réale *et al.*, 2007). The tested pig, i.e. the resident, encounters in its home pen an intruder (that should be of slightly smaller size) and the latency until attack by the resident is recorded. Recording aggressive behavior among pigs in a group is difficult since the pigs must be identified individually. Aggressiveness can be investigated by direct or video observations, either by recording the total number of initiated and received attacks for each pig or by recording the identity and outcome of each dyadic encounter.

In general, pigs search for positive and close interactions with humans. Fear of humans is an indicator of low welfare and different methods are used to record fear. One way, used by Velie *et al.* (2009) and others, is to let a person unfamiliar to the pigs enter the pen and stand there motionless, and the latency for pigs to approach and touch the human is recorded. The success of this test relies on the pigs' motivation to voluntarily approach the human. The trait "easy to handle" can be measured during routine work, e.g., when pigs are moved between pens or weighed.

Tail Biting

Pigs' behavioral response to chronic stressors in the environment may translate into abnormal behavior such as tail biting. It could be a redirected behavior related to the need to explore and forage. Even so, tail biting can also be observed in outdoor production (Walker and Bilkei, 2006). Pigs performing tail biting with a high frequency also perform much ear-biting and belly-nosing. Brunberg *et al.* (2011) identified three types of pigs: biters, victims, and neutral pigs that never perform nor receive tail biting even during an ongoing outbreak. The frequency of biters and the severity of tail biting are highly influenced by the environment. The number of injured tails is easy to record, but to identify the biters is quite complicated. Breuer *et al.* (2003) used pigs' motivation to chew a rope as an indirect measure of their biting tendency.

Sexual Behavior

In most countries males raised for slaughter are castrated, but there are exceptions like the U.K. Due to welfare reasons, the European Union wants to ban surgical castration from 2018. Entire males are more likely to show aggressive behavior and are more active than castrates (Cronin *et al.*, 2003). They also display more sexual behavior such as mounting (Rydhmer *et al.*, 2010). In boars used for mating and semen collection, there is genetic variation in libido traits, with heritabilities around 0.15 (Rothschild and Bidanel, 1998).

By means of pheromonal communication, gilts and sows grouped together synchronize estrus. During estrus, the female shows standing reflex and there is a genetic variation in the ability to show this sexual behavior. The heritability of this categorical trait was estimated at 0.3 on the underlying scale in a research herd (Rydhmer *et al.*, 1994). Recording of estrus symptoms was later performed in Norwegian nucleus herds, but almost no genetic variance was found in that environment (Holm, 2004).

Maternal Behavior

The pig is an exception among the ungulates regarding prolificacy; it has more in common with rodents and many carnivores. The sow produces a very large litter for being such a large animal. At the approach of farrowing, the pregnant sow becomes intolerant of group mates (Stolba and Wood-Gush, 1989). In the wild, the sow leaves the group 3 days before farrowing and searches for a suitable site where she builds a nest with materials from the vegetation. In the farrowing stable, the sow also builds a nest if given building material and space. Sows spend around 15 hours on nest-building activities (Thodberg *et al.*, 1999) and in the absence of material, sows direct their activity at the floor, walls, and bars (Wischner *et al.*, 2009). They humidify the floor with their snout and paw insistently with their front legs. The nest-building activities are very similar in wild boar sows and domestic sows housed indoors (Stolba and Wood-Gush, 1989).

When the first piglet is born the sow reduces the number of postural changes and lies still until most piglets are born. The farrowing takes on average 4 hours (Wallenbeck *et al.*, 2009). Sows differ in their maternal skills, which influence the probability of survival of the piglets. Before lying down, some sows carefully check if there are any piglets, in order to avoid crushing them. Others do what Wechsler and Hegglin (1997) describe as "flopping straight down" which is a great risk for the piglets. An attentive sow reacts to piglet screams by changing posture and thus many piglets are saved. Sows' reactions to the sound of screaming piglets have been used as a measurement of maternal behavior (Grandinson *et al.*, 2003). Unlike the ewe, the sow does not lick her young. Instead, the piglets initiate frequent nose-to-nose contact with their mother. Most sows can bond to alien piglets, which is convenient in high-producing herds since it allows cross-fostering as a means to optimize piglet survival.

The nursing is based on a sophisticated interaction between the sow and its piglets (Canario, 2006). After vigorous competition among littermates during the first days, each piglet gets its own teat. The sow nurses the piglets every hour, day and night. She initiates the nursing event by grunting and lying down on her side. The piglets respond by massaging the udder for a few minutes, which stimulates oxytocin release resulting in the milk let-down. After the milk ejection, which is around 20 seconds, the piglets continue to massage the udder until the sow interrupts nursing by standing up or rolling on her belly, or until the piglets leave the udder or fall asleep. The total nursing event takes, on average, 5 minutes but the variation is large (Wallenbeck *et al.*, 2008).

FAWC'S FIVE FREEDOMS

The Farm Animal Welfare Council's definition of animal welfare (FAWC, 1992) lists five freedoms. These freedoms define ideal states and form a comprehensive framework for analysis and discussions of animal welfare within the proper constraints of an effective pig production unit. The five freedoms are:

- Freedom from hunger and thirst—by ready access to fresh water and a diet to maintain full health and vigor.
- Freedom from discomfort—by providing an appropriate environment including shelter and a comfortable resting area.
- Freedom from pain, injury, or disease—by prevention or rapid diagnosis and treatment.
- Freedom from fear and distress—by ensuring conditions and treatment which avoid mental suffering.
- Freedom to express normal behavior—by providing sufficient space, proper facilities, and company of the animal's own kind.

Although these freedoms, as stated by FAWC, do not refer directly to genetics and breeding, we use them as a structure for this chapter.

FREEDOM FROM HUNGER AND THIRST

Are hunger and thirst welfare problems for pigs? In many countries, young pigs during the growing–finishing phase as well as lactating sows are given *ad libitum* access to feed and water. There is sometimes competition between pen mates at the feeder or water nipple. Such competition is, however, probably more related to "fear and distress" or "discomfort" than to "hunger and thirst".

Some newborn piglets do, in fact, starve, and most piglets have a restricted access to milk. A piglet ingests around 40 g of milk per meal and the piglet's total milk intake can be estimated as 25% of the increase in body weight (Noblet and Etienne, 1989). Piglets have a very high capacity to eat (or rather to drink) and it is not known whether the difference between milk intake capacity and realized milk intake is related to hunger, but piglets starving to death are of course a welfare issue. Piglet mortality and growth are governed by both the genes of the sow and the genes of the piglet (Grandinson, 2003). Sows' milk production (recorded as litter growth) also has a genetic background (Lundgren, 2011).

A lactating sow producing as much as 15 liters of milk per day cannot eat enough to fulfill its nutritional needs during lactation. One way to overcome this restriction and increase milk production could be to select for increased appetite or voluntary feed intake during lactation. Pig breeders seldom think of appetite as a behavioral trait, but in humans food intake is often discussed in relation to behavior. Selection for increased voluntary feed intake of lactating sows was proposed by Eissen *et al.* (2000). Problems with low appetite increase with increasing temperature. Bergsma and Hermesch (2012) have recently shown that breeding for reduced thermal sensitivity of feed intake is possible. Lundgren *et al.* (2013) found genetic correlations between appetite and piglet growth rate (higher appetite – heavier piglets) and between appetite and body condition at weaning (higher appetite – better body condition). Selection for high voluntary feed intake may thus improve welfare of both piglets and lactating sows, but what would the consequences of selection for larger appetite in lactating sows be for the welfare of dry sows?

Dry sows never have *ad libitum* access to feed and the high-energy, grain-based feed used for sows is quickly digested and results in long-term periods of hunger (EFSA, 2007). One way of assessing total welfare impact is to take the number of animals, severity, and duration into account, and sows are lactating for only a quarter of the year. The genetic correlation between appetite during different phases of the sow's life is not known, but it seems likely that selection for high voluntary feed intake of lactating sows will result in more hungry dry sows. The goal conflict between high voluntary feed intake during lactation and restricted feed provision during pregnancy could be solved by giving free access to roughage or straw in addition to selection for high appetite. Straw not only reduces hunger, it also provides an occupation. Feed restriction and a boring environment may result in stereotypies. Spoolder *et al.* (1995) showed that straw reduces the development of excessive bar manipulation in sows on restricted feeding. In poultry, it has been proposed that feather pecking is a redirected foraging behavior (Brunberg, 2011) and there are genetic differences in predisposition for feather pecking. Likewise, there may be genetic differences between sows in how they respond to environmental enrichment such as straw, but in general it can be assumed that straw and other edible rooting material improves the welfare of all pigs.

There are genetic differences in eating behavior of young, growing pigs. When fed with an automatic feeding station, the size and duration of each meal, distribution of meals over time, etc., can be recorded (Figure 11.2). The duration of meals and feeding frequency are heritable traits ($h^2 = 0.4–0.5$, Labroue *et al.*, 1997). Fernández *et al.* (2011) found specific feeding strategies for different breeds. The number of meals per day was negatively correlated with sizes of meals and duration of meals. Large White pigs were "nibblers and fast eaters", Pietrain pigs were "nibblers and slow eaters", Duroc pigs were "meal and slow eaters" and Landrace pigs were "meal and fast eaters". According to Fernández *et al.* (2011), the "meal and fast eater" strategy is related to higher growth rate. Several genes or markers with a significant effect on feeding behavior have been found. The MC4R gene, which codes for a melanocortin receptor, has a positive effect on daily feed intake, probably mediated through the central control of appetite (Kim *et al.*, 2000). Zhang *et al.* (2009) identified a QTL on chromosome 7 for the number of visits of pigs to the feeder.

Aggressive interactions often occur during feeding (Rydhmer *et al.*, 2006). In a study by Jonsson and Jørgensen (1989) where feed was a

FIGURE 11.2 Pigs often eat together, but when individual feed intake should be recorded, they have to eat one by one in an automatic feeding station. *Photograph: Nils Lundeheim*

restricted resource, social rank was correlated to growth rate (high rank – high growth rate). Growth rate is included in the genetic evaluation of all pigs and in many breeding programs young pigs are tested in systems with automatic feeders where only one pig at a time can eat. The genetic correlation between feeding behavior in this system and feeding behavior when the pigs are group fed in a trough is not known.

Residual feed intake (RFI) is a measure of feed efficiency that accounts for the animal's energy requirements for production and maintenance. It is calculated as the difference between observed and predicted feed intake. The RFI concept was proposed as a selection trait for several species in the 1990s. With today's increasing awareness of the climate impact of animal production, breeders' interest in RFI has increased. Low RFI in the growing pig is genetically correlated to high leanness and low feed intake. Selection for decreased RFI is, however controversial from a welfare standpoint, given that the genetic variation in RFI reflects the genetic variation in activity and response to stress (Luiting and Urff, 1991). In fact, RFI indirectly measures the quantity of buffer resource available for activity, immune system, some metabolic processes and stress response. Sadler *et al.* (2011) found a genetic correlation between activity and RFI (high RFI – high activity level). Reduction of RFI by selection increases feed efficiency at the expense of reduced feeding activity, owing to a positive genetic correlation between RFI and feeding duration (von Felde *et al.*, 1996). Young and Dekkers (2011) compared a line selected for reduced RFI with a control line in pens where pigs from both lines were mixed, and fed with a single-space electronic feeder from 3 months until slaughter. Pigs selected for low RFI had less meal per day than control pigs, especially during peak eating times. The authors conclude that feed efficiency may be affected by feeding behavior because selection for decreased RFI results in pigs that spend less time eating and eat faster.

FREEDOM FROM DISCOMFORT

According to FAWC (1992), animal welfare includes "Freedom from discomfort—by providing an appropriate environment including shelter and a comfortable resting area". What is an appropriate environment for a pig, and could the environment be more or less appropriate depending on the pig's genotype? In FAWC's report on welfare of pigs kept outdoors (FAWC, 1996) it says:

Breeding companies, and those responsible for the selection of breeding stock to be kept on outdoor enterprises, must ensure that only those strains of pig with the genetic potential to thrive in the conditions provided are used.

Thus, FAWC assumes there are important genotype by environment interactions (G × E) with regard to welfare. One example could be genetic differences in coat color, where pigs with alleles for white color probably have a higher risk of getting burnt by the sun if there is no shade. In the tropics, cattle breed differences have been observed with regard to seeking shadow (Hernández *et al.*, 2002). We are not aware of any pig studies describing genetic variation in behavior outdoors. Maternal behavior is, however, often mentioned as a trait of extra importance outdoors (Wallenbeck, 2009).

The variation in temperature is larger outdoors than indoors, and both very cold and very hot weather can lead to thermal stress. High temperature is a welfare problem also indoors in many countries, during summer in tempered areas, and all year round in tropical areas. Selection for increased lean growth has led to higher total heat production in modern pigs, which increases the pressure to maintain homeothermy when it is hot. Activity increases heat production and pigs spend more time lying and less time eating as temperature increases (Brown-Brandl *et al.*, 2001). There are breed differences in heat stress susceptibility and Gourdine *et al.* (2006) found signs of G × E when comparing Creole and Large White sows. Several genes related to heat stress have been found in pigs (Maak *et al.*, 1998). Kanis *et al.* (2004) use thermal discomfort as a model trait when discussing the ability to breed pigs for increased robustness and thereby increased welfare. Based on physiological and behavioral responses of pigs to changing temperature, different welfare zones can be identified and the individual transition points between these zones can serve as a selection trait, aiming for a broader welfare zone. An alternative is to test pigs under harsh conditions and select those animals with the least problems associated with coping (Kanis *et al.*, 2004).

In intensive, large-scale indoor production systems, a varying environment is probably not the main welfare problem. On the contrary, the uniformity and lack of events may be a large challenge for growing pigs. Thus we have to stress the question "Appropriate environment for whom?" Are there any lines of pigs with the genetic ability to "thrive" in the ordinary indoor conditions where pigs are usually kept during the growing–finishing period? The pig is an explorative animal, but there is not much to explore in pens for growing pigs. Breed differences in explorative behavior have been found; in a comparison between young Large White, Landrace and Duroc pigs, the Durocs were more exploratory (Breuer *et al.*, 2003). Meishan sows perform more explorative behavior than Large White sows (Canario *et al.*, 2009). In the Code of good practice for farm animal breeding and reproduction organizations (EFAB, 2013) the industry has agreed that breeding organizations must maintain "the intrinsic characteristics of domesticated species". The members of the Federation of Veterinarians of Europe are concerned about breeding resulting in "animals with unnatural... behavioral characteristics" (FVE, 1999). The meaning of natural and unnatural for domestic animals will be discussed later, but uncurious pigs could be regarded as unnatural.

"Freedom from Discomfort" also implies providing "a comfortable resting area". No matter how well designed the pen or the floor is, there will be little rest if the pigs are continuously disturbing each other. Entire males perform more social behavior than castrates and too much social behavior seems to decrease animal welfare even when it is not painful (Rydhmer *et al.*, 2010). Nosing, pushing, sniffing, and nibbling are social behaviors normally performed by pigs. Bench and Gonyou (2007) found that breed and also sire within breed had a significant effect on belly-nosing and sucking behaviors in young weaned pigs. In a breed comparison by Breuer *et al.* (2003), Duroc pigs performed less belly-nosing than Landrace and Large White, suggesting a genetic background to these behaviors. If rearing of entire males for slaughter would become the dominating production system, selection for an optimal frequency of both aggressive and non-aggressive social behaviors could be relevant.

FREEDOM FROM PAIN, INJURY OR DISEASE

Tail Biting

Tail biting is difficult to prevent because its occurrence has a multi-factorial origin. The most extreme preventive treatment is to cut off the tails as a means of improving pig welfare. According to the European Food Safety Authority (EFSA, 2007), over 90% of the pigs within the European Union are tail docked, even though routine tail docking is prohibited (EU Directive 91/630 EEC). In Sweden, where no tail docking is performed, tail biting or other tail damage is observed in 7% of the pigs at the slaughter plant

(Keeling *et al.*, 2012). Is selection against tail biting possible, as a complement to environmental improvements?

The Royal Society for the Prevention of Cruelty to Animals, which is the organization behind "Freedom food", knows that tail biting has a genetic background. They write on their home page that "the genetics of the pigs on a unit can affect the likelihood of tail biting occurring" (RSPCA, 2012). But it is a difficult trait to study because it is hard to identify the biter. Breuer *et al.* (2005) defined biters as pigs involved in more than 50% of the tail-biting incidences within 10 minutes of observation. Tail biting is not performed by all pigs, almost 90% are never observed to perform tail bites (Brunberg *et al.*, 2011).

Breed differences in the tendency to tail-bite group mates have been found by Breuer *et al.* (2003). Nordic studies show that Landrace and Duroc pigs tail-bite more often than Yorkshire and Hampshire pigs. Yorkshire pigs are tail-bitten more often whereas Hampshire pigs are tail-bitten less often than the other breeds (Sinisalo *et al.*, 2012; Westin, 2000). However, Lund and Simonsen (2000) found no clear breed effect. The genetic influence on this behavior can be more relevant within than between breeds.

Around 3% of the pigs in the study of Breuer *et al.* (2005) were classified as biters. The heritability of this "yes or no trait" was estimated at 0.3 (after transformation to an underlying continuous scale) in Landrace pigs. The corresponding heritability was zero in Large White pigs. At the genetic level, tail biting is unfavorably correlated with growth rate and leanness. A genome-wide association study was done on records of tail biting from Norwegian farms. The first results showed that different chromosomal regions on the pig genome are associated with the delivery and receipt of tail-biting behaviors (Wilson *et al.*, 2012).

To get a better understanding of the molecular mechanisms of tail biting, Emma Brunberg studied gene expression in the brain (hypothalamus and prefrontal cortex) in her PhD project (Brunberg, 2011). She used pigs from a Finnish herd and analyzed the abundance of RNA molecules from specific genes. There were differences in gene expression between biters and victims and also between these types and neutral pigs not involved in tail biting although they were housed together with biters. Many of the genes differed between neutral pigs and the other categories. The EGF gene that codes for an epidermal growth factor and has an important function in the dopaminergic system was less expressed in neutral pigs. This gene is involved in novelty-seeking in humans (Keltikangas-Järvinen *et al.*, 2006). Furthermore, the GTF2I gene which is involved in sociability is up-regulated in the hypothalamus of neutral pigs. Also, the GHRL gene which codes for a hormone related to appetite and the PDK4 gene related to fat content were up-regulated in these pigs. The results give further grounds to the suspicion that selection for high production is followed by some abnormal behaviors directed towards group mates.

Aggressive Behavior in Young Pigs

The Code of good practice for farm animal breeding and reproduction organizations (EFAB, 2013) states that:

Breeding Organisations ensure the health and welfare of the animals they keep and select, so that pain and suffering are minimized; this may include selection against aggressive behavior between animals.

Aggressive behavior is performed especially by pigs that meet for the first time and are similar in body weight. Again, identifying the recipient is much easier than identifying the aggressive pig initiating the fight. In an attempt to evaluate the genetic background to aggressive temperament, Velie *et al.* (2009) used the resident–intruder test on growing pigs. They used a simple scoring based on the cumulative number of attacks over two tests with a score between 0 (the pig never attacked) and 2 (the pig attacked in both tests). The trait was analyzed as categorical using a threshold model. According to this study, attacking an intruder has a low heritability ($h^2 = 0.1$) and the latency to the first attack is not heritable.

To record skin lesions is easier than to study aggressive behavior. The number of fresh skin lesions can be counted, e.g., immediately before and 24 h after mixing. Lesions located in the front of the body are associated with a high level of reciprocal fighting. Those located in the rear are the result of receipt of non-reciprocated aggression (i.e. being bullied whilst being chased or attempting to retire from a conflict). The total count of lesions had a heritability of 0.2 in the study by Turner *et al.* (2006b). When distinguishing between different parts of the body, the heritability for lesions in the rear was lower both when recorded the day after mixing and 3 weeks after mixing (Turner *et al.*, 2009). Again, this indicates that the genotype of victims is different from the genotype of the attacking pigs.

Nucleus herds have been used to study aggressive behavior at mixing. Being involved in fights has a moderate heritability ($h^2 = 0.4–0.5$) and so has the delivery of non-reciprocated aggression ($h^2 = 0.3–0.4$). The genetic correlation between number of lesions and being involved in fights were all strongly positive in one study (0.7) but ranged from close to zero (0.1, rear) to 0.7 (front) in another study (Turner *et al.*, 2008; 2009). The genetic correlations between bully delivery and receipt were inconsistent between these studies, maybe due to differences in aggression patterns. Turner *et al.* (2009) suggest selection against pigs which fight and bully others by recording lesions on different parts of the body as different traits. They conclude, based on low and negative genetic correlations between lesions and growth and backfat depth that selection against lesions would not induce correlated changes in production traits.

In theory, pigs who fight allocate less energy to growth. Torrey *et al.* (2001) showed that Landrace gilts selected for large loin-eye area were involved in more aggressive interactions after mixing than gilts with smaller loin-eye area. Social rank, recorded during feeding, was highly heritable when estimated on entire males at a test station and the genetic correlation between position in social rank order and growth rate was high (Jonsson and Jorgensen, 1989). Thus, selection for high growth rate could result in increased aggressiveness, which also has been proposed by Schinkel *et al.* (2003). We recently found that pigs with a high direct breeding value for growth rate are more successful during social contests (Canario *et al.*, 2012). They have a higher genetic merit for both initiating and winning fights and bullying than other pigs. Furthermore, these pigs are less frequently bullied and have fewer lesions in the rear.

Individual behavior is adapted to the group around the individual and the performance of the individual is dependent on the group mates. Selection for individual performance maximizes the individual's results but not necessarily that of the group the animal is raised in. Bill Muir became interested in aggressive behavior because of its negative influence on production results. His research has shown alternative methods of selection, improving both production and welfare. Muir developed a model to handle an unfavorable correlation between growth rate and competitive behavior without any need for behavioral observations. The model is called "the group model", "the social model", or "the competitive model". The quantitative relation between individual and group productivity was first presented by Griffing (1967) who extended classic population genetic models to include social effects. Based on Griffing's theory, Muir (2005) presented a social model that was further developed by Bijma *et al.* (2007) to estimate social genetic parameters. The social model for genetic evaluation of growth rate includes not only the "ordinary" direct genetic effect on an individual's growth rate, but also the social genetic effect on all group members' growth rates (Figure 11.3).

With this approach, two breeding values are estimated for each animal, one describing the animal's genetic ability to grow and the other describing the animal's genetic ability to influence the growth of other animals in the pen, the so called social breeding value. The correlation between the direct (D) genetic effect and the social (S) genetic effect (r_{DS}) is negative (unfavorable) if the animals compete for limited resources such as food and space. If that is the case, it cannot be recommended to select the animal with the highest breeding value for the direct effect since fast-growing animals are best at the expense of their group members. Instead, animals with rather high breeding values for both the direct and the social effect should be selected. If the genetic correlation between the direct genetic effect and the social genetic effect is positive, i.e. favorable, selection of the animal with the highest breeding value for the direct effect will not harm group members. Even so, using both direct and social breeding values in the selection can lead to a

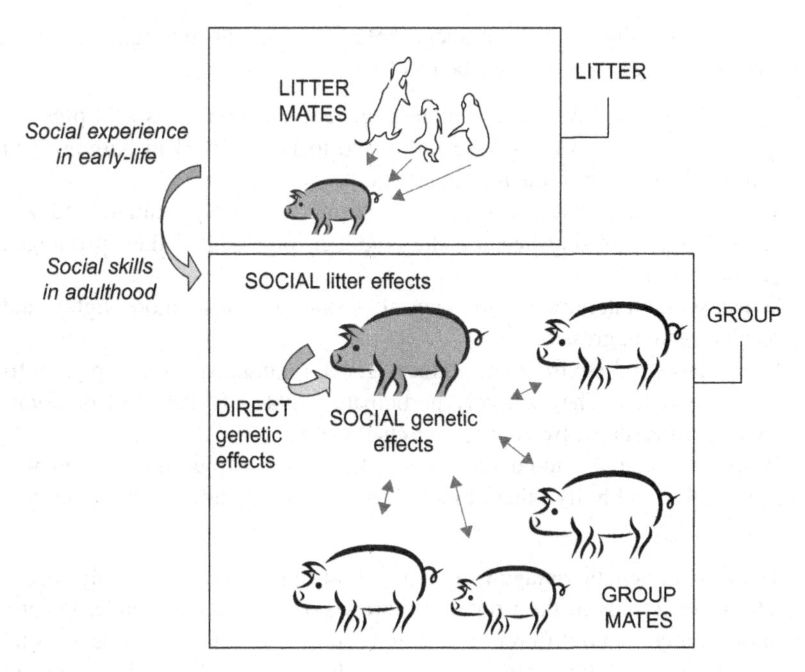

FIGURE 11.3 The direct genetic effect and the social genetic effect on the performance of pigs can be used to estimate direct and social breeding values. The social skills of a pig that influence its social interactions with other group mates are determined by its social experience in early life (achieved in contact with litter mates) and its genotype. *Illustration: Laurianne Canario*

larger genetic progress than only using the direct breeding value (Rodenburg *et al.*, 2010).

Bergsma *et al.* (2008) used the social model to study growth rate. The results indicated that pigs cooperate rather than compete ($r_{DS} = 0.2$). In another study the correlation was estimated at -0.1 (Canario *et al.*, 2010). Chen *et al.* (2009a) found a range from negative to positive social interactions in different American populations ($r_{DS} = -0.4$ to 0.7). This confirms that the genetic background of aggressive behavior varies between pig populations.

What characterizes pigs with high breeding values for the social genetic effect? From a Swedish data set, we have both behavioral and lesion score data on the animals and can compare pigs with high and low breeding values for direct and social effects on growth rate (Canario *et al.*, 2012). We simulated different situations in which the pigs displayed negative social interactions (genetic antagonism between own growth and the growth of group mates; $r_{DS} < 0$), neutral social interactions ($r_{DS} \sim 0$) or positive social interactions (genetic mutualism between own growth and the growth of group mates; $r_{DS} > 0$). Pigs with high social breeding value at mixing, referred to here as social pigs, can be described as follows:

- When the social environment induces genetic antagonism, social pigs are genetically predisposed to lose fights and to be involved in bullying and being bullied more frequently than others.
- When the genetic antagonism is very strong, social pigs initiate and win fewer fights, probably because the dominant pigs defend their privileged access to food.
- When social interactions are neutral, social pigs lose more fights and receive more aggression.
- When the social environment induces genetic mutualism, social pigs initiate more fights. They actively participate in the establishment of dominance relationships, by being more challenging.
- When the genetic mutualism is very strong, social pigs initiate and win more fights and bully others more. That means they adopt a more aggressive strategy.

Under both genetic antagonism and neutral interactions, social pigs accumulate fewer lesions in the front and more lesions in the rear. Under genetic mutualism, there is no difference in lesion scores between more or less social pigs. Three weeks later, social pigs do not differ from others with regard to lesions in the front but if the social environment induces genetic mutualism they have more lesions in the rear, which is a sign of involvement in both reciprocal fighting and receipt of bullying. Social pigs spend less time standing, which indicates that they are calmer. (Canario *et al.*, 2012).

If pigs are selected for both the direct and social effects on growth rate in a situation of neutral or positive interactions, social pigs will initiate more bullying immediately after mixing. When the dominance order has been set, they will be less willing to fight and at greater risk of being attacked (Canario *et al.*, 2012). Aggressive behavior at mixing is a necessary adaptation to the rearing environment which does not hamper pig result and is probably not correlated to aggressive behavior later on, in stable groups. Aggression among pigs is of two broad kinds: intensive fighting during a brief period when unfamiliar pigs are mixed and longer-term competition over feed or other limited resources to maintain hierarchy. This stresses the importance of considering an appropriate time window in pig life if one intends to select directly on a behavioral trait.

Unfavorable correlations between the direct and the social genetic effect have been found in trees, quails, and several laying hen populations (Bijma, 2011). Bouwman *et al.* (2010), on the contrary, found no direct social correlation on early growth in pigs. The group model assumes that an animal's social effect is similar on all other pigs in the group, regardless of whether they are relatives or not. Only cooperation between relatives fits well into Dawkins' story of *The Selfish Gene* (Dawkins, 1976), but pigs rely on familiarity rather than genetic relatedness and they are not able to recognize relatives if they have not been raised together (Bekoff, 1981). Are there "helpful" pigs, and if so, why? Natural selection would result in selfish individuals that are successful at the expense of others, not in successful groups. This is observed, e.g., in wild plants which compete for light, but breeding is not equal to natural selection. Maybe "helpful" pigs are merely pigs not disturbing other pigs? Pigs genetically predisposed for eating and staying calm without spending time and energy on bothering group mates would get high breeding values for own growth and for pen mates' growth. That strategy is probably only successful if feed is not a limited resource.

It would be interesting to study how general the breeding values for social effects are in different environments with more or less limited resources. With a group size ranging from five to 15 animals per pen, the social genetic effects were diluted in larger groups (Canario *et al.*, 2010). Turner *et al.* (2001) suggest that in very large groups, social strategies depend more and more on no social rules. It can be assumed that the behavioral response to selection for growth will be quite different with different pen and group size and different feeding systems.

Aggressive Behavior in Sows

Sows that have been housed in farrowing pens fight during the first 24 hours after weaning, when mixed with other sows. Sows should be loose housed for welfare reasons but fighting can be detrimental for low-ranked sows. FAWC's report on the Welfare Implications of Animal Breeding and Breeding Technologies in Commercial Agriculture (2004) state that:

> *FAWC is aware that selection for temperament is becoming increasingly important ... for species such as pigs and laying hens where a move away from close confinement systems, driven by either legislation or market forces, has revealed the importance of behavioral traits such as reduced levels of aggression.*

It would be interesting to use the group model on loose-housed sows, to estimate direct and social effects on reproduction traits like interval from weaning to estrus and pregnancy rate.

In a Danish field study of pregnant sows, all aggressive interactions during the first 30 minutes after mixing were recorded (Lövendahl *et al.*, 2005). The identity of sows delivering and receiving attacks were noted. The model included two genetic effects, one describing the capacity to attack and the other describing the predisposal for being attacked. Repeated measures from all dyads (all pairs) of sows in the pen were analyzed. Performing aggressive behavior was heritable ($h^2 = 0.2$), whereas being a victim had a heritability close to zero. A simpler way of recording (and analyzing) is to just record the sum of attacks given by each sow and base the genetic evaluation on that sum. This gives a higher heritability and would work well in practice. In a nucleus herd of 100 sows, approximately 12 sows are weaned every third week. One hour's work every third week would allow the farmer to mark the sows and record aggressive behavior and then the breeding organization could include aggressive behavior of sows in the genetic evaluation and select against it, in order to improve welfare. Hellbrügge *et al.* (2008) recorded aggressive behavior of pregnant sows when mixed in small groups during washing, before entering the farrowing unit. Eighteen per cent of the sows were aggressive and the heritability was estimated with a threshold model at 0.3. They also found favorable genetic correlations between aggression in groups and maternal behavior (less aggressive – stronger reaction to the sound of a screaming piglet).

Marker-assisted selection is proposed as a method of decreasing aggression in pigs. Several markers and some genes governing aggressiveness have been found in rodents and with a candidate gene approach based on genes with a known impact on the hypothalamic–pituitary–adrenal (HPA) axis, associations between skin lesions and the gene NR3C coding for a glucocorticoid receptor as well as the gene AVPR1B coding for a vasopressin receptor were found in pigs (Muráni *et al.*, 2010). Terenina *et al.* (2012) used the same data set and found markers indicating that both dopamine- and serotonin-related genes are involved in the genetic variation of aggressiveness.

Maternal Behavior

Piglet mortality is a severe problem in pig production related to the ongoing selection for larger litters and leanness. Genetic studies on behavior of newborn piglets are scarce, but several genetic studies on maternal behavior have been performed. The long-term goal of most of these studies is to decrease piglet mortality by breeding. The International Coalition for Animal Welfare (ICFAW) represents non-governmental animal welfare organizations from all over the world and contributes to the decision-making at the World Organisation for Animal Health (OIE). ICFAW recommends:

> *Selective breeding should focus on improving the welfare of pigs by selecting for, e.g., breeds which make good mothers and are less likely to crush their young, smaller litters of healthier, more robust piglets and resistance to stress and disease.*
> (ICFAW, 2010)

Maternal behavior starts already before farrowing, with nest building. This behavior makes the sow different from other ungulates. Nowak *et al.* (2000) wrote "The fact that nest building has been retained in spite of domestication indicates how robust this maternal behaviour is." Nest building is performed to provide the piglets with shelter and comfort, and particularly to keep them warm. This is especially important for small piglets due to the greater surface to body mass ratio resulting in higher heat losses. Nest building is a heritable trait in mice (Lynch, 1980).

The duration of farrowing influences piglet survival and the risk of dying during or shortly after farrowing is greater for piglets born late (Canario *et al.*, 2009). Farrowing duration is therefore a potential selection trait, but it is very difficult to record on a large scale. For a proper assessment, the observer should stay beside the sow and at the same time avoid disturbing the farrowing. There is a genetic variation in farrowing duration (but the heritability is low when recorded in nucleus herds; $h^2 \leq 0.1$) and it is correlated with litter size and number of stillborn (Holm *et al.*, 2004). No genetic relationships between farrowing duration and behavior are reported in the literature.

Sows are sometimes aggressive towards their newborn piglets. Infanticide (also called savaging) is associated with endocrine changes during farrowing, farrowing experience, environment, and genotype. The heritability for savaging piglets is high, but it is a tricky trait to analyze and the range of heritability estimates is large ($h^2 = 0.1-0.9$; Knap and Merks, 1987; Van der Steen *et al.*, 1988). Although rarely included in the genetic evaluation, there is an ongoing selection against this trait at the farm level. Besides, natural selection helps the breeders; often there are no piglets alive to select in affected litters. Single genes with a large influence on this defect have been found and many of them seem to interact. Several QTL have been detected on different chromosomes, among which a promising QTL on the X chromosome has been found in a Large White Duroc × Erhualian population (Chen *et al.*, 2009b). Some QTL regions found in pigs have corresponding regions in humans and rodents that harbor genes controlling anxiety, bipolar disorder, or coping behavior (Quilter *et al.*, 2007), and some mitochondrial genes are differentially expressed when savaging and not savaging sows are compared. This is interesting because the mitochondria contain DNA that is inherited solely through the mother. Several of these genes are related to neural pathways.

Crushing by the sow is one of the most common causes of piglet mortality. The introduction of farrowing crates was primarily aimed at reducing crushing. Crates also facilitate good hygiene and protect the farmer from aggressive sows. There is growing contest against this production system which induces higher stress in sows as compared to farrowing pens without crates (Jarvis *et al.*, 1997). In a recent EFSA report, the experts recommended that

> *... the use of loose farrowing systems should be considered as a serious alternative to conventional crated systems, as good performance results can be obtained in such systems if sows are kept in sufficiently large pens.*
>
> Spoolder *et al.*, 2011

Since the crate is very stressful for the sow, we must use alternative ways to prevent crushing, such as selection for improved maternal behavior.

Half of the crushed piglets are probably healthy piglets, but to identify the cause of death accurately on a large scale, e.g., in nucleus herds, is very difficult. The heritability for percentage of crushed piglets was estimated at 0.06 on data from a research herd (Grandinson *et al.*, 2002) and at 0.04 on data from nucleus herds (Grandinson *et al.*, 2003). The sow's reaction to the sound of a screaming piglet has been studied by several researchers. The postural reaction of the sow is heritable (Grandinson *et al.*, 2003). Using a questionnaire to the farmers, Vangen *et al.* (2005) estimated the heritability of sow carefulness in two populations at 0.1 and 0.2.

The sow reaction to a screaming piglet is genetically correlated to piglet mortality; stronger reaction – lower mortality ($r_g = -0.2$; Grandinson *et al.*, 2003). But Hellbrügge *et al.* (2008) found a stronger correlation between piglet mortality and sows' reaction to pop music ($r_g = -0.3$) than to piglet scream. Our interpretation is that a test of this kind reveals the sow's maternal ability, regardless of whether it believes the sound comes from its own piglets or not.

The relationship between a sow and its piglets seem to rely upon piglets' bond to their teats rather than strong sow–piglet bonds. Even so, there is a variation in sows' reactions to being separated from their piglets. Hellbrügge *et al.* (2008) estimated the heritability for this trait at 0.1 both at farrowing and three weeks later. The reaction was stronger at three weeks and the genetic correlation to piglet mortality was rather high ($r_g = -0.4$). Gäde *et al.* (2008) evaluated maternal ability 3–5 days after farrowing on a scale from 1 to 5 by combining several aspects: behavior during farrowing, behavior during lying down (carefulness), nursing behavior (position during nursing), reaction to screaming piglets, and number of crushed piglets. They estimated the heritability for this combined trait at 0.05. Canario *et al.* (2007) have observed breed differences in nursing behavior between Large White and Meishan sows, and Vangen *et al.* (2005) found some genetic variance in nervousness during suckling and willingness to expose the udder.

Many sows are kept in crates during farrowing and lactation, but in some European countries farrowing crates are forbidden due to welfare reasons. Furthermore, sows in outdoor production are never in crates. A frequently asked question is whether there are important G × E for maternal behavior in systems with crates, farrowing pens without crates, or huts outdoors. Baxter *et al.* (2011) compared two indoor sow lines, one selected for high survival from birth to weaning, and a control line selected for average piglet survival. All studied animals were born and reared on outdoor commercial units. They were later compared in two environments, indoors in pens without crates and outdoors. In both environments, the control gilts crushed more piglets than the high-survival gilts. Thus, selection for improved piglet survival limits the crushing behavior regardless of environment. Indoors, the high-survival gilts did, however, show more aggression towards the piglets and savaged more of them as compared to the control gilts. No such difference was observed outdoors. Probably the high-survival gilts reacted more strongly to the new indoor environment than the control gilts, and this reaction was expressed as aggressiveness towards the piglets. Thus, when selecting animals for improved maternal behavior, the production system should be taken into account.

Piglet survival and growth depend both on the genotype of the piglets and the genotype of the sow. Leenhouwers *et al.* (2001) compared newborn piglets with high and low direct breeding values for survival. The difference in breeding values was not reflected in early postnatal behavior. However, sows with high maternal breeding values for piglet survival gave birth to piglets that took shorter time to reach the udder and suckle for the first time, as compared to sows with low maternal breeding values (Knol *et al.*, 2002). It could be assumed that this difference reflects differences in sow behavior and calmness during farrowing. Engelsma *et al.* (2011) estimated correlations between the breeding value for mothering ability (based on percentage of liveborn piglets that survived) and various behavioral traits. Around farrowing, sows with high breeding values tended to perform less postural changes and be less active, but these correlations were not significant. During lactation (week 1–2), sows with high breeding values spent more time lying laterally and less time sitting, thus facilitating piglets' access to the udder. They also performed less postural changes which may limit crushing.

Questionnaires in which the farmers summarize their observations of sows' behavior during farrowing and lactation can be an alternative to time-consuming behavior tests or video recordings under field conditions. Vangen *et al.* (2005) used a questionnaire where farmers were asked to assess maternal behavior of Norwegian and Finnish Landrace sows on a scale from 1 to 7. Most of the results were very consistent across countries. The highest heritabilities were found for the sow's reaction to piglet screaming when handled ($h^2 = 0.1-0.2$) and how often the sow showed fear during routine management ($h^2 = 0.1-0.2$).

There is variation in maternal behavior between breeds. At farrowing, Meishan sows spend more time standing, manipulating straw, and exploring the floor, and they have more nose contact with the piglets than Large White sows (Canario *et al.*, 2007). Meishan piglets show higher vitality at birth than Large White piglets. When crossbred F1-piglets of these breeds are compared (i.e. similar piglet genotype for different maternal genotypes), piglets born from Meishan sows spend more time sleeping in contact with the sow's udder but are less eager to suckle than piglets born from Large White sows (Figure 11.4). This emphasizes the interaction between maternal and offspring behavior.

A study design with both purebred and crossbred piglets in the same litter further highlights the mechanisms of interactions between generations. For this purpose, purebred animals were inseminated with a mixture of semen from both breeds (Dauberlieu *et al.*, 2011). At birth, both purebred and

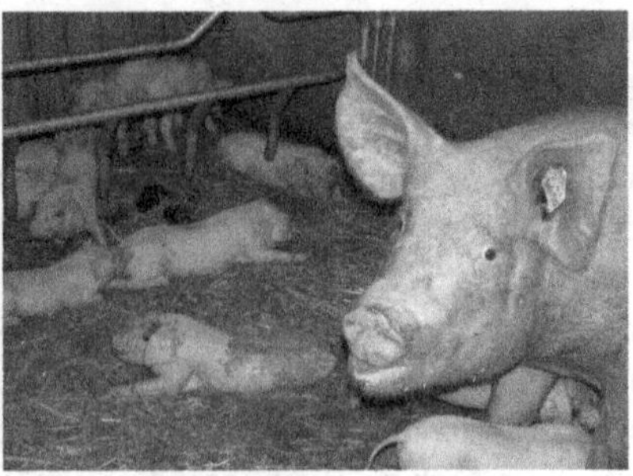

FIGURE 11.4 Comparison of maternal behavior between Meishan and Large White sows with crossbred F1-litters. *(Photograph: Laurianne Canario)*

crossbred piglets born from Meishan sows take less time to reach the udder than piglets born from Large White sows, for which purebred Large White piglets are the slowest. The purebred Large White piglets are quicker to reach the udder when raised by Large White dams which lie laterally for longer times at farrowing. Purebred and crossbred piglets born from Large White sows differ more than those born from Meishan sows in behavior and growth over the first days after birth.

Sows selected for high lean-tissue growth rate are less active at farrowing and more responsive to piglets than sows selected for low lean-tissue growth rate (McPhee *et al.*, 2001). At first sight, this may be interpreted as good for piglet welfare because of a decreased risk of crushing. However, the low activity reveals a poor welfare of the sow with negative consequences for the piglets. The probability of stillbirth is higher in piglets born from sows that are lying for much of the time, because low activity is associated with more farrowing difficulties (Canario, 2006).

Only weak genetic trends could be found in maternal behavior in French Large White sows that have been selected for prolificacy and lean growth rate for decades. By using frozen semen, two types of sows could be compared. Crosses between old-type or modern-type sires and modern-type sows were used to create two populations that were studied in the same environment and at the same time for several generations (Canario, 2006). The lines did not differ with regard to cause of piglet neonatal mortality (crushing by sow and piglet starvation) and infanticide was rare in both lines. The reaction of the sow when separated from the litter did not differ between the two lines. The sow's response to nose contact initiated by piglets was stronger for modern-type sows than for old-type sows. The largest difference was observed for postural activity of the sow at farrowing. Modern-type sows were lying for a longer period after the onset of farrowing, which was associated with greater farrowing difficulties. (Canario, 2006). In line with this difference, stillbirth and piglet weight were both higher in modern-type litters (Canario *et al.*, 2007).

A complicating factor when selecting for improved maternal capacity is that cross-fostering is widely practiced, even in nucleus herds. The breeding value for maternal capacity can be calculated as the maternal genetic effect on piglet survival related to the sow that fosters the piglets, regardless of whether it is the biological mother or not. It is, however, not obvious whether piglets dying shortly after being moved should be regarded as a result of the biological or the foster mother. An advanced model can include three genetic effects; the maternal effects of the biological and the foster mother and the direct effect of the piglet. Such a model puts very high demands on data structure and computer capacity and is difficult to use in routine genetic evaluation.

FREEDOM FROM FEAR AND DISTRESS

High levels of fear and anxiety are disadvantageous for pigs as well as farmers. Fearful and aggressive animals are more difficult to handle and they can be a risk for the handler. Fortunately, pigs are not very aggressive towards humans. All caretakers know they should never turn their back to a boar, but we are not aware of any genetic studies on human-directed aggressive behavior of boars. We have, however, noted that the word "aggressive" is used in the marketing of sire lines. A sire line can, e.g., be described as "aggressive AI boar excellent for semen collection", which could reflect behavioral differences between lines.

Fearfulness

Domestic pigs meet humans every day, and there is usually no escape. Fearfulness thus could have a large effect on pig welfare, but in general pigs seek contact with the caretaker rather than avoid him/her, unless the caretaker behaves "badly". There is an ongoing threshold selection against fear at the herd level which is partly a natural selection and partly an active selection performed by the breeder. This selection has always been an important part of the domestication process. In case of unfavorable genetic correlations between fear and production traits, natural selection may however not be enough to hinder a negative development. There are indications that selection for lean growth rate results in a more excitable temperament in pigs (Grandin, 1992). Leaner pigs are more stressed at transportation, balk more, and are more difficult to drive through races at the slaughter plant than fatter pigs (Grandin, 1998). Scott *et al.* (2000) compared a lean and a fat Landrace line. The lean animals were more reluctant to voluntarily approach a human.

The heritability for fear of human has been estimated based on growing pigs, gilts, and sows. The response of pigs to humans tends to be more heritable when the trait is recorded on animals isolated from the group. The heritability of voluntary approach to a human (time to first contact) was estimated at 0.4 based on isolated gilts (Hemsworth *et al.*, 1990) but the heritability was only 0.1 when based on gilts in groups (Hellbrügge *et al.*, 2007) and 0.0 when based on growing pigs of both sexes in groups (Velie *et al.*, 2009). The comparison must be done with caution since the animals belong to different populations and also differ in age.

Easy to Handle

In growing–finishing pigs, the assessment of "easy to handle" can be performed in association with a routine procedure that takes more time if the pigs are not co-operative. Variation between breeds exists in this trait. Terlouw *et al.* (1997) showed that Large White pigs react more strongly to human presence than Duroc pigs. Lepron (2006) compared three genotypes: a Large White line, a line intensively selected for loin and ham muscle development (Meat line), and a line with Meishan and Large White genes. The Meat line pigs took longer time to exit the pen, run more frequently at weighing and required more human interventions when moved out of the pen, as compared to the other lines.

Moving pigs is a common task on any pig farm. Pigs are moved between pens, to the scale at weighing, to the boar at mating and on/off the trailer when transported. This work is time consuming and there might be an optimal tameness of pigs. Very flighty animals run off in the wrong direction, jump over barriers, and can create chaos. Very tame animals move slowly because they want to come close and interact with the caretaker instead of walking in front of him/her. Day *et al.* (2002) found that pigs exposed to a pleasant-handling treatment took more time to move out of a pen because of reduced fear of humans. Grandin (1989) also reported that handling becomes difficult when pigs are overly tame. The Meat line pigs in Lepron's study (2006) also exhibited behaviors indicative of fear, stress, and excitation, such as bouts and U-turns. Thus, they seemed more nervous than the other two lines. Large White pigs also showed some distinctive reactions to handling. They vocalized more and their heart rate increased more than that of the Meat line pigs. Pigs from the line with Meishan genes appeared to be the least stressed when moved out of their pen.

Recent studies have shown that the response to weighing is a heritable trait in young pigs: the heritability for score for restlessness in the scale was estimated at 0.2 (Holl *et al.*, 2010) and for time to exit the scale at 0.1–0.2 (Jones *et al.*, 2009). The genetic correlations between restlessness and growth rate was negative ($r_g = -0.4$, more restless – lower growth rate) and the corresponding correlation with leanness was positive ($r_g = 0.1-0.2$, more restless – leaner). The genetic correlations between exit time and growth rate was positive ($r_g = 0.1$, longer time – higher growth rate) and the corresponding correlation with leanness was negative ($r_g = -0.2$, longer time – leaner). Selection for growth rate would thus lead to less excited animals whereas selection for leanness would lead to more excited animals. D'Eath *et al.* (2009) recorded handling scores during weighing. Aggressive behavior at mixing at 10 weeks was genetically associated with handling scores at 22 weeks; pigs that fought and delivered non-reciprocal aggression were more active during weighing whereas pigs that were bullied needed more encouragement to enter the scale. When sow behavior was recorded with a questionnaire in two populations, protests by the sow when moved to the farrowing pen was heritable in one of them ($h^2 = 0.2$) but not in the other (Vangen *et al.*, 2005).

Lactating Sows' Aggression Toward Humans

Aggressive behavior of sows is more often observed when lactating sows are kept outdoors or in family pens than indoors (Wallenbeck *et al.*, 2009). Aggressiveness may be less of a problem for the caretaker if sows are kept in stalls or crates. Even so, aggressive behavior of sows is not regarded as a big problem in Sweden, where stalls for dry sows and farrowing crates have been forbidden for many years. Less than 10% of the sows reacted aggressively in a piglet-handling test where the breeder entered the farrowing pen, lifted a 4-day-old piglet and squeezed it gently until it screamed (Grandinson *et al.*, 2003). Avoidance of humans was also recorded in the piglet-handling test and 70% of the sows neither moved forward nor withdrew from the breeder. The heritabilities of aggression toward humans and avoidance were estimated at 0.1 in that field study. Piglet mortality was not genetically correlated to aggression but to avoidance ($r_g = 0.4$, more avoidance – higher mortality). It must be noted that the genetic analyses included only 1100 observations and the standard deviations of these correlation estimates were very large.

When aggression of lactating sows towards human was recorded with a questionnaire to farmers, the heritability was estimated at 0.1 (Vangen *et al.*, 2005). The heritability of fear of humans was estimated at 0.1–0.2 in the same study. We do not know if these two traits are genetically correlated, but admit that the thought of meeting a large sow who has no fear of humans can be somewhat alarming, especially in outdoor production. In the study by Baxter *et al.* (2011), sows selected for higher piglet survival were both more aggressive towards their piglets and towards the humans handling their piglets. They were also more reactive animals, which suggests that farrowing was a period of acute stress for these sows.

Pigs' Ability to Handle Stress

Many studies of stress in pigs have been performed, but the motives have often been more related to meat quality than to pig welfare. The genetic selection for high production has probably contributed to a reduction of HPA axis activity and consequently to a decreased capacity of modern, high-producing animals to maintain their production despite of environmental changes (Mormède and Terenina, 2012). The acute response of the HPA axis can be studied by monitoring the release of ACTH, and cortisol levels are often used to measure chronic stress when assessing animal welfare (Wiepkema and Koolhaas, 1993). The stress response, measured by the response to ACTH, is heritable. Gene expression studies (Hazard *et al.*, 2008; Jouffe *et al.*, 2009) have indicated candidate genes for differences in ACTH response. But the association between ACTH response and behavior is seldom investigated. Désautés *et al.* (1997) showed that Large White pigs had higher post-stress ACTH levels than Meishan pigs after an open-field test. They also showed significant correlations between vocalization and locomotion scores and post-stress ACTH levels, suggesting that these measures reflect the level of reactivity to the environmental challenge, and that they may have a common genetic background.

Engelsma *et al.* (2011) used the open-field test and found that sows with higher breeding values for piglet survival were more explorative; they manipulated the floor and the walls for a longer time than sows with low breeding values. In a small study where Meishan (MS) and Large White (LW) gilts were compared, a test series including the reaction in an open-field test (5 minutes of exploration), the reaction to a novel object (a bucket descended from the ceiling and left on the floor for 5 minutes), and finally the reaction to an unfamiliar human entering the open field and standing there motionless for 5 minutes was used (Canario *et al.*, 2009). In the first challenge, the MS females that stand motionless were the ones with the highest litter growth later on. The LW females that explore less were the ones with the lowest stillbirth rate and piglet mortality later on. In the last challenge, the MS gilts which remained standing all the time had higher piglet growth and the LW gilts that showed higher latency to first move had higher piglet survival. Thus, gilts' ability to handle stressful situations determines their maternal ability later in life but these relations differ between breeds.

Low ability to handle stress, so called Porcine Stress Syndrome (PSS) is caused by a mutation in the locus for the ryanodyne receptor gene RYR(1) and the mutated allele, n, is recessive. It is associated with high leanness and high susceptibility to the development of stress. Stressful stimuli trigger a potentially lethal condition known as malignant hyperthermia in homozygous nn pigs and the n allele has been eradicated from several populations. In some sire-line populations, the n allele is still regarded as a resource, due to its association with leanness. Since all crossbred pigs raised for slaughter are heterozygote the problems of stress sensitivity and mortality are mainly seen in the purebred sire line. Even so, according to Murray and Johnson (1998) Nn pigs also die more often during transport than NN pigs. Furthermore, Nn pigs are less active than NN pigs in an open-field test (Fàbrega *et al.*, 2004). Under commercial conditions, no behavioral differences were found between NN and Nn pigs of various ages in traits like frequency of resting, non-suckling activity, suckling/eating activity and social interactions (Fàbrega *et al.*, 2005). The example of the halothane gene is interesting, because it was one of the first major genes that could be used in breeding programs. Some breeding organizations decided to refrain from the n allele and its high leanness due to welfare reasons as long ago as the 1980s, while other breeding organizations still use the n allele without ethical concerns.

Terlouw *et al.* (1997) and her colleagues have compared different breeds with regard to behaviors that reflect the ability to feel or cope with stress as well as stress hormone levels and meat quality. Duroc pigs touched humans more often than Large White pigs did in a human-exposure test. However, when doing so, the Duroc pigs had higher heart rates than the Large Whites. Touching the man or not in the test depends on the fear the man induces in the pig and the pig's motivation to have contact with the man. It is not known whether the Durocs were less fearful or more motivated to touch the man as compared to Large Whites. Anyway, the response in this test (heart rate) had less effect on meat quality in Durocs than in Large Whites. Terlouw (2005) concludes that the impact of stress response on meat quality is breed dependent. Maybe the same is true for the impact of stress response on welfare?

FREEDOM TO EXPRESS NORMAL BEHAVIOR

In the fifth freedom of FAWC (1992) it says that animals should have the freedom to express "normal behavior". In line with that, demands for the ability to express normal or natural behavior are found in animal welfare laws, and in organic farming natural behavior is a key issue (Lund *et al.*, 2004). The Swedish Animal Welfare Act (1988), for example, states that "Animals shall be accommodated and handled ... in such way as to ... permit natural behavior." The international standards for organic animal husbandry state that "systems that change from conventional to organic production require a conversion period to develop natural behavior" and that "organic livestock husbandry is based on ... respect for physiological and behavioral needs of livestock" (IFOAM, 2005). In a resolution adopted in 1999 by the Federation of Veterinarians of Europe (FVE) it says that selected animals "may be unable to express their natural behavior". The definition of natural behavior of domestic animals is, however, problematic. Are all behaviors performed by wild boars "natural" for domesticated pigs? Segerdahl (2007) proposes that natural behavior of domesticated animals is a philosophical rather than a biological concept. During domestication, animals adapt to agricultural environments via interaction with caretakers and this genetic change will influence their behavior.

The normal or natural behavior concept is further complicated by the fact that such behaviors can be "bad behaviors", leading to goal conflicts when aiming for improved welfare. Aggressive behavior has a genetic background, as described above. To attack a foreign pig could be regarded as a normal behavior in confined situations where contact cannot be avoided. The consequence of the freedom to express such behavior has only been studied from one perspective, i.e. the perspective of the attacked pig whose welfare is decreased. Whether having the freedom to express an aggressive behavior is related to higher welfare has not been studied. Assessment of positive emotions in animals is a complex but growing research field (Boissy *et al.*, 2007).

One example of a normal or natural behavior is mounting. Mounting is more frequent in entire males than castrates and females, but the entire males seem to mount other pigs regardless of sex (Rydhmer *et al.*, 2006). According to Hemsworth and Tilbrook (2007) there are breed differences in male sexual behavior and the heterosis observed for libido in crossbred males also indicates a genetic background. In boars used for mating or for semen collection, sexual behavior is, of course, a wanted behavior. In entire males raised for slaughter, on the contrary, mounting is an unwanted behavior. Excessive mounting leads to injuries (Rydhmer *et al.*, 2006). It would probably be possible to decrease the frequency of mounting by selection, but selection against sexual behavior can hardly be included in any breeding program claiming to be sustainable in a long-term perspective.

CONCLUDING REMARKS

As this review has shown, many behavioral traits important for welfare are heritable and could thus be modified by selection. Furthermore, the ongoing selection for high production seems to influence pigs' behavior. For example, selection for leaner pigs is risky since it increases reactivity to stressors, and should not be done without monitoring pig behavior (as well as health and reproduction). The responses to both acute and chronic stressors need to be considered when analyzing the capacity of an animal to adapt behaviorally to the environment, because they refer to different mechanisms of adaptation.

The lowest welfare in pig production is probably found among growing–finishing pigs kept indoors. These animals live in an environment that is extremely boring, especially for such an explorative animal as the pig. Are pigs too smart for their own good? George Orwell knew about pigs' mental capacity when he wrote *Animal Farm*: "The work of teaching and organizing fell naturally upon the pigs, who were generally recognized as being the cleverest of animals" (Orwell, 1945). Could and should we select for calmer, more passive pigs; pigs that are less explorative and therefore less bored pigs?

Let us assume it would be possible to select pigs for less internally motivated curiosity, and thus less need for a rich environment. Would such a selection be a good way to improve welfare? In the Code of good practice for farm animal breeding and reproduction organizations (EFAB, 2013), it says that breeding organizations must maintain "the intrinsic characteristics of domesticated species". We interpret intrinsic as "belonging to", by its very nature. An uncurious pig could be regarded as an unnatural pig, even if its welfare would be better. We agree with the Code of good practice about maintaining the intrinsic characteristics of pigs and thus we would not recommend such a selection. Instead, management systems should be developed and housing of growing pigs should be improved.

We have described some genetic studies on how easy the pigs are to handle. A question we need to ask ourselves is: Do we select for "easy-care animals" so that we do not have to care? There is a risk that we select animals that can stand a bad environment instead of improving the environment, which is thoroughly discussed by Sandöe *et al.* (1999) in the article "*Staying Good while Playing God*." There are no easy answers to these ethical questions. We have to work with them again and again, together with all stakeholders, as long as we perform animal breeding.

In order to increase welfare and at the same time breed for efficient animals, we need to improve breeding programs further. This development should be done together with ethologists, to develop standardized, relevant, and accurate measurements of behavior or other traits related to welfare. Molecular surveys on behavior will greatly improve our understanding of the genes and biological pathways behind behavioral expressions. Genomic selection (where many markers along the whole genome are used for the genetic evaluation) opens up new possibilities to select for traits that are difficult and thus expensive to record, like many behavioral traits are. It is, however, important to remember that high-quality recording of phenotypic data is the base for all breeding work, regardless of method.

We recommend that breeding organizations should define a broader breeding goal, aiming for a sustainable pig production, and they should reflect on the possibility of including pig behavior in the genetic evaluation. Direct selection for behavioral traits is, of course, not the only solution to improve welfare. Often it can be more rewarding to select for improved functional traits. With regard to behavior, we conclude:

- Group housing of sows will be the standard housing system in the future. Selection for less aggressive behavior in sows when kept in groups is feasible and it could be worthwhile, for welfare reasons.

- Selection against fear of humans could be relevant, but it is probably more important to choose the caretakers carefully.
- Maternal behavior is crucial for piglet production, but it may be more efficient (and much easier) to select for piglet survival and growth, including both direct and maternal effects. Even so, further studies on genetic and environmental effects on sow and piglet behavior are important in order to better understand the biological background of piglet survival.
- It is still too early to use the group model in the genetic evaluation of growth rate. We need more knowledge about pigs' behavior in groups and consequences of selection for social effects before using the group model on a large scale. In the meantime, the consequences of a direct selection against aggressiveness in young pigs ought to be further investigated.

REFERENCES

Arey, D.S., 1999. Time course for the formation and disruption of social organisation in group-housed sows. Appl. Anim. Beh. Sci. 62, 199–207.

Arey, D.S., Franklin, M.F., 1995. Effects of straw and unfamiliarity on fighting between newly mixed growing pigs. Appl. Anim. Beh. Sci. 45, 23–30.

Baxter, E.M., Jarvis, S., Sherwood, L., Farish, M., Roehe, R., Lawrence, A.B., et al., 2011. Genetic and environmental effects on piglet survival and maternal behaviour of the farrowing sow. Appl. Anim. Beh. Sci. 130, 28–41.

Bekoff, M., 1981. Mammalian sibling interactions: genes, facilitative environments, and the coefficient of familiarity. In: Gubernick, D.J., Klopfer, P.H. (Eds.), Parental Care in Mammals. Plenum Press, New York, pp. 307–346.

Bench, C.J., Gonyou, H.W., 2007. Effect of environmental enrichment and breed line on the incidence of belly nosing in piglets weaned at 7 and 14 days-of-age. Appl. Anim. Beh. Sci. 105, 26–41.

Bergsma, R., Hermesch, S., 2012. Exploring breeding opportunities for reduced thermal sensitivity of feed intake in the lactating sow. J. Anim. Sci. 90, 85–98.

Bergsma, R., Kanis, E., Knol, E.F., Bijma, P., 2008. The contribution of social effects to heritable variation in finishing traits of domestic pigs (*Sus scrofa*). Genetics 178, 1559–1570.

Bijma, P., 2011. Breeding for social interaction, for animal welfare. In: Meyers, R.A. (Ed.), Encyclopedia of Sustainability Science and Technology. Springer Science & Business Media LLC.

Bijma, P., Muir, W.M., Van Arendonk, J.A.M., 2007. Multilevel selection 1: quantitative genetics of inheritance and response to selection. Genetics 175, 277–288.

Boissy, A., Manteuffel, G., Jensen, M.B., Moe, R.O., Spruijt, B., Keeling, L.J., et al., 2007. Assessment of positive emotions in animals to improve their welfare. Physiol. Behav. 92, 375–397.

Bouwman, A.C., Bergsma, R., Duijvesteijn, N., Bijma, P., 2010. Maternal and social genetic effects on average daily gain of piglets from birth until weaning. J. Anim. Sci. 88, 2883–2892.

Breuer, K., Sutcliffe, M.E.M., Mercer, J.T., Rance, K.A., Beattie, V.E., Sneddon, I.A., et al., 2003. The effect of breed on the development of adverse social behaviours in pigs. *Appl. Anim. Behav. Sci.* 84, 59–74.

Breuer, K., Sutcliffe, M.E.M., Mercer, J.T., Rance, K.A., O'Connell, N.E., Sneddon, I.A., et al., 2005. Heritability of clinical tail-biting and its relation to performance traits. Livest. Prod. Sci. 93, 87–94.

Broom, D.M., Sena, H., Moynihan, K.L., 2009. Pigs learn what a mirror image represents and use it to obtain information. Anim. Behav. 78, 1037–1041.

Brown-Brandl, T.M., Eigenberg, R.A., Nienaber, J.A., Kachman, S.D., 2001. Thermoregulatory profile of a newer genetic line of pigs. Livest. Prod. Sci. 71, 253–260.

Brunberg, E., 2011. Tail biting and feather pecking: using genomics and ethology to explore motivational backgrounds. Acta Universitatis Agriculturae Sueciae, Doctoral Thesis no. 2011:76. Swedish University of Agricultural Sciences, Uppsala, Sweden.

Brunberg, E., Wallenbeck, A., Keeling, L.J., 2011. Tail biting in fattening pigs: associations between frequency of tail biting and other abnormal behaviours. Appl. Anim. Behav. Sci. 133, 18–25.

Canario, L., 2006. Aspects génétiques de la mortalité des porcelets à la naissance et en allaitement précoce: relations avec les aptitudes maternelles des truies et la vitalité des porcelets. Thèse doctorale en génétique animale et comportement de l'Institut National Agronomique Paris-Grignon.

Canario, L., Moigneau, C., Billon, Y., Bidanel, J.P., 2007. Comparison of maternal abilities of Meishan and Large White breeds in a loose-housing system. 58th Annual Meeting of the European Association for Animal Production, Dublin, Irlande. Book of abstracts, p. 148.

Canario, L., Billon, Y., Mormède, P., Poirel, D., Moigneau C., 2009. Temperament, adaptation and maternal abilities of Meishan and Large White sows kept in a loose-housing system during lactation. 60th Annual Meeting of European Association of Animal Production, Barcelona, Italy. Book of abstracts, p. 282.

Canario, L., Lundeheim, N., Bijma, P., 2010. Pig growth is affected by social genetic effects and social litter effects that depend on group size. 9th World Congress on Genetics Applied to Livestock Production. Leipzig, Germany. CD-ROM Communication ID170.

Canario, L., Turner, S.P., Roehe, R., Lundeheim, N., D'Eath, R.B., Lawrence, A.B., et al., 2012. Genetic associations between behavioral traits and direct-social effects of growth rate in pigs. J. Anim. Sci. 90, 4706–4715.

Chen, C.Y., Johnson, R.K., Newman, S., Kachman, S.D., van Vleck, L.D., 2009a. Effects of social interactions on empirical responses to selection for average daily gain of boars. J. Anim. Sci. 87, 844–849.

Chen, C., Guo, Y., Yang, G., Yang, Z., Zhang, Z., Yang, B., et al., 2009b. A genome wide detection of quantitative trait loci on pig maternal infanticide behavior in a large scale White Duroc × Erhualian resource population. Behav. Genet. 39 (818), 213–219.

Cronin, G.M., Dunshea, F.R., Butler, K.L., McCauley, I., Barnett, J.L., Hemsworth, P.H., 2003. The effects of immuno- and surgical-castration on the behaviour and consequently growth of group-housed, male finisher pigs. Appl. Anim. Beh. Sci. 81, 111–126.

Curtis, S.E., Edwards, S.A., Gonyou, H., 2001. Ethology and psychology. In: Pond, W.G., Mersmann, H.J. (Eds.), The Biology of the Domestic Pig. Comstock Publishing Associates, pp. 41–78.

Dauberlieu, A., Billon, Y., Bailly, J., Launay, I., Lagant, H., Liaubet, L., *et al.*, 2011. Neonatal mortality in piglets: genetics to improve behavioural vitality. 62nd Annual Meeting of the

European Association of Animal Production, Stavanger, Norway. Book of abstracts, session 34, p. 209.

Dawkins, M.S., 2004. Using behaviour to assess animal welfare. Anim. Welfare 13, S3–S7.

Dawkins, R., 1976. The Selfish Gene. Oxford University Press.

Day, J.E.L., Spoolder, H.A.M., Burfoot, A., Chamberlain, H.L., Edwards, S.A., 2002. The separate and interactive effects of handling and environmental enrichment on the behaviour and welfare of growing pigs. Appl. Anim. Beh. Sci. 75, 177–192.

de Jonge, F.H., Boleij, H., Baars, A.M., Dudink, S., Spruijt, B.M., 2008. Music during playtime: using context conditioning as a tool to improve welfare in piglets. Appl. Anim. Beh. Sci. 115, 138–148.

D'Eath, R.B., Roehe, R., Turner, S.P., Ison, S.H., Farish, M., Jack, M.C., et al., 2009. Genetics of animal temperament: aggressive behaviour at mixing is genetically associated with the response to handling in pigs. Animal 3, 1544–1554.

Désautés, C., Bidanel, J.-P., Mormède, P., 1997. Genetic study of behavioral and pituitary–adrenocortical reactivity in response to an environmental challenge in pigs. Physiol. Behav. 62 (9), 337–345.

Docking, C.M., Van de Weerd, H.A., Day, J.E.L., Edwards, S.A., 2008. The influence of age on the use of potential enrichment objects and synchronisation of behaviour of pigs. Appl. Anim. Behav. Sci. 110, 244–257.

Dzieciolowski, R.M., Clarke, C.M.H., Frampton, C.M., 1992. Reproductive characteristics of feral pigs in New Zealand. Acta Theriol. 37, 259–270.

Edwards, G.P., Pople, A.R., Saalfeld, K., Caley, P., 2004. Introduced mammals in Australian rangelands: future threats and the role of monitoring programmes in management strategies. Aust. Ecol. 29, 40–50.

EFAB, 2013. Code of Good Practice for Farm Animal Breeding and Reproduction organizations. <www.code-efabar.org>.

EFSA, 2007. Scientific report of the panel on animal health and welfare on animal health and welfare in fattening pigs in relation to housing and husbandry. Report by European food safety authority. EFSA J. 564, 1–100.

Eissen, J.J., Kanis, E., Kemp, B., 2000. Sow factors affecting voluntary feed intake during lactation. Livest. Prod. Sci. 64, 147–165.

Engelsma, K.A., Bergsma, R., Knol, E.F., 2011. Phenotypic relations between mothering ability, behaviour and feed efficiency of sows peri- and post partum. In Genetic aspects of feed intake in lactating sows (R. Bergsma). Thesis, Wageningen University, Wageningen, The Netherlands.

Enquist, M., Leimar, O., 1983. Evolution of fighting behaviour: decision rules and assessment of relative strength. J. Theor. Biol. 102, 387–410.

Fàbrega, E., Font, J., Carrión, D., Velarde, A., Ruiz-de-la-Torre, J., Diestre, A., et al., 2004. Differences in open field behaviour between heterozygous and homozygous negative gilts for the RYR(1) gene. J. Appl. Anim. Welfare Sci. 7, 83–93.

Fàbrega, E., Font, J., Carrion, D., Velarde, A., Luis Ruiz-de-la-Torre, J., Diestre, A., et al., 2005. Behavioural patterns and performance in heterozygous or halothane free suckling piglets and growing gilts. Anim. Res. 54, 95–103.

FAWC, 1996. Report on the welfare of pigs kept outdoors. <www.fawc.org.uk>.

FAWC, 1992. Farm animal welfare council, 1992. Vet. Rec. 131, 357.

FAWC, 2004. FAWC Report on the Welfare Implications of Animal Breeding and Breeding Technologies in Commercial Agriculture. <www.fawc.org.uk>.

Fernández, J., Fàbrega, E., Soler, J., Tibau, J., Ruiz, J.L., Puigvert, X., et al., 2011. Feeding strategy in group-housed growing pigs of four different breeds. Appl. Anim. Beh. Sci. 134, 109–120.

Fraser, D., Kramer, D.L., Pajor, E.A., Weary, D.M., 1995. Conflict and cooperation: sociobiological principles and the behavior of pigs. Appl. Anim. Behav. Sci. 44, 139–157.

FVE, 1999. FVE policy on genetic modifications in animals. <www.fve.org>.

Gäde, S., Bennewitz, J., Kirchner, K., Looft, H., Knap, P.W., Thaller, G., et al., 2008. Genetic parameters for maternal behaviour traits in sows. *Livest. Sci.* 114, 31–41.

Gourdine, J.L., Bidanel, J.P., Noblet, J., Renaudeau, D., 2006. Effects of season and breed on the feeding behavior of multiparous lactating sows in a tropical humid climate. J. Anim. Sci. 84, 469–480.

Grandin, T., 1989. Effect of rearing environment and environmental enrichment on the behavior of neural development of young pigs. Doctoral Dissertation. University of Illinois, USA.

Grandin, T., 1992. Effect of genetics on handling and CO2 stunning of pigs. Meat Focus Int. July, 124–126.

Grandin, T., 1998. Genetics and behavior during handling, restraint, and herding. In: Grandin, T., Deesing., M.J. (Eds.), Genetics and the Behavior of Domestic Animals. Academic Press.

Grandinson, K., 2003. Genetic aspects of maternal ability in sows. Doctoral thesis. Acta Universitatis Agriculturae Sueciae. Agraria 390. Swedish University of Agricultural Sciences, Uppsala, Sweden.

Grandinson, K., Lund, M.S., Rydhmer, L., Strandberg, E., 2002. Genetic parameters for the piglet mortality traits crushing, stillbirth and total mortality, and their relation to birth weight. Acta Agric. Scand., Sect. A, Anim. Sci. 52, 167–173.

Grandinson, K., Rydhmer, L., Strandberg, E., Thodberg, K., 2003. Genetic analysis of on-farm tests of maternal behaviour in sows. Livest. Prod. Sci. 83, 141–151.

Griffing, B., 1967. Selection in reference to biological groups. I. Individual and group selection applied to populations of unordered groups. Aust. J. Biol. Sci. 20, 127–139.

Hazard, D., Liaubet, L., Sancristobal, M., Mormède, P., 2008. Gene array and real time PCR analysis of the adrenal sensitivity to adrenocorticotropic hormone in pig. BMC Genomics 9, 101.

Hellbrügge, B., Tölle, K.H., Bennewitz, J., Henze, C., Presuhn, U., Krieter, J., 2007. A note on genetic parameters for human-approach tests on gilts. In: Genetic aspects of piglet losses and the maternal behaviour of sows. *Schriftenreihe des Instituts fur Tierzücht und Tierhaltung der Christian-Albrechts-Universität zu Kiel*, Heft 158.

Hellbrügge, B., Tölle, K.H., Bennewitz, J., Henze, C., Presuhn, U., Krieter, J., 2008. Genetic aspects regarding piglet losses and the maternal behaviour of sows, Part 2. Genetic relationship between maternal behaviour in sows and piglet mortality. Animal 2, 1281–1288.

Hemsworth, P.H., Barnett, J.L., Treacy, D.T., Madgwick, P., 1990. The heritability of the trait fear of humans and the association between this trait and subsequent reproductive performance of gilts. Appl. Anim. Beh. Sci. 25, 85–95.

Hemsworth, P.H., Barnett, J.L., 1992. The effects of early contact with humans on the subsequent level of fear of humans. Appl. Anim. Behav. Sci. 35, 83–90.

Hemsworth, P.H., Tilbrook, A.J., 2007. Sexual behavior of male pigs. Horm. Behav. 52, 39–44.

Hernández, A.I., Cianzio, D., Olson, T.A., 2002. Physiological performance and grazing behavior of Senepol, Brahman and Holstein heifers in Puerto Rico. Senepol Symposium, St. Croix, USVI November 8–10, 2002.

Holm, B., 2004. Genetic Analysis of Sow Reproduction and Piglet Growth: Including Genetic Correlations to Production Traits. Doctoral thesis 2004:47. Norwegian University of Life Sciences, Aas, Norway.

Holm, B., Bakken, M., Vangen, O., Rekaya, R., 2004. Genetic analysis of litter size, parturition length, and birth assistance requirements in primiparous sows using a joint linear-threshold animal model. J. Anim. Sci. 82, 2528–2533.

Holl, J.W., Rohrer, G.A., Brown-Brandl, T.M., 2010. Estimates of genetic parameters among scale activity scores, growth, and fatness in pigs. J. Anim. Sci. 88, 455–459.

ICFAW, 2010. OIE recommendations for the on-farm welfare of pigs. Submission by the International Coalition for Animal Welfare (ICFAW). <www.icfaw.org>.

IFOAM, 2005. The International Federation of Organic Agriculture Movements (IFOAM) basic standards for organic production and processing. <www.ifoam.org>.

Jarvis, S., Lawrence, A.B., McLean, K.A., Deans, L., Chirnside, J., Calvert, S.K., 1997. The effect of environment on behavioural activity, ACTH, beta-endorphin and cortisol in pre-parturient gilts. Anim. Sci. 65, 465–472.

Jensen, P., 1982. An analysis of agonistic interaction patterns in a group-housed dry-sows aggression regulation through an "avoidance order". Appl. Anim. Ethol. 9, 47–61.

Jones, R.M., Hermesch, S., Crump, R.E., 2009. Evaluation of pig flight time, average daily gain and backfat using random effect models including growth group. Proc. Assoc. Adv. Anim. Breed. Genet. 18, 199–202.

Jonsson, P., Jørgensen, J.N., 1989. Selection of fattening pigs under consideration of social ranking. Archiv. Tierzucht. 32, 147–154.

Jouffe, V., Rowe, S.J., Liaubet, L., Buitenhuis, B., Sancristobal, M., Mormède, P., et al., 2009. Using microarrays to identify positional candidate genes for QTL: the case study of ACTH response in pigs. BMC Proc. 3 (S4), S14.

Kanis, E., van den Belt, H., Groen, A.F., Schakel, J., de Greef, K.H., 2004. Breeding for improved welfare in pigs: a conceptual framework and its use in practice. Anim. Sci. 78, 315–329.

Keeling, L.J., Wallenbeck, A., Larsen, A., Holmgren, N., 2012. Scoring tail damage in pigs: an evaluation based on recordings at Swedish slaughterhouses. Acta. Vet. Scand. 54, 32 (Epub ahead of print.).

Keltikangas-Järvinen, L., Puttonen, S., Kivimäki, M., Rontu, R., Lehtimäki, T., 2006. Cloninger's temperament dimensions and epidermal growth factor A16G polymorphism in Finnish adults. Genes Brain Behav. 5, 11–18.

Kim, K.S., Larsen, N., Short, T., Plastow, G., Rothschild, M.F., 2000. A missense variant of the porcine melanocortin-4 receptor (MC4R) gene is associated with fatness, growth, and feed intake traits. Mamm Genome 11, 131–135.

Kittawornrat, A., Zimmerman, J.J., 2011. Toward a better understanding of pig behavior and pig welfare. Anim. Health Res. Rev. 12, 25–32.

Knap, P.W., Merks, J.W.M., 1987. A note on the genetics of aggressiveness of primiparous sows towards their piglets. Livest. Prod. Sci. 17, 161–167.

Knol, E.F., Verheijen, C., J.I. Leenhouwers, T. van der Lende, T., 2002. Genetic and biological aspects of mothering ability in sows. 7th World Congress on genetics Applied to Livestock Production. Montpellier, France.

Labroue, F., Guéblez, R., Sellier, P., 1997. Genetic parameters of feeding behavior and performance traits in group-housed Large White and French Landrace growing pigs. Gen. Sel. Evol. 29, 451–468.

Lawrence, A.B., Terlouw, E.M.C., 1993. A review of behavioural factors involved in the development and continued performance of stereotypic behaviors in pigs. J. Anim. Sci. 71, 2815–2825.

Leenhouwers, J.I., De Almeida Junior, C.A., Knol, E.F., van der Lende, T., 2001. Progress of farrowing and early postnatal pig behavior in relation to genetic merit for pig survival. J. Anim. Sci. 79, 1416–1422.

Lepron, E., 2006. Comportement de trois lignées génétiques de porcs à l'engrais et relations avec la consommation d'énergie résiduelle. Université de Laval. Mémoire électronique. <http://archimede.bibl.ulaval.ca/archimede/fichiers/23439/23439.html>.

Lindberg, C.A., 2001. Group life. In: Keeling, L.J., Gonyou, H.W. (Eds.), Social Behaviour in Farm Animals. Wallingford CBI Publ., pp. 37–58.

Luiting, P., Urff, E.M., 1991. Optimization of a model to estimate residual feed consumption in the laying hen. Livest. Prod. Sci. 27, 321–338.

Lund, V., Anthony, R., Röcklinsberg, H., 2004. The ethical contract as a tool in organic nimal husbandry. J. Agr. Environ. Ethics 17, 23–49.

Lund, A., Simonsen, H.B., 2000. Aggression and stimulus-directed activities in two breeds of finishing pig. Pig J. 45, 123–130.

Lundgren, H., 2011. Genetics of sow performance in piglet production. Doctoral thesis. Acta Universitatis Agriculturae Sueciae 91. Swedish University of Agricultural Sciences, Uppsala, Sweden.

Lundgren, H., Fikse, W.F., Grandinson, K., Lundeheim, N., Canario, L., Vangen, O., *et al.*, 2013. Genetic parameters for feed intake, litter weight, body condition and reproduction in Norwegian Landrace Sows. (Submitted).

Lövendahl, P., Damgaard, L.H., Nielsen, B.L., Thodberg, K., Su, G., Rydhmer, L., 2005. Aggressive behaviour of sows at mixing and maternal behaviour are heritable and genetically correlated traits. Livest. Prod. Sci. 93, 73–85.

Lynch, C.B., 1980. Response to divergent selection for nesting behavior in mus musculus. Genetics 96, 757–765.

Maak, S., Petersen, K., Von Lengerken, G., 1998. Association of polymorphisms in the porcine HSP 70.2 gene promoter with performance traits. Anim. Genet. 29 (Suppl.), 71.

Mauget, R., 1981. Behavioural and reproductive strategies in wild forms of Sus scrofa (European wild boar and feral pigs). In: Sybesma, W. (Ed.), The Welfare of Pigs. Martinus Nijhoff, The Hague, pp. 3–13.

McPhee, C.P., Kerr, J.C., Cameron, N.D., 2001. Peri-partum posture and behaviour of gilts and the location of their piglets in lines selected for components of efficient lean growth. Appl. Anim. Behav. Sci. 71, 1–12.

Mormède, P., Terenina, E., 2012. Molecular genetics of the adrenocortical axis and breeding for robustness. Domest. Anim. Endocrinol. 43, 116–131.

Muir, W.M., 2005. Incorporation of competitive effects in forst tree or animal breeding programs. Genetics 170, 1247–1259.

Muráni, E., Ponsuksili, S., D'Eath, R.B., Turner, S.P., Kurt, E., Evans, G., et al., 2010. Association of HPA axis-related genetic variation with stress reactivity and aggressive behaviour in pigs. BMC Genetics 11, 74.

Murray, A.C., Johnson, C.P., 1998. Impact of the halothane gene on muscle quality and pre-slaughter deaths in Western Canadian pigs. Can. J. Anim. Sci. 78, 543–548.

Noblet, J., Etienne, M., 1989. Estimation of sow milk nutrient output. J. Anim. Sci. 67, 3352–3359.

Nowak, R., Porter, R.H., Lévy, F., Orgeur, P., Benoist Schaal, B., 2000. Role of mother–young interactions in the survival of offspring in domestic mammals. Reviews Reprod 5, 153–163.

Orwell, G. 1945. Animal Farm.

Quilter, C.R., Blott, S.C., Wilson, A.E., Bagga, M.R., Sargent, C.A., Oliver, G.L., et al., 2007. Porcine maternal infanticide as a model for puerperal psychosis. Am. J. Med. Genet. B. Neuropsychiatr. Genet. 144, 862–868.

Réale, D., Reader, S.M., Sol, D., McDougall, P., Dingemanse, N., 2007. Integrating animal temperament within ecology and evolution. Biol. Rev. 82, 291–318.

Rodenburg, T.B., Bijma, P., Ellen, E.D., Bergsma, R., de Vries, S., Bolhuis, J.E., et al., 2010. Breeding amiable animals? Improving farm animal welfare by including social effects in breeding programmes. Anim. Welfare 19 (S), 77–82.

Rothschild, M.F., Bidanel, J.P., 1998. Biology and genetics of reproduction. In: Rothschild, M.F., Ruvinsky, A. (Eds.), The Genetics of the Pig. Wallingford CAB International, pp. 313–343.

RSPCA, 2012. The Royal Society for the Prevention of Cruelty to Animals. <www.rspca.org.uk>.

Rydhmer, L., Eliasson-Selling, L., Johansson, K., Stern, S., Andersson, K., 1994. A genetic study of estrous symptoms at puberty and their relationship to growth and leanness in gilts. J. Anim. Sci. 72, 1964–1970.

Rydhmer, L., Zamaratskaia, G., Andersson, H.K., Algers, B., Guillemet, R., Lundström, K., 2006. Aggressive and sexual behaviour of growing and finishing pigs reared in groups, without castration. Acta Agric. Scand., Sect. A, Animal Sci. 56, 109–119.

Rydhmer, L., Lundström, K., Andersson, K., 2010. Immunocastration reduces aggressive and sexual behaviour in male pigs. Animal 4, 965–972.

Sadler, L.J., Johnson, A.K., Lonergan, S.M., Nettleton, D., Dekkers, J.C.M., 2011. The effect of selection for residual feed intake on general behavioral activity and the occurrence of lesions in Yorkshire gilts. J. Anim. Sci. 89, 258–266.

Sandøe, P., Nielsen, B.L., Christensen, L.G., Sorensen, P., 1999. Staying good while playing god—the ethics of breeding farm animals. Anim. Welf. 8, 313–328.

Schinkel, A.P., Spurlock, M.E., Richert, B.T., Muir, W.M., Weber, T.E., 2003. 54th Annual Meeting of the European Association for Animal Production, Rome, Italy.

Scott, K.A., Torrey, S., Stewart, T., Weaver, S.A., 2000. Pigs selected for high lean growth exhibit increased anxiety response to humans. Society for Neurosciences's 30th Annual Meeting, New Orleans, LA, USA.

Segerdahl, P., 2007. Can natural behavior be cultivated? The farm as local human/animal culture. J. Agr. Environ. Ethics 20, 167–193.

Signoret, J.P., Baldwin, B.A., Fraser, D., Hafez, E.S.E., 1975. The behaviour of swine. In: Hafez, E.S.E. (Ed.), The Behaviour of Domestic Animals, third ed. Baillibre Tindall, London, pp. 295–329.

Sinisalo, A., Niemi, J.K., Heinonen, M., Valros, A., 2012. Tail biting and production performance in fattening pigs. Livest. Sci. 143, 220–225.

Spitz, F., 1986. Current state of wild boar biology. Pig News Inf. 7, 171–175.

Spoolder, H.A.M., Burbidge, J.A., Edwards, S.A., Simmins, H.P., Lawrence, A.B., 1995. Provision of straw as a foraging substrate reduces the development of excessive chain and bar manipulation in food restricted sows. Appl. Anim. Beh. Sci. 43, 249–262.

Spoolder, H., Bracke, M., Mueller-Graf, C., Edwards, S., 2011. Preparatory work for the further development of animal measures for assessing the welfare of pigs. Technical report submitted to EFSA.

Stolba, A., Wood-Gush, D.G.M., 1989. The behaviour of pigs in a semi-natural environment. Anim. Prod. 48, 419–425.

Swedish Animal Welfare Act. 1988. The Animal Welfare Act. Ministry of Agriculture Art. No. Jo 09.021.

Terenina, E., Bazovkina, D., Rousseau, S., Salin, F., D'Eath, R., Turner, S., et al., 2012. Association entre polymorphisms de genes candidats et comportements agressifs chez le porc. J. Rech. Porcine Fr. 44, 45–46.

Terlouw, C., Rybrarczyk, P., Fernandez, X., Blinet, P., Talmant, A., 1997. Comparaison de la réactivité au stress des porcs de races Large White et Duroc. J. Rech. Porcine Fr. 29, 383–390.

Terlouw, E.M.C., 2005. Stress reactions at slaughter and meat quality in pigs: genetic background and prior experience. A brief review of recent findings. Livest. Prod. Sci. 94, 125–135.

Thodberg, K., Jensen, K.H., Herskin, M.S., Jorgensen, E., 1999. Influence of environmental stimuli on nest building and farrowing behaviour in domestic sows. Appl. Anim. Behav. Sci. 63, 131–144.

Torrey, S., Pajor, E.A., Weaver, S., Kulhers, D., Stewat, T.S., 2001. Effect of genetic selection for loin-eye area on behavior after mixing in Landrace pigs. 35th International Congress International Society Applied Ethology, Davis, USA. p. 98.

Turner, S.P., Horgan, G.W., Edwards, S.A., 2001. Effect of group size on aggressive behaviour between unacquainted domestic pigs. Appl. Anim. Beh. Sci. 74, 203–215.

Turner, S.P., Farnworth, M.J., White, I.M.S., Brotherstone, S., Mendl, M., Knap, P., et al., 2006a. The accumulation of skin lesions and their use us a predictor of individual aggressiveness in pigs. Appl. Anim. Beh. Sci. 96, 245–259.

Turner, S.P., White, I.M.S., Brotherstone, S., Farnworth, M.J., Knap, P.W., Penny, P., et al., 2006b. Heritability of post-mixing aggressiveness in grower-stage pigs and its relationship with production traits. Anim. Sci. 82, 615–620.

Turner, S., Roehe, R., Mekkawy, W., Farnworth, M., Knap, P., Lawrence, A., 2008. Bayesian analysis of genetic associations of skin lesions and behavioural traits to identify genetic components of individual aggressiveness in pigs. Behav. Genet. 38, 67–75.

Turner, S.P., Roehe, R., D'Eath, R.B., Ison, S.H., Farish, M., Jack, M.C., et al., 2009. Genetic validation of post-mixing skin injuries in pigs as an indicator of aggressiveness and the relationship with injuries under more stable social conditions. J. Anim. Sci. 87, 3076–3082.

Vangen, O., Holm, B., Valros, A., Lund, M.S., Rydhmer, L., 2005. Genetic variation in sows' maternal behaviour, recorded under field conditions. Livest. Prod. Sci. 93, 63–71.

Van der Steen, H.A.M., Schaeffer, L.R., de Jong, H., de Groot, P.N., 1988. Aggressive behaviour of sows at parturition. J. Anim. Sci. 66, 271–279.

Velie, B.D., Maltecca, C., Cassady, J.P., 2009. Genetic relationships among pig behavior, growth, backfat, and loin muscle area. J. Anim. Sci. 87, 2767–2773.

von Felde, A., Roehe, R., Looft, H., Kalm, E., 1996. Genetic association between feed intake and feed intake behavior at different stages of growth of group-housed boars. Livest. Prod. Sci. 47, 11–22.

Walker, P.K., Bilkei, G., 2006. Tail-biting in outdoor pig production. Vet. J. 171, 367–369.

Wallenbeck, A., 2009. Pigs for organic production. Studies of sow behaviour, piglet-production and G×E interactions for performance. Doctoral thesis. Acta Universitatis Agriculturae Sueciae 37. Swedish University of Agricultural Sciences, Uppsala, Sweden.

Wallenbeck, A., Rydhmer, L., Thodberg, K., 2008. Maternal behaviour and performance in first-parity outdoor sows. Livest. Sci. 116, 216–222.

Wallenbeck, A., Gustafson, G., Rydhmer, L., 2009. Sow performance and maternal behaviour in organic and conventional herds. Acta Agric. Scand. A 59, 181–191.

Wechsler, B., Hegglin, D., 1997. Individual differences in the behaviour of sows at the nest-site and the crushing of piglets. Appl. Anim. Beh. Sci. 51, 39–49.

Westin, R., 2000. Svansbitning hos gris relaterat till individuell tillväxt och ras. Examensarbete 2000:46. Swedish University of Agricultural Sciences.

Wiepkema, P.R., Koolhaas, J.M., 1993. Stress and animal welfare. Anim. Welfare 2, 195–218.

Wischner, D., Kemper, N., Krieter, J., 2009. Nest-building behaviour in sows and consequences for pig husbandry. Livest. Sci. 124, 1–8.

Wilson, K., Zanella, R., Ventura, C., Johansen, H.L., Framstad, T., Janczak, A., et al., 2012. Identification of chromosomal location associated with tail biting and being a victim of tail-biting behaviour in the domestic pig (SUS Scrofa domesticus). J. Appl. Genetics. 53, 449–456.

Wood-Gush, D.G.M., Vestergaard, K., 1991. The seeking of novelty and its relation to play. Anim. Behav. 42, 599–606.

Wyckoff, A.C., Henke, S.E., Campbell, T.A., Hewitt, D.G., VerCauteren, K.C., 2009. Feral swine contact with domestic swine: a serologic survey and assessment of potential for disease transmission. J. Wildl. Dis. 45, 422–429.

Young, J.M., Cai, W., Dekkers, J.C., 2011. Effect of selection for residual feed intake on feeding behavior and daily feed intake patterns in Yorkshire swine. J. Anim. Sci. 89, 639–647.

Zhang, Z., Ren, J., Ren, D.R., Ma, J.W., Guo, Y.M., Huang, L.S., 2009. Mapping quantitative trait loci for feed consumption and feeding behaviors in a White Duroc x Chinese Erhualian resource population. J. Anim. Sci. 87, 3458–3463.

Chapter 12

Genetics and Animal Welfare

Temple Grandin* **and Mark J. Deesing**[†]
**Department of Animal Sciences, Colorado State University, Fort Collins, Colorado, USA;*
[†]Grandin Livestock Handling Systems, Inc., Fort Collins, Colorado, USA

INTRODUCTION

The productivity of domestic livestock and poultry has almost tripled in the last 100 years through the use of both improved feeding methods and genetic selection. Freeman and Lindberg (1993) state that enhancements in genetic selection in the dairy industry have contributed more to increased milk production than improvements in management. The National Milk Producers Federation (1996) reports that during a 30-year period, milk production in Holstein cows more than doubled. Since the first edition of this book was published in 1998, milk production has had an additional 14% increase from 1999 until 2009 (USDA/NASS, 2010). Data from the United Kingdom shows that from 1995 until 2012, milk production increased approximately 30%. Similar gains have been made in poultry (Gordy, 1974; Maudlin, 1995). Due to genetic selection, the ability of a chicken to gain weight has increased phenomenally. In 1923, it took 16 weeks to produce a broiler chicken. In 1993, only 6.5 weeks were required.

Adverse effects on bird welfare are an inevitable and unavoidable consequence of selective breeding for high-production traits. The ancestors of the modern-day chicken laid about 25 eggs a year; today's laying hens produce more than 250 eggs. After about a year they are considered "spent." In the meat sector, broiler chickens can convert 3 kg of food into 2 kg of meat (Robbins and Phillips, 2011). To put this growth rate into perspective, Boersma (2001) calculates; "If you grew as fast as a chicken, you'd weigh 349 pounds at age 2". Selection for rapid growth rate has resulted in decreased heart and lung size relative to the rest of the body (Decuypere et al., 2007), and in skeletal defects that affect walking ability (Corr *et al.*, 2003; Knowles *et al.*, 2008). Although many breeding companies now increasingly incorporate health and welfare goals alongside economic goals, from a welfare point of view, growth rate is still seen as a problem (Cooper and Wrathhall, 2010).

For years, both authors have been concerned that in the future, the most serious animal welfare problems may be caused by over-selection for

Genetics and the Behavior of Domestic Animals. DOI: http://dx.doi.org/10.1016/B978-0-12-394586-0.00012-3
© 2014 Elsevier Inc. All rights reserved.

production traits such as rapid growth, leanness, and high milk yield. Unfortunately, some of these concerns have materialized. Many scientists are worried about indiscriminate selection for production traits that seriously compromise animal welfare. Presently, the dairy cow is producing more milk than its ancestor would have produced. The amount is 10-times the beef cattle average of 1000–2000 kg (Webster, 1993). Due to this high metabolic drain, many Holstein cows are only "productive" for two years and are slaughtered around the age of four when profitability drops. Selection for increased milk production has also led to epidemics of so-called "production-related diseases," such as lameness and mastitis, the two leading causes of dairy cow mortality in the United States (USDA, 2008).

The improvements in plant and animal production is largely a result of genetic selection. Turner (2011) warns that selection for economically important traits may be a factor in the expression of behaviors detrimental to welfare. Some examples are aggression in pigs and feather pecking in laying hens. The use of genomic information has the potential to help breeders select against harmful behavior, but if it is used indiscriminately, it could accelerate selection of harmful behaviors. Prunier *et al.* (2010), Jensen *et al.* (2008), and Rodenburg and Turner (2012) all state that new breeding tools could be used to improve animal welfare by selecting against harmful behaviors.

Several long-term selection studies using a variety of small mammals have clearly shown that over-selection for a single trait may have adverse or unexpected effects on other traits (Belyaev, 1979; Dobzhansky, 1970; Lerner, 1954; Wright, 1978). Belyaev (1979) and Belyaev and Borodin (1982) demonstrated the effects of over-selection for a single trait in long-term selection experiments in foxes. Selecting foxes for a single behavioral trait (tameness) caused unexpected changes in coat color, breeding cycles, hormonal profiles, and subsequent changes in body traits. One theory is that the genes guiding the animals' behavior do so by altering chemicals in their brains. Changes to those neurochemicals have "downstream" influences on the animals' physical appearance (Kukekova, Chapter 10).

In domestic pigs, selection for meat production traits has resulted in animals with lowered reproductive ability (Dickerson, 1973). To compensate, modern pork producers often use maternal sow lines selected for high reproduction ability, bred to sire lines selected for high meat production. The resulting offspring are terminal cross pigs fattened for market. Buchanan (1987) stated that the use of crossbreeding in the swine industry has increased in the last 60 years. In the 1990s, most commercial market pigs were crossbred. The use of hybrid sire and dam genetic lines is still a standard practice in 2012. Genetic selection of highly inbred lines with extremely fast growth rates is associated with biochemical disorders. Some of the metabolic diseases associated with genetic selection are Mulberry heart disease (Shin *et al.*, 2011), porcine stress syndrome (Leman *et al.*, 1992), and osteochondrosis, a disease of bone and cartilage growth (Stern *et al.*, 1995). Of particular interest, fast-growing pigs display patterns of genetically determined pathologic alterations similar to those observed for human populations afflicted by obesity (Cummings and Schwartz, 2003).

With the exception of a short section on biotechnology at the end of this chapter, all welfare problems discussed here were created with conventional breeding methods. The animals are not GMOs (genetically modified organisms) or products of biotechnology.

WELFARE PROBLEMS CAUSED BY HARMFUL SOCIAL BEHAVIORS

Since the year 2000, harmful social behavior has been thoroughly researched. In hens, selection for production traits such as greater egg production is associated with increased feather pecking (Muir, Chapter 9). In pigs, selection for production traits is associated with an increase in pig aggression, tail biting, and savaging of piglets (reviewed by Turner, 2011). Much of the current research has been motivated by legislation in Europe to move away from practices such as beak trimming or tail docking. The ban on gestation stalls in Europe, and pressure from retailers in the United States to move to group housing provides further impetus to develop less aggressive animals that still have high levels of productivity.

Aggression in Sows

Practical experience shows that certain genetic lines of sows produce efficiently in stalls but have high levels of fighting injuries when housed in groups. Pig production companies that have successfully moved from gestation stalls to group housing have done so by changing genetics and selecting for less aggressive sows. When sows were kept in individual stalls, there was no informal selection by producers to eliminate the most aggressive animals.

Research clearly shows that aggression in sows is heritable (d'Eath *et al.*, 2009a). Chapter 11 of this book, by Lotta Rydhmer, covers the genetic factors that affect sow aggression. Aggression interacts with other behavioral traits in odd ways. Purebred Yorkshires and Yorkshire × Landrace pigs that stood still on a scale during weighing were less aggressive, and took longer to move in and out of the scale (d'Eaath *et al.*, 2009b). Fortunately, selecting for less aggressive sows does not affect maternal ability (Lovendahl *et al.*, 2006). On-farm observations indicate that pigs bred for lower aggression must never be mixed with more aggressive pigs. Low-aggression pigs suffer many attacks and may become severely injured.

Tail Biting in Pigs

Breuer *et al.* (2004) found that tail biting was heritable in Landrace pigs and not in Large Whites. Breuer *et al.* (2004) found that genetic selection for lean growth and low back fat is associated with tail biting in Landrace pigs. This agrees with the first author's observations of thousands of market pigs at many slaughter plants. A much higher percentage of bitten tails is seen in certain genetic lines of pigs. Pigs from these lines are more excitable and chew harder on a person's boots compared to other genetic lines. Taylor *et al.* (2010) has an interesting hypothesis and suggests three types of tail biting, each with different motivations. The differences are referred to as; "sudden," "forceful," and "obsessive" (Taylor *et al.*, 2010). The first author speculates that intense, persistent boot chewing by these pigs may be the "obsessive" motivation. Other forms of tail biting may be misdirected foraging. Biting tails and fighting may have two totally different motivations. Tail-biting pigs spend more time manipulating enrichment devices (Zonderland *et al.*, 2011), and victims of tail biting are more aggressive than pigs that bite tails. Misdirected foraging is very different than aggression (Zonderland *et al.*, 2011). On a novel object test, tail biters had lower anxiety compared to victims (Zupan *et al.*, 2012). Sinisalo *et al.* (2012) report breed differences in tail biting and found that when Yorkshire and Landrace pigs are mixed, the Yorkshires are more likely to be victims of tail biting. When the first author visited Japan, she observed confinement-reared pigs that did not have to have their tails docked to prevent tail biting (Figure 12.1). Whether tail biting is motivated by aggression or misdirected foraging, Van der Weerd *et al.* (2008) found that the use of straw and chewable enrichment devices may help to prevent severe tail biting. In this study, three enrichment procedures were used; straw, a rootable feed dispenser, and a feed dispenser providing flavorful water when chewable rods were manipulated. Results show that a full bed of straw was the most successful way of occupying the

FIGURE 12.1 These Japanese pigs seldom bite tails even though they live in an environment that is barren. Genetic factors have an influence on the occurrence of tail biting.

pigs. Pigs could manipulate the straw even while lying down. Age may also influence aggression and tail biting. Li *et al.* (2012) suggest that young sows are vulnerable in group-gestation systems when mixed with mature sows. Sorting sows by the number of times they have given birth improves welfare and performance by reducing severe injuries and fighting.

Savaging in Sows

Sows killing their piglets is an example of an extreme welfare problem that may interact with environmental conditions. Sows that savage are more restless at farrowing. Ahlstrom *et al.* (2002) suggest that aberrant maternal behavior during farrowing or an individual's inability to cope with restrictive environments may be contributing factors in savaging. Chen *et al.* (2008) compared behaviors such as grinding teeth, rearing, lying down with care, and prepartum nest-building behaviors. These behaviors were not a predictor of savaging. However, Chen *et al.* (2008) reported that sows that made more postured changes during farrowing were more likely to savage. This suggests that savaging is part of a more generalized behavioral pathology that includes increased excitability—not specifically piglet directed. Baxter *et al.* (2011) found that modifying the farrowing environment affected maternal behavior in sows, and suggest that it is essential to evaluate sows in the environment they will be housed in. This is especially important when sows from different genetic lines are being switched from indoor, intensive environments to outdoor environments. Sows attacking and killing their own newborn offspring is a serious welfare concern that requires further study.

Feather Pecking in Hens

In the egg-production sector, feather pecking is a serious welfare and economic problem. Several recent studies have focused on behaviors of genetic lines of hens selected for either high or low feather pecking. Evidence suggests that feather pecking in laying hens may be misdirected foraging behavior. Rodenburg *et al.* (2010) investigated the underlying motivation of feather pecking in genetic lines of chickens bred for high and low feather pecking. Results suggests that fear may not be the primary motivator, since the two genetic lines had similar reactions on the novel object test and human-approach test (Rodenburg *et al.*, 2010). It has also been suggested that dietary factors could play a role in feather pecking, however, hens selected for high feather pecking ate more of a high-fiber diet (Kalmendal and Bessei, 2012). Lines selected for increased feather pecking showed increased motivation for locomotor activity (Kjaer, 2009). de Haas et al. (2010) found that a high feather-pecking line ate worms faster. This is further evidence that feather pecking is displaced foraging. High-producing animals require vast quantities of food. The first author speculates that selection for high egg production and increased appetite are related. Fisikowski *et al.* (2008) found that two sub-haplotypes of the DRD4 gene are associated with feather pecking. Dopamine genes are associated with exploratory behavior in many species. This finding is additional evidence that foraging behavior and/or increased appetite is related to high feather pecking. Behavioral observations also show that hens perform gentle pecks during dust bathing and vigorous pecks when foraging. Pecking behavior directed toward other birds resembles pecking behavior during foraging (Dixon *et al.*, 2009). In a subsequent study, the same research team showed that providing environmental enrichment opportunities for foraging are effective for reducing feather pecking (Dixon *et al.*, 2010). Kjaer and Jorgensen (2011) found that birds from a high-feather-pecking line had higher autonomic nervous system arousal during physical restraint compared to those from a low-feather-pecking line. Brunberg *et al.* (2011) suggest that feather pecking may serve as a model for human obsessive–compulsive disorder. This study provides a gene list that may be useful in further research on the mechanisms behind feather pecking.

EFFECTS OF OVERSELECTION IN CATTLE AND PIGS

Since 1971, the first author has observed hundreds of thousands of animals at slaughter plants, farms, and feedlots in the United States, Canada, Europe, Australia, and New Zealand. In the early 1990s she began to observe an increased number of highly excitable, nervous pigs and cattle. These animals were more difficult to handle and more likely to panic and become extremely agitated when subjected to sudden novel experiences. For example: a light tap on the rear causes squealing in pigs with an excitable temperament, but a tap on the rear has little effect on pigs with a calmer temperament. The appearance of highly excitable and difficult-to-handle animals appeared to coincide with genetic selection for both rapid growth and high, lean meat yield (Grandin, 1994). In cattle, the most reactive animals are primarily crossbreeds from the European continent. In the United States, these cattle became popular when producers started selecting for lean beef (Grandin, 1994). Along with welfare problems and handler safety issues, highly excitable animals have an increase of meat-quality problems. Cattle that become highly agitated during handling are more likely to have tough meat and more dark cutters (Voisinet *et al.*, 1997). Many studies now show an association with poorer meat quality and excitable temperaments in cattle (del Campo *et al.*, 2010; Muchenje *et al.*, 2009; Turner *et al.*, 2011), and pigs (Barton-Gade, 1984; Sayre *et al.*, 1964). Pigs that become excited just before slaughter have more PSE and lower meat quality. Since the first edition of this book, an industry-wide acceptance of this problem has motivated producers to select intensively for calmer cattle (see Chapter 4 on cattle temperament testing). The results are calmer, easier to handle cattle with better meat quality (Ribeiro *et al.*, 2012). However, increased selection for calm temperaments is swinging the pendulum the other way and concerns have now shifted to problems associated with a loss of beneficial foraging and maternal traits in calm cattle.

In both the United States and Australia, cattle are grain-fed in large outdoor feedlot pens. During the 1970s, the first author never observed tongue rolling or other abnormal behaviors in grain-fed cattle. In a 1996 visit to several large feedlots, she noticed an increase of "stereotypic" (abnormal) tongue rolling in fed Holsteins. Some Holstein steers obsessively lick fences, gates or other surfaces in the feedlot pen. This behavior is expressed even though cattle were fed (*ad lib.*) grain and corn. Licking is so obsessive that Holstein steers learned to lick open gate latches. Beef-breed cattle in feedlots never engage in constant licking or tongue rolling.

Both authors speculate that continuously selecting for high milk production and increased appetite may explain the increase in licking and tongue rolling. A large amount of food is required for Holstein cows to produce such large amounts of milk, and the effects of increased appetite are also revealed in male offspring of these cows. We further hypothesize that licking fences and gates may be a precursor to more serious problems if genetic selection for the highest production is continued. For example, the increased incidence of "weaver" condition in Holsteins is possibly related to increased selection for high milk production. Weaver is an inherited degenerative mycloencephalopathy (Freeman and Lindberg, 1993).

From a welfare standpoint, grain-fed Holstein steers are probably not compromised, but from a health standpoint, welfare problems already exist. The first author has observed that Holstein steers on a high brain ration bloat more often compared to beef breeds on similar rations. In addition to bloat, grain-fed Holsteins have more sudden death than beef cattle. Feed yard employees have observed healthy-appearing Holsteins suddenly fall over and die while standing at a feed trough.

MUSCLE GROWTH AND WELFARE

In the first edition of this book we discussed the compromised welfare of poultry by genetic selection for rapid growth and high meat production. Growth of the skeleton and the internal organs can not kept up with the growth of muscle mass in broiler chickens selected to grow very quickly, resulting in lameness and leg deformities (Muir, Chapter 9). Today, a combination of selective breeding and changes in feeding practices has reduced leg problems. For example, a lower-energy feed is given to young chicks to slow down growth slightly and allow the leg bones to develop. However, broiler chickens continue to have reduced cardiopulmonary capacity in relation to their muscle mass, and cannot withstand much physical exertion (Broom, 1987, 1993; Julian, 1993; Julian *et al.*, 1986). Heavy broilers have little stamina, and lie down after walking a few meters. Reproductive fertility has also been reduced by extensive selection for rapid growth—muscle growth in the breast can make mating difficult. Turkeys selected for large breasts are unable to breed naturally (Dinnington *et al.*, 1990). Another example is Belgian Blue cattle. This breed is so heavily muscled that a high percentage of their calves must be delivered by cesarean section (Broom 1987, 1993). Not only do births of such calves require surgical intervention, their tongue muscles are sometimes too enlarged to suckle, leading to death (Lips *et al.*, 2001; Uystepruyst *et al.*, 2002). Fortunately, some Belgian Blue genetic lines are now available that have much easier calving. The genetics of the double muscle trait have been extensively studied. Selection for this trait is associated with the inactivation of the myostatin gene, coinciding with reduced organ mass (Fiems, 2012). A mutation causes the myostatin gene to become inactive, resulting in muscle hypertrophy, and coinciding with a susceptibility to respiratory disease, kidney stones, lameness, heat and nutritional stress, and dystosia. In addition, Belgian Blue cattle require a diet high in nutrients, due to a reduced feed intake capacity (Fiems, 2012). In 2012, the first author visited a farm in the United Kingdom where Belgian Blue sires were bred to Devon × Limousin cows enabling the production of heavily muscled calves that required no growth promotants. The huge Devon × Limousin cross cows required a high-nutrient diet that worked well in the extensive, lush pastures in the United Kingdom, but some nutrient-poor pastures in the United States would not be able to support these cattle. High lean-meat yield is desirable, but it is important to find a balance between good performance in double-muscled animals and acceptable welfare.

Porcine Stress Syndrome and Leg Problems

In the past, some pork producers deliberately used boars positive for the PSS (Porcine Stress Syndrome) gene. The crossbred offspring from these boars have a higher percentage of lean meat and larger loin eyes (Aalhus *et al.*, 1991). PSS is inherited in a classical Mendelian manner. Pigs that are either homozygous negative or heterozygous with one PSS gene (carrier state) will not display the symptoms of PSS when they become excited. Furthermore, when a homozygous-positive PSS boar is bred to a sow that is free of the PSS gene, none of the offspring will be homozygous-positive for PSS. To maximize kilograms of lean meat, swine-breeding companies sell PSS homozygous-positive boars for breeding to PSS homozygous-negative sows. A welfare problem is created if a producer breeds PSS-heterozygous sows (carrier state) to PSS-homozygous boars. Unfortunately, this practice is quite common because a producer can breed terminal cross gilts from his own herd instead of buying new gilts that are homozygous-negative and completely free of the PSS gene.

The first author has observed that some of the lean hybrid pig lines developed in the 1990s often had five-times as many death losses, compared to older genetic lines of pigs with more back fat. In the 1990s, truck drivers transporting pigs often reported one or two dead pigs on a truckload of hybrid market animals. These deaths were likely to occur even when pigs were transported under good conditions. Older genetic lines often had no dead pigs on a truck. Many of the truckloads with high death losses contained pigs sired by PSS-positive boars. Murray and Johnson (1998) reported higher death losses in market pigs with the halothane PSS gene. In the 2000s, the use of PSS genetics declined and death losses are now lower. Recently, USDA figures show that dead pigs arriving at the slaughter plants dropped from 0.29% in 2000 to 0.17% in 2010.

Although the PSS problem had been reduced and the U.S. Pork Industry has now almost eliminated PSS genetics, other problems associated with a demand for lean meat and larger loin eyes are occurring. Beta agonist feed additives (ractopamine) are now used to enhance the growth of lean muscle. High doses of ractopamine can cause weakness, downed pigs, hoof cracking, and increased aggression (Grandin, 2010; Marchant-Forde *et al.*, 2003; Poletto *et al.*, 2011). Feeding the beta agonist ractopamine can suppress the expression of serotonergic genes in the brains of the pigs (Poletto *et al.*, 2011). Problems corrected by eliminating PSS genetics may now be repeated by using drugs.

As the U.S. industry was phasing out PSS genetics, new problems with poor leg conformation and lameness started to occur. Single-minded selection for large loin eye size, thin back fat, and rapid growth resulted in a neglect of selecting breeding stock with good leg conformation. In the early 2000s, the first author observed approximately 50% lame market-weight pigs in some herds. Today, breeders have added standards for structural correctness of feet and legs, which has reduced lameness.

High death loss in ultralean hybrids was a more serious problem in the United States than in Europe. Death losses in Europe were much lower for the same hybrid pig lines from the same commercial breeding company. The reason for this may be that European pigs are fed a limited grain ration and are grown more slowly. Slower growth results in older market-weight pigs that have a more mature skeleton. In the United States, market pigs are fed *ad lib.* grain and attain market weight at an earlier age. Another factor influencing increased death losses and non-ambulatory pigs in the United States is higher slaughter weights compared to pigs in Europe. The first author has observed that heavier animals are more likely to collapse during physical exertion.

PROBLEMS CAUSED BY GENETIC SELECTION FOR APPETITE

Animals genetically selected for rapid weight gain or production of large quantities of milk, eggs, or meat require a tremendous appetite and food intake in order to produce large amounts of these products. Selection for rapid muscle growth correlates highly with selection for increased appetite.

Research on chickens shows that birds selected for egg production stop eating when their metabolic needs are met, but broiler chickens selected for meat production do not stop eating until their gut is completely full (Nir *et al.*, 1978).

From an appetite standpoint, the welfare of broiler chickens and market pigs being fattened for slaughter is good. These animals are allowed to eat until satiated. However, a welfare problem related to increased appetite occurs in breeder hens and sows of the fast-growing offspring. Breeders have to be maintained on a calorie-restricted diet (Close, 1996). For example, broiler breeders on a restricted diet produce more eggs (Robinson *et al.*, 1991), but if allowed to eat until satiated, they develop reproductive problems (Yu *et al.*, 1992a). Furthermore, if sows are fed all they can eat, they become over-fat, which can result in leg problems and difficulties farrowing. Problems occur because breeder animals' appetites far exceed their basic metabolic needs.

Broiler breeders used for egg production are fed 60–80% less than they would eat if fed *ad lib.* to prevent them from becoming overweight (Karunajeewa 1987; Hocking 1993; Hocking *et al.*, 1993; Yu *et al.*, 1992a,b; Zuidhoff *et al.*, 1995). When hens go into egg production to produce broiler chicks, they are restricted to 25–50% of what they would eat if they ate until satisfied. Feed restriction in sows is slightly less extreme compared to broiler chickens. Gestating sows are fed approximately 60% of their *ad lib.* intake of a standard grain concentrate diet (Lawrence *et al.*, 1989). A sow nursing piglets is allowed to eat all she wants, but during gestation, she is kept on a calorie-restricted diet to prevent her from becoming too fat.

A review of a number of studies shows that feed restriction in breeding sows and chickens results in many abnormal behaviors, such as stereotypies (Lawrence and Terlouw, 1993). Appleby and Lawrence (1987) and Tolkcomp *et al.* (2005) found that stereotypies developed in sows only when their feed intake was restricted. In a subsequent experiment, the same research team concluded that the amount of feed sows are fed on commercial farms is not high enough to satisfy their motivation for feeding (Lawrence *et al.*, 1989). They further concluded that hunger resulting from a restricted diet may be a major cause of stress in confined housing systems. Bergeron and Gonyou (1997) found that stereotypies developed when the diet did not have sufficient energy to prevent hunger. Similar results have been found in other animals. Savory *et al.* (1992) found that feed restriction in broiler breeder hens also results in stereotypies. It is becoming increasingly clear that feed restriction in animals selected for increased appetite is a serious welfare concern.

Importance of Roughage Feeds

It is a common practice in the poultry and pork industries to feed highly concentrated diets consisting of mostly grain to breeding sows and broiler breeders. This does not compromise the welfare of market animals that are fed *ad lib.*, but it may increase welfare problems in calorie-restricted breeding animals. Studies in poultry, pigs, and cattle show that the incidence of abnormal behavior can be reduced by providing roughage in the diet (Lawrence *et al.*, 1989; Redbo and Nordblad, 1997; Robert *et al.*, 1993; Zuidhoff *et al.*, 1995). Roughage helps to reduce stereotypies and other abnormal behavior because diets high in roughage can be restricted in calories and fill the animal's gut. Diets with added roughage also take more time to eat and satisfy the animal's motivation for mouth activities. However, artificially increasing food intake time by placing hanging chains in the feed trough failed to reduce the development of stereotypies (Bergeron and Gonyou, 1997). The wild ancestors of chicken and pigs spent many hours pecking and rooting to obtain their food. Stolba and Wood-Gush (1989) report that domestic swine reared in woods and grasslands display most of the same behaviors as their ancestor, the European wild boar. Behaviors used for obtaining food, such as grazing and rooting, occupied large portions of each day. Even though the adult domestic sows used in this study came out of an intensive housing system, they quickly reverted to ancestral foraging patterns.

It is important to provide sows with roughage feed that provides both gut fill and mouth activities. Feeding roughage such as straw, sugar beet pulp, or oat hulls can improve the welfare of breeding animals kept on a calorie-restricted diet (Close *et al.*, 1985). Zuidhoff *et al.* (1995) found that feeding concentrates diluted with 15% oat hulls increased the time required to consume the feed and reduced stress. Stress was measured with a heterophil/lymphocyte test. Diets high in fiber (lignocellulose) were effective in reducing feeding motivation (DeJong and Guemene, 2011; Souza de Silva *et al.*, 2012). Chopped straw was added to the ration and hunger was measured by counting how many times boars would press a panel to obtain feed rewards. Whole straw provided to pigs is more effective for reducing abnormal behavior than chopped straw. Stereotypies in sows housed in individual stalls can be prevented by feeding small amounts of straw (Fraser 1975). Compared to baled hay, pelleting, or wafering, roughage feeds decrease bulking by about 75% (Haenlein *et al.*, 1966). Willard *et al.* (1977) found that horses fed pellet diets spent more time chewing wood and eating manure compared to horses fed hay.

Some sectors of the industry have been reluctant to feed roughage because roughages are too bulky to move through some types of automated feeding systems. Roughages are also more expensive to transport and increase the load on manure-handling systems. Management issues aside, adding roughage to the diet improves productivity. Broiler breeder hens fed a bulky diet of 15% oat hulls had higher egg production than hens fed grain concentrates (Zuidhoff *et al.*, 1995).

While feeding roughage can reduce hunger motivation in some highly productive breeding animals, it may not always be sufficient to maintain welfare. Tina Widowski, an animal welfare specialist at the University of Guelph in

Canada, has extensive experience of raising both breeding sows and broiler breeder chickens. She told the first author that "broiler breeders are so highly selected for appetite that they do not work behaviorally as animals." For example, when broiler breeders were fed *ad lib.*, they spent 50% less time pecking objects (Sandilands *et al.*, 2005). There is also some evidence that sows that produce large litters have more stereotypic behavior (von Borell and Hurnick, 1990). D'Eath *et al.* (2009a) and Lawrence *et al.* (1989) both warn that feeding a bulkier, less nutrient-dense diet may not solve all the welfare problems associated with feeding motivation. Further research is needed on the effects of "metabolic hunger." There may be a point where continued breeding of animals for appetite will cause grave problems.

MOVEMENT RESTRICTION VERSUS FEED RESTRICTION

The general public is often most concerned about movement restriction when animal welfare is being discussed. Some examples are egg-laying hens housed in cages, veal calves housed in crates, and gestating cows kept in stalls where they are unable to turn around. Research clearly shows that animals are motivated for movement (Dellmeier *et al.*, 1985; McFarlane *et al.*, 1988). However, movement restriction may be less stressful to the breeding animals than feed restriction (Rushen, 1995; Rushen *et al.*, 1993). This could be especially true in animals genetically selected for high appetite. In one experiment, gilts were housed in crates that made turning around difficult. In the first treatment, food and water were positioned at opposite ends of the crates so the gilts had to turn around in order to eat and drink. In the second treatment, the water and feeder were placed at the same end of the crate. Gilts that did not have to turn around did so just as many times as gilts that had to turn in order to eat or drink (McFarlane *et al.*, 1988).

In another experiment, Dellmeier *et al.* (1985) found that veal calves housed singly in small stalls for six-and-a-half weeks responded with greater activity when tested in an open-field arena, compared to calves housed in groups in a yard. The calves housed in small stalls ran around and kicked up their heels when turned loose in the test arena. Calves raised in large group pens did not do this. Pasille *et al.*, (1995) questioned the results of Dellmeier *et al.* (1985), due to the fact that the open field was more novel for the calves housed in stalls. Pasille *et al.* (1995) suggest that Dellmeier *et al.* (1985) did not control for the novelty factor in the test arena. However, the first author has observed that cattle kept in a large, open feedlot pens and cattle housed singly in small "dog run" pens behave differently when released. Large, fat cattle raised in small "dog runs" ran up and down a 3.5-meter-wide alley when released. The increased activity cannot be explained by the effects of novelty. Both groups of cattle were raised outside on the same farm and the alley was visible from their pens. Our informal observations support the findings of Dellmeier *et al.* (1985).

NERVOUS SYSTEM ABNORMALITIES DUE TO MUTATIONS OR SELECTIVE BREEDING

The literature is full of examples of neurological or behavioral abnormalities in dogs, pigeons, mice, rabbits, and rats. In most cases, the abnormalities occurred as spontaneous mutations in laboratory stocks, which were then continuously selected for research purposes. An example of spontaneous mutation is muscular hypertrophy in Dorset sheep (Cockett *et al.*, 1996). One of the most well-researched examples is nervous pointer-breed dogs. Pointers are dogs selected to freeze and point when they see game birds hidden in the bushes. Pointing behavior is similar to an orienting response that becomes frozen. The selection of pointers for the pointing trait may be selection for an abnormal orienting response. Breeders have known for years that some dogs are so nervous that they are useless as hunting dogs (Dykman *et al.*, 1979) The pointing trait and nervousness may be linked. A neurologically normal dog will orient toward a bird or other prey, then either chase it or return to its original activity. A normal dog does not stay frozen in an orienting posture. Breeders of pointers have recognized that a fine line exists that divides a good pointer from a bad one. Dogs with a heightened pointing response are generally too nervous to make reliable hunting dogs. Dykman *et al.* (1979) and Peters *et al.* (1967) found obvious differences in the behavior of normal pointers versus nervous pointers. Furthermore, visitors to the laboratory could easily differentiate which dogs were from the nervous genetic line. Nervous pointers cowered in their cages and failed to approach visitors, whereas the normal pointers approached and wagged their tails.

Geneticists have selectively bred both normal pointers, with the characteristic pointing behavior, and abnormal, nervous pointers. Nervous pointers are excessively timid and display a hyperstartle response, avoidance of people, and catatonic freezing in the close presence of people (Klien *et al.*, 1988). Some nervous pointers display strikingly bizarre behavior. One dog became "frozen" in a point and fell over when accidently bumped by another dog (McBryde and Murphee, 1974). It appears that with nervous pointers, a continuum of neurological defects exists. Further research by Klien *et al.* (1988) revealed that many nervous pointers were also deaf. Inherited deafness in lines of pointers selected for excessive nervous behavior is an autosomal recessive trait, and the nervous trait may be inherited dominantly (Steinberg *et al.*, 1994).

The welfare of normal pointers (ones without the nervous trait) is acceptable. However, nervous pointers may have very poor welfare unless raised under specialized conditions. For example: training and environmental modification can help nervous pointers act more like normal dogs. McBryde and Murphee (1974) found that training nervous pointers alongside normal pointers made them less timid. The dogs from the nervous genetic line become less timid and followed a normal pointer. After a period of training, the nervous dogs no longer needed to be kept with a normal dog. McBryde and Murphee (1974) concluded that breeding an animal continuously for a certain genetic trait often resulted in the occurrence of other less desirable traits. In the 1990s the Arkansas pointer lines were terminated. However, they contributed to our understanding of how genes and environment influence behavioral difference between dogs, including the development of potential therapies for shy or nervous dogs.

Inherited Neurological Defects

Arman (2007) states that the prevalence of inherited defects in dogs is too high. Breed associations should encourage selecting dogs for their functionality and allow the introduction of new genetics into the registered breeds. The list of neurological problems in dogs is huge (Ackerman, 2011). Many examples of neurological defects in other animals can be found in the genetics literature. For example, roller-tumbler pigeons are bred to roll in flight. Pigeon fanciers select for birds with an intermediate expression of this trait. Birds with an excessive expression of this trait continue rolling in flight until they crash into the ground (Entrikin and Erway, 1972). Pigeons that roll in flight have acceptable welfare, but when they hit the ground, welfare is severely compromised. In comparison to pointer dogs, a slightly abnormal orienting response makes a good bird dog, but an excessive amount of the trait results in a dog that is a nervous wreck.

NERVOUS DEFECTS IN RODENTS AND SMALL MAMMALS

For hundreds of years, the Japanese have bred "fancy" mice as a hobby. In Europe, there is also a long tradition of mouse fanciers (Morse, 1978). Some fancy mice were selected for abnormalities of movement. Most of the fancy mouse mutations can be found in today's laboratory mice. One is called the "Japanese Waltzer." Cools (1972a,b) reported that these animals have an abnormality of the inner ear, are also hyperactive and behave like mice injected with amphetamines (Cools, 1972a,b).

Researchers have bred rodents with many abnormalities such a seizure-prone rats (Seyfried, 1982; Serikawa and Yamada, 1986; Wimmer and Wimmer, 1985). Some rats have epileptic seizures triggered either by lowering their body temperature or by sound. Mutant, seizure-prone rats are so sensitive that a minor disturbance, such as changing drinking bottles, can cause convulsions (Serikawa and Yamada, 1986). Others are generally prone to seizures, and many different types of stimuli can trigger a seizure. There are also strains of rodents with juvenile onset or adult onset susceptibility to seizure. Rabbits have also been selectively bred for susceptibility to sound-induced seizures (Hohenboken and Nellhaus, 1970). A preference for alcohol has also been genetically selected (Wimmer and Wimmer, 1985). McClearn and Rogers (1961) discovered a preference for alcohol over water in an inbred mouse line. Crabbe (1983) found two distinct genetic effects and two separate behavioral effects of alcohol. The first effect of alcohol on mice is on activity levels after a dose. The second is "hypothermic sensitivity" which includes a long sleep time after a dose of alcohol.

In rodents and small mammals used for research, welfare problems can appear unexpectedly. In a German laboratory, workers reported an outbreak of self-mutilation in checkered cross rabbits bred for a high resistance to infection (Inglauer et al., 1994). The rabbits bit off their own toes. The researchers achieved the goal of increased disease resistance, but the price was rabbits that injured themselves. This experience is just one more example of behavioral and physiological traits linked in ways that are difficult to predict.

DEPIGMENTATION PATTERNS AND THE NERVOUS SYSTEM

For centuries, animal breeders have recognized that a lack of eye and body pigmentation may signal neurological defects. Animals with extensive depigmented areas on the body are more likely to have developmental or nervous system abnormalities (Cowling *et al.*, 1994; Searle, 1968; Webb and Cullen, 2010). The degree of depigmentation could possibly serve as an early warning for welfare-related nervous system abnormalities in animals highly selected for production traits.

It is important, however, to differentiate between depigmentation and animals with white hair and fully pigmented skin. Arab horses and Brahman cattle have light hair and dark skin. Depigmented areas of the body are characterized by both white hair and pink skin. The Holstein cow is depigmented in the white areas of the body. Light-colored cattle, such as the Charolais, are not depigmented and have brown eyes and some hair pigment. Most beef breeds do not have large areas of depigmentation. In addition, some crossbred white pigs with Duroc genetics have white hairs with a very faint reddish tint. The Yorkshire and Landrace pig breeds have dark-colored eyes and white hair and skin.

Coat color and patterns are controlled by a common genetic mechanism in many different mammals (Murray, 1988). For example, many domestic animals have a white blaze on the forehead and a white tail tip. Dogs, cats, and cattle are all found to have this pattern. Schaible (1969) and Schaible and Brumbaugh (1976) explain that migration of pigment-producing cells occurs during embryonic growth, and common patterns between species frequently occur. For example: mice can be selected for a white belt on a mostly black body. This pattern is very similar in appearance to that of Hampshire pigs and Belted Galloway cattle. The common white areas in all three species are characterized by a lack of pigment, both in the hair and in the skin beneath the hair. In domestic horses, depigmented facial and leg markings were not present, or were very rare in their wild ancestors (Guerts, 1977). Woolf (1990, 1991, 1995) suggests that white markings on the face or legs provide less color protection from predators. However, flashy white markings are favored by some breeders, and Forbis (1976) reports that common white markings in the Arabian have been present since ancient times. Attractive white markings on the legs and forehead of horses are often deliberately selected for. In contrast, depigmentation in production livestock and poultry is possibly a secondary effect of selection for production traits. Many domestic animals are partially depigmented, especially high-producing animals. A completely depigmented animal is albino with pink eyes. A good example is white rats and mice. Several food-producing animals are partially depigmented, and have either colored eyes or some areas of the body that are not white. The highest-producing pigs, milk cows, and chickens are either white or partially white. The maternal sow lines of commercial pigs are mostly white. High-producing commercially bred turkeys and broiler chickens have white feathers and white skin.

Depigmentation is both good and bad. Partial depigmentation in varying degrees is highly correlated with production traits. Partial depigmentation is also linked to behavior. The two calmest breeds of cattle, Herefords and Holsteins, have depigmented white areas on their heads. However, large areas of depigmentation often appear in conjunction with serious developmental or neurological problems. Many veterinarians report that mostly white Holstein dairy cows are more difficult to handle and more nervous than Holsteins with larger pigmented areas of hair and skin. Also, a genetic disorder called "white heifer disease" is commonly found in white heifers of the Shorthorn breed, or in breeds related to them (Rendel, 1952; Spriggs, 1946a,b). White Shorthorn heifers are usually sterile (Spriggs, 1946a,b). The ovaries and external genitalia are normal but the uterus and the vagina are not completely developed.

Depigmentation, Behavior, and Defects

There is definitely a relationship between depigmentation and behavior. When Belyaev (1979) selected foxes for tameness, the animals developed a piebald coat pattern with areas of depigmented white fur. Depigmentation of skin, hair, and eyes is related to the development of the nervous system. Searle (1968) found a relationship between depigmentation and deafness in several rodent species. In deer mice, Cowling *et al.* (1994) report a relationship between deafness and the amount of depigmentation on the head. The mice also had ataxia (staggering) and retinal problems. Mice with the most extensive white areas were most likely to be deaf. Also in mice, the Variant-Waddler is a neurological mutant strain where individuals are hyperactive, deaf, and display an abnormal circling behavior. A remarkable relationship was found between asymmetrical spotted areas of fur and the preferred direction of circling (Cools, 1972a,b). This author postulated that asymmetry in the fur pattern and laterality in circling are expressions of the same basic neurological disturbance.

It is well known that white cats with blue eyes are often deaf (Bergsma and Brown, 1971; Schaible and Brumbaugh, 1976). In both dogs and cats, a white or piebald coat color is also related to an increased incidence of deafness (Schaible and Brumbaugh, 1976; Sorsby and Davey, 1954). Bergsma and Brown's (1971) review of the literature discussed a continuum in the relationship between the depigmentation and deafness. White cats with dark eyes or dark-pigmented areas of hair and skin on the head are less likely to be deaf, in comparison to white cats with blue eyes. Hudson and Rubin (1962) and Schaible and Brumbaugh (1976) both discuss deafness in Dalmatians, a heavily depigmented breed. Thirty per cent of Dalmatians are deaf (Strain *et al.*, 2009). Dalmations with the most extensive white areas are most likely to be deaf. In addition to Dalmations, other extensively depigmented dog breeds, such as pale colored Australian shepherds have a high incidence of brain, ear, and eye abnormalities (Schaible and Brumbaugh, 1976; Sponenberg and Lamoreaux, 1985; Strain *et al.*, 2009; Strain, 2011). A higher incidence of depigmentation and deafness is also found in albino cattle (Leipold and Hutson, 1962). Paint horses can also have neurological abnormalities (Figure 12.2). The authors speculate that selection for a moderate amount of depigmentation appears to provide production advantages, but excessive selection will cause serious welfare problems.

Depigmentation is also related to abnormalities in the visual system. Guillery (1974) reports that many albino mammals have visual pathways with crossed connections to the brain. Siamese cats, which are partially albino, have crossed visual pathways (Guillery *et al.*, 1971). The crossed

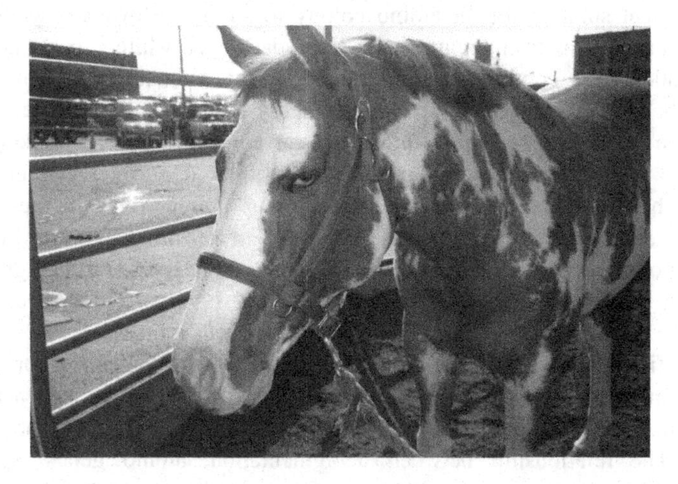

FIGURE 12.2 A paint horse with blue eyes has a nervous switch that resembles Tourettes (the disorder) in humans. Overselection for appearance traits is sometimes linked to neurological problems.

visual pathways may not compromise welfare of Siamese cats kept as pets because they adapt well to their miswired visual systems (Guillery, 1974). However, visual tests that were more demanding than shape discrimination revealed that albinos have subtle visual perception deficiencies. Sheridan (1965) found that black and white hooded rats with pigmented eyes performed better than albinos on a visual discrimination task with one eye covered. Compared to the albinos, the hooded rats performed the task better when they had to switch eyes. This principle is also vividly shown in mice. Guillery *et al.* (1971) found that flecked mice with variegated areas of both pigmented and depigmented areas of skin and fur have normal visual pathways to the brain, even though they are roughly 50% albino. The white areas of skin and hair are albino and the dark areas are pigmented. The welfare of laboratory albino rats is not compromised because they eat, drink, and mate normally. However, abnormalities of the visual system and auditory systems may cause functional problems and could possibly confound the results of research on vision or hearing. Albino rats have more unmyelinated nerve fibers in the optic system compared to hooded rats (Lund, 1973). In albinos, the visual cortex is also thinner (Lund, 1973). This is a further indication of a correlation between nervous system abnormalities and depigmentation.

Highly depigmented or albino animals are rare in nature. When they do occur, the animals generally have difficulty surviving in the wild. One brief reference in *Science* by Minckler and Pease (1938) refers to a colony of albino rats living under feral conditions. These rats inhabited an area at a local dump in Montana. The exact source of these animals is unknown, but it was presumed that students from the local university released them. Abundant food, water, shelter, and a lack of predators created a sheltered environment suitable for an albino colony to survive. However, white cats depend on humans to maintain them. Even dark-eyed white cats do not survive under feral conditions. Furthermore, many animals with genetically based behavioral problems are often highly depigmented. For example, Viennese white rabbits are known to have seizures (Hohenboken and Nellhaus, 1970). In the paper on nervous pointer dogs by Dykman *et al.* (1969), both authors noticed that five nervous pointer dogs in the photographs were almost all white. These dogs had greater areas of white fur than typical pointers. It would be interesting to measure the amount of depigmented areas on the bodies of both normal and nervous pointers and then compare these to Dykman *et al.*'s (1969) continuum of nervous traits.

The first author has observed possible signs of over-selection for depigmentation in high-producing, commercially bred hybrid pigs. The nervous, excitable pigs discussed earlier are usually very lean and are white or mostly white. The relationship between depigmentation, albino genes, spotting genes, and nervous system problems is documented extensively in mammals (Arman, 2007; Silversides and Smyth, 1986). The link in poultry is much weaker. Harper *et al.* (1988) found that the "bobbler" defect in turkeys is not linked to feather color. The "bobbler" defect involves abnormalities in the inner ear and balancing difficulties. However, Silversides and Smyth (1986) found a relationship between lighter feather color and congenital tremor. The affected birds had abnormal myelin in the cerebellum of their brain and a reduction in melanin pigment in the eyes and feathers. A possible explanation for less of an effect in poultry might be due to the fact that baby chicks of the white poultry breeds are born with yellow down and adult white poultry have yellow-pigmented legs. The mutation that makes some of the feathers white in bronze colored Leghorns is associated with weakness and the homozygous mutant does not survive (Somes, 1979). Photos accompanying this paper show that the partially white mutants also have very pale (almost white) feet and beaks as adults. Albinism and depigmentation in most mammals causes abnormalities in the rods in the retina that are used for night vision. However, albino birds do not have retinal abnormalities (Jeffery, 1997). Unlike most mammals, birds have a cone (color-vision)-dominant retina and the development of the cones is unaffected. A few albino mammals, such as the squirrel, have relatively normal retina development because cones predominate (Jeffery, 1997).

More recently, Oskina *et al.* (2009) studied the involvement of glucocorticoid hormones in the appearance of white spots during embroygenesis. A delay in the migration and development of melanoblasts (pigment cells) in the embryos of fully pigmented gray rats was caused either by restraint stress (emotional) or by administering dexamethasone in drinking water. Dexamethasone is a potent synthetic member of the glucocorticoid class of steroid drugs. At days 12–14 of gestation, pregnant females were assigned to a treatment plan that consisted of being restrained in a small plastic cage for 45 minutes three times a day, or in the second treatment dexamethasone was given to the pregnant rats in drinking water. In both treatment groups, three times more offspring were born with large areas of depigmented skin and hair on the ventral side of the body compared to the untreated control groups. It was also demonstrated that in adult offspring of mothers treated with dexamethasone, the response of the hypothalmo–pituitary–adrenal axis to emotional stress was lower. Prenatal emotional stress in fully pigmented mother rats caused a rise in glucocorticoid (stress hormone) which is associated with digmented areas of skin and hair of the offspring. There is also a down-regulation of the HPA-axis leading to a reduced response to stress in adult offspring. The same effects were shown by artificial administration dexamethasone (Oskina *et al.*, 2009).

FLUCTUATING ASYMMETRY AND GENETIC PROBLEMS

Another possible warning sign of both genetic problems and detrimental effects of the environment on development is greater fluctuating asymmetry of an animal's body (Markow, 1992). Fluctuating symmetry (FA) refers to random deviations from bilateral asymmetry, and can be measured in the body—as in bilateral symmetry of finger or foot lengths—or in a particular organ. FA is a key concept in evolution and development and underlies the theory of developmental stability—the ability to maintain normal development under stress. Animals with high levels of FA have reduced developmental stability. Deviations from perfect bilateral symmetry can be caused by environmental stress, or genetic problems during development, leading to developmental instability (Graham *et al.*, 2010). Conservation biologists use measurements of body symmetry to assess detrimental effects of environmental pollution in rainbow trout (Leary and Allendorf, 1989), chickens (Buijs *et al.*, 2012; Erikson *et al.*, 2003; Moller *et al.*, 1995), and horses (Manning and Ockenden, 1994). Graham *et al.* (2010) discuss the importance of taking body measurements on several different body locations to get an accurate measure.

Fluctuating asymmetry is also affected by purely genetic causes. Parsons (1990) states that inbreeding increases the degree of asymmetry in a variety of animals. The teeth of hybrid mice are more symmetrical compared to inbred mice (Bader, 1965). Symmetrical trout have greater heterozygosity and gene variation (Leary and Allendorf, 1989). Asymmetry also affects an animal's performance and adaptability. Manning and Ockenden (1994) measured the differences in the width and length of knees, teeth, ears, coronet bands, upper legs, and six measurements of the heads of racehorses. More symmetrical horses performed better on the racetrack. Symmetry is also involved in sexual selection. Moller (1992, 1993) found that female swallows preferred males with long, symmetrical tails.

There also appears to be a relationship between domestication and fluctuating asymmetry. Domestic animals tend to be more asymmetrical than wild animals (Parsons, 1990). Fluctuating asymmetry and minor physical abnormalities are signs of greater problems. Human children with developmental abnormalities such as mental retardation and autism have a greater percentage of minor physical abnormalities than the general population (Links *et al.*, 1980; Steg and Rapport, 1975). They are also more likely to have greater asymmetry (Malina and Buschang, 1984). Assessments of body asymmetry may be one simple way to provide an early warning of potential problems with a selective breeding program. Intuitively, dairy cow breeders for years have evaluated various body parts for symmetry. For example: dairymen routinely look at traits such as a well-formed, symmetrical udder. All major dairy bull studs do this because they intuitively know that symmetry is a good trait.

An important conclusion can be made. Over-selection for a single trait ruins the animal. In breeding animals, people must be aware of the complex interaction between traits that do not appear to be related.

WELFARE AND GENETIC SELECTION

At what point has genetic selection for a certain trait gone too far? Changing an animal's diet or housing conditions may provide decent welfare under some conditions, but not others. There may be a point at which the animal is so defective that it will suffer even when it is fed and housed in an ideal environment. If genetic selection for milk production goes too far, it is likely that Holstein cows may develop serious health problems which cannot be compensated for by improving their nutrition or housing. The excessive metabolic drain on dairy cows, and the amount of work required during peak lactation is so great that a human doing a comparable amount of work would have to exercise vigorously for 6 hours every day (USDA, 2008).

Animals can be altered by genetic selection to such an extent that serious structural or neurological defects develop that can cause great discomfort. For example, animals with bowed legs or other structural abnormalities may have difficulty walking. Both authors agree that animal welfare is not acceptable if high-producing meat animals are chronically lame due to weak legs. It is also our opinion that an adequate level of animal welfare is not possible if selected traits becomes so extreme that obvious mobility problems occur, or if the condition is known to be painful in humans. A good example of this is arthritis.

One may argue that measuring how an animal feels is very difficult. However, the nervous system of pigs and cattle has the same basic design as the human nervous system. Careful studies of different mammals reveal that surgical procedures do cause pain in animals (Molony *et al.*, 1995; Short and Poznak, 1992). Noxious stimuli that cause pain in people are also likely to cause pain in animals (Molony *et al.*, 1995; Short and Poznak, 1992). Both chickens and rats self-medicate to relieve pain by eating or drinking a bitter-tasting painkiller (Colpaert *et al.*, 2001; Danbury *et al.*, 2000). When a lesion in their joint heals, they stop ingesting the painkiller and switch to plain feed or water. This indicates that animals definitely feel pain (Danbury *et al.*, 2000).

Both authors suggest guidelines for determining the ethical limit for genetic selection. Animals should have freedom from pain due to structural abnormalities. Over-developed meat-producing muscle that causes fatigue, or an animal becoming non-ambulatory during handling is not acceptable. Welfare is also seriously compromised in females unable to give birth naturally and forced to endure severe dystocias. There should also be low levels of behavior detrimental to welfare, such as hyperexcitability during handling or high levels of aggression. Some examples include heart attacks in pigs homozygous for PSS, hyperexcitable pigs that constantly back up in races and bunch and pile up at the slaughter plant, and nervous pointers that have difficulty adapting to normal dog environments. Another example is an inbred strain of mice that does not habituate to repeated exposure to a novel environment (Boleij *et al.*, 2012).

Since these mice lack normal habituation, their welfare may be severely compromised during standard laboratory procedures. To maintain an acceptable level of welfare, these mice need to be housed in conditions where exposure to sudden novelty is minimized. Severely compromised welfare can occur in these high-fear mice if not slowly trained to tolerate novelty. Selecting for excessive appetite motivation may cause stress, stereotypies, and other abnormal behaviors, especially in feed-restricted breeding animals. If stereotypies develop in large numbers of breeding animals housed and fed under the best conditions, genetic selection may have gone too far in the wrong direction. In some situations, a genetic line with low levels of repetitive stereotypies had higher-fear temperaments (Hansen and Jeppensen, 2006; Svendsen *et al.*, 2007). More recent research has shown that relationships between temperament and high and low stereotypies are not simple. Minks showing low fear when fear was measured by willingness to explore a stick stuck in their cage may become very frightened when fear is measured with a different method. Minks with high levels of stereotypies had higher fecal cortisol metabolites after they were immobilized (Malmkwist *et al.*, 2011). High stereotyping minks had more kits, but the low stereotyping minks had fewer, better-quality kits (Meagher *et al.*, 2012). Maybe the high stereotyping minks were high "seekers" and more willing to approach novelty (see Chapter 1). Malmkwist *et al.* (2011) conclude that minks with high stereotypies may be more sensitive to stress. An animal demonstrating continuous stereotypies has severe difficulty with basic functioning.

Siamese cats with mildly crossed eyes have good welfare, but animals bred to have constant epileptic seizures may have poor welfare if not housed in an environment free of seizure triggers. It is clear that animal welfare problems have occurred in some animals as a result of overzealous selection for a single trait. Many dog breeds have structural problems that compromise welfare (Ackerman, 2011; Arman, 2007; Ott, 1996). Some examples are eye problems in collies, back problems in dachshunds, breathing problems in bulldogs, and hip problems in many breeds. Many of these traits can be selected without compromising welfare if selection is done in moderation. Animal breeders need to take a more holistic view when selecting breeding stock, and think about optimizing the whole animal instead of selecting for a few traits. This would help prevent the mistakes made by pig breeders discussed earlier in this chapter. Pig breeders selected for production traits and forgot to examine animals for structural soundness, and failed to see the leg problems until half of the pigs became lame.

The science of genetic engineering now makes it possible to delete genes or swap genes between species. A major difference between conventional selective breeding and genetic engineering is that the process of selection is greatly sped up (Fox, 1989, 1992; Grandin, 1991; Rollin, 1995). The speed of change is accelerated to a point where less time is available to take corrective action if selection mistakes are made.

USE BIOTECH CAREFULLY

Genetic engineering is used to make animal models for various human diseases by inserting genes into animals. Today there are hundreds of genetically engineered mice for use as animal models for human neurological conditions (Crawley, 2007; McFarlane *et al.*, 2007; Wellberg, 2011; Walberg, 2011). For example, mice that have autistic-like behaviors can be purchased from a commercial company. The "autistic" mice have increased repetitive behavior, such as self-grooming (McFarlane *et al.*, 2007). To avoid a possible welfare problem in this animal model, the mice should be carefully monitored to ensure that excessive self-grooming does not turn into self-injurious behavior. Mice altered by removing a gene involved in the regulation of serotonin display behavior that resembles obsessive–compulsive disorder (OCD) in humans. The mice altered by genetic engineering methods ran around repeatedly in the same alley of a maze. Normal mice will quickly try a new alley (Chen *et al.*, 1994). Hooper *et al.* (1987) created a mouse model for Lesch-Nyhan syndrome. Children afflicted with Lesch-Nyhan syndrome self-mutilate and bite themselves. If this model is developed to the point at which the syndrome is fully expressed, the welfare of the mice will be severely compromised (Rollin, 1995). Transgenic cattle and pigs are also being used to produce substances useful in human medicine (Murray *et al.*, 2010). For readers interested further in this topic, there is an excellent illustrated book to help technicians monitor health and welfare of mice, rats, and rabbits (Pritchett-Corning *et al.*, 2011).

It is beyond the scope of this chapter to discuss all the ethical arguments about genetic engineering, but there is a greater ethical justification to cause some suffering in a mouse if the mouse model makes it possible to cure Lesch-Nyhan syndrome. This syndrome is one of the most horrible genetic conditions that can affect a child. It is the opinion of both authors that causing pain in order to make a pig grow slightly faster is absolutely not acceptable.

Unexpected Linked Traits

Genetically altering an animal by knocking out selected genes can affect the behavior of the animal in unexpected ways. Researchers and commercial companies using biotechnology should be careful and not forget that changing a targeted piece of genetic code can have unexpected effects. Researchers are cautioned to avoid repeating the mistakes we reviewed in the first edition of this book. For example, Silva *et al.* (1992a,b) used genetic engineering methods to knock out the gene that encodes for a substance called calcium/calmodulin-dependent kinase II. This substance is involved in the mechanisms of learning. The "knock-out" mutant mice displayed impaired spatial learning when tested in a water maze that required them to find a platform hidden under the surface of milky-colored water. The impaired spatial learning was similar to that of normal mice with hippocampus lesions. The hippocampus is a brain structure crucial for forming memories. However, Silva *et al.* (1992a,b) found that knocking out this gene affected more than just learning: the knock-outs also had an abnormally enhanced acoustic startle response, and were more likely to jump when they heard a sudden loud noise. In 1994, researchers in another lab made a startling discovery of these same knock-out mutant mice. Mice were sometimes found dead in their cages with broken backs. The fighting behavior of these mice was so extreme that they suffered broken backs (Chen *et al.*, 1994). This is another example of an unexpected effect. Knocking out the gene for calcium/calmodulin kinase II affected more than just learning: it also completely removed fear causing uninhibited fighting. Calcium/calmodulin kinase II is also involved in the regulation of the neurotransmitter serotonin. The knock-outs had normal behavior except in the area of defensive aggression. Defensive aggression was measured by quantifying attacks by a resident mouse against a mouse that was placed in its cage. Offensive aggression was measured by determining how hard an intruder mouse fought a resident. Since the knock-outs had no fear, they continuously attacked a resident mouse when placed in its cage. The heterozygote knock-outs with one knock-out gene had abnormally high defensive aggression (Chen *et al.*, 1994). Offensive aggression, which is attack aggression, was attenuated in the genetically engineered mutants.

Researchers in another laboratory were also surprised by unexpected results when a specific gene was blocked. Giros *et al.* (1996) found that blocking the effects of the dopamine (neurotransmitter) gene in mice had both predicted and unexpected results. Blocking the effect of the gene made the mice produce an overabundance of dopamine in their brains. Dopamine excess is believed to be responsible for some of the symptoms of schizophrenia in people. As expected, homozygous mice had two copies of the blocked gene were hyperactive and ran around more compared to the heterozygous (one copy of the gene), or normal, wild-type mice. Neither type of genetically altered mice had stereotypies. An unexpected finding was that injections of amphetamines had no effect on the homozygotes, and an eight-fold increase in locomotor activity in the heterozygotes and normal, wild-type mice (Giros *et al.*, 1996). This experiment provided important insights into how pharmaceuticals work, and showed that the effects of drugs on behavior are not simple and straightforward. Similar effects of fear may operate in mice bred and selected by conventional methods. Male mice that were bred to have heightened reactivity to stimulation were less aggressive (Gariepy *et al.*, 1988). Since the first edition of this book, researchers have developed more sophisticated behavioral tests for use with all the new mouse models (Crawley, 2007).

SPECULATIONS OF GENETIC ENGINEERING

What if, through the use of genetic engineering, an animal was created that would feel no pain and have no fear? What would be the ethical implications of this? The work of Chen *et al.* (1994) showed very clearly that knock-out mice with no fear when mixed with unfamiliar mice caused serious welfare problems due to fighting and injury.

In contrast, selecting for extreme docility with conventional breeding methods could cause similar problems. Ranchers have observed that placid Hereford bulls spend more time fighting with other bulls to determine their social rank compared with flighty, excitable Salers bulls. The more excitable animals are too fearful to fight. The second author has made similar observations in horses. The calmest horse on a pasture is more likely to bully the other horses. Timid horses avoid a confrontation that could lead to injury. There may be an optimum temperament between extreme docility and flightiness. Motivation is complex. In Chapter 8, Faure's research shows that fear is not a simple trait. Fear behaviors may look similar but have different motivations. For example, feather pecking in hens is probably misdirected foraging behavior instead of true aggression. Cattle and pig breeders need to select for a calm temperament and cull individuals that panic when confronted with novelty, but it may be a mistake to select for the absolutely calmest animals. An extremely docile cow may not take care of her calf (see Chapter 5). In the Chen *et al.* (1994) experiment, increased defensive aggression only occurred in mice heterozygous for the knock-out gene. Mice with both calcium/calmodulin kinase II genes knocked out had overall attenuated aggression and many other abnormal behaviors. The behavior in these mice differed depending on whether one copy, or both copies of the gene were knocked-out. This research clearly shows that knocking out genes is not going to provide clear-cut results. As with natural breeding, traits are linked. The behavior of an animal is the sum of a complex interaction between inherited traits and those traits' interactions with the environment.

GENETIC DIVERSITY

Another serious concern brought up by ethicists and geneticists is the problem of restricting the gene pool. This problem is not limited to genetic engineering. Many scientists are concerned about the loss of genetic diversity (Notter, 1996; Cundiff *et al.*, 1996). Larry Cundiff, a researcher who works for the U.S. Department of Agriculture, stated that pig breeds in China are more heterozygous. Preserving the genetic diversity of Chinese pig breeds is important.

The loss of genetic variability in agricultural plants and animals could result in disaster if disease struck a susceptible high-producing genetic line. Nothing is free. If one selects for just one trait there is usually a price to be paid by weakening another trait. In 1977, the American Livestock Breeds Conservatory was founded to preserve rare breeds of domestic livestock and poultry. However, some agriculturists do not see the wisdom of keeping older and lower-producing breeds of pigs and chickens. The reason for conserving older breeds is they retain desirable traits such as disease resistance, fertility, and hardiness (Thomas, 1995; Sponenberg, 1995).

The more specialized a domestic animal becomes, the more specialized an environment it will require. A Holstein cow requires more environmental support by humans than a beef cow. Beef cows and horses can still go feral and survive under wild conditions. The first author has observed feral cows on mountain ranches that were able to avoid being rounded up for up to 10 years. The Holstein milk cow, with her huge udder, would have difficulty living under natural conditions. In addition, the first author has observed that Holstein calves are weaker and take longer to walk through races and chutes compared to beef-breed calves.

In the plant kingdom, corn is an example of a totally domestic plant. It cannot grow without assistance from humans. Modern corn is radically different from its ancestor, teosinte. Teosinte had hard kernels covered with an inedible shell, whereas modern corn has more kernels and no hard shell over them. Research by Jane Dorweiller at the University of Minnesota showed that converting teosinte into modern corn involved changing just a small stretch of DNA. Plants have been modified more by domestication than animals.

ETHICAL QUESIONS

One may speculate, what if a farm animal breed was manipulated as much as corn has been? Modern corn breeds look like a different species compared to wild corn. Would it be ethical to create microcephalic cows with almost no brain? Dr Mike Fox (1989, 1992) was the first to ask this question. If the nervous system was modified so the mutant would not suffer, would it be ethical? Philosophers can fight over this one.

Being practical people with years of hands-on experience with animals, both authors agree that decisions on the ethical use of biotechnology should be based on the concept of ethical cost. Invasive or painful experiments or the creation of animals with chronic pain should not be done for frivolous reasons. It may be justified to cause some pain or discomfort in an animal to find a cure for multiple sclerosis, but it would not be morally justified to make animals suffer in order to grow larger amounts of meat or produce a few more kilograms of milk.

Furthermore, it is our opinion that extremely fearful or aggressive agricultural animals are not acceptable from an animal welfare standpoint. We are also very concerned about the welfare of breeding animals selected for increased appetite. In the laboratory, genetically altered animals can provide tremendous knowledge about both genetics and the nervous system if researchers take a few simple precautions, most mutants can be maintained in a laboratory with good welfare. The "no-fear" mice would have an adequate level of welfare if caretakers made sure not to mix unfamiliar mice together. However, homozygous mutants with hyper-locomotor activity may be more compromised. Breeding large numbers of these mice may not be ethically justified, but breeding in small numbers may be justified to learn about mechanisms of neurotransmitter pathways and that could lead to development of new medications and treatments for many serious disorders. A researcher sensitive to the welfare of the animals in his laboratory could maintain mutant breeding stock with large numbers of heterozygotes, which have more normal behavior, and keep just enough homozygotes for ongoing experiments.

The Farm *Versus* the Laboratory

A big difference exists between hundreds of mice in a laboratory and thousands of pigs or chickens used for the commercial production of animal protein. Since animals on commercial farms often are not managed as carefully as research animals, genetic characteristics causing minor welfare problems in a controlled laboratory setting could cause horrible suffering on a farm. This would be especially true if the farm was poorly managed. If a "no-fear" pig was developed, on farm welfare could be deplorable because the animals might seriously injure each other during fights. New research shows that fear is not a single trait. Different types of fear are controlled by different brain circuits (Gross and Canteras, 2012). Selective removal of different fear pathways could have different effects. On the other hand, if aggression was totally removed to eliminate fighting, new problems might occur. Maybe a fearless, totally nonaggressive cow or sow would fail to take care of her babies, or totally non-emotional mutants would be too lazy to fully graze a pasture. The authors conclude that one should be more cautious when genetically altering an agricultural animal compared to a laboratory animal. Even when animals are bred naturally, one must be careful to ensure that the animal fits into its environment. LeNeindre *et al.* (1996) found that cattle and sheep been bred for intensive conditions, may have behavioral problems when housed extensively. For example: gentle cows under intensive conditions may become aggressive toward humans when reared on large pasture. This raises the question: is it preferable to engineer animals to fit industrial systems, or rather to engineer systems that fit the animals in the first place?

In 1991, the first author was asked to present a paper on biotechnology and meat production (Grandin, 1991). In this paper, a brief history of science was reviewed. Scientists in the past were persecuted and sometimes killed for discussing and discovering forbidden knowledge. Yesterday's forbidden knowledge is today's accepted fact. Galileo was persecuted for writing that the Earth was not the center of the universe. One should remember, in the past medical knowledge was held up for a thousand years due to prohibition against dissection. The editor of *Science* suggested that we must proceed with biotechnology, but we should proceed with caution (Koshland, 1989). Biotechnology has proceeded. Genetically modified corn and soybeans are routinely used around the world, but biotechnology is still controversial. Many people fear GMOs (genetically modified organisms) in their food. Golden Rice, which has been genetically modified to biosynthesize beta-carotine could help prevent blindness in Africa, and genetically engineered food animals are still not being used in the food supply. While editing the last part of this chapter, we removed a sentence we wrote in 1998, which read, "Today, genetic engineering is controversial, but tomorrow it may be routine." Fifteen years later, it is still controversial. In the U.S., research funding for genetically modified food animals has ground to a halt (Maxmen, 2012). In China and Brazil research on genetically modified animals is continuing on projects such as fast-growing carp, cows with less allergenic milk, and goats with milk that can be used to treat diarrhea. On the other hand, genetic technologies could also be used to improve welfare. An example is using genetic engineering to remove horns from dairy cows so farmers will not have to perform painful removal (Maxmen, 2012). Colorado State University Professor Bernard Rollin, professor of animal sciences, biomedical sciences, and philosophy, has introduced as a guiding principle the concept of "conservation of welfare": when genetically engineering animals, the transgenic animals should be no worse off afterwards than their parents were (Pew Initiative on Food and Biotechnology, 2005).

We end this chapter with the last two sentences of the first author's paper given in 1991.

We should proceed cautiously, but we should definitely proceed. Biotechnology can be used for noble, frivolous, or evil purposes. Decisions on the ethical use of this powerful new knowledge must not be made by extremists or people motivated purely by profit.

REFERENCES

Aalhus, J.L., Jones, S.D.M., Robertson, A.K.W., Tong, K.W., Sather, A.P., 1991. Growth characteristics and carcass composition of pigs with known genotypes for stress susceptibility over a weight range of 70 to 120 kg. Anim. Prod. 52, 347–353.

Ackerman, L.A., 2011. The genetic connection: a guide to health problems in purebred dogs. Am. Anim. Hosp. Assoc.

Ahlstrom, S., Jarvis, S., Lawrence, A.B., 2002. Savaging gilts are more restless and responsive to piglets during expulsive phase of parturition. Appl. Anim. Behav. Sci. 76, 83–91.

Appleby, M.C., Lawrence, A.B., 1987. Food restriction as a cause of stereotypic behavior in tethered gilts. Anim. Prod. 45, 103–110.

Arman, K., 2007. A new direction for kennel club regulations and breed standards. Can. Vet. J. 48, 953–965.

Bader, R., 1965. Fluctuating asymmetry in the dentition of the house mouse. Growth 29, 291–300.

Barton-Gade, P., 1984. Influence of Halothane genotype on meat quality in pigs subjected to various preslaughter treatments. Proceeding of the International Congress on Meat Science and Technology 30th Bristol, England pp. 8–9.

Baxter, E.M., Jarvis, S., Sherwood, L., Parrish, M., Roebe, R., Lawrence, A.B., et al., 2011. Genetic and environmental effects on piglet survival and material behavior of the furrowing sow. Appl. Behav. Sci. 130, 28–41.

Belyaev, D.K., 1979. Destabilizing selection as a factor in domestication. J. Hered. 70, 301–308.

Belyaev, D.K., Borodin, P.M., 1982. The influence of stress on variation and its role in evolution. Biol. Zentralbl. 100, 705–714.

Bergeron, R., Gonyou, H.W., 1997. Effects of increasing energy intake and foraging behaviors on the development of stereotypics in pregnant sows. Appl. Anim. Behav. Sci. 53, 259–270.

Bergsma, D.R., Brown, K.S., 1971. White fur, blue eyes, and deafness in the domestic cat. J. Hered. 62, 171–185.

Boersma, S., 2001. Managing rapid growth rate in broilers. World Poultry 17 (8), 1388–3119.

Boleij, H., Salomons, A.R., Sprundel, M., Arndt, S., Ohi, F., 2012. Not all mice are created equal. Welfare implications of behavioral habituation profiles in four 129 mouse substrains. PLOS One. 7 (18), e42544.

Breuer, K., Sutcliffe, M.E.M., Mercer, J.T., Rance, K.A., O'Connell, N.E., Sneddon, A., et al., 2004. Heritability of clinical tail biting and its relation to performance traits. Livest. Sci. 93, 87–94.

Broom, D.M., 1987. Applications of neurobiological studies to farm animal welfare. Curr. Top, Vet. Med. Anim. Sci. 42, 101–110.

Broom, D.M., 1993. A usable definition of animal welfare. J. Agric. Environ. Ethics 6, Spec. (Suppl. 2), 15.

Brunberg, E., Jensen, P., Isaksson, A., Keeling, L., 2011. Feather pecking behavior in laying hens, Hypothalamic gene expression in birds performing and receiving pecks. Poult. Sci. 90, 1145–1152.

Buchanan, D.S., 1987. The crossbred sire: experimental results for swine. J. Anim. Sci. 65, 117–127.

Buijs, S., Van Poucke, E., Van Dongent, S., Lens, L., Baert, J., Tuyttens, F.A.M., 2012. The influence of stocking density on broiler chicken bone quality and fluctuating asymmetry. Poult. Sci. 8, 1759–1767.

Chen, C., Rainnie, D.G., Greene, R.W., Tonegawa, S., 1994. Abnormal fear response and aggressive behavior in mutant mice deficient for a calcium-calmodi kinease. II. Science 266, 291–294.

Chen, C., Gilbert, C.L., Yang, G., Gup, Y., Segonels-Pichon, A., Ma, J., et al., 2008. maternal infanticide in sows: incidence and behavioral comparisons between savaging and nonsavaging sows at parturition. Appl. Anim. Behav. Sci. 109, 238–248.

Close, W.H., 1996. Nutritional management of swine is an ever-evolving challenge. Feedstuffs. January 22, pp. 16–19, 46–47.

Close, W.H., Noblet, J., Heavens, R.M., 1985. Studies on the energy metabolism of the pregnant sow. 2. The partition and utilization of metabolizable energy intake in pregnant and nonpregnant animals. Br. J. Nutr. 53, 267–279.

Cockett, N.E., Jackson, S.P., Shay, T.L., Farnir, F., Berghmans, S., Snowder, G.D., et al., 1996. Polar overdominance at the *Ovine callipyge* locus. Science 273, 236–238.

Colpaert, F.C., Taryre, J.P., Alliga, M., Kock, W., 2001. Opiate self-administration as a measure of chronic nocioceptive pain in arthritic rats. Pain 91, 33–34.

Cools, A.R., 1972a. Neurochemical correlates of the waltzing-shaker syndrome in the variant-waddler mouse. Psychopharmacology (Berlin) 24, 384–396.

Cools, A.R., 1972b. Asymmetrical spotting and the direction of circling in the Variant-Waddler mouse. J. Hered. 63, 167–171.

Cooper, M.D., Wrathall, J.H.M., 2010. Assurance schemes as a tool to tackle genetic welfare problems in farm animals: broilers. Anim. Wel. 19, 51–56.

Corr, S.A., Gentle, M.J., McCorqoudale, C.C., Bennett, D., 2003. The effect of morphology on walking ability in the modern broiler: a gait analysis study. Anim. Wel. 12, 159–171.

Cowling, K., Robins, R.J., Haigh, G.R., Teed, S.K., Dawson, W.D., 1994. Coat color genetics of Peromyscus. IV. Variable white, a new dominant mutation in the deer mouse. J. Hered. 85, 48–52.

Crabbe, J.C., 1983. Sensitivity to ethanol in inbred mice. Genetic correlations among several behavioral responses. Behav. Neurosci. 97, 280–289.

Crawley, J.E., 2007. What's wrong with my mouse? Behavioural Phenotyping of Transgenics and Knock-Out Mice, second ed. Wiley, Hoboken, New Jersey.

Cummings, D.E., Schwartz, M.W., 2003. Genetics and pathophysiology of human obesity. Annu. Rev. Med. 54, 453–471.

Cundiff, L., Young, L., Leymaster, K., 1996. Genetic utilization from the breed to the gene level in livestock. Paper Presented at the National Animal Germplasm Program Symposium, American Society of Animal Science Rapid City, South Dakota, July 24, 1996.

Danbury, T.C., Weeks, C.A., Chambers, J.P., Waterman-Peareson, A.E., Kestin, S.C., 2000. Self selection of the analgesic drug carprofen in lame broilers. Vet. Rec. 146, 307–311.

d'Eath, R.B., Roehe, R., Turner, S.P., Ison, S.H., Farish, M., Jack, M., et al., 2009a. Genetics of animal temperament Aggressive behavior at mixing is genetically associated with handling in pigs. Animal 3, 1544–1554.

d'Eath, R.B., Tolkamp, B.J., Kyriazykis, I., Lawrence, A.B., 2009b. Freedom from hunger and preventing obesity: the animal welfare implications of reducing food quantity and quality. Anim. Behav. 77, 275–288.

deHaas, E.N., Nielsen, B.L., Buitenhuis, A.J., Rodenburg, T.B., 2010. Selection for feather pecking affects responses to novelty and foraging behavior in laying hens. Appl. Anim. Behav. Sci. 124, 90–96.

Dejong, L.C., Guemene, D., 2011. Major welfare issues in broiler breeders. Worlds Poult. Sci. J. 57, 73–82.

del Campo, M., Brito, G., Soares de Lima, J., Hernandez, P., Montossi, F., 2010. Finishing diet, temperament and lariage time effects on carcuss and meat quality traits in steers. Meat Sci. 86 (4), 908–914.

Dellmeier, G.R., Friend, T.H., Gbur, E.E., 1985. Comparison of four methods of calf confinement, II. Behavior. J. Anim. Sci. 60, 1102–1109.

Dickerson, G.E., 1973. Inbreeding and heterosis in animals. In: Proceedings of the Animal Breeding and Genetics symposium in Honor of Dr. J.L. Lush. American Society of Animal Science, Champaign IL.

Dinnington, E.A., Siegel, P.B., Anthony, N.B., 1990. Reproductive fitness in selected lines of chickens and their crosses. J. Hered. 81, 217–218.

Dixon, L., Duncan, I.J.H., Mason, G.L., 2009. Using movement to determine motivation comparison of the motor patterns involved in feather pecking dust bathing and foraging. Br. Poult. Sci. Abstr. 5, 26–27.

Dixon, L.M., Duncan, I.J.H., Mason, G.J., 2010. The effects of four kinds of enrichment on feather pecking behavior in laying hens housed in barren environments. Anim. Welf. 19, 429–435.

Decuypere, E., Onagbesan, O., Swennen, Q., Buyse, J., Bruggeman, V., 2007. The endocrine and metabolic interface of geno-type nutrition interactions in broilers and broiler breeders. Worlds Poult. Sci. J. 63, 115–128.

Dobzhansky, T., 1970. Genetics of the Evolutionary Process. Columbia University Press, New York.

Dykman, R.K., Murphee, O.D., Peters, J.E., 1969. Like begets like. Ann. N.Y. Acad. Sci. 159, 976–1007.

Dykman, R.K., Murphee, O.D., Reese, W.G., 1979. Familial anthropophobia in pointer dogs. Arch. Gen. Psychiatry 36, 988–993.

Entrikin, R.K., Erway, L.C., 1972. A genetic investigation of roller and tumbler pigeons. J Hered 63, 351–354.

Erikson, M.S., Haug, A., Torjesen, P.A., Bakkan, M., 2003. Prenatal exposure to corticosterone impairs embryonic development and increases fluctuating asymmetry in chickens (*Gallus gallus domesticus*). Br. Poult. Sci 44 (5), 690–697.

Fiems, L.O., 2012. Double Muscling in cattle: genes, husbandry, carcasses and meat. Animals 2 (3), 472–506.

Fisikowski, K., Schwarzenbacher, H., Wysocki, M., Weigend, S., Preisinger, R., Kjaer, J.B., et al., 2008. Variation in neighboring genes of the dopaminergic and serofonergic systems affects feather pecking in hens. Anim. Genet. 40, 192–199.

Forbis, J., 1976. The Classical Arabian Horse. Liveright, New York.

Fox, M., 1989. Genetic engineering and animal welfare. Appl. Anim. Behav. Sci. 22, 105–113.

Fox, M., 1992. Super Pigs and Wondercorn. Lyons and Buford, New York.

Fraser, D., 1975. The effect of straw on the behavior of sows in either stalls. Anim. Prod. 21, 59–68.

Freeman, A.E., Lindberg, G.L., 1993. Challenges to dairy cattle management: genetic considerations. J. Dairy. Sci. 76, 3143–3159.

Gariepy, J.L., Hood, K.E., Cairns, R.B., 1988. A developmental–genetic analysis of aggressive behavior in mice (*Mus Musculus*): III Behavior mediation by heightened reactivity or increased immobility?. J. Comp. Physiol. 102, 392–399.

Giros, B., Jaber, M., Jones, S.R., Wightman, R.M., Caron, M.G., 1996. Hyperlocomotion and indifference to cocaine and amphetamine in mice lacking the dopamine transporter. Nature (London) 379, 606–612.

Gordy, J.F., 1974. American Poultry History 1823–1973. American Printing and Publishing, Madison, WI.

Graham, J.H., Shmuel, R., Hagit, H.-O., Eviatar, N., 2010. Fluctuating asymmetry: methods, theory, and applications. Symmetry 2, 466–540.

Grandin, T., 1991. Biotechnology and animal welfare. In: Fiems, L.O., Cottyn, B.G., Demyer, D.I. (Eds.), Animal Biotechnology and the Quality of Meat Production. Elsevier, Amsterdam, pp. 145–147.

Grandin, T., 1994. Solving livestock handling problems. Vet. Med. 89, 989–998.

Grandin, T., 2010. Improving Animal Welfare: A Practical Approach. CABI Publishing, Wallingford, Oxfordshire, UK.

Gross, C.T., Canteras, N.S., 2012. The many paths to fear. Nat. Rev. Neurosci. 13, 651–658.

Guerts, R., 1977. Hair Color in the Horse. Allen, London.

Guillery, R.W., 1974. Visual pathways in albinos. Sci. Am. 230 (5), 44–54.

Guillery, R.W., Amoron, C.S., Eighmy, B.B., 1971. Mutants with abnormal visual pathways: an explanation of anomalous geniculate laminae. Science 174, 831–832.

Haenlein, G.F.W., Holdren, R.D., Yoon, Y.M., 1966. Comparative response of horses and sheep to different physical forms of alfalfa hay. J. Anim. Sci. 25 (3), 740–743.

Hansen, S.W., Jeppensen, L.L., 2006. Temperament, stereotypies and anticipatory behaviour as a measure of welfare in mink. Appl. Animal. Behav. Sci. 99, 172–182.

Harper, J.A., Bernier, P.E., Savage, T.F., West, D.J., 1988. Bobber: a cervical ataxic mutant in the domestic turkey responsive to sunlight. J. Hered. 79, 155–159.

Hocking, P.M., 1993. Welfare of broiler breeders and layer females subjected to food and water control during rearing, Quantifying the degree of restriction. Br. Poult. Sci. 34, 53–64.

Hocking, P.M., Maxwell, M.H., Mitchell, M.A., 1993. Welfare assessment of broiler breeder and layer females subjected to food restriction and limited access to water during rearing. Br. Poult. Sci. 34, 443–458.

Hohenboken, W.D., Nellhaus, G., 1970. Inheritance of audiogenic seizures in the rabbit. J. Hered. 61, 107–114.

Hooper, M., Hardy, K., Handyside, A., Hunter, S., Monk, M., 1987. HPRT-deficient (Lesch-Nyhan) mouse embryos derived from germline colonization by cultured cells. Nature (London) 326, 292–295.

Hudson, W.R., Ruben, R.J., 1962. Hereditary deafness in the Dalmatian dog. Arch. Otolaryngol. 75, 213–219.

Inglauer, F., Beig, C., Dimigen, J., Gerold, S., Gocht, A., Seeburg, A., et al., 1994. Hereditary compulsive self-mutilating behavior in laboratory rabbits. J. Hered. 85 (3), 126–135.

Jeffery, 1997. The albino retina: an abnormality which provides insight into normal retinal development. Trends Neurosci. 20, 165–169.

Jensen, P., Builerhuis, B., Kjaar, J., Zanella, A., Mormede, P., Pizzar, T., 2008. Genetics and genomics of animal behavior and welfare: challenges and possibilities. Appl. Anim. Behav. Sci. 113, 383–403.

Julian, R.J., 1993. Ascites in poultry. Avian Pathol. 22, 419–454.

Julian, R.J., Friars, G.W., French, H., McQuinton, M., 1986. The relationship of right ventricular hypertrophy right ventricular failure and ascites to weight gain in broiler and rooster chickens. Avian Dis. 31 (1), 130–135.

Kalmendal, R., Bessei, W., 2012. The preference for high fiber feed in laying hens selected divergently on feather pecking. Poult. Sci. 51, 1785–1789.

Karunajeewa, H., 1987. A review of current poultry feeding systems and their potential acceptability to animal walfarists. Worlds Poult. Sci. 43, 20–32.

Kjaer, J.B., 2009. Feather pecking in domestic fowl is generally related to locomotor activity: implications for a hyperactivity disorder model in feather pecking. Behav. Genet. 39, 564–570.

Kjaer, J.B., Jorgensen, H., 2011. Heartrate variability in domestic chicken lines genetically selected on feather pecking behavior. Genes. Brain. Behav. 10, 747–755.

Klien, E., Steinberg, S.A., Weiss, S.R.B., Mathews, D.M., Uhde, T.W., 1988. The relationship between genetic deafness and fear-related behaviors in nervous pointer dogs. Physiol. Behav. 43, 307–312.

Knowles, T.G., Kestin, S.C., Haslam, S.M., Brown, S.N., Green, L.E., Butterworth, A., et al., 2008. Leg disorders in broiler chickens: prevalence, risk factors, and prevention. PLos One 3 (2), e1514.

Koshland, D., 1989. The engineering of species. Science 244, 1233.

Lawrence, A.B., Appleby, M.C., Illius, A.E., Macleod, H.A., 1989. Mearuring hunger in the pig using operant conditioning: the effect of dietary bulk. Anim. Prod. 48, 213–220.

Lawrence, A.B., Terlouw, E.M.C., 1993. A review of behavioral factors involved in the development and continued performance of stereotypic behaviors in pigs. J. Anim. Sci. 71, 2815–2825.

Leary, R.F., Allendorf, F.W., 1989. Fluctuating asymmetry as an indictor of stress: implications for conservation biology. Trends Ecol. Evol. 4, 214–217.

Leipold, H.W., Huston, K., 1962. A herd of glass-eyed albino cattle. J. Hered. 53, 179–182.

LeNeindre, P., Boivin, X., Boissy., A., 1996. Handling extensively kept animals. Appl. Anim. Behav. Sci. 49, 73–81.

Leman, A.D., Straw, B.E., Mengenling, W.L., 1992. Diseases of Swine. Iowa State University Press, Ames Iowa City. (p. 763).

Lerner, I.M., 1954. Genetic Homeostasis. Oliver & Boyd, London.

Li, Y.Z., Wang, L.H., Johnson, L.J., 2012. Sorting by parity to reduce aggression towards first parity sows in group-gestation housing systems. J. Anim. Sci. 90, 4514–4522.

Links, P.S., Stockwell, M., Abichandani, F., Simeon, J., 1980. Minor physical abnormalities in childhood autism. J. Autism. Dev. Disord., 273–285.

Lips, D., De Tavernier, J., Decuypere, E., Van Outryve, J., 2001. Ethical Objections to Caesareans: Implications on the Future of the Belgian White Blue. Preprints of EurSafe 2001: Food Safety, Food Quality, Food Ethics, Florence, Italy.

Lovendahl, P., Damgaard, L.H., Nielsen, B.L., Thodberg, K., Su, G., Rydhmer, L., 2006. Aggressive behavior in sows at mixing and maternal behavior are heritable and genetically correlated traits. Livest. Sci. 93, 73–85.

Lund, R.D., 1973. Uncrossed visual pathways of hooded and albino rates. Science 149, 1506.

Malina, P.M., Buschang, P.H., 1984. Anthropometric asymmetry in normal and mentally retarded males. Am. Hum. Biol. 11, 515–531.

Malmkwist, J., Jeppensen, L., Palme, R., 2011. Stress and stereotypic behavior in mink (mustela vision): A focus on adrenocortical activity. Stress 14, 312–329.

Manning, J.T., Ockenden, L., 1994. Fluctuating asymmetry in race horses. Nature (London) 370, 185–186.

Marchant-Forde, J.N., Lay, D.C., Pajor, J.A., Richert, B.T., Schninckel, A.P., 2003. The effects of ractopamine on the behavior and physiology of finishing pigs. J. Anim. Sci. 81, 416–422.

Markow, T.A., 1992. Genetics and developmental stability: an integrative conjecture on aetiology and neurobiology of schizophrenia. Psychol. Med. 22, 295–305.

Maudlin, J.M., 1995. Behavior and Management of Broiler Breeder Chickens. American Society of Agricultural Engineering, Chicago [(ASAE Pap. No. 954512)].

Maxmen, A., 2012. Politics holds back animal engineers. Nature 490, 318–319.

McBryde, W.C., Murphee, O.D., 1974. The rehabilitation of genetically nervous dogs. Pavlovian. J. Biol. Sci. 9, 76–84.

McClearn, G.E., Rogers, D.A., 1961. Genetic factors in alcohol preference of laboratory mice. J. Comp. Physiol. Psychol. 54, 116–119.

McFarlane, J.M., Boe, K.E., Curtis, S.E., 1988. Turning and walking by gilts in modified gestation crates. J. Anim. Sci. 66, 3267–3333.

McFarlane, H.G., Kusek, G.K., Yang, M., Phoenix, J.L., Bolivar, V.J., Crawley, J.N., 2007. Autism-like behavioral phenotypes in BTBR T + tf/J mice. Genes Brain Behav. 7, 152–163.

Meagher, R.K., Bechard, A., Mason, G.L., 2012. Mink with divergent activity levels have divergent reproductive strategies. In: Davies, M.B., Krebs, J.R., West, S.A. (Eds.), An Introduction to Behavioral Ecology, fourth ed. Wiley Blackwell, pp. 543–554.

Minckler, J., Pease, F.D., 1938. A colony of albino rats existing under feral conditions. Science 87, 460–461.

Moller, A.P., 1992. Female swallows preference for asymmetrical male sexual ornaments. Nature (London) 357, 238–240.

Moller, A.P., 1993. Symmetrical male sexual ornaments, paternal care, and offspring quality. Behav. Ecol. 5, 188–194.

Moller, A.P., Santora, G.S., Vestergaard, K.S., 1995. Developmental stability in relation to population density and breed of chickens *Gallus gallus*. Poult. Sci. 74 (11), 1761–1771.

Molony, V., Kent, J.E., Robertson, I.S., 1995. Assessment of acute and chronic pain after different methods of castration of calves. Appl. Anim. Behav. Sci. 46, 33–48.

Morse, H.C. (Ed.), 1978. Origins of Inbred Mice. Academic Press, New York.

Muchenje, V., Dzama, K., Chimonyo, M., Strydom, P.E., Raats, J.G., 2009. Relationship between pre-slaughter stress responsiveness and beef quality in three cattle breeds. Meat. Sci. 81 (4), 653–657.

Murray, A.C., Johnson, C.P., 1998. Importance of the halothane gene on musch quality and pre-slaughter death in Western Canadian pigs. Can. J. Anim. Sci. 78, 543–548.

Murray, J.D., 1988. How the leopard gets its spots. Sci. Am. 3–9.

Murray, J.D., Mohamad-Fauzi, N., Cooper, C.A., Maya, E.A., 2010. Current status of transgenic animal research for human health application. Acta Scientiae. Vet. 38 (Suppl. 12), 5267–5632.

National Milk Producers Federation, 1996. Dairy Producers Highlights. NMPF, Arlington, VA.

Nir, I., Nitsan, Z., Dror, Y., Shapira, N., 1978. Influence of overfeeding on growth, obesity, and intestinal tract in young chicks of light and heavy breeds. Br. J. Nutr. 39, 27–35.

Notter, D., 1996. The importance of genetic diversity I livestock populations in the future. Paper Presented at the National Germplasm Program Symposium. American Society of Animal Science, Rapid City, SD, July 24, 1996.

Oskina, I.N., Prasolova, L.A., Plyusnina, I.Z., Trur, L.N., 2009. Role of glucocorticoids in coat depigmentation in animals selected for behavior. J. Cytol. Genet. 44 (5), 286–293.

Ott, R.S., 1996. Animal selection and breeding techniques that create diseased populations and compromise welfare. J. Am. Vet. Med. Assoc. 208, 1969–1974.

Parsons, P.A., 1990. Fluctuating asymmetry: an epigenetic measures of stress. Biol. Rev. Camb. Philos. Soc. 65, 131–145.

Pasille, A.M., Rushen, J., Martin, F., 1995. The behavior of calves in an open-field test: factor analysis. Appl. Anim. Behav. Sci., Sci. 45, 201–213.

Peters, J.E., Morphee, O.D., Dykman, R.O., 1967. Genetically-determined abnormal behavior in Dogs, Cond. Reflex 2, 206–215.

Pew Initiative on Food and Biotechnology, 2005. Proceedings of Exploring the Moral and Ethical Aspects of Genetically Engineered and Cloned Animals, Rockville, MD.

Poletto, R., Cheng, H.W., Meisel, R.L., Richard, B.T., Marchant-Forde, J.N., 2011. Gene expression of serotonin and dopamine receptors and monoamine oxidase-A in the brain of dominant and subordinate pubertal domestic pigs (sus serofa) fed B-adrenoreptor agonist. Brain Res. 1381, 11–20.

Pritchett-Corning, K.R., Girod, A., Avellaneda, G., Fritz, P.E., Chou, S., Brown, M.J., 2011. Handbook of Clinical Signs in Rodent And Rabbits. Charles Rivis Laboratories, Wilmington, MA.

Prunier, A., Heinonen, M., Quesnel, H., 2010. High physiological demands on intensively raised pigs: impact on health and welfare. Animal 4, 886–898.

Redbo, I., Norblad, A., 1997. Stereotypics in heifers are affected by feeding regime. Appl. Anim. Behav. Sci. 53, 193–202.

Rendel, J.M., 1952. White Heifer disease in a herd of dairy shorthorns. J. Gene. 51, 89–94.

Ribeiro, J.d.S., Goncalves, T.d.M., Ladeira, M.M., Tullio, R.T., Campos, F.R., Bergman, J.A.G., 2012. Reactivity, performance, color and tenderness of meat from Zebu cattle finished in feedlot. R. Bras. Zootech. 41, 4.

Robbins, A., Phillips, C.J.C., 2011. International approaches to the welfare of meat chickens. Worlds Poult. Sci. J. 67, 351–369.

Robert, S., Matte, J.J., Farmer, C., Girard, C.L., Martineau, G.P., 1993. High giber diets for sows. Effects on stereotypics and adjunctive drinking. Appl. Anim. Behav. Sci. 37, 297–309.

Robinson, F.E., Robinson, N.A., Scott, T.A., 1991. Reproductive performance, growth rate and body composition of full-fed versus feed-restricted broiler breeder hens. Can. J. Anim. Sci. 71, 549–556.

Rodenburg, T.B., deHaas, E.N., Nielsen, B.L., Buttenuis, A.J., 2010. Fearfulness in laying hens divergently selected for high and low feather pecking. Appl. Annl. Ischaun. Sci. 121, 91–96.

Rodenburg, T.B., Turner, S.P., 2012. The role of breeding and genetics in the welfare of farm animals. Anim. Front. 2, 16–21.

Rollin, B.E., 1995. The Frankenstein Syndrome. Cambridge University Press, Cambridge, UK.

Rushen, J., 1995. The coping hypothesis of all stereotypic behavior. Anim. Behav. Sci. 45, 613–615.

Rushen, J., Lawrence, A.B., Terlouw, E.M.C., 1993. The motivational basis of stereotypies. In: Lawrence, A.B., Rushen, J. (Eds.), Stereotypic Animal Behavior: Fundamentals and Applications for Animal Welfare. CABI International, Wallingford, UK, pp. 41–64.

Sandilands, V., Tolkamp, B.J., Kyriazakis, L., 2005. Behaviour of food restricted broilers during rearing and lay—effects of an alternative feeding method. Physiol. Behav. 85 (2), 115–123.

Savory, C.E., Seawright, E., Watson, A., 1992. Stereotyped behavior in broiler breeders in relation to husbandry and opioid receptor blockade. Appl. Anim. Behav. Sci. 32, 349–360.

Sayre, R.N., Kiernat, B., Briskey, E.J., 1964. Processing characteristics of porcine muscle related to pH and temperature during rigor mortis development and to gross morphology, 24 hours port-mortem. J. Food. Sci. 29, 175.

Schaible, R.H., 1969. Clonal distribution of melanocytes in piebald-spotted and variegated mice. J. Food. Sci. 172, 181–200.

Schaible, R.H., Brumbaugh, J.A., 1976. Electron microscopy of pigment cells in variegated and non-variegated piebald spotted dogs. Pigm. Cell 3, 191–200.

Searle, A.G., 1968. Comparative Genetics of Coat Color in Mammals. Logos Press, London.

Serikawa, T., Yamada, J., 1986. Epileptic seizures in rats homozygous for two mutations, zitter and tremor. J. Hered. 77, 441–444.

Seyfried, T.M., 1982. Developmental genetics of audiogenic seizure susceptibility in mice. In: Anderson, V.E., Hauder, W.A., Penry, J.K., Sing, C.F. (Eds.), Genetics Basis of the Epilepsies. Raven Press, New York, pp. 199–210.

Sheridan, C.L., 1965. Interocular transfer of brightness and pattern discriminations in normal and corpus-callosum-sectioned rates. J. Comp. Physiol. Psychol. 59, 292–294.

Shin, H., Thomas, P.R., Ensley, S.M., Kim, W.-I., Loynachan, A.T., Halbur, P.G., et al., 2011. Vitamin E and Selenium levels are within normal range in pigs diagnosed with Mullberry Heart Disease and evidence for viral involvement in the syndrome is lacking. Transboundry and Emerging Diseases 58 (6), 483–491.

Short, C.E., Poznak, A.V., 1992. Animal Pain. Churchill-Livingstone London.

Silva, A.J., Stevens, C.F., Tonegawa, S., Wang, Y., 1992a. deficient hippocampal long-term potentiation in a calcium–calmodulin kinase II mutant mice. Science 257, 201–206.

Silversides, F.G., Smyth Jr, J.R., 1986. Fader shaker, a lethal pigment and neurological mutation in the chicken. J.Hered. 77, 295–300.

Sinisalo, A., Niemi, J.K., Heinonen, M., Valros, A., 2012. Tail biting and production in fattening pigs. Livest. Sci. 143, 220–225.

Somes, R.G., 1979. White-wing, a lethal feather achromatic mutant in the fowl. J. Hered. 79, 373–378.

Sorsby, A., Davey, J.B., 1954. Occular associations of dappling (or merling) in the coat color of dogs, I. Clinical and genetical data. J. Gene. 52, 425–440.

Souza de Silva, C., Van Der Borne, J.J., Gerrits, W.J., Kemp, B., Bolhois, E., 2012. Effect of dietary fibers and different physiochemical properties on feeding motivation in adult female pigs. Physiol. Behav. 107, 218–230.

Sponenberg, D.P., Lamoreux, M.L., 1985. Inheritance of tweed, a modification of merle, in Australian Shepherd dogs. J. Hered. 76, 303–304.

Sponenberg, D.P., 1995. Livestock genetic resource conservation, practical field application of theoretical aspects. J. Anim. Sci. 73 (Suppl. 1), 123.

Spriggs, D.N., 1946a. White heifer disease. Vet. Rec. 58, 405–409.

Spriggs, D.N., 1946b. White heifer disease part II. Vet. Rec. 58, 415–418.

Steg, J.P., Rapport, J.L., 1975. Minor physical abnormalities in normal, neurotic, learning disabled, and severely disturbed children. J. Autism Child. Schizophr 5, 299–307.

Steinberg, S.A., Klien, E., Killens, R.L., Udhe, T.W., 1994. Inherited deafness among nervous pointer dogs. J. Hered. 85, 56–59.

Stern, S., Lundeheim, N.K., Johansson, K., Andersson, K., 1995. Osteochondrosis and leg weakness in pigs selected for lean tissue growth rate. Livest. Prod. Sci. 44, 45–52.

Stolba, A., Wood-Gush, D.G.M., 1989. The behavior of pigs in a semi-natural environment. Anim. Prod. 48, 419–425.

Strain, G.M., 2011. Deafness in Dogs and Cats. CABI Publishing, Wallingford, Oxon, UK.

Strain, G.M., Cark, L.A., Wahl, J.M., Turner, A.E., Murphy, K.E., 2009. Prevalence of deafness in dogs, Heterozygous or homozygous for the Merle Allelle. J. Vet. Intern. Med. 23, 282–286.

Svendsen, P.M., Hansen, B.K., Melmkwist, J., Hansen, S.W., Palme, R., Jeppensen, L.L., 2007. Selection against stereotypic behavior may have contradictory consequences for the welfare of farm mink (Mustela vision). Appl. Anim. Behav. Sci. 107, 110–119.

Taylor, N.R., Main, D.C., Mendi, M., Edwards, S.A., 2010. Tail biting: a new perspective. Vet J. 186, 137–147.

Thomas, D.L., 1995. Importance of maintaining genetic diversity among livestock populations. J. Anim. Sci. 73 (Suppl. 1), 123.

Tolkomp, B.J., Kyriazakis, I., Sandilands, V., 2005. Behavior of food restricted broilers during rearing and laying, Effects of an alternative feeding method, physiology, and behavior, 85:115–123.

Turner, S.P., 2011. Breeding against harmful social behaviors in pigs and chickens: state of the Art and The Way Forward. Appl. Anim. Behav. Sci. 134, 1–8.

Turner, S.P., Navajas, E.A., Hyslop, J.J., Ross, D.W., Richardson, R.J., Prieto, N., et al., 2011. Association between response to handling and growth and meat quality in frequently handled *Bos tauras* beef cattle. J. Anim. Sci. 89 (2), 4239–4248.

United States Department of Agriculture (USDA) 2008. Changes in the U.S. Dairy Cattle Industry, 1991–2007, IN: Dairy 2007 Part II.

USDA-National Agricultural Statistics Service, 2010. Overview of the United States Dairy Industry.

Uystepruyst, C., Coghe, J., Dorts, T., Harmegnies, N., Delsemme, M., Art, T., et al., 2002. Optimal timing of elective caesarean section in Belgian white and blue breed of cattle. Vet. J. 163 (3), 267–282.

Van der Weerd, H.A., Docking, C.M., Day, J.E.L., Brever, K., Ewards, S.A., 2008. Effect of species relevant environmental enrichment of the behavior and productivity of finishing pigs. Appl. Anim. Behav. Sci. 99, 230–242.

Voisinet, B.D., Grandin, T., Tatum, J.D., O'Conner, S.F., Deesing, M.J., 1997. *Bos indicus*-cross feedlot cattle with excitable temperaments have tougher meat and a higher incidence of borderline dark cutters. Meat Sci. 46, 367–377.

Von Borell, E., Hurnick, J.F., 1990. Stereotypic behavior and productivity in sows. Can. J. Anim. Sci. 70, 953–956.

Walberg, L., 2011. Neurodevelopmental disorders: mice that mirror autism. Nat. Rev. Neurosci. 12, 615.

Webb, A.A., Cullen, C.L., 2010. Coat color and coat color pattern related neurologic and neuro-ophthalmic disease. Can. Vet. J. 51, 653–657.

Webster, J., 1993. Understanding the Dairy Cow, second ed. Blackwell, Oxford, UK.

Wellberg, L., 2011. Neurodevelopmental disorders; Mice that mimic autism. Nat. Rev. Neurosci. 12, 615.

Willard, J.G., Willard, J.C., Wolfram, S.A., Baker, J.P., 1977. Effect of diet on cecal PH and feeding behavior of horses. J. Anim. Sci. 45, 87–93.

Wimmer, R.E., Wimmer, C.C., 1985. Animal behavior genetics: a search for the biological foundations of behavior. Ann. Rev. Psychol. 36, 171–218.

Woolf, C.M., 1990. Multifactoral inheritance of common white markings in the Arabian horse. J. Hered. 81, 250–256.

Woolf, C.M., 1991. Common white facial markings in bay and chestnut Arabian horses and their hybrids. J. Hered. 82, 167–169.

Woolf, C.M., 1995. Influence of stochastic events on the phenotypic variation of common white leg markings in the Arabian horse. Implications for various genetic disorders in humans. J. Hered. 86, 129–135.

Wright, S., 1978. The relation of livestock breeding to theories of evolution. J. Anim. Sci. 46, 1192–1200.

Yu, M.W., Robinson, F.E., Robbler, A.R., 1992a. Effect of feed allowance during rearing and breeding on female broiler breeders, I. Growth and carcass characteristics. Poult. Sci. 71, 1739–1749.

Yu, M.W., Robinson, F.E., Charles, R.G., Weingardt, R., 1992b. Effect of feed allowance during rearing and breeding on female broiler breeders, 2. Ovarian morphology and production. Poult. Sci. 71, 1750–1761.

Zonderland, J.J., Schepers, F., Bracke, M.B.M., Hartog, L.A., den Kemp, B., Spoolder, H.A.M., 2011. Characteristics of biter and victim piglets apparent before tail biting outbreak. Animal 5, 767–775.

Zuidhoff, M.J., Robinson, F.E., Feddes, J.J.R., Hardin, R.T., 1995. The effect of nutrient dilution on the well-being performance of female broiler breeders. Poult. Sci. 74, 441–456.

Zupan, M., Janczak, A.M., Framstad, T., Zanella, A.J., 2012. The effect of biting tails and having tails bitten in pigs. Physiol. Behav. 106.

FURTHER READING

Gebhart, G.F., 1997. Quoted in Mukerjee, M. (1997) Trends in animal research. Sci. Am. 276 (2), 86–93.

King, H.D., 1930. Life processes in gray Norway rats during fourteen years of captivity. Am. Anat. Mem. 17, 1–72.

Lambooij, E., Garssen, G.J., Walstra, P., Mateman, G., Merkus, G.S.M., 1985. Transport by car for two days: some aspects of watering and loading density. Livet. Prod. Sci. 13, 289–299.

Index

Note: Page numbers followed by "*f*" refer to figures, respectively.

C

D

G

H

Q

R

S

U

V

W

Printed in the United States
By Bookmasters